Research and Discovery

Landmarks and Pioneers in American Science

Volume Three

Edited by

Russell Lawson

SHARPE REFERENCE

an imprint of M.E. Sharpe, Inc.

SHARPE REFERENCE

Sharpe Reference is an imprint of M.E. Sharpe, Inc.

M.E. Sharpe, Inc.
80 Business Park Drive
Armonk, NY 10504

Library of Congress Cataloging-in-Publication Data

Research and discovery: landmarks and pioneers in American science / Russell Lawson, editor.
    p. cm.
Includes bibliographical references and index.
ISBN 978-0-7656-8073-0 (hc: alk. paper)
1. Science—United States—History—Encyclopedias. 2. Research—United States—History—
Encyclopedias. I. Lawson, Russell.

Q127.U6R45 2007
509.73—dc22   2006014012

Cover images (clockwise from top left corner) provided by: Keystone/MPI/Hulton Archive/Getty Images;
North Wind Picture Archive; NASA/Time & Life Pictures/Getty Images; Hulton Archive/Getty Images;
Hulton Archive/Getty Images; Alfred Eisenstaedt/Time & Life Pictures/Getty Images; Nina Leen/Time & Life
Pictures/Getty Images; NASA/Buyenlarge/Time & Life Pictures/Getty Images.

Printed and bound in the United States of America

The paper used in this publication meets the minimum requirements of
American National Standard for Information Sciences
Permanence of Paper for Printed Library Materials,
ANSI Z 39.48.1984.

(c) 10 9 8 7 6 5 4 3 2 1

Publisher: Myron E. Sharpe
Vice President and Editorial Director: Patricia Kolb
Vice President and Production Director: Carmen Chetti
Executive Editor and Manager of Reference: Todd Hallman
Executive Development Editor: Jeff Hacker
Project Editor: Laura Brengelman
Program Coordinator: Cathleen Prisco
Editorial Assistant: Alison Morretta
Text Design: Carmen Chetti and Jesse Sanchez
Cover Design: Jesse Sanchez

Contents

iv Contents

Section 12: Mathematics and Computer Science

Section 13: Applied Science

Section 14: History and Philosophy of Science

Topic Finder

Note: Numbers in parentheses indicate field of study; see the complete Contents in Volume 1.

Natural Phenomena and Features

**Notable Figures: Exploration,
Natural History, and Philosophy**

Notable Figures: Physical Sciences

Processes, Techniques, and Treatments

Projects, Experiments, and Expeditions

Theories, Concepts, and Philosophical Perspectives

Tools, Inventions, and Technological Achievements

Research and Discovery

Volume Three

Section 10

PHYSICS

ESSAYS

Aristotelian Physics in Colonial America

When the premier scientific institution of early America, Harvard College, was founded in 1636, the Aristotelian worldview was dominant in thought and science in Europe and America. The trials of building new communities and educating clergy to take the intellectual helm of the Massachusetts Bay Colony distracted colonial Americans from focusing on the "new science" of Nicolaus Copernicus, Galileo Galilei, Johannes Kepler, René Descartes, Pierre Gassendi, and Robert Boyle that was in the process of transforming European scientific theory. Hence, Harvard's science curriculum of arithmetic, geometry, physics, astronomy, and botany continued to be beholden to an Aristotelian worldview.

Late in the seventeenth century, however, the works of the "new science" as well as the publications of Isaac Newton's *Principia Mathematica* led to the erosion of Aristotelian science. This transition made Harvard, in the words of historian Louis B. Wright, "a center for the promulgation of Copernican and presently Newtonian science."

Aristotelian Science

Aristotelian science was a composite of the work and theories of a variety of ancient Greek scientists, such as the sixth- and fifth-century B.C.E. philosophers Thales, Anaximander, Anaximenes, Anaxagoras, Archelaus, Socrates, and Plato, Aristotle's teacher; fourth-century B.C.E. atomists such as Epicurus and Zeno; and Hellenistic scientists of the final three centuries B.C.E. and first two centuries C.E. such as Eratosthenes, Hipparchus, Apollonius of Perga, and Claudius Ptolemy. All except Aristarchus of Samos agreed that the universe was geocentric: that is, Earth is an immovable force in the center of a finite universe, and the heavenly bodies orbit Earth in the order of the moon, Mercury, Venus, the sun,

Mars, Jupiter, and Saturn. All planets and the sun are perfect spheres, god-like. The outer rim of the universe is the realm of the fixed stars, limited in number, forming a stellar shell to the universe.

The planets were thought to move in a stellar substance called "ether," a fifth element Aristotle added to the traditional four elements of ancient Greek science: earth, air, fire, water. Aristotle's physics did not allow for inertial movement of bodies. In his theory, Earth is the heaviest of bodies, because it is at the center of the universe. All things tend toward the center, which explains gravity, as well as why fire leaps upward, water flows to the center, air (being lighter) forms the atmosphere, and the planets orbit through ether in their own celestial spheres.

Two World Systems

During the mid-seventeenth century, the writings of European proponents of the "new science" crossed the Atlantic to Boston and Philadelphia, changing ideas about the Aristotelian model of the universe. Europeans such as the logician Peter Ramus challenged Aristotelian logic by arguing for a more simple method. Gassendi, the atomist, along with Descartes, the French philosopher and mathematician, challenged the Aristotelian assumption of the existence of ideal forms and truths that are not supported by observation and experiment. Francis Bacon's *Novum Organum* (1620) presented the empirical point of view.

Galileo's studies of motion contradicted the geocentric universe and disproved the Aristotelian argument that the heavenly bodies are perfect and unchanging and that the universe is finite. Galileo's universe is potentially infinite. Robert Boyle expanded on Gassendi's ideas to hypothesize a mechanistic universe based on corpuscles. Johannes Kepler showed that the orbit of planets is elliptical rather than spherical.

The conflict between the Aristotelian worldview and the new science of Copernicus, Galileo, Boyle, and Bacon is illustrated by Charles Morton, who was a Puritan minister, a noted dissenter in England against the royalist Anglican Church, a natural philosopher, and a physicist who believed firmly that natural law depends on divine law, or the will of God. In 1686, Morton emigrated to Massachusetts, where he became pastor at the First Parish at Charlestown, Massachusetts, and a teacher at Harvard College. His *Compendium Physicae* (*Compendium of Physics*), published in 1683, was adopted by the Cambridge scientific community in 1687, becoming the sole physical science text used at Harvard until 1728. The *Compendium* was also used at Yale College.

Morton's *Compendium* is an odd mixture of Aristotelian and Copernican science. His physics rests on the ancient Greek four elements, which was quickly being replaced in Europe by the corpuscular system of Gassendi and Boyle. Yet Morton accepted most of the ideas of the "new science," such as the heliocentric universe, Galileo's laws of motion, the empirical method, and the material basis of the universe.

Morton was not alone in holding conflicting views. He and Cotton Mather were two of the greatest advocates of science in colonial America, and both assumed the reality of witchcraft.

Russell Lawson

Sources

Cohen, I. Bernard. *The Birth of a New Physics.* New York: W.W. Norton, 1985.

Daniels, George H. *Science in American Society: A Social History.* New York: Alfred A. Knopf, 1971.

Stearns, Raymond Phineas. *Science in the British Colonies of North America.* Urbana: University of Illinois Press, 1970.

Newtonian Physics and Early American Science

The publication of Isaac Newton's *Principia Mathematica* (*Mathematical Principles of Natural Philosophy*) in 1687 laid to rest the Aristotelian universe and brought to fruition the work of Nicolaus Copernicus, Galileo Galilei, Johannes Kepler, René Descartes, Francis Bacon, and Robert Boyle into one system of explaining the universe—the Newtonian paradigm.

The *Principia* relies on empiricism over speculation; it brings sophisticated mathematics—calculus—to bear on questions of force and motion. The *Principia* explores the universe, as it were, using Kepler's laws of planetary motion, Galileo's laws of falling bodies, Bacon's empirical methods, Descartes's mechanistic view of the universe, and Boyle's corpuscle theory of matter.

Newton's Laws

The *Principia* outlines three laws of motion on which Newton's physics and worldview were based. According to the first law, "Every body perseveres in its state of being at rest or of moving uniformly straight forward, except insofar as it is compelled to change its state by forces impressed upon it." According to the second law, "A change in motion is proportional to the motive force impressed and takes place in the direction of the straight line along which that force is impressed." According to the third law, "For every action, there is an equal and opposite reaction."

Newton's three laws of motion resulted in his discovery of the general law of universal gravitation, which explains the mutual forces at work in the universe, specifically the multiple forces of attraction at work in the solar system, where the moon, sun, Earth, and all the other planets exert proportional forces on each other. The law of universal gravitation explains the elliptical orbits of planets, the positions of planets and satellites in the solar system, why Earth rotates on its axis inclined to the plane of the ecliptic (the precession of the equinoxes), why tides exist, and why things do not fly off Earth into space.

The Newtonian universe is, in short, a universe that is rational and benevolent, fashioned by a benevolent and rational creator; a universe that operates according to natural laws that make sense and accommodate human reason; a universe that is constant and predictable; a universe that operates like a machine of absolute perfection that is eternal and without error; a universe of matter in motion that contains the subtle signature of the divine. To understand Newton's universe, scientists needed scientific instruments such as telescopes, microscopes, electrostatic machines, air pumps, theodolites, sextants, longitudinal instruments, and chronometers to study time, space, gravity, the heavens, the unseen, friction, motion, electric charges, static, energy, pressure, inertia, elasticity, vacuums, and hydrostatics—all that involves cause and effect in the movement of matter.

Newtonian Science in America

Isaac Newton never traveled to America, but his ideas did by means of his various publications, in particular the *Principia* and the *Opticks* (1704). The *Principia* is written in Latin and features extremely complex mathematics; few Americans could master it. The *Opticks* is written in English and is more for the average intelligent person interested in science.

Newton's reputation arrived in America before his books did, so that when copies of the *Principia* began to arrive, there was much excitement among American scientists. Jeremiah Dummer brought one of the first copies to Yale College in 1715. Not everyone could probe Newton's mind through the medium of his books, however. Benjamin Franklin, for example, complained of the difficulty of the *Principia* and preferred the *Opticks* to inform him of Newton's theories. The few who could understand the *Principia*, such as James Logan of Philadelphia, were important patrons and advocates of the Newtonian system.

Although Newton's works were not required reading at Harvard and Yale, the colleges hoped to teach students about Newton's laws of motion if they could not yet read them firsthand. The Hollis professorship at Harvard was established to inculcate the Newtonian paradigm in the minds of students. The Hollis Chair in the eighteenth century was directed to engage Newtonian science as follows: "The lectures which shall be delivered in the Philosophy Chamber to the resident Bachelors and the two Senior Classes, between the twenty-first of March and twenty-first of June annually, shall contain a complete course of Experimental Philosophy, in the various branches of it; and in the progress of the experiments, the principles and construction of the various machines made use of, shall be explained."

John Winthrop IV, the Hollis Professor of Mathematics and Natural Philosophy at Harvard from 1738 to 1779, relied on experimental science to impress on his students the Newtonian universe. His course was devised "to instruct students in a system of natural philosophy and a course of experimental, in which to be comprehended pneumatics, hydrostatics, mechanics, statics, optics and in the elements of geometry, together with the doctrine of proportions, the principle of algebra, conic sections, planes and solids, in the principles of astronomy and geography, viz. the doctrine of the sphere, the use of the globes, the motions of the heavenly bodies according to the different hypotheses of Ptolemy, Tycho Brahe and Copernicus, with the general principles of Dialling, the division of the world into its various kingdoms, with the use of the maps, &c."

Newton was thought to be an unparalleled savant, rather like Albert Einstein today. As Alexander Pope, the author of *Essays on Man*, wrote at the time:

Nature, and Nature's Laws lay hid in Night.
God said, *Let Newton be!* and All was *Light*.

Russell Lawson

Sources

Cohen, I. Bernard. *The Birth of the New Physics*. New York: W.W. Norton, 1985.

Daniels, George H. *Science in American Society: A Social History*. New York: Alfred A. Knopf, 1971.

Stearns, Raymond Phineas. *Science in the British Colonies of North America*. Urbana: University of Illinois Press, 1970.

Benjamin Franklin, American Physicist

Americans were on the frontier of science during Benjamin Franklin's lifetime, which spanned much of the eighteenth century. Franklin, although one of America's leading scientists, found the theories of Isaac Newton daunting and Newton's *Principia Mathematica* (1687) difficult to understand because of the sophisticated mathematics and esoteric concepts. More to Franklin's liking was Newton's more elementary treatise *Opticks* (1704), which explained the same concepts covered in *Principia* but according to experiment and observation rather than theory.

Franklin was a representative American scientist because of his ability to take a commonsense, experimental approach to science rather than a metaphysical one—that is, to take Newton's difficult concepts and apply them to practical, realistic situations. For example, what if the strange phenomenon of lightning were produced by conditions present not just in a thunderstorm but all around, inherent in nature? Was lightning a consequence of movement and change? Franklin realized that the power that produces lightning, that charged the key in his famous kite experiment, is the same power produced by rubbing two materials together, or when one person touches another and feels the shock of static electricity. That being the case, how was this static electricity produced? Could anyone create such a charge?

Franklin, using as his basis Newton's first law of motion respecting universal conservation of energy, argued that an electric charge, such as that of static electricity, is not created but is displaced from one conductor to another. Electric charges are universal in nature, and charge is never created or destroyed, just transferred from one host to another. This was Franklin's law of conservation of charge.

For the kite experiment, Franklin constructed a silk kite with a small iron rod protruding from the top. A brass key was attached to the silk kite string that Franklin held. Flying the kite in a storm, Franklin stood on a small wooden platform within a shelter. The conduction of "electric fluid" from the charged clouds was detectable by

means of the key: When Franklin touched it, he received a sharp shock, which indicated that the key was conducting electric current.

According to Franklin's law of conservation of charge, the electric current must go somewhere, displaced from one host to another. Had Franklin been standing on the ground, he would have become a conductor of the current to the earth. (When one French experimenter made this mistake, it was his last.) By standing on the wooden platform, Franklin was not a conductor; instead, the current made its way to the key, as brass is a strong conductor of electricity. Whereas previous theory had it that electricity was comprised of two distinct forces, Franklin saw it as a unified force with positive and negative charges.

Franklin used the same theories to invent the lightning rod, a long iron rod attached to the roof of a house, typically alongside the chimney. An iron filament, attached to the bottom of the rod, made its way through the house to the basement and to the earth. This way, the current was drawn from the clouds through the iron conductor to the earth, without harming the house.

Franklin, who loved gadgets, contrived an extension of his lightning rod into his study. Attached to the rod was a bell. When the strong electrical current of a thunderstorm was outside his house, the scientist knew by the ringing of the bell. One night, he heard a harsh crackling sound. Going to his study, he found it bathed in light; the bell was white hot with electric current. More than a century before Thomas Edison's inventions, Franklin accidentally discovered artificial light using electric current.

Franklin's practical approach to electricity helped to stir interest and inspire others to engage in experimentation to discover the true essence of the "electric fluid." Joseph Priestley of England, a good friend of Franklin's who would emigrate to America in 1794, showed how an electric force is comparable to a gravitational force, where the force is proportional to the inverse square of the distance. In Europe, Charles Augustin Coulomb illustrated how forces exist between the positive and negative poles of an electric charge according to the same inverse

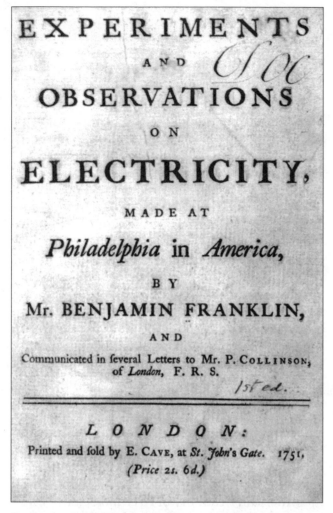

EXPERIMENTS
AND
OBSERVATIONS
ON
ELECTRICITY,

MADE AT

Philadelphia in *America,*

BY

Mr. BENJAMIN FRANKLIN,

AND

Communicated in feveral Letters to Mr. P. COLLINSON,
of *London*, F. R. S.

1st ed.

———————————————

L O N D O N:
Printed and fold by E. CAVE, at *St. John's Gate.* 1751,
(Price 2s. 6d.)

Benjamin Franklin's highly influential *Experiments and Observations of Electricity* (1751) explains and supports his theories on the nature of electricity—perhaps his greatest contribution to pure science. *(Library of Congress, LC-USZ62–58219)*

square ratio. Hans Christian Oersted discovered electromagnetism, the unification of electrical and magnetic phenomena. Andre-Maria Ampere in 1820 described the force and movement of electric current. Michael Faraday in 1821 demonstrated the conversion of electrical energy into mechanical energy. Faraday had a major impact on James Clark Maxwell, who introduced the concept of "electro tonic function," which is the relationship between electromagnetic induction, force between current-carrying wires, and magnetic action.

Franklin also focused on the movement of air and the relationship between masses of hot and cold air, exploring theories of meteorological phenomena as well as principles of heating air by means of fireplaces. Using theories about the interaction of hot and cold air, Franklin addressed the practical problem of heating a home in winter. He discussed his ideas in *An Account of the New Invented Pennsylvanian Fire-Place* (1744).

The fireplace that Franklin devised was actually a wood-burning iron stove connected to a flue through which smoke ascended out the chimney. Franklin's aim was to displace cool air in the room with heated air. A draft brought cool air from the room through an in-flow vent at the bottom of the stove. The heated air flowed out through vents on the sides of the stove. Smoke was carried under the stove in a pipe connected to the flue.

The Franklin stove experienced mix success: On extremely cold days, the smoke tended to back up into the stove and then the room. Franklin's work on the properties of heating air influenced Benjamin Thompson, also known as Count Rumford, who invented the highly effective Rumford fireplace.

Russell Lawson

Sources

Cohen, I. Bernard. *Benjamin Franklin's Science.* Cambridge, MA: Harvard University Press, 1990.

Franklin, Benjamin. *An Account of the New Invented Pennsylvanian Fire-Place.* Philadelphia, 1744.

———. *The Autobiography of Benjamin Franklin.* New York: Macmillan, 1962.

———. *Experiments and Observations in Electricity, Made at Philadelphia in America.* London: E. Cave, 1751.

Van Doren, Carl. *Benjamin Franklin.* New York: Viking, 1938.

ALVAREZ, LUIS
(1911–1988)

A Nobel Prize–winning scientist and physics professor, Luis Walter Alvarez was one of the most prolific American scientists of the twentieth century. His contributions include the invention of the hydrogen bubble chamber technique, the design and development of radar systems, the invention of the detonators used in nuclear weapons, and research in archeology and paleontology.

Alvarez was born in San Francisco, California, on June 13, 1911. After receiving his undergraduate degree in physics in 1932, master's degree in 1934, and Ph.D. in 1936 from the University of Chicago, he joined the Radiation Laboratory of the University of California as a research fellow. Except for several sabbaticals, including leaves to conduct research aimed at the development of microwave beacons and antennas at the Massachusetts Institute of Technology from 1940 to 1943 and to work on nuclear weapons detonators at Los Alamos from 1944 to 1945, Alvarez spent his entire career at the University of California, Berkeley.

In applied science, Alvarez, in collaboration with Swiss physicists and fellow Nobel laureate Felix Bloch in 1939, made the first measurement of the "magnetic moment" of the neutron, which determined the direction and strength of its magnetic field. Alvarez's avocational interest in aviation led him to develop what is now called the ground-controlled approach, an instrument-based navigational system for landing military and civilian airplanes at night and in conditions of poor visibility. Alvarez was also responsible for the design and construction of Berkeley's 40 foot proton linear accelerator in 1947 and, in the 1960s, he used muon detectors to look for hidden chambers inside the second pyramid of Giza in Egypt.

Alvarez received the 1968 Nobel Prize in Physics for his development of the hydrogen bubble chamber technique, a device for tracking the trajectories of ionized particles by photographing their trails of bubbles through super-heated hydrogen, and for the development of high-speed measurement and analysis devices needed to process the millions of photographic images produced by the bubble chamber experiments. His efforts in this area made it possible to record and study the short-lived subatomic particles created by particle accelerators.

In 1980, Alvarez collaborated with his son Walter, a geology professor at Berkeley, to develop the asteroid-impact theory of mass extinction in an attempt to explain the presence of a worldwide layer of the element iridium in the boundary between rocks laid down in the Cretaceous and Tertiary geological periods. According to the Alvarez theory, a giant asteroid struck Earth about 65 million years ago, killing off the dinosaurs and raising a huge cloud of iridium-bearing dust that eventually settled over the globe and into the geologic record.

In addition to receiving the Nobel Prize, Alvarez was named to the National Academy of Sciences, the National Academy of Engineering, the American Academy of Arts and Sciences, and he was a fellow of the American Physical Society. He died on September 1, 1988, in Berkeley.

Todd A. Hanson

Sources

Alvarez, Luis W. *Adventures of a Physicist.* New York: Basic Books, 1987.
Trower, Peter, ed. *Discovering Alvarez: Selected Works of Luis W. Alvarez with Commentary by His Students and Colleagues.* Chicago: University of Chicago Press, 1987.

ANDERSON, CARL DAVID
(1905–1991)

Experimental physicist and Nobel laureate Carl David Anderson is best known for his discovery of the positron, a subatomic particle.

He was born on September 3, 1905, to Swedish immigrant parents in New York City. Almost all of

Anderson's education and professional career was connected to the California Institute of Technology. He received both his bachelor of science degree (1927) and doctorate in physics (1930) at the Pasadena institution. He worked there as a research fellow from 1930 to 1933, then as an assistant professor of physics from 1933 to 1939, and as a full professor from 1939 through his retirement in 1976. During World War II, Anderson also served with the National Defense Research Committee and the Office of Science Research and Development, working in the field of rocketry.

Anderson's early work was in X-ray research. His doctoral dissertation examined the distribution of photoelectrons ejected from various gases when subjected to X-rays. As a research fellow, he turned to the study of cosmic rays and, initially under the supervision of physics professor Robert Andrew Millikan, made the discovery of the positron that led to the 1936 Nobel Prize in Physics, an award he shared with Austrian scientist Victor Hess. (Anderson and Hess did not collaborate, but both were studying the properties of cosmic rays.)

The first form of antimatter ever discovered, the positron is the antiparticle of the electron. Like the electron, it is a subatomic particle that orbits the nucleus of the atom, but the positron has a positive charge rather than a negative one. When positrons and electrons collide, as in the nuclear furnace of a star, they produce a form of energy known as gamma rays. Anderson made his discovery by shooting gamma rays produced by thorium carbide through a device that combined a gas chamber, a lead plate, and a magnet, which bent differently charged particles in different directions, allowing for the separation and identification of the physical properties of each particle. The positron is critical to the working of positron emission tomography (PET), a form of nuclear imaging developed in the 1970s and used to make three-dimensional renderings of the body.

The same year that he won the Nobel Prize, Anderson discovered, along with graduate student Seth Neddermeyer, the muon (originally the mu-meson), another subatomic particle. The muon, with a mass over 200 times that of an electron, is critical to the understanding of the strong nuclear force.

Anderson continued his research on subatomic particles as well as teaching physics at Cal Tech for another forty years. Among his many academic honors are the Gold Medal of the American Institute of the City of New York (1935), the Elliott Cresson Medal of the Franklin Institute (1937), the Presidential Certificate of Merit (1945), and the John Ericsson Medal of the American Society of Swedish Engineers (1960). He died in San Marino, California, on January 11, 1991.

James Ciment

Sources

Anderson, Carl D. "The Positive Electron." *Physical Review* 43:6 (1933): 491–94.

Kevles, Daniel J. *The Physicists: The History of a Scientific Community in Modern America.* New York: Vintage Books, 1979.

BETHE, HANS ALBRECHT (1906–2005)

The German American physicist Hans Bethe contributed to the development of the hydrogen bomb in the early 1950s. He is known for his work on the theory of atomic nuclei and on the nuclear reactions that supply energy in the stars and sun—work for which he was awarded a Nobel Prize in 1967.

Bethe was born on July 2, 1906, in Strasbourg, Alsace-Lorraine. In 1912, the family moved to Kiel, Germany, where his father hired a tutor for Bethe, who by age fourteen was reading books on trigonometry and calculus. At the Goethe Gymnasium in Frankfurt, Bethe immersed himself in physics and mathematics. In 1924, he enrolled in the University of Frankfurt but transferred two years later to the University of Munich, where, in 1928, he received a Ph.D. in physics.

In 1930, he published in the journal *Annalen der Physik* what he regarded as his best paper, "The Theory of the Passage of Swift Corpuscular Rays through Matter." In this paper, he quantified the energy released in the collision between an atom and a subatomic particle.

A recipient of a Rockefeller Foundation fellowship, Bethe went on to study at Cambridge University in England in 1930. The following year, he studied at the University of Rome in Italy under the tutelage of Enrico Fermi. In 1932, Bethe became assistant professor in theoretical physics at the University of Tübingen in Württemberg, Germany.

Anti-Semitism in Nazi Germany drove Bethe, whose mother and maternal grandparents were Jewish, from his post at the University of Tübingen and to the University of Manchester in England in 1933. In 1935, Bethe accepted a position at Cornell University in Ithaca, New York.

Bethe's interest in nuclear physics stemmed from the proposal of physicists Robert Atkinson and Friedrich Georg Houtermans in 1929 that the sun derives its energy from nuclear reactions. Stimulated by a conference on the production of energy in the sun, Bethe in April 1938 began work to determine which reactions between elements yield enough energy to supply the sun. Working down the periodic table, he caused nuclear reactions from increasingly heavier elements, arriving at the reaction between carbon and nitrogen as giving the best estimate of the energy in the sun. Bethe, using his own work and that of others, charted the evolution of stars from the initial hydrogen reactions to the helium-carbon reactions of a star's old age.

The outbreak of World War II pitted the United States against Germany in a race to build an atomic bomb. Bethe's research in nuclear physics brought him to the attention of J. Robert Oppenheimer, the director of the Manhattan Project, who, in 1942, invited Bethe to the University of California, Berkeley, to begin work on a bomb. Upon establishing his headquarters in Los Alamos, New Mexico, Oppenheimer appointed Bethe director of the theoretical division in 1943. At the time, Bethe seems to have had few qualms about the morality of using an atomic bomb, declaring such questions the province of philosophy rather than physics.

Soviet detonation of a uranium bomb in 1949 led physicist Edward Teller to approach Bethe about joining him in building a hydrogen bomb. In his research on the sun, Bethe had outlined the fusion reaction that would detonate such a bomb, but he doubted that such a weapon could be built. When President Harry S. Truman authorized the building of a hydrogen bomb in 1949, Bethe joined several members of the American Physical Society in a press conference to warn the public of the dangers of thermonuclear weapons. Between February and September 1952, however, he acquiesced in working with Teller on the bomb.

Bethe's relationship with Teller broke in 1954 over the revocation of Oppenheimer's security clearance by the Atomic Energy Commission. Bethe defended Oppenheimer and urged Teller to do the same, and he felt betrayed when Teller testified to the commission against Oppenheimer.

Increasingly uneasy about the dangers of nuclear weapons, in 1960, Bethe urged the U.S. military not to develop an intercontinental ballistic missile, and he supported the 1963 Partial Test Ban Treaty between the United States and the Soviet Union. In 1967, he won the Nobel Prize in Physics for his work on the thermonuclear reaction in the sun.

Bethe retired from Cornell University in 1975. He continued to speak out on public issues until his death on March 6, 2005.

Christopher Cumo

Sources

Bernstein, Jeremy. *Hans Bethe, Prophet of Energy.* New York: Basic Books, 1980.

Marshak, Robert E., ed. *Perspectives in Modern Physics: Essays in Honor of Hans A. Bethe on the Occasion of His 60th Birthday, July 1966.* New York: Interscience, 1966.

COLD FUSION

Cold fusion is the nuclear reaction that would occur if two atomic particles, or nuclei, were brought close enough together to fuse and form another heavier element at low temperatures. If it is ever developed, cold fusion could provide a clean and virtually inexhaustible source of energy without the extremely high temperatures required by nuclear fusion. But most scientists believe that cold fusion is impossible, or at least improbable, despite significant popular belief to the contrary. In 1989, two researchers claimed to have achieved cold fusion, causing what many in the science world consider the greatest scientific fiasco of the twentieth century.

On March 23, 1989, University of Utah chemistry professor Stanley Pons and Martin Fleischman, a research professor of electrochemistry at England's University of Southampton, held a press conference in Salt Lake City, Utah, to announce that their attempt to achieve cold fusion had been successful. The duo had constructed a simple electrolytic cell containing deuterium oxide, or heavy water, into which they inserted palladium and platinum electrodes. Their intent was to cause the deuterium nuclei from the heavy water to fuse together using electrical current by forcing the nuclei into the palladium atomic lattice. The researchers asserted that a chemical reaction in the device had produced energy, in the form of heat, with an output that was four times greater than the energy input.

The Pons-Fleischman announcement received considerable attention from the international news media, generating widespread public interest in cold fusion as a potential solution to world energy needs. Public anticipation of a limitless energy supply rose to a frenzy. In the weeks and months following the announcement, scientists at several other institutions around the world claimed to have achieved similar results when replicating the experiment. Few, however, were bold enough to claim with any certainty that it was the result of cold fusion. Many other scientists who tried to replicate the experiment reported no such results. Eventually, the claim that this experiment resulted in cold fusion was proven false.

Some scientists suggested that the Pons-Fleischman experiment was an inadvertent replication of a 1924 experiment by University of Berlin researchers Fritz Paneth and Kurt Peters and Swedish scientist John Tandberg. Paneth, Peters, and Tandberg had claimed that a similar device could be used to produce helium. They, too, were proven wrong. All scientists make mistakes, and Pons and Fleischman were no different. In rushing to announce their discovery, they had circumvented the traditional peer review process that most likely would have pointed to the flaws in their conclusion and would have avoided the confusion and embarrassment that resulted from their premature announcement.

Despite the doubts about cold fusion held by mainstream science, some scientists and nonscientists continue to believe that it is possible. Current knowledge in particle physics suggests that it is not.

Todd A. Hanson

Sources

Huizenga, John R. *Cold Fusion: The Scientific Fiasco of the Century.* Rochester, NY: University of Rochester Press, 1992.

Taubes, Gary. *Bad Science: The Short Life and Weird Times of Cold Fusion.* New York: Random House, 1993.

COMPTON, ARTHUR HOLLY (1892–1962)

Arthur Holly Compton was an American physicist and Nobel laureate whose studies of X-ray scattering led to the particle concept of electromagnetic radiation. While he is best known for this physics work, Compton's life combined work in science with service to both academia and the nation at large.

He was born on September 10, 1892, in Wooster, Ohio. His father, Elias Compton, was a professor of philosophy and dean of the College of Wooster. Young Compton studied at Wooster and received his Bachelor of Science degree in 1913. He received a Ph.D. from Princeton University in 1916 for his studies of the angular distribution of X-ray reflection from crystals as a means of studying atomic structure.

After a year of teaching physics at the University of Minnesota, Compton became a research engineer at the Westinghouse Lamp Company in Pittsburgh, Pennsylvania. In 1919, he was appointed a National Research Council Fellow at Cambridge University's Cavendish Laboratory. The following year, he was hired as the head of the Department of Physics and the Wayman Crow Professor of Physics at Washington University in St. Louis, Missouri.

At Washington University, Compton continued the X-ray reflection studies he had begun in his dissertation. The research would lead to a Nobel Prize in 1927 for his discovery of the effect created when photons collide with electrons—a simultaneous increase in wavelength and decrease in energy. The discovery introduced the particle concept of electromagnetic radiation, which later become known as the Compton effect. It proved crucial to understanding the absorption of shortwave electromagnetic radiation and the newly discovered phenomenon of cosmic rays, which Compton also studied with great interest.

In 1923, Compton took a job as professor of physics at the University of Chicago, where he remained for the next two decades. There, he also began a lifelong commitment to public service in government and academia. In 1934, he served as president of the American Physical Society.

He was appointed chair of the National Academy of Sciences Committee to Evaluate Use of

Atomic Energy in War in 1941. In this position, Compton appointed J. Robert Oppenheimer as the committee's lead theorist; Oppenheimer would go on to lead the Manhattan Project. Compton also directed the University of Chicago's Metallurgical Laboratory, or Met Lab. In December 1942, he worked with Enrico Fermi, another Manhattan Project luminary, to help create the first sustainable nuclear chain reaction. That same year, Compton was named president of the American Association for the Advancement of Science.

In 1946, Compton returned to Washington University as the university's ninth chancellor, and he served in that capacity until 1953. He remained on the Washington University faculty as Distinguished Service Professor of Natural Philosophy until his retirement in 1961. Compton died in Berkeley, California, on March 15, 1962.

Todd A. Hanson

Sources

Compton, Arthur H. *Atomic Quest: A Personal Narrative.* New York: Oxford University Press, 1956.
———. *Cosmos of Arthur Holly Compton.* New York: Alfred A. Knopf, 1967.

CYCLOTRON

The cyclotron is a device for accelerating subatomic particles to nearly the speed of light. In early 1930, American physicist Ernest Orlando Lawrence began constructing a device at the University of California, Berkeley, that he believed could use magnetic fields in a vacuum to bend the path of charged particles emitted from a radioactive source around a semicircular trajectory.

By pairing two hollow D-shaped electrodes, called "dees," back to back under a spinning magnet and then reversing the magnetic field just as the particles completed traveling one of the half circles, Lawrence's device accelerated the particles across the gap into the second dee and a second field to make a complete circle. As the particles cycled through the same constant magnetic fields over and over again, they accelerated in a spiral path, until they emerged from the cyclotron at speeds close to the speed of light.

During the course of 1930, graduate student Milton Stanley Livingston carried out much of the actual hardware construction under Lawrence's direction. In January 1931, the duo had their first success. Using a device 4.5 inches (11.4 cm) in diameter, Lawrence and Livingston used a charge of 1,800 volts to accelerate hydrogen ions to energies of 80,000 electron volts (the standard units for measuring particle energy). The following year, using a larger cyclotron with an 11 inch (28 cm) diameter magnet, they achieved an output of more than 1 million volts-electrons.

In the decades that followed, cyclotrons became one of the basic tools of nuclear physics research, as more were built at universities and laboratories around the world. In 1939, Lawrence received the Nobel Prize in Physics for the invention and development of the cyclotron and for research results on artificial radioactive elements obtained with it.

As physicists built larger cyclotrons based on Lawrence's designs, they discovered a speed limit at which the mechanics of classical physics no longer worked as well. The machines were limited by a constraint described by Einstein's theory of relativity, the idea that mass, length, and time change with velocities near the speed of light. To adapt to the relativistic speed of particles, larger cyclotrons had to be modified so that the frequency of the accelerating voltage changed as the particles accelerated. These frequency-modulated cyclotrons, or synchrocyclotrons, evolved into the current generation of synchrotrons, in which the magnetic fields and electric fields are carefully synchronized to produce particle beam energies of several hundred million electron volts.

Synchrotrons are used in basic nuclear physics research to further understanding of atomic nuclei. Synchrotrons are also used in material analysis and to make special and rare isotopes for scientific research and medical uses. Work with cyclotrons has resulted in the positron emission tomography (PET) scans used by hospitals to create three-dimensional images of biochemical activity.

Todd A. Hanson

Sources

Baron, E., and M. Lieuvin. *Cyclotrons and Their Applications 1998.* Proceedings of the Fifteenth International Conference on Cyclotrons and Their Application. Bristol, UK: Institute of Physics Publishing, 1999.
Wilson, Edmund. *An Introduction to Particle Accelerators.* Oxford, UK: Oxford University Press, 2001.

EINSTEIN, ALBERT
(1879–1955)

Widely characterized for nearly a century as the quintessential theoretical physicist, Albert Einstein is an icon of American science. His influence has surpassed that of giants of previous centuries, even as theorists continue to debate the precise implications of his theories and models.

He was born in Ulm, Germany, on March 14, 1879, to Hermann and Pauline Einstein. Shortly after his birth, the family moved to Munich, where he attended school at the Luitpold Gymnasium. In 1894 his family moved to Pavia, Italy; Einstein joined them a year later without finishing school or getting his secondary school certificate. When he failed the entrance examination for the Swiss Federal Institute of Technology, or Zurich Polytechnic, his parents sent him to the Swiss city of Aarau to finish secondary school. After receiving his diploma in 1896, he was finally able to enroll in the Zurich Polytechnic and graduated with a teaching diploma in 1900.

In 1901, Einstein became a Swiss citizen; the following year, he took a job as a technical assistant examiner at the Swiss Patent Office in Bern. On January 6, 1903, he married Mileva Maric, a Serbian mathematician he had met in college; a little more than a year later, the couple's first son was born. In 1905, Einstein received his doctorate for his thesis "A New Determination of Molecular Dimensions," but he remained employed as a patent clerk. That would soon change.

Annus Mirabilis

While working at the patent office, Einstein wrote and submitted four papers to the physics journal *Annalen der Physik* (Annals of Physics) in 1905. These papers would have such a profound effect on the course of twentieth-century physics that they came to be known as the "Annus Mirabilis Papers" (from the Latin *annus mirabilis*, meaning "year of wonders"). In the first, "On a Heuristic Viewpoint Concerning the Production and Transformation of Light," Einstein proposed the idea of "light quanta" (now called photons) and theorized how the concept could be used to explain the photoelectric effect. The article

The work of theoretical physicist Albert Einstein (left) was nothing less than a revolution in the human understanding of physical reality. He is seen here with American Nobel laureate Arthur Compton in 1940. *(Keystone/Hulton Archive/Getty Images)*

marked a momentous break with classical physics and would become one of the seminal papers of the fledgling field of quantum physics.

In the second paper, "On the Motion—Required by the Molecular Kinetic Theory of Heat—of Small Particles Suspended in a Stationary Liquid," Einstein put forth his theories of Brownian motion, the physical phenomenon named after Scottish botanist Robert Brown, who noticed in 1827 how minute pollen particles floating in water seemed to move around randomly. Relying on the kinetic theory of fluids, which asserted that gases are made up of molecules in constant random motion, Einstein theorized that molecules of water also move at random and that, in any short period of time, a small particle would receive any number of random impacts from random directions. These impacts would cause a sufficiently small particle to be in motion the way Brown described. Einstein's theory provided the first empirical evidence for the reality of atoms.

Einstein's third paper of 1905, "On the Electrodynamics of Moving Bodies," introduced the theory of relativity for which he would become widely known. The paper provided a new way of understanding the relationship between time and space that came to be called the "theory of special relativity," in order to distinguish it from his later "theory of general relativity." The paper was not perfect, and the paradoxes it contained earned Einstein a considerable amount of ridicule from the scientific world. Eventually, he worked out the apparent contradictions and the theory gained general acceptance.

Titled "Does the Inertia of a Body Depend upon Its Energy Content?" Einstein's fourth paper was the one that would earn him the most recognition with the general public, putting forth the famous $E=mc^2$ equation—or, the energy (E) of a body at rest equals its mass (m) times the speed of light (c) squared. Building on the theory of special relativity, the paper contended that mass and energy are interchangeable.

In 1911, Einstein was appointed extraordinary professor of theoretical physics at the University of Zurich. The following year, he was named full professor at the Zurich Polytechnic, and, in 1914, he adopted German citizenship after taking a position as a professor in the University of Berlin and the director of the Kaiser Wilhelm Physical Institute (now the Max Planck Institute). Einstein and his wife separated when he moved to Berlin; in 1919, they divorced, and he married his cousin Elsa Löwenthal.

Berlin Years

In Berlin, Einstein built his reputation and produced a series of stunning physics theories. In 1916, he published his paper on the general theory of relativity. Many historians of physics believe that this paper was probably Einstein's greatest intellectual achievement. In 1917, he wrote a paper in 1917 on the stimulated emission of light that laid the foundation for the invention of the laser.

For his discovery of the law of the photoelectric effect, Einstein received the 1921 Nobel Prize in Physics. Recognition of the paper's true seminal influence had come in 1919, when British astrophysicist Arthur Eddington's measurements during solar eclipses demonstrated how the light emanating from a distant star was bent by the sun's gravity as it passed by. Despite the apparent evidence, many scientists were still unconvinced of the theory's validity, for reasons ranging from simple disagreement with interpretation to intolerance for Einstein's lack of an absolute frame of physical reference.

In 1922, Einstein published his first work on the unified field theory, which would be an enduring intellectual quest for the rest of his life, as he looked for a classically based unifying theory of gravity and electromagnetism. He also lectured on his work in venues around the world.

In 1925, he received a paper from a young physicist from India named Satyendra Nath Bose. The paper described light as a gas of photons. When Einstein realized that Bose's theory could be applied to atoms in a gas, he published an article that incorporated Bose's model and explained its implications. Thus, "Bose-Einstein statistics" are now used to describe assemblies of elementary particles called bosons.

Einstein remained in Berlin until 1933, when he renounced his German citizenship after Adolf Hitler came to power. Emigrating to the United States, Einstein took a position at the new Institute for Advanced Study at Princeton University in New Jersey.

Princeton Years

Einstein's years at Princeton were filled with as many political issues as scientific ones. He considered himself a pacifist and a humanitarian and had moved to the United States to be free from the nascent Nazi regime in Germany.

To guarantee that Hitler did not build the first atomic bomb, Einstein signed a letter to President Franklin Roosevelt on August 2, 1939, urging Roosevelt to initiate a vigorous atomic bomb research program. Roosevelt responded to Einstein's appeal by setting up a committee charged with investigating the use of uranium in a weapon. The committee's investigation resulted in the creation of the Manhattan Project.

In later years, Einstein opposed nuclear weapons development and pushed for nuclear disarmament. At the July 1955 Pugwash Conferences on Science and World Affairs in London, he and co-author Bertrand Russell, the British philosopher, released the Russell-Einstein Manifesto, which called for all nations to renounce nuclear weapons.

Einstein wrote a number of books, the most important of which include *Special Theory of Relativity* (1905), *General Theory of Relativity* (1916), *Relativity* (1920), *Investigations on Theory of Brownian Movement* (1926), and *The Evolution of Physics*

(1938). He also received a number of honorary degrees from major American and European universities.

Einstein spent the last decade of his life focused on the unification of the laws of physics, which he referred to as the unified field theory, while trying to develop a generalized theory of gravitation. His legacy to the world will always be his science, but, in many ways, he was more than a great physicist. To many, Albert Einstein embodied genius, intellect, compassion, and the triumph of reason over the unknown. He died on April 18, 1955, in Princeton, New Jersey.

Todd A. Hanson

Sources

Bolles, Edmund Blair. *Einstein Defiant: Genius Versus Genius in the Quantum Revolution.* Washington, DC: Joseph Henry, 2004.

Einstein, Albert. *Relativity: The Special and General Theory.* Trans. Robert W. Lawson. New York: Routledge, 2001.

Highfield, Roger, and Paul Carter. *The Private Lives of Albert Einstein.* London: Faber and Faber, 1993.

Pais, Abraham. *Subtle Is the Lord: The Science and the Life of Albert Einstein.* Oxford, UK: Oxford University Press, 1982.

Stachel, John. *Einstein's Miraculous Year: Five Papers That Changed the Face of Physics.* Princeton, NJ: Princeton University Press, 1998.

FERMI, ENRICO (1901–1954)

Enrico Fermi was one of America's most brilliant and best-known physicists. He received the 1938 Nobel Prize for his work on developing the experimental proof of the role of neutrons in nuclear fission and artificial radioactivity. He also played a critical role in the Manhattan Project and the development of atomic energy.

Fermi was born on September 29, 1901, in Rome, Italy. At the age of seventeen, he began studies at the University of Pisa, graduating in 1922 with a doctorate in physics. After doing post-doctoral studies with Max Born in Göttingen, Germany, and then Paul Ehrenfest in Leiden, the Netherlands, Fermi returned to Italy in 1925 and spent the next two years as a lecturer in mathematical physics and mechanics at the University of Florence. In 1927, he became a professor of theoretical physics at the University of Rome.

During the 1930s, Fermi worked on radioactivity, methodically going through the periodical chart, bombarding elements with neutrons to see the result. In 1938, he discovered the phenomenon of slow neutrons, in which neutron bombardment actually increases radioactivity. Also in 1938, Fermi went to Sweden to accept the Nobel Prize for his work with artificial radioactivity produced by neutrons and nuclear reactions caused by slow neutrons. Immediately after the award ceremony, Fermi and his family left for the United States to escape Italy's fascist dictatorship. In New York, Fermi worked as a professor of physics at Columbia University. In 1942, Fermi left Columbia for the University of Chicago, where he supervised a team constructing the world's first nuclear reactor. On December 2, 1942, he and his team initiated the world's first controlled, self-sustaining nuclear chain reaction, which operated for twenty-eight minutes and produced roughly 200 watts of power before being shut down.

Fermi became an American citizen in July 1944; shortly thereafter, he moved from Chicago

Enrico Fermi fled Fascist Italy in 1938, the year he won the Nobel Prize in Physics. His theoretical and practical work led to the first sustained nuclear reaction in 1942 and the construction of the first atomic bomb in 1945. *(Hulton Archive/Getty Images)*

to Los Alamos, New Mexico, to lead the F (for "Fermi") Division of the Manhattan Project. This division conducted both theoretical and experimental physics work in support of the development of the atomic bomb.

After World War II, Fermi became a professor at the University of Chicago. There, he turned his attention to high-energy physics and cosmic rays. He taught there until his death on November 29, 1954.

On May 11, 1974, the National Accelerator Laboratory in Batavia, Illinois, was renamed the Fermi National Accelerator Laboratory in his honor. Also named for him are the fermion, an atomic particle, and the element fermium.

Todd A. Hanson

Sources

Cooper, Dan. *Enrico Fermi and the Revolutions of Modern Physics.* Oxford, UK: Oxford University Press, 1998.
Fermi, Laura. *Atoms in the Family: My Life with Enrico Fermi.* Albuquerque: University of New Mexico Press, 1988.

FEYNMAN, RICHARD
(1918–1988)

Richard Feynman made a variety of notable contributions to theoretical physics and was a participant in the Manhattan Project during World War II. A prolific physicist, Feynman used mathematics to explain liquid helium at extremely cold temperatures, provided an explanation for electrons in high-energy collisions, and discovered the cause of the 1986 *Challenger* space shuttle tragedy.

Richard Phillips Feynman was born in Far Rockaway, Long Island, New York, on May 11, 1918. Although his parents were not college educated, they recognized the value of a scientific education. His father encouraged him to play with objects and arrange them in sets or patterns, always stressing the importance of math. Richard was encouraged to experiment with chemicals, electricity, and mechanical inventions. He built a motor to rock his sister's cradle and a burglar alarm for the house.

In high school, he found algebra too easy. In geometry and trigonometry, he amused himself by challenging traditional concepts and devising his own sets of symbols. By the age of fifteen,

Feynman had already mastered what most graduating seniors find a struggle. When the Great Depression forced his family to move from their home into a small apartment, Feynman found a new project in collecting broken radios or typewriters to fix. At graduation time, his classmates named him "the mad genius."

Feynman earned his B.A. from the Massachusetts Institute of Technology in 1935 and his Ph.D. in physics from Princeton in 1942. While at Princeton, working with the isotron project, he was recruited for the Manhattan Project at Los Alamos, New Mexico. During the course of that endeavor, Feynman discovered that government bureaucracy had created dangerous conditions in the workplace. He systematically examined the rooms, making note of which chemicals were being stored without proper safety precautions. He reported his findings to senior U.S. Army officers, thereby averting a potential disaster in the building. At the same time, Feynman displayed a unique talent—making play of work.

Unfortunately, at a time when his professional life was burgeoning, his personal life was devastated in 1945 by the death of his wife, Arline, who had been confined to a sanatorium for tuberculosis. In 1957, Feynman met Gweneth Howarth, an English librarian, and they were married shortly thereafter. They had one son, Carl, in 1962; six years later, they adopted a daughter, Michelle Catherine.

After World War II, Feynman, went to Cornell University, where he studied antimatter: particles with equal mass but opposite electric charges. His diagrams demonstrated the relationship between the negatively charged electron in matter and the positively charged positron in antimatter. Intrigued by physicist Paul Dirac's call for "some essentially new physical ideas" in the field of quantum physics, Feynman examined the behavior of atomic particles and discovered the infinite energy of interaction between electrons. The resulting "Feynman Diagram" demonstrates the interactions of particles in both space and time. In 1950, Robert Bacher, a friend from the Atomic Energy Commission at Los Alamos, recruited Feynman for the physics department at the California Institute of Technology, where he used the Feynman Diagram to teach his classes.

In 1965, Feynman shared the Nobel Prize in Physics with Shin-Ichiro Tomonaga and Julian

Schwinger; Feynman was cited for his work on quantum electrodynamics (QED). Although QED could not predict what would happen in a given experiment, it could predict the statistical probability of what would happen. In addition, Feynman was interested in the role that subatomic particles, specifically protons, played in QED.

His research with superconductors involved materials that would conduct electricity without offering resistance. This would improve the efficiency of systems where heat and light, the result of resistance, would be minimized or nonexistent. He also worked with superfluidity, the twin of superconductivity. Superfluidity is the tendency of a substance, specifically a fluid, to resist viscosity. Dry water, which sounds like an oxymoron, would be such a substance. Feynman pointed out that superfluid helium resembled such a substance.

Between 1965 and 1985, he worked on a variety of projects at Cal Tech, including quantum chromodynamics, a synthesis of field theories and quark jets. During this time, many physicists came to work with him; these included Murry Gell-Mann, with whom Feynman would collaborate for the rest of his life. In 1972, at the annual meeting of the American Physical Society, Feynman was given the Oersted Medal for contributions to the teaching of physics.

On January 28, 1986, the space shuttle *Challenger* burst into flames shortly after take-off, killing all seven of the astronauts onboard. Feynman was recruited to serve on the commission established to determine the cause of the disaster. After examining photographs and drawings, and meeting with engineers, Feynman turned his attention to the O-rings, made of synthetic rubber, and the seals on the field joints. The solid rocket boosters were in sections held together by joints that had to be sealed to prevent the escape of hot gasses.

As Feynman questioned the members of the commission, he found that there had been a long history of problems with the O-rings. In a simple demonstration with cold water and an O-ring, he showed that lowering the temperature of the water caused the O-ring to crack. Cold weather on the day of the launch had diminished the effectiveness of the seals. The problem, it turned out, had been known for a long time but ignored. Feynman's presentation was as succinct

as it was powerful. He concluded his findings by stating, "For a successful technology, reality must take precedence over public relations, for nature cannot be fooled."

Feynman's health declined during the last decade of his life. He experienced myxoid liposarcoma, a rare gastrointestinal cancer, during the late 1970s and early 1980s. In the mid-1980s, a second rare cancer, Waldenstroms macroblobulinemia, attacked his body, further weakening him. He died on February 15, 1988.

Lana Thompson

Sources

Feynman, Richard P., with Ralph Leighton. *Surely You're Joking Mr. Feynman.* New York: W.W. Norton, 1985.
Gleick, James. *Genius: The Life and Science of Richard Feynman.* New York: Vintage Books, 1992.

FISSION

Nuclear fission is a reaction in which the nucleus of a heavy, unstable atom splits, or fissions, into two or more lighter nuclei, releasing massive amounts of energy. The discovery of fission by European scientists paved the way for technological application by American scientists, notably the development of atomic weapons.

The science of fission began in Europe in the early twentieth century. Scientists such as Marie Curie, Ernst Rutherford, and Niels Bohr discovered the structure of the atom as well as the unique atomic nuclei of heavy elements such as radium and uranium. Nuclear fission occurs in nature slowly and spontaneously as radioactive decay.

In 1932, English scientist James Chadwick discovered the neutron of the atom, which has a neutral charge (in contrast to the positively charged proton and the negatively charged electron). In 1934, Hungarian scientist Leo Szilard theorized that a neutron colliding with an atom of a heavy element would produce a change in the atom and the release of energy. The reaction also releases two or more additional neutrons. If the path of these neutrons is not blocked, or controlled, some of the neutrons hit other atomic nuclei, causing them to fission also. If there is enough fission fuel available—a critical mass—for the production of free neutrons by fission to be greater than those lost, the result is a self-sustaining nuclear chain reaction. An uncontrolled self-sustaining nuclear chain reaction results in an explosive release of energy. In 1938, German scientists Otto Hahn and Lise Meitner recognized and named the concept of a nuclear fission reaction. They used the "liquid drop model," in which a spherical nucleus undergoing fission begins to elongate into a dumbbell shape until it reaches a point of no return and splits at the neck into two equal or nearly equal fragments, releasing energy in the process.

The fission research of the 1930s made it possible for a team led by Enrico Fermi—who had moved from Italy to the United States in 1939—to explore fission's potential as a source of nuclear energy. Using uranium as a neutron source and 400 tons of graphite as a neutron moderator, and working in a room that had formerly been a squash court, Fermi's team initiated a self-sustaining nuclear chain reaction on December 2, 1942. Fermi's fission research in Chicago, and then at the Manhattan Project in Los Alamos, New Mexico, led to the development of the atomic bomb and the world's first uncontrolled nuclear fission explosion in the New Mexican desert in July 1945.

Fission research after World War II focused primarily on making smaller, more efficient, fission nuclear weapons and on the development of nuclear reactor technologies. Although the nuclear fusion weapons developed during the early years of the Cold War replaced most of the fission-based weapons, the thermonuclear designs continued to use a nuclear fission device called a "primary" to initiate the fusion reaction.

In 1953, President Dwight D. Eisenhower proposed his Atoms for Peace program for research and development of the use of nuclear energy for electrical power generation. By 1957, the Atomic Energy Commission began operation of the Shippingport nuclear reactor in Pennsylvania. In the decades that followed, many countries built nuclear fission reactors to generate electrical power. Fission also has had many other civilian, military, and space applications.

Today, fission science research in the United States continues to make important, if incremental, advances in knowledge. In 2001 physicists at Los Alamos National Laboratory and the Japanese Atomic Energy Research Institute developed a more comprehensive understanding of the mechanisms underlying nuclear fission, using a computer model that defined critical shapes of elongation, neck diameters, fragment deformation, and mass division in the liquid drop model. The research allowed scientists to draw a number of new conclusions about the fission process.

One recent discovery is that, for some lighter actinide elements, two fission paths exist: one where the fissile particle divides into unequal fragment masses, and another with equal fragment mass divisions. Such discoveries made by ongoing fission research illustrates that even a well-known, established scientific concept can be open to new and different interpretations.

Todd A. Hanson

Sources

Cowan, George A. "A Natural Fission Reactor." *Scientific American* (July 1976): 36–47.

Graetzer, Hans, and David Anderson. *The Discovery of Nuclear Fission.* New York: Van Nostrand Reinhold, 1971.

Mackintosh, Ray, ed. *Nucleus: A Trip into the Heart of Matter.* Baltimore: Johns Hopkins University Press, 2002.

Rhodes, Richard. *The Making of the Atomic Bomb.* New York: Simon and Schuster, 1995.

FUSION

Nuclear fusion is the reaction that occurs when two lightweight atomic nuclei are brought together under extremely high temperatures to fuse and form a heavier element particle. The union of the atomic nuclei releases vast quantities of energy. Fusion is the underlying principle of thermonuclear weapons and stellar burning.

English physicists Ernest Rutherford, Marcus Oliphant, and Paul Harteck discovered the hydrogen fusion reaction at Cambridge University in 1934 after accelerating deuterium nuclei into deuterium in the form of heavy water. Driven by the energy of acceleration, hydrogen nuclei fuse to create helium, hydrogen's heavier neighbor in the periodic table of elements. The fusion process creates energy in the form of thermal, neutron, and gamma radiation. Because the Cambridge fusion reaction was not self-sustaining, it was not useful as an energy source. However, the discovery laid the foundation for the development

Tokamak is a device use to produce a doughnut-shaped magnetic field in plasma physics research. The long-term goal of researchers is controlled nuclear fusion as a viable source of energy. *(Yale Joel/Time & Life Pictures/Getty Images)*

of the hydrogen bomb, which uses the energy of a fission-based nuclear device to rapidly compress and then ignite a mass of fusion fuel to create a massive uncontrolled thermonuclear reaction.

Nuclear fusion research in the United States has involved research and development of uncontrolled fusion found in thermonuclear weapons. Uncontrolled nuclear fusion research in the United States began with a design proposed by Stanislaw Ulam and Edward Teller in 1951. Called the Teller-Ulam Configuration, the design used a fission nuclear device called a "primary" as a trigger to compress and heat a mass of fusion fuel—usually an isotope of hydrogen—through a process called radiation implosion. The fusion of the hydrogen nuclei releases large quantities of energy in the form of an explosion. The Teller-Ulam design was modified and refined during the Cold War through a series of thousands of above-ground and underground nuclear weapons tests.

In 1993, the United Nations General Assembly began negotiations for the creation of a comprehensive test-ban treaty prohibiting all nuclear explosions in all environments. Since that time, nuclear weapons fusion research in the United States has been restricted to subcritical nuclear tests—explosions driven by conventional explosives where no critical mass is formed and no self-sustaining nuclear chain reaction occurs.

Controlled nuclear fusion has become the holy grail of nuclear physics research in the United States because of its potential as an energy source. Scientists are studying magnetic confinement fusion and inertial confinement fusion as possible ways to achieve a sustained nuclear fusion reaction. Technical difficulties lie in the inability to confine or contain the extremely high temperatures (in excess of 100 million degrees Celsius) needed to achieve fusion.

Controlled nuclear fusion's association with nuclear weapons and its confusion with cold fusion have hindered research. Nuclear fusion is not cold fusion, which most scientists consider technically improbable and perhaps impossible. If nuclear fusion can be harnessed, only small amounts of material would be needed to produce vast amounts of energy. Experts estimate that a mere 10 grams of deuterium, extracted from seawater, and 15 grams of tritium, produced from lithium, would produce enough energy to meet the lifetime electricity needs of an individual.

Todd A. Hanson

Sources

Bromberg, Joan Lisa. *Fusion: Science, Politics, and the Invention of a New Energy Source.* Cambridge, MA: MIT Press, 1982.

Herman, Robin. *Fusion: The Search for Endless Energy.* New York: Cambridge University Press, 2006.

Rhodes, Richard. *Dark Sun: The Making of the Hydrogen Bomb.* New York: Simon and Schuster, 1995.

Zirker, Jack B. *Journey from the Center of the Sun.* Princeton, NJ: Princeton University Press, 2001.

GELL-MANN, MURRAY (1929–)

The American theoretical physicist Murray Gell-Mann is known as the discoverer of the quark, an elementary particle comprising protons and neutrons, which, in turn, form the nucleus of the atom. Gell-Mann also proposed the "eightfold way," a simplification of the hierarchy of all subatomic particles, and developed a system of particle classification according to their isotopic spin (or isospin) and strangeness. This approach brought order to the more than 100 subatomic particles found lurking in the nucleus of the atom. However, unlike Dmitri Mendeleev's periodic table for chemistry, based on visible properties such as mass and chemical behavior, Gell-Mann's system has no counterparts in the world of everyday experience. In fact, for many years Gell-Mann believed that his beloved quarks were nothing more than a mathematical construct to be used to explain subatomic particle behavior.

Born in New York City on September 15, 1929, Gell-Mann was recognized as a child prodigy. He obtained a Ph.D. from the Massachusetts Institute of Technology in 1951. His first significant theory was that particles exhibit "strangeness," an explanation of why some unstable particles disintegrate more slowly than others. This idea of a fundamental new property led to the notion that all subatomic particles fall into orderly patterns, defined by how much strangeness and electric charge they possess.

Like Mendeleev's periodic table, used to predict new elements, Gell-Mann's "eightfold way" proposal of 1962 suggested that missing particles would be discovered. He calculated the theoretical characteristics of the missing particles, including their mass. When physicists discovered the particle omega-minus in 1964, the mass of the particle was within one-half of 1 percent of Gell-Mann's estimate. Soon after, Gell-Mann predicted the existence of even more fundamental particles in various combinations, which he called "quarks." Experiments bombarding hydrogen atoms with electrons soon detected three quarks of two different types comprising both protons and neutrons.

In 1969, Gell-Mann was awarded the Nobel Prize in Physics for his work on subatomic particles; it was one of the few unshared Nobel awards in physics. Gell-Mann also played a key role in explaining how quarks and the force among them prevent the protons, neutrons, and atomic nucleus from pulling apart: the strong force permanently confines the quarks through the exchange of gluons. From this, he and many other physicists developed the field theory of quantum chromodynamics.

Robert Karl Koslowsky

Sources

Gell-Mann, Murray. *The Quark and the Jaguar.* New York: W.H. Freeman, 1994.

Johnson, George. *Strange Beauty: Murray Gell-Mann and the Revolution in 20th-Century Physics.* New York: Alfred A. Knopf, 1999.

Koslowsky, Robert. *A World Perspective Through 21st Century Eyes.* Victoria, Canada: Trafford, 2004.

Stehle, Philip. *Physics: The Behavior of Particles.* New York: Harper and Row, 1971.

GRAVITY

In the terms of the general theory of relativity, gravity is the curve of space and time that results from a concentration of mass or energy. European theorists over the space of several millennia developed an understanding of gravity, the particulars of which American physicists have illustrated through experimentation.

The European search to understand gravity began with the Greek philosopher Aristotle, who believed that each thing has a natural place and motion associated with it. The natural place for objects containing large amounts of earth was at the center of things—Earth. If an Earth object was displaced from its natural spot, it attempted to return to the center. Aristotle's concept of natural motion held until Italian scientist Galileo Galilei's investigations toward the end of the sixteenth century. Galileo showed through experimentation that all falling objects accelerate at the same rate and that the velocity of a falling object is proportional to the square of the fall time.

The observations of Galileo and others enabled English physicist Isaac Newton to determine his law of gravity in the seventeenth century. Newton proposed that any mass attracts any other mass along a line drawn between their centers. Further, the strength of attraction is proportional to the size of two masses and inversely proportional to the square of the distance between them. In the early twentieth century, Albert Einstein's general theory of relativity understood gravity according to the distortion of space-time, which causes an object's normally straight path to curve along with space-time.

The European theory of gravity required experimental confirmation. In 1916, Einstein proposed three tests of his general theory of relativity: the variations in the perihelion of Mercury, the deflection of light by a gravitational field, and the red-shifting of radiation by gravity. American scientists were involved in the latter two tests. The first attempt to demonstrate that light was deflected by a gravitational field occurred during a total eclipse of the sun in 1919. English scientists Arthur Eddington, Frank W. Dyson, and Charles R. Davidson measured the position of certain stars in the sky at night and then measured the position of the same stars again during the eclipse. Presumably, the presence of the sun would cause a deflection in the light and an apparent shift in the position of the stars. The results of these observations agreed with the predictions of general relativity, but many physicists questioned the accuracy of the experiment.

The controversy continued until 1976, when Americans Edward B. Fomalont and Richard A. Sramek published the results of their work at the National Radio Astronomy Observatory in West Virginia. The pair used radio interferometry to measure the shift in apparent position of distant quasars during an eclipse. Further confirmation of the bending of light in a gravitational field was found in the form of gravitational lensing. As light from distant stars passes massive objects, its path bends in the same manner that light is bent by a lens. The result can be a shift in apparent position or even the formation of a new image of the distant object. The first observed instant of gravitational lensing was found by Dennis Walsh, Bob Carswell, and Ray Weymann in 1979 at the Kitt Peak National Observatory in Arizona.

Red-shifting occurs when a photon's wavelength is lengthened by some process. Gravitational red-shifting of photons was verified by Robert V. Pound and Glen A. Rebka, Jr., in 1959 at the Lyman Laboratory at Harvard University. In the experiment, the two scientists monitored the frequency of photons emitted from unstable iron atoms located at different distances from the surface of Earth. They found that the photons emitted closer to the surface, where there is a stronger gravitational field, experienced the red-shift predicted by general relativity.

One of the most important and common applications of general relativity is found in the global positioning satellites (GPS). The GPS system, developed by the U.S. Department of Defense, is a set of twenty-four satellites in Earth orbit. Each satellite emits a time signal that is synchronized with the other satellites. A receiver that detects the signal from at least four of these satellites is able to pinpoint its longitude, latitude, and altitude with a margin of error of less than one meter. To achieve this accuracy, the atomic clocks on board the satellites must be synchronized within about four nanoseconds. Such agreement requires the extensive use of general relativity to correct for gravitational effects such as gravitational time dilation and gravitational red-shift. Without correction for general relativistic effects, positioning could drift by as much as ten kilometers per day.

As is indicated by its use with the GPS system, general relativity is one of the best tested and most accurate theories of modern physics. However, the theory of general relativity is apparently incompatible with quantum physics, another modern and extremely well tested theory. It is currently not possible to produce a quantum theory of gravity, but this remains one of the most heavily investigated areas of physics. Reconciliation of gravity and quantum theory would almost certainly produce important new understandings of the world.

R. Dwayne Ramey

Sources

Feynman, Richard P., Robert B. Leighton, and Matthew Sands. *The Feynman Lectures of Physics.* Vol. 1. Reading, MA: Addison-Wesley, 1977.

Schutz, Bernard. *Gravity from the Ground Up: An Introductory Guide to Gravity and General Relativity.* Cambridge, UK: Cambridge University Press, 2003.

GREENWOOD, ISAAC (1702–1745)

Colonial mathematician and physicist Isaac Greenwood held the first Hollis professorship at Harvard.

Born in Boston on May 11, 1702, the son of a merchant and shipbuilder, Greenwood attended Harvard between 1717 and 1721, studying under mathematics professor Thomas Robie, who introduced him to Newtonian science and calculus. In 1722, Greenwood became involved in the controversy in Boston over smallpox inoculation; his pamphlet *A Friendly Debate* supported inoculation.

A Puritan minister, Greenwood traveled to England in 1723 to preach. There, he became acquainted with members of the Royal Society of London. His ability to interlace scientific knowledge with religious belief won him influential admirers, including Thomas Hollis, a London merchant who had been making donations to Harvard since 1719 and who founded a chair of divinity in 1722. Hollis established a professorship in mathematics and natural philosophy at Harvard, and Greenwood became the first Hollis Professor in 1728.

Greenwood was a prolific author in physics and mathematics. In 1726, he published *An Experimental Course of Mechanical Philosophy*, a textbook used at Harvard. He also contributed observations on meteorology, dampness in wells, and the aurora borealis to the *Philosophical Transactions of the Royal Society*. His 1729 textbook *Arithmetick, Vulgar and Decimal* was used at both Harvard and Yale. In 1731, Greenwood wrote *Philosophical Discourse Concerning the Mutability and Changes of the Material World*, a deist approach to Newton's ideas, in which he examined the commensurability of science and religion.

Greenwood also cataloged the scientific apparatus Hollis donated to Harvard. Hollis's grateful nephew, also named Thomas, sent more equipment to Harvard, including an orrery (a mechanical model of the solar system), which, in part, inspired Greenwood to write *Explanatory Lectures on the Orrery, Armillary Sphere, Globes and Other Machines* in 1734. Four years later, Greenwood published his lecture series *A Course of Mathematical Lectures and Experiments*, followed in 1739 by *A Course of Philosophical Lectures*.

An alcoholic, Greenwood was warned by the Harvard Corporation to get sober numerous times over the course of sixteen months before he was dismissed in 1738. He could not support his household on his public subscription lectures alone, even after he sold his house and land, so he created a traveling scientific show in 1740 that was advertised by his friend Benjamin Franklin.

In desperation, Greenwood served as a Royal Navy chaplain from 1742 to 1744. After his discharge, he died from the effects of alcoholism in Charleston, South Carolina, on October 12, 1745.

Amy Ackerberg-Hastings

Further Reading

Leonard, David C. "Harvard's First Science Professor: A Sketch of Isaac Greenwood's Life and Work." *Harvard Library Bulletin* 29 (1981): 135–68.

Shipton, Clifford K. *Biographical Sketches of Graduates of Harvard University.* Vol. 6. Boston: Massachusetts Historical Society, 1937.

Simons, Lao Genevra. "Isaac Greenwood, First Hollis Professor." *Scripta Mathematica* 2 (1934): 117–24.

HOLLIS PROFESSORSHIP

Thomas Hollis was a wealthy London merchant who endowed chairs of theology, mathematics, and natural philosophy at Harvard College. A Baptist, he was disinclined to support the Anglican universities at Oxford and Cambridge. Believing that people in a young and rustic colony such as Massachusetts had the potential to educate boys into strong, moral men, he donated £5000 in books, scholarship funds, scientific apparatus, and professorial endowments to Harvard between 1719 and 1731. A chair designated as the Hollis Professor of Divinity was first filled in 1722 by Edward Wigglesworth, Sr.

By 1727, Hollis had become convinced that education in the physical sciences was necessary and appropriate for furthering the Christian faith of leaders. He asked for suggestions about establishing a professorship from the hymn writer Isaac Watts as well as from David Neal, Jeremiah Hunt, and Isaac Greenwood.

Greenwood was a twenty-five-year-old Harvard graduate who had impressed members of the Royal Society with his comprehension of experimental philosophy. He became a protégé of Jean Théophile Desaguliers and explained Newtonian physics to general audiences. Hollis appreciated the younger man's intellect but thought his behavior irresponsible, as Greenwood was in debt when he left London for America. Greenwood had influential friends in Boston who lobbied on his behalf, however; in 1728, he was installed at Harvard as the first Hollis Professor of Mathematics and Natural Philosophy.

Early holders of the Hollis professorships made Harvard the leading center for maintaining a connection with European developments in science. Greenwood taught Newtonian science, wrote two textbooks and several treatises, and delivered popular public lectures, before he was removed for intemperance in 1738. His successor, John Winthrop IV, was a productive astronomer. He rebuilt Harvard's instrument collection and library after a devastating fire in 1764, in part, with funds from Hollis's nephew, a lawyer also named Thomas. By the time of his death in 1779, Winthrop was the most notable scientist in the United States.

Samuel Williams continued Winthrop's astronomical excursions between 1780 and 1788, although teaching standards fell somewhat under his leadership and that of Samuel Webber, who held the chair from 1789 to 1806. Webber did compile a widely used mathematics compendium in 1801.

Instruction was revitalized by John Farrar, who during this tenure from 1807 to 1836, replaced Webber's textbook with a series of translations of late eighteenth-century French treatises on mathematics and physics. Farrar and his successor Joseph Lovering collaborated on an 1842 publication, *Electricity, Magnetism, and Electro-Dynamics*, which included "A Course of Natural Philosophy" by Farrar and "A New Course of Physics" by Lovering, who served as Hollis professor from 1838 to 1888. Benjamin Osgood Peirce, who succeeded Lovering and was Hollis professor from 1888 to 1914, published *Elements of the Theory of the Newtonian Potential Function*, as well as *A Short Table of Integrals* for use in mathematics classes.

Recent Hollis professors have been Andrew Gleason, from 1969 to 1992, and Bertrand I. Halperin, who has served in this position since 1992. Papers for each Hollis professor are found at the Harvard Archives.

Amy Ackerberg-Hastings

Sources

Birkhoff, Garrett. "Mathematics at Harvard, 1836–1944." In *History of Mathematics: A Century of Mathematics in America.* Providence, RI: American Mathematical Society, 1989.

Quincy, Josiah. *The History of Harvard University.* 2 vols. Cambridge, MA, 1840.

"Some of Harvard's Endowed Professorships." *Harvard Alumni Bulletin* 29 (1926–1927): 65–69, 145–50, 387–93, 1026–28; 30 (1927–1928): 138–40.

KINNERSLEY, EBENEZER (1711–1778)

Ebenezer Kinnersley was the foremost freelance scientific lecturer in mid-eighteenth-century America.

Born in 1711, Gloucester, England, he immigrated with his family to America in 1714. A Baptist clergyman in the 1730s, Kinnersley was forced from his Philadelphia pulpit when he attacked the emotionalism of the New Lights (who professed a modified Calvinism) during the Great Awakening of the mid-eighteenth century.

Kinnersley's desire to have his polemics printed led him to make the acquaintance of the publisher of the *Pennsylvania Gazette*, Benjamin Franklin. Kinnersley became part of Franklin's intellectual circle. With help and encouragement from Franklin, he commenced a tour of the American colonies as an electrical demonstrator, lecturer, and showman that lasted from 1749 to 1753.

Franklin and Kinnersley operated in many ways as a team. In addition to performing the entertaining tricks with electricity that were the mainstay of the iterant electrical demonstrator, Kinnersley disseminated Franklin's electrical theories. Franklin, who drew up the outline of some of Kinnersley's presentations, helped promote the course of lectures on electricity Kinnersley offered in Philadelphia in 1751.

Kinnersley's presentations, which were advertised in colonial newspapers, featured such standard tricks as the electrified woman, the charge from whose lips would purportedly discourage anyone from kissing her, and "Electrified Money, which scarce any Body will take when offer'd to them," as well as discussions of electrical theory. Kinnersley also offered to kill animals instantaneously with electricity, if the audience provided the animals. More importantly, in the 1750s, Kinnersley ceaselessly promoted Franklin's new invention, the lightning rod, using miniature buildings to show how lightning rods protected them from electricity. Most colonial American electrical demonstrators patterned their demonstrations and advertisements on Kinnersley's, although few could rival the quality of his equipment.

Kinnersley also performed independent research into such questions as the electrical charge of clouds and the relative abilities of different substances to conduct electricity. He was the most active electrical researcher in the American colonies other than Franklin, publishing scientific articles and inventing an "electrical air thermometer." This device tested Kinnersley's theory that electrical current generated heat. In 1758, one of Franklin's enemies accused Franklin of having stolen his electrical theories from Kinnersley, a charge Kinnersley, then Professor of English and Oratory at Philadelphia College, indignantly denied.

In 1774, in a show of patriotism at a Philadelphia demonstration against the enemies of Franklin and the independence movement, Kinnersley used electricity to set fire to effigies of unpopular Tories, such as the governor of Massachusetts, Thomas Hutchinson. Kinnersley died on July 4, 1778.

William E. Burns

Sources

Heilbron, J.L. *Electricity in the 17th and 18th Centuries: A Study in Early Modern Physics.* 2nd ed. Mineola, NY: Dover, 1999.

Lemay, J.A. Leo. *Ebenezer Kinnersley: Franklin's Friend.* Philadelphia: University of Pennsylvania Press, 1964.

LAWRENCE, ERNEST
(1901–1958)

The American nuclear physicist Ernest Lawrence, a Nobel laureate, was famous for inventing the cyclotron, founding the Lawrence Berkeley National Laboratory, and developing what has come to be known as "big science."

Born in Canton, South Dakota, on August 8, 1901, to Carl and Gunda Lawrence, Ernest Orlando Lawrence received his bachelor's degree in chemistry from the University of South Dakota in 1922. The following year, he received his master's degree from the University of Minnesota; he received his Ph.D. from Yale University in 1925.

In 1928, Lawrence was hired as an associate professor of physics at the University of California, Berkeley. Two years later, he was made a full professor, the youngest at Berkeley at the time. By 1931, he had acquired an unused civil engineering laboratory building to accommodate his research. The building would later become the university's Radiation Laboratory and serve as the precursor to today's Lawrence Berkeley National Laboratory. Lawrence remained at Berkeley as a professor and director of the Radiation Laboratory until his death on August 27, 1958, in Palo Alto.

Lawrence received the Nobel Prize in Physics in 1939 for research on creating artificial radioactive elements using a cyclotron. With the help of graduate student Milton Stanley Livingston, Lawrence constructed at Berkeley the first cyclotron, a device for accelerating subatomic particles to near the speed of light by using magnetic fields to bend the path of the particles around a semicircular trajectory. Using the cyclotron, Lawrence and other scientists propelled subatomic particles into certain elements, breaking up their atomic structures and creating isotopes of the elements. In the decades that followed, cyclotrons became one of the basic tools of nuclear and particle physics research.

During World War II, Lawrence contributed to the development of the atomic fission bomb as part of the Manhattan Project. Along with scientists working at Princeton University, he devised a method for the large-scale separation of uranium isotopes by electromagnetic means, a method that was later used at the Y-12 laboratory at Oak Ridge, Tennessee. In September 1945, after witnessing the test of the atomic bomb and seeing evidence of the devastation at Hiroshima and Nagasaki, Lawrence joined some of his physics colleagues in opposing, on moral grounds, the development of the hydrogen "superbomb." Lawrence later served as a member of the U.S. delegation to the Geneva Conference of 1958, seeking a ban on nuclear weapons tests.

Lawrence has been called the "father of big science" for his role in developing large-scale scientific projects grouped around a central research goal or instrument and involving numerous scientists from a number of universities or government laboratories. Lawrence employed this model at Berkeley for the use of the cyclotron in isotope research. Often, these big science teams are multidisciplinary, requiring the expertise of researchers in several different technical or scientific fields. This model of large-scale scientific projects has become the norm in many areas of physics research, particularly high-energy physics.

Todd A. Hanson

Sources

Childs, Herbert. *An American Genius: The Life of Ernest Lawrence.* New York: E.P. Dutton, 1968.

Heilbron, J.L., and Robert W. Seidel. *Lawrence and His Laboratory: A History of the Lawrence Berkeley Laboratory.* Berkeley: University of California Press, 1989.

Herken, Gregg. *Brotherhood of the Bomb: The Tangled Lives and Loyalties of Robert Oppenheimer, Ernest Lawrence, and Edward Teller.* New York: Henry Holt, 2002.

LIGHT

Light is a wave of energy. It is the visible portion of the electromagnetic spectrum, the full range of which extends from high-energy cosmic radiation to low-energy radio waves.

Light waves range in size from about 400 to 700 billionths of a meter (or about 0.4 to 0.7 microns). Not unlike an ocean wave, light waves vibrate at right angles to each other, carrying a variety of energy sources. Light travels at a speed of approximately 186,282.3959 miles or 299,792.4574 kilometers per second. When matter and energy interact, light is often the by-product, as when heated steel glows or a wood fire produces light. The source of the majority of light on Earth is the sun, which projects 1,370 watts of light per mile on Earth and nearly 6,000 kelvin of energy.

The English mathematician Isaac Newton first used a prism in 1666 to reveal that white light (sunlight) in fact comprises the colors of the spectrum: violet, blue, green, yellow, orange, and red. He was also able to recombine the spectrum colors back into white light. In 1704, Newton began to publish his historic four-volume *Opticks,* which presented explanations of the properties of light, including color, reflection, and refraction.

In 1800, English astronomer William Herschel discovered infrared light by heating red light and then heating a dark region and seeing the temperature rise much higher. The word *infrared* means "below red." Infrared light is not visible to the human eye, because its wavelength is longer than that of white light.

In 1877, working for the U.S. Navy as a research scientist in Washington, D.C., Albert Michelson revised a previously conducted experiment to measure the speed of light. Michelson used improved optics and equipment to determine the most accurate measure of the speed of light, documenting his findings in the 1878 article "Experimental Determination of the Velocity of Light." For his work, Michelson became the first American to receive the Nobel Prize in Physics, in 1907.

Albert Einstein, while working at the Swiss Patent Office in 1905, theorized that light is comprised of "quanta," indivisible entities later called photons. He studied the photoelectric effect—a quantum electronic phenomenon in which electrons are emitted from matter after it absorbs light energy. For his work, Einstein was awarded the Nobel Prize in Physics in 1921.

The increased understanding of the nature of light led to vital innovations in science, medicine, and technology. Once the relationship between electromagnetism and light was understood, for example, it became possible to produce artificial light using electricity. The inventor Thomas Edison, in Menlo Park, New Jersey, created the first electricity-powered incandescent light bulb in 1879 as a practical means of providing artificial light.

X-rays were put to practical use by a number of physicists, including the Serbian American inventor Nikola Tesla at his lab in New York City in the 1890s. Tesla's contribution to X-ray technology was the single electrode tube, in which charged particles (electrons) were passed inside a cathode tube with tungsten to emit radiation, or what he called "radiant energy." X-rays were eventually used for making medical images of the human body and in the treatment of cancer, killing the specific cells that cause the disease.

The use of radio waves in long-distance communication represented another understanding in the spectrum of light. In 1906, American scientist Lee De Forest invented the audion—a vacuum tube with a third electron—which could amplify radio waves. A pioneer in the modern electronics industry, De Forest was later forced by bankruptcy to sell his design to Bell Laboratories for $50,000.

During the 1950s, American physicists bombarded atoms with electromagnetic radiation, forming light composed of wavelengths that formed a concentrated beam known as a laser beam. Gordon Gould, a graduate student at Columbia University in New York, came up with the design concept in 1957, and in 1960, Theodore Maiman built a laser at the Hughes Research Laboratories in Malibu, California. The laser is now widely used in medicine and communications.

Discovery of the properties of light also led to advances in astronomy and the understanding of the universe. In 1912, Henrietta Swan Leavitt, an astronomer at Harvard College Observatory in Cambridge, Massachusetts, discovered a means to calculate the distance to stars based on their magnitude (brightness). This method ultimately was applied by Edwin Hubble, an astronomer working at the Mount Wilson Observatory in Pasadena, California, in 1929. Hubble had noticed that the light from distant galaxies was shifting toward the red portion of the spectrum. This "red shift" indicated that objects emitting light are moving away from the observer and that the universe is expanding.

In 1965, Bell Laboratory physicists Arno Penzias and Robert Wilson detected long wavelength microwave radiation coming from all directions in space. Working out of the lab in Murray Hill, New Jersey, they were at first annoyed by the interference, then realized that this radiation represented light left over from the "big bang" that had red-shifted to the point of becoming nonvisible microwave radiation.

In 1970, Corning Glass Works in Corning, New York, manufactured the first wire that could carry information via light. Set to a 17-decibel optic attenuation per kilometer, this glass, plastic, and titanium cord was later called optical fiber. Fiber optics would change communications forever by allowing high-speed transmission of voice and data over great distances.

James Fargo Balliett and Ron Davis

Sources

Brill, Thomas. *Light: Its Interaction with Art and Antiquities.* New York: Plenum, 1980.
Sobel, Michael I. *Light.* Chicago: University of Chicago Press, 1989.
Wagner, David. *Light and Color.* New York: Wiley, 1982.
Waldman, Gary. *Introduction to Light: The Physics of Light, Vision, and Color.* New York: Dover, 2002.

LIGHT, SPEED OF

The phenomenon of light was a source of intrigue for centuries among European thinkers and scientists who tried to understand its nature and determine its velocity. After repeated European attempts to determine the speed of light, it was an American, Albert Michelson, who in the late nineteenth century provided the most accurate estimate. Michelson also teamed with Edward Morley to use the speed of light to disprove the existence of ether.

For centuries, scientists believed that light traveled from one point to another instantaneously. It was not until the seventeenth century that Galileo Galilei challenged conventional thinking that the speed of propagation of light could not be measured; in 1630, he used terrestrial distances and flashing lanterns to try to determine its speed. An assistant was placed on a distant hill and flashed light in response to Galileo's lantern signals. This test was repeated on a second set of hills, located farther apart, so that the difference in interval (factoring out the assistant's reaction time) might provide an indication of the speed of light. Galileo noted no extra time in either test, which meant that only the assistant's reaction time was being measured. He concluded that light traveled extremely fast and could not be measured using terrestrial distances.

Other Europeans following in Galileo's footsteps, such as the Danish astronomer Olaus Roemer in the seventeenth century, used celestial distances in their calculations of the speed of light. Roemer recorded the time delay between one of Jupiter's moons producing a shadow on the planet's surface at different points in its orbit, and he calculated the speed of light at 132,000 miles per second.

During the nineteenth century, the French physicist Armand Hippolyte Fizeau used a technique involving reflecting mirrors, instead of flashing lanterns, and a rotating toothed wheel to measure the time it took for reflected light to return to its source. This approach marked the return to terrestrial distances for calculating light speed. Using remote equipment instead of human assistants, Fizeau estimated the speed of light at 196,000 miles per second.

Another nineteenth-century French scientist, Jean Bernard Leon Foucault, adopted Fizeau's method but replaced the toothed wheel with a second mirror. His innovation enabled him to conduct light-speed experiments in a laboratory. Foucault's adjustment allowed the reflected light from the second mirror to be deflected to a spot on a screen. Based on the displacement of a second spot on the screen, resulting from the second mirror rapidly spinning at a predetermined rate, Foucault in 1862 reported a speed of light of 185,000 miles per second. Foucault extended his laboratory setup to measure the speed of light not only in air but also in water and a variety of other materials. The higher the refraction of light moving through a medium, he demonstrated, the slower light travels.

The early experiments in Europe were key to the development of the interferometer by American physicist Albert Abraham Michelson in 1881. An interferometer is a device that splits a light beam in two and sends the resulting beams down different paths until they are brought back together, so that the interference pattern is used to calculate speed. By splitting a beam of light in two and transmitting it across a 700 meter distance, Michelson measured the light's speed at 186,355 miles or 299,895 kilometers per second, very close to James Clerk Maxwell's predicted speed of 300,000 kilometers per second.

Scientists such as Michelson believed that light propagates in an all-pervasive medium of space called ether, and that there must be a way of measuring the motion of the ubiquitous ether relative to Earth. Michelson believed that measuring the speed of light by means of an interferometer would allow him to determine the existence of the ether. In 1887, Michelson teamed with another American physicist, Edward Morley, to detect the ether by measuring relative changes in the speed of light as Earth orbits the sun. They believed their sensitive experimental setup would separate incoming light into two paths, which would allow them to compare the distance light travels in the path moving parallel to the ether with the distance light traverses in the path perpendicular to the ether. The result would appear as a delay in the light beam moving perpendicular to the ether manifest in the interference pattern of the recombined beams. Much to the puzzlement of Michelson and Morley, however, no interference fringes were found.

In 1905, Albert Einstein solved the contradiction raised by the Michelson-Morley experiments. With his theory of special relativity, Einstein showed that light travels at one speed in a vacuum, regardless of how or where it is measured. He postulated that absolute, uniform motion cannot be detected, and that the speed of light is independent of the source's motion. If Earth and the entire testing apparatus are considered to be at rest, no time differences should be found (all directions are equivalent). Thus, there is no ether in which light propagates; rather, light and all other forms of electromagnetic radiation propagate through space without a medium. One of the implications of Einstein's special theory of relativity is that time intervals shorten as a frame of reference moves. This means that a traveler leaving Earth at the speed of light for many years would return having experienced an elapsed time on Earth of only a few months.

The invention in the 1960s of the laser, a light source with a single wavelength, allowed for more accurate measurements of the speed of light by determining the precise number of waves produced in one second. Using this technique, American physicist Kenneth Evenson, in 1972, calculated the speed of light at 186,282.3959 miles or 299,792.4574 kilometers per second. This work led to the National Institute of Standards and Technology adopting the speed of light in the redefinition of the meter.

Einstein's theories are predicated on the speed of light as a fundamental constant of nature. This suggests that nothing in the universe can travel faster. Some physicists postulate, however, that at the moment of the big bang, faster-than-light particles may have existed in a subatomic process. A hypothetical particle, called the tachyon, may never have possessed speeds below the speed of light. Consequently, the speed of light becomes a two-way speed barrier: it prevents slower-moving particles from acquiring enough energy to reach the speed of light and denies faster-moving particles from releasing enough energy to decelerate to the speed of light.

Robert Karl Koslowsky

Sources

Hughes, Thomas Parke. *Science and the Instrument-maker.* Washington, DC: Smithsonian Institution, 1976.

Magueijo, Joao. *Faster than the Speed of Light.* New York: Perseus, 2003.

Perkowitz, Sidney. *Empire of Light.* New York: Henry Holt, 1996.

MAGNETISM

Magnetism as it relates to electricity, which fascinated European theorists during the eighteenth and nineteenth centuries, became the focus of American applied science during the late nineteenth and early twentieth centuries.

Europeans who inaugurated the modern study of electromagnetism include the sixteenth-century English physician William Gilbert, who wrote *De magnete* (1600), the first treatise on magnetism, in which he discussed his work with electricity and argued that Earth is a gigantic magnet. A quantitative experiment by eighteenth-century French physicist Charles Coulomb showed that the forces between magnetic poles vary inversely with the square root of the distance between them. The Danish physicist Hans Christian Oersted, building on Gilbert's contributions to magnetism and Coulomb's work on electricity, developed the concept of electromagnetism, unifying the connection between electricity and magnetism.

Practical applications finally appeared during the 1820s, when the English scientist William Sturgeon invented the electromagnet by wrapping copper wire around iron and applying electric current to magnetize the iron. As long as current was applied, the electromagnet could lift metal objects and move them about. Sturgeon found that by bending the iron into a horseshoe shape, effectively bringing the opposite poles closer together, the magnetic strength and hence the lifting power of the electromagnet was greatly increased.

Joseph Henry, in the nineteenth century, was the first American to contribute significantly to the field of electromagnetism. In addition to discovering the phenomenon of inductance, the magnetic strength of an electric field, Henry's main innovation was to wind insulated wires around iron to produce powerful electromagnets. He built the largest electromagnet in the world, one that could lift 2,300 pounds, and observed large sparks when the circuit was disrupted. From this, he deduced the property of electric inductance. During his experiments, Henry found that inductance is defined by the circuit layout, especially the coiling of wire. His investigations also led to the making of noninductive windings by folding wire back on itself.

The Serbian American inventor and researcher Nikola Tesla built on the ideas of Henry and discovered the rotating magnetic field in 1883, during his experiments on generators. His discovery established alternating current as an alternative to direct current for the growing electric industry. Tesla then built the induction motor, a crucial step in the spread of alternating current around the world.

The trend during the twentieth century was a doubling about every decade of the maximum energy product of magnetic materials, a measure of a magnet's ability to produce work for a given size. This progress spawned a whole new series of alloys and ceramic materials used to produce magnets for a diverse range of new applications.

The American mathematical physicist John H. van Vleck shared the Nobel Prize in Physics in 1977 for his lifetime of research on the magnetic properties of these materials, which provided essential knowledge for the solution of practical problems and technological applications. Laser devices and magnetic resonance imaging, which produces diagnostic images of the body using a magnetic field, are just some of the products derived from his work.

Robert Karl Koslowsky

Sources

Lehrman, Robert L. *Physics—The Easy Way.* Hauppauge, NY: Barron's, 1998.

Russell, Colin A. *Michael Faraday: Physics and Faith.* Oxford, UK: Oxford University Press, 2000.

Verschuur, Gerrit L. *Hidden Attraction: The Mystery and History of Magnetism.* Oxford, UK: Oxford University Press, 1993.

MAYER, MARIA GOEPPERT (1906–1972)

Maria Goeppert Mayer, a Nobel laureate in physics, was born on June 28, 1906, in Katowice, a city that at the time was part of Germany but is now in Poland. Her parents were academics Friedrich Goeppert and Maria Wolff Goeppert, and through her father she represented a seventh generation of university professors.

She grew up in Göttingen, where her father was a professor of pediatrics. In 1924, she entered the University at Göttingen, receiving her doctorate in theoretical physics in 1930. Also that year, she married Joseph Edward Mayer, an American physicist on a Rockefeller fellowship. Shortly thereafter, the couple settled in Baltimore, Maryland. Maria Goeppert Mayer became a U.S. citizen in 1932.

Mayer was appointed to positions in a number of institutions, including Johns Hopkins University, Columbia University, and Sarah Lawrence College, and she worked at the Los Alamos Laboratory in New Mexico during World War II. In 1946, she joined the University of Chicago as an associate professor and later became a full professor. At the same time, she worked as a senior physicist in the Theoretical Physics Division of Argonne National Laboratory, established by the federal government outside Chicago in July 1946. From 1960 to 1972, she was a professor at the University of California, San Diego.

In her early career at Johns Hopkins, Mayer did research on the color of organic molecules. Later experiments focused on the separation of isotopes of uranium and nuclear shell structure. In 1955, she coauthored a book with Hans Jensen, *Elementary Theory of Nuclear Shell Structure.* She studied elements having the "magic numbers" of 2, 8, 20, 28, 50, 82, and 126 protons or neutrons: elements with these particular nucleon numbers are unusually stable, as the nucleons move in stable orbits around the nucleus. Elements with the first three magic numbers—2, 8, and 20—had already been explained, but Mayer gave a convincing explanation for the stability of elements with higher magic numbers; her model of their orbits explored new facts on the structure of atomic nu-

German-born Maria Goeppert Mayer of the University of Chicago won the 1963 Nobel Prize in Physics for the theory that protons and neutrons are arranged in a shell in the atomic nucleus, much as electrons are outside it. *(Library of Congress, LC-USZ62–118262)*

clei. Mayer shared the 1963 Nobel Prize in Physics with Eugene Paul Wigner and Hans Jensen. She was the first woman to receive the prize for work in theoretical physics and the second after Marie Curie to win in physics.

Mayer, who was honored with membership in the National Academy of Sciences, contributed to physics through her research in areas such as the phenomenon of changes to nucleons, atomic properties of transuranic elements (those with an atomic weight greater than 92), and opaqueness in substances. She died on February 20, 1972, in San Diego.

Patit Paban Mishra and Sudhansu S. Rath

Sources

Gabor, Andrea. *Einstein's Wife.* New York: Viking, 1995.

McGrayne, Sharon Bertsch. *Nobel Prize Women in Science.* New York: Birch Lane, 1993.

Nobel Lectures. *Physics 1963–1970.* Amsterdam, The Netherlands: Elsevier, 1972.

MICHELSON-MORLEY EXPERIMENT

Designed and carried out by Albert A. Michelson (1852–1931) and Edward Williams Morley (1838–1923), the Michelson-Morley Experiment was designed to determine the velocity of the planet Earth through space. The experiment,

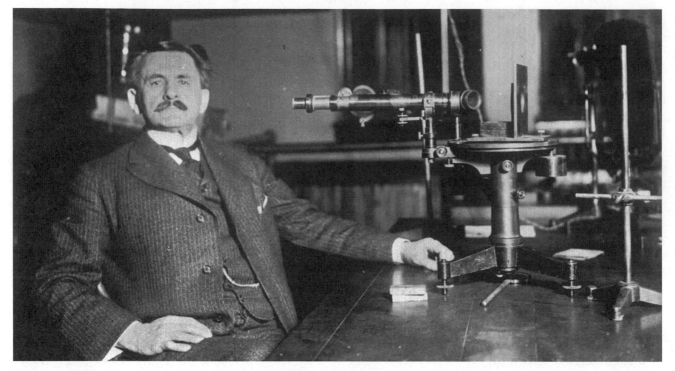

Albert Michelson developed a device called the interferometer, used in his classic 1887 experiment with Edward Morley on the motion of Earth through space and the ether. Michelson's Nobel Prize in 1907 was the first for an American in the sciences. *(Boyer/Roger Viollet/Getty Images)*

according to its presuppositions, resulted in the conclusion that Earth was not moving at all.

In the nineteenth century, scientists widely believed that because light has a wavelike (undulatory) character, it had to be transmitted through a substance, in the same way that sound waves are carried through the air. This substance was called the luminiferous ("light-bearing") ether.

Therefore, several possibilities presented themselves with regard to the movement of Earth. Earth might drag the ether along with it or pass through the ether, or there might be some combination of these two motions. Or the ether might drift past Earth. If that were so, then the speed of light would vary according to direction.

The reason for the variation in light's velocity can be understood by thinking of two boats crossing a stream with significant current. One boat goes across the stream and back again, and the other boat goes up stream and back, both boats traversing the same distance at the same relative speed. The boat going across the stream and back will take less time than the one going up and back, since the current of the stream works against the boat's speed at a 90 degree angle, rather than directly against and then with the boat's speed.

Michelson, a U.S. Naval Academy graduate, combined an interest in optics with his nautical expertise, which entailed the need to determine the relationship between wind and current and the movement of a ship. Morley, a Congregationalist minister and professional chemist, provided necessary technical expertise in building the device used in their experiment.

The apparatus Michelson designed, later called an interferometer, consisted of a silvered mirror that would split a beam of light; two tunnel-like "arms" set at 90 degree angles and containing a series of mirrors; and an observation eyepiece, to which both beams of light would ultimately be directed. The device could be rotated 360 degrees. As it turned, one of the arms would, at some point along the compass, turn into the "stream," or ether "wind." Each of the two beams of light split by the silvered mirror would then be traveling at different speeds—one at 90 degrees to the ether, the other against the "stream." Through the eyepiece, the observer would be able to detect the different speeds by means of different patterns of alternating bands of light and darkness. The change in spacing between the light and dark bands was called the "fringe shift."

To everyone's astonishment, the experiment, conducted in July 1887, failed to detect any significant fringe shift. The implication was that Earth is not moving—a geocentric conclusion repugnant to the majority of modern scientists.

The experiment's result was a significant factor in the later acceptance by most physicists of Albert Einstein's special theory of relativity, which dismissed the existence of the ether and which maintained that the speed of light is an absolute standard. In addition, the interferometer, besides its value in astrophysics, has proved to be of tremendous use in making highly accurate measurements of microscopic distances and in high-resolution spectroscopy.

Frank J. Smith

Sources

Aspden, Harold. *Modern Aether Science.* Southampton, UK: Sabberton, 1972.

Swenson, Lloyd. *The Ethereal Aether: A History of the Michelson-Morley-Miller Aether-Drift Experiments.* Austin: University of Texas Press, 1972.

MILLIKAN, ROBERT A. (1868–1953)

One of the most influential American physicists of the twentieth century, Robert Andrews Millikan was an author, teacher, scientist, and university president. He inspired generations of American physicists and was a Nobel laureate for his work on the elementary charge of electricity and the photoelectric effect.

Millikan was born on March 22, 1868, in Morrison, Illinois, the son of a Congregational minister. After attending high school in Iowa, he entered Oberlin College in Ohio in 1886. Following his graduation in 1891, he taught physics for two years until being appointed a fellow in physics at Columbia University in New York. Millikan received his Ph.D. in 1895 for research on the polarization of light emitted by incandescent surfaces.

After a year of postdoctoral study in Germany, in 1896, he was hired as an assistant at the newly established Ryerson Laboratory at the University of Chicago. Millikan showed an outstanding aptitude for teaching physics and, in the process, poured substantial amounts of time and energy into writing textbooks and improving methods of physics instruction. In 1902, he married Greta Erwin Blanchard, with whom he would have three sons. Millikan was made a professor in 1910, and, in the decades that followed, he would go on to author, or co-author with other scientists, more than a dozen influential physics texts.

In 1917, Millikan joined the World War I effort as vice chair of the National Research Council of the National Academy of Sciences in Washington, D.C., where he conducted research on the detection of submarines. His experience in Washington would be a turning point in his career, as it would introduce him to the astronomer George Ellery Hale. In 1921, Hale persuaded Millikan to join the faculty of the fledgling California Institute of Technology, where he became director of the Norman Bridge Laboratory of Physics and Cal Tech's first president. Under Millikan's leadership, Cal Tech would become one of the leading scientific research centers in America. Also in 1921, Millikan served as the prestigious American delegate to the Solvay Congress—the International Congress of Physics at Brussels.

Millikan received the Nobel Prize in Physics in 1923 for his work on the photoelectric effect and determining the atomic structure of electricity. The latter research included a determination of the charge carried by an electron using the "falling-drop method," which Millikan developed to provide experimental proof that the charge was a constant for all electrons. His work on the photoelectric effect, whereby matter subjected to electromagnetism absorbs photons and releases electrons, experimentally verified Einstein's theories on the photoelectric effect and provided further evidence of the wave and particle behavior of matter. Millikan also researched cosmic radiation, the charged particles entering Earth's atmosphere.

During the course of his working life, Millikan received honorary degrees from twenty-five universities. He served as vice president of the American Association for the Advancement of Science and as president of the American Physical Society. In addition to receiving the Nobel Prize, he was the recipient of the American Institute of Electrical Engineers Edison Medal and the Comstock Prize from the National Academy of Sciences. Millikan died in San Marino, California, on December 19, 1953.

Todd A. Hanson

Sources

Kargon, Robert. *The Rise of Robert Millikan: Portrait of a Life in American Science.* Ithaca, NY: Cornell University Press, 1982.

Millikan, Robert A., and I. Bernard Cohen. *Autobiography of Robert A. Millikan.* North Stratford, NH: Ayer, 1980.

PARTICLE PHYSICS

Particle physics, the study of subatomic particles, is a relatively new science that began in the early 1900s. Europeans such as Niels Bohr, Marie Curie, Ernest Rutherford, and James Chadwick made initial discoveries on the nature of the atom and the particles that compose it, such as the electron and proton. Chadwick discovered the neutron in 1932, the same year that American physicist Carl David Anderson, using a device called a cloud chamber, discovered the positive electron, or positron. In 1936, Anderson was awarded the Nobel Prize for his discovery of the positron; also in 1936, he and a graduate student, Seth Neddermeyer, discovered the muon, part of the subatomic particle family of mesons.

Physicists throughout the world discovered more particles in subsequent years, and a classification system, the Standard Model, emerged based on work by researchers in quantum mechanics. The Standard Model states that particles are either fermions or bosons. Fermions are matter particles and include protons, electrons, and neutrons. Bosons include massless particles such as photons and gravitons. The four universal physical forces—gravitational, electromagnetic, strong, and weak—are produced through a mediation process among these particles. For example, electromagnetism causes electrons to orbit an atom's nucleus, and the weak force causes radioactivity. These four forces are responsible for all fermion interactions.

In 1964, Murray Gell-Mann and George Zweig were key contributors in identifying quarks and how the strong nuclear force holds the different types of quarks together in forming the proton and neutron within the atom's nucleus. In 1973, David J. Gross, Frank Wilczek, and H. David Politzer explained why quarks could never be seen apart from one another. In 2004, the three Americans received the Nobel Prize in Physics for their work in clarifying how the strong force binds the atomic nucleus together. Their efforts led to the theory of quantum chromodynamics, or QCD, whereby quarks come in six "flavors" and three "colors." The colors interact through an exchange of energy bundles called gluons, which develop a color charge to ensure particle stability.

The bubble chamber, invented and developed by 1960 Nobel laureate Donald Arthur Glaser in 1952 to uncover the existence of particles, became indispensable in tracking the paths of high-energy subatomic particles. Superheated liquid is expanded within the bubble chamber just before particles are sent through. The interactions produced by the streaming particles ionize atoms in the superheated liquid and create a bubble path along the particle trajectory. The bubbles reveal the particles tracks, which are photographed during their trip for further analysis.

Extensive progress in the realm of particle physics has led to the view that all matter comprises three types of foundational objects: quarks, leptons, and bosons. Quarks are particles within the atomic nucleus. Leptons are particles outside the atom. Bosons provide the bases for forces in the universe.

Robert Karl Koslowsky

Sources

Johnson, George. *Strange Beauty: Murray Gell-Mann and the Revolution in 20th-Century Physics.* New York: Alfred A. Knopf, 1999.

Koslowsky, Robert. *A World Perspective Through 21st Century Eyes.* Victoria, Canada: Trafford, 2004.

Stehle, Philip. *Physics: The Behavior of Particles.* New York: Harper and Row, 1971.

PAULI, WOLFGANG
(1900–1958)

Nobel laureate Wolfgang Pauli was born in Austria on April 25, 1900, to Berta and Wolfgang Joseph Pauli. His mother was an author and his father was a medical doctor and professor at the University of Vienna.

After his early education in Austria, Pauli studied at the Ludwig Maximilian University of Munich, Germany, earning a Ph.D. in 1921. He did postdoctoral work with the Danish physicist Niels Bohr, who developed the quantum theory, before spending five years as a lecturer at the University of Hamburg. In 1928, Pauli was appointed professor of theoretical physics at the Federal Institute of Technology in Zurich, Switzerland.

Pauli spent much of the decade prior to World War II working and lecturing in the United States. In 1931, he was hired as a visiting professor at the University of Michigan; in 1935, he became a visiting professor at the Institute for Advanced Study at Princeton, New Jersey, where he met and worked with Albert Einstein.

Following the outbreak of the war in Europe, Pauli decided to stay in the United States. In 1940, he was named the chair of theoretical physics at Princeton. He returned to the University of Michigan the following year and spent time at Purdue University in 1942. At the end of the war, Pauli became a naturalized U.S. citizen.

Pauli made significant contributions to the field of quantum mechanics. The Exclusion (or Pauli) Principle states that no two identical particles of an atom (electrons) can exist in the same quantum energy state at the same time. The reason that electrons do not congregate together is the electron spin, with each electron having a different quantum number. Electron spin was experimentally confirmed in 1926. Pauli was also the first to theorize the existence of a neutral, massless particle that accounts for the energy discrepancy that results when a nucleus of an atom loses an electron. Enrico Fermi would later name the particle a "neutrino." Its observation for the first time in 1956 verified Pauli's theory.

In addition to the 1945 Nobel Prize in Physics, Pauli received a number of awards and honors during his career. In 1930, he was awarded the Lorentz Medal from the Royal Dutch Academy of Sciences, and he received the Max Planck Medal in 1958. Wolfgang Pauli died in Zurich on December 15, 1958, of pancreatic cancer.

Todd A. Hanson

Sources

Enz, Charles P. *No Time to Be Brief: A Scientific Biography of Wolfgang Pauli.* Oxford, UK: Oxford University Press, 2002.

Laurikainen, K.V. *Beyond the Atom: The Philosophical Thought of Wolfgang Pauli.* Berlin, Germany: Springer-Verlag, 1989.

PRINCE, JOHN
(1751–1836)

The master scientific instrument maker John Prince was born in Boston on July 22, 1751, was educated at Harvard College, and became a Congregational clergyman serving the First Parish of Salem in 1779. He was at the Salem parish for forty-five years, and, in his spare time, he collected books on science and discussed science with other clergy, such as the Reverend Manasseh Cutler, from nearby Hamilton, who along with Prince and the Reverend Thomas Barnard formed the core of the Salem Philosophical Library.

Prince fully embraced the eighteenth-century view that science and religion were complementary, reflecting in a 1796 letter that science is involved "in promoting a knowledge of the works of nature among men, and leading their minds through these footsteps up to their Divine Author: in making the best and noblest use of Philosophy, that of expanding the idea of the Supreme Being in the minds of men, and impressing them with proper sentiments of piety towards him."

Prince corresponded with scores of scientists and clergy on both sides of the Atlantic, and he was involved in scientific societies, such as the Salem Philosophical Library, American Academy of Arts and Sciences, American Philosophical Society, and Massachusetts Historical Society. He had a library of more than 3,000 volumes pertaining to science and philosophy and developed friendships with other scientists such as Nathaniel Bowditch, the Reverend William Bentley, Harvard President Joseph Willard, Benjamin Silliman, physician Edward Holyoke, and Alexander Wilson. He also contributed to science through his work as a scientific instrument maker, one of the best of the late eighteenth and early nineteenth centuries.

Prince made sophisticated scientific instruments through much of his long life, using a workshop adjacent to the parsonage. He was an expert with various materials, such as iron, bronze, glass, and wood. Patronized by colleges and intellectuals throughout America and England, Prince made telescopes, microscopes, surveying and navigational instruments, electrostatic machines, electrometers, and "magic lanterns" to project enlarged images for viewing.

He gained the greatest fame for his air pump, invented in the 1780s and described in the *Memoirs of the American Academy of Arts and Sciences*. Prince's design was simple and efficient, as Jefferson noted in a letter from 1788: "A considerable improvement in the Air pump has taken place in America. You know that the valves of that machine are it's [sic] most embarassing parts. A clergyman in Boston has got rid of them in the simplest manner possible."

Air pumps were used by eighteenth-century scientists to investigate the qualities of air—its various volumes in different conditions—and how air pressure impacted other substances. Although Prince and many other contemporary instrument makers tried and failed to devise an air pump that created a vacuum, Prince's air pump came closest to achieving this goal.

Russell Lawson

Source

Schechner, Sara J. "John Prince and Early American Scientific Instrument Making." *Publications of the Colonial Society of Massachusetts* 59 (1982): 431–503.

PRINCIPIA MATHEMATICA

The *Philosophiae Naturalis Principia Mathematica* (1687), commonly known as the *Principia Mathematica* or *Principia*, is considered one of the greatest treatises in the history of science. Completed by the English physicist and mathematician Isaac Newton in eighteen months, it was originally published in Latin in three books, with the financial help of Edmond Halley. Although Newton laid the foundations of calculus, most of the proofs in the *Principia* are geometrical arguments. It would be the French physicist and mathematician Pierre Laplace in his *Celestial Mechanics* (1799–1825) who would translate the geometrical arguments of the *Principia* into calculus or physical mechanics.

Book 1 explores the mathematics of the motion of bodies. Book 2 examines motion in resistant media (physical reality). And Book 3 describes the cosmology of a physical reality based on laws Newton proposed. Newton established the validity of his formulations and conclusions by calculating the masses of the sun and of planets having satellites, the density of Earth, and the trajectory of a comet. Similarly, he explained the variations in the moon's motion, the precession of the equinoxes, the motion of the tides, and the variation in gravitational acceleration depending on latitude.

Though Newton had first begun to develop his theories of mechanics as a Cambridge University student, it was in the *Principia* that he stated his three universal laws of motion: (1) every object continues in its state of rest or of uniform motion in a straight line, unless it is compelled to change that state by forces impressed upon it; (2) the acceleration of an object is directly proportional to the net force acting on the object, is in the direction of the net force, and is inversely proportional to the mass of the object; and (3) whenever one object exerts a force on a second object, the second object exerts an equal and opposite force on the first.

Newton used the term *gravitas* (weight) in the *Principia*, where he first stated the law of universal gravitation: gravitational attraction is directly dependent upon the masses of both objects and inversely proportional to the square of the distance that separates their centers. Newton demonstrated that gravity is universal, extending beyond Earth to the whole of physical reality.

In the *Principia*, Newton presented the first analytical determination (based on Boyle's law) of the speed of sound in air as 968 feet per second. Knowing the true value to be approximately 1,116 feet per second, Newton attempted to reconcile the difference by postulating a number of nonideal effects. Laplace's application of calculus to the problem resolved the discrepancy, however, without the ancillary postulations.

The title *Mathematical Principles of Natural Philosophy* summarizes the intent of the work to apply mathematics to natural philosophy, synthesizing into a single construct cosmology, history, and theology. The publication of the

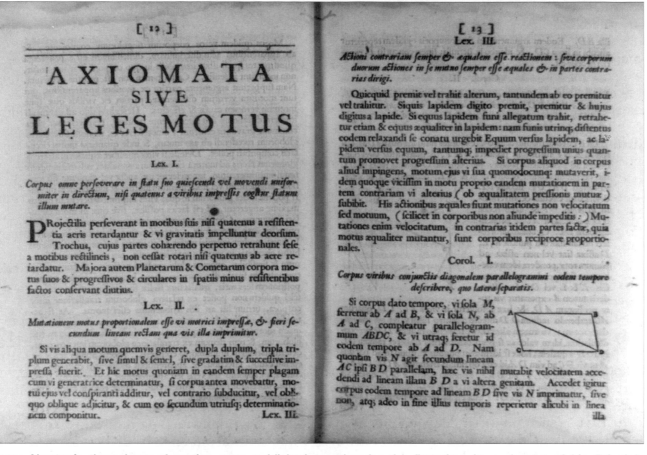

Isaac Newton's three laws of motion were published together for the first time in a chapter of his *Principia Mathematica* (1687), titled *"Axiomata sive leges motus."* The *Principia* is counted among the greatest achievements in the history of Western science. *(Library of Congress, LC-USZ62-95173)*

Principia was the culmination of Newton's scientific contributions. He lost interest in scientific inquiry and suffered from depression that ended in a nervous breakdown. Therafter, Newton became a university representative to the British parliament, was appointed warden and then master of the Royal Mint, was elected president of the Royal Society, and was knighted.

Richard M. Edwards

Sources

Butterfield, Herbert. *The Origins of Modern Science.* New York: Free Press, 1997.

Cohen, I. Bernard. *Introduction to Newton's Principia.* Cambridge, MA: Harvard University Press, 1971.

Gribbin, John. *The Scientists: A History of Science Told Through the Lives of Its Greatest Inventors.* New York: Random House, 2003.

Newton, Isaac. *Newton's Principia: The Central Argument: Translation, Notes, and Expanded Proofs.* 1687. Santa Fe, NM: Green Lion, 1995.

QUANTUM PHYSICS

Quantum physics is the area of science that predicts the behavior of objects at molecular, atomic, and subatomic levels, focusing on the particle-wave duality of matter and energy. Although it was initially developed by European physicists in the late nineteenth and early twentieth centuries, American physicists began to take the lead in quantum physics research in the mid-twentieth century. This research examines the microscopic rather than macroscopic world.

The macroscopic world behaves according to classical or Newtonian physics, which is based primarily on Newton's three laws of motion. These laws describe how and why objects move. First attempts to predict the behavior of very small objects were based on the same classical principles, but scientists encountered an impasse in the nineteenth century, because classical theory predicted that a black body, which absorbs electromagnetic radiation, should radiate vastly more energy at high frequencies than could be experimentally verified. From this and other seemingly small discrepancies in classical theory emerged quantum physics.

In response to the problems of black body radiation, the German physicist Max Planck proposed in 1900 that the atoms forming a black body could absorb and emit energy only in discrete packets called quanta. Further, the amount of energy present in each quanta was proportional to its frequency.

In 1905, Albert Einstein used Plank's quantum theory to posit how electric current is generated in certain light-sensitive materials—the photoelectric effect. Central to Einstein's model was the thesis that quantization, the packaging of energy into the discrete quanta, is not a special property of a black body. All electromagnetic radiation, he argued, is naturally quantized and correctly viewed as being composed of either waves or quanta.

In 1924, French physicist Louis de Broglie attempted to unite the classical view of matter as formed from particles and the new quantum view of radiation. He proposed that just as radiation could be viewed either as a wave or a particle, so matter could be viewed as either a particle or a wave. In 1926, the Americans Clinton Davisson and Lester Germer verified the wave-particle duality of matter for electrons. In 1928, the Danish physicist Niels Bohr summarized the current understanding of quantum physics in the "complementarity principle," which states that the physical properties of matter and energy cannot be described by either a particle or a wave model alone but must incorporate both.

Mathematical formulations, somewhat analogous to Newton's laws of motion for classical physics and known as quantum mechanics, were developed in terms of matrix algebra by the German theoretician Werner Heisenberg in 1925, and in terms of wave motion by the Austrian Erwin Schrödinger in 1926. Quantum mechanics describes how matter and energy behave and interact in quantum physics.

Using the wave description of quantum physics, Max Born, another German, showed in 1926 that the wave associated with a particle represents the probability that the particle will be found in a given region of space. Quantum physics then presents a world in which the outcome of physical events is probabilistic in nature rather than explicitly determined. This statistical view of the universe has profound philosophical implications and provoked a skeptical Einstein to counter, "God does not play dice with the universe." After a prolonged debate with Bohr, Einstein eventually admitted the logical consistency of quantum physics. Quantum physics has been extensively tested since that time and no part of the theory has been disproved, though many philosophical questions remain.

It was expected that quantum physics would be capable of predicting how matter and energy interact, necessitating a quantum theory of forces—a quantum field theory. Such a theory was developed for the electromagnetic field—called quantum electrodynamics, or QED—by American physicists Richard Feynman and Julian Schwinger and the Japanese physicist Shinichiro Tomonaga. QED details the interactions between matter and radiation and enables predictions about areas as diverse as atomic interactions, the properties of light, and the masses of certain subatomic particles. The trio received the 1965 Nobel Prize in Physics for their work in QED, which is one of the most thoroughly tested theories of modern physics.

American physicists Sheldon Glashow and Steven Weinberg and Pakistani physicist Abdus Salam developed a quantum theory for the weak nuclear force and showed that at high energies this force and the electromagnetic force are aspects of the same object. The three won the 1979 Nobel Prize for their explanation of electroweak theory.

Following the path laid out in the development of QED, a theory detailing the strong nuclear force was soon developed. Quantum chromodynamics, or QCD, detailed the quark and its behavior in mediating the strong nuclear force. David Gross, Frank Wilczek, and David Politzer showed that interactions between quarks could become extremely weak under certain conditions. This discovery allowed the use of QCD for predictive calculations and won these three American physicists the 2004 Nobel Prize in Physics.

With the development of QCD, the only fundamental force without a quantum field theory is gravitation. The most popular of current attempts lies in the area of string theory, according to which the basic building blocks of the universe are one-dimensional, vibrating strings rather than zero-dimensional, point particles. Originally proposed in 1970 by Yoichiro Nambu of the University of Chicago and Stanford University professor Leonard Susskind, along with Danish physicist Holger Bech Nielsen, to explain phenomena now covered by QCD, string theory was seen as the potential basis of a unified quantum theory of all four fundamental forces and the fundamental particles.

Initially, there were a large number of string theories, all apparently sound, but there was no logically consistent method for selecting a specific theory. Edward Witten of the Institute for Advanced Study showed that all competing theories are actually specific cases of an eleven-dimensional theory now known as M-theory. The extra dimensions are posited to be "rolled up" with such small cross-sections that they are undetectable in normal circumstances. While M-theory shows promise, no convincing tests have yet been proposed, let alone performed, to determine its accuracy.

R. Dwayne Ramey

Sources

Feynman, Richard P., Robert B. Leighton, and Matthew Sands. *The Feynman Lectures of Physics.* Vol. 1. Reading, MA: Addison-Wesley, 1977.

Zee, Anthony. *Quantum Field Theory in a Nutshell.* Princeton NJ: Princeton University Press, 2003.

Rabi, I.I.
(1898–1988)

Isadore Isaac Rabi was a Nobel Prize–winning physicist. His discovery of resonances within a single molecule led to a greater understanding of the internal structure of molecules, atoms, and atomic nuclei, as well as to the development of magnetic resonance imaging (MRI). He also played a key role in helping create two of the world's most eminent physics laboratories.

Rabi was born in Raymanov, Austria, on July 29, 1898, to David Rabi and Janet Teig. He moved with his family to the United States in 1899 and grew up in New York City. He graduated from Cornell University with a degree in chemistry in 1919 and received his Ph.D. from Columbia University in 1927 for studies on the magnetic properties of crystals. After working for two years in Europe with the eminent physicists Werner Heisenberg, Niels Bohr, and Wolfgang Pauli, Rabi returned to Columbia in 1929 as a lecturer in theoretical physics and became a professor in 1937.

Rabi took a sabbatical from Columbia in 1940 to work at the Massachusetts Institute of Technology on the development of radar, which he thought was crucial to an Allied victory in World War II. In 1943, Rabi helped J. Robert Oppenheimer recruit physicists for the Manhattan Project, and he later worked as a consultant at the Los Alamos, New Mexico, laboratory. For his work in developing molecular beam magnetic resonance, a method for recording the magnetic properties of a molecule, Rabi received the Nobel Prize in 1944.

When Rabi returned to Columbia in 1945 after the war, he was named executive officer of the Physics Department. In that position, he was instrumental in organizing nine northeastern universities into a nonprofit organization that would build a nuclear science laboratory devoted to research on the peaceful uses of atomic energy. By 1947, Brookhaven National Laboratory for Atomic Research was being built at the former site of Camp Upton on Long Island, New York.

In June 1950, Rabi was a U.S. delegate to a meeting of the United Nations Educational, Scientific, and Cultural Organization (UNESCO) in Florence, Italy. Working on behalf of the U.S. State Department, and in consultation with leading European physicists, Rabi helped get a resolution passed that called on the UN to assist and encourage the formation of regional research centers and laboratories to increase scientific collaboration among Western European countries. By pooling human and financial resources, nations could build and acquire together many of the expensive, large-scale scientific research instruments that no nation alone could afford.

The resolution resulted in the Conseil Européen pour la Recherche Nucléaire (European Council for Nuclear Research), or CERN, and the construction of the CERN laboratory on the border of France and Switzerland, near Geneva. Later renamed the European Organization for Nuclear Research, it is still widely known by its original acronym and remains one of the world's largest particle physics laboratories. Rabi died on January 11, 1988, in New York.

Todd A. Hanson

Sources

Rabi, I.I. *My Life and Times as a Physicist.* Claremont, CA: Friends of the Colleges at Claremont, 1960.

Rhodes, Richard. *The Making of the Atomic Bomb.* New York: Simon and Schuster, 1995.

Rigden, John S. *Rabi: Scientist and Citizen.* Cambridge, MA: Harvard University Press, 2000.

RAMSEY, NORMAN
(1915–)

Nobel laureate Norman Foster Ramsey, known for his work on the Manhattan Project and in developing the atomic clock, was born August 27, 1915, in Washington, D.C., the son of a U.S. Army officer and a university mathematics instructor. When he graduated from high school at age fifteen, his parents expected him to follow his father to West Point, but he was too young to be admitted, so he entered Columbia College in 1931. He later wrote, "Though I started in engineering, I soon learned that I wanted a deeper understanding of nature than was then expected of engineers so I shifted to mathematics."

After graduating from Columbia in 1935 with a bachelor's degree in physics, he was awarded a university fellowship that allowed him to earn a second bachelor's degree at Cambridge University. At Cambridge, he worked at the Cavendish Laboratory, then one of the world's leading physics research facilities. Becoming interested in molecular beams, he returned to Columbia to study with I.I. Rabi, who developed the atomic beam magnetic resonance (ABMR) method of measuring the atomic oscillation of electromagnet radiation (for which Rabi would win the 1944 Nobel Prize in Physics).

In 1940, Ramsey joined the faculty of the University of Illinois. World War II led him to the Radiation Laboratory of the Massachusetts Institute of Technology, where he consulted with the government on radar. He also worked on the Manhattan Project, which built the first atomic bombs.

After the war, he returned to Columbia and became one of the founders of Brookhaven National Laboratory on Long Island, a research center for particle physics. He served as the first head of its physics department, before joining Harvard University in 1947, where he taught and did research until his retirement in 1987. Ramsey took time away from Harvard to serve as the first science adviser (assistant secretary general for science) to the North Atlantic Treaty Organization and as a visiting professor at several colleges and universities.

He continued studying electromagnetic radiation of atoms but, unable to solve the problem of maintaining uniform magnetic fields in ABMR, he invented the separated oscillatory field method, which made possible much higher resolution atomic spectroscopy. This was a key step in developing the cesium atomic clock, the most accurate chronometer yet developed. With a graduate student, Daniel Kleppner, Ramsey invented the hydrogen maser (similar to a laser, it uses microwaves instead of visible light).

Ramsey, the author of five books and more than 300 academic papers, has received numerous honors, including the E.O. Lawrence Award (1960), Davisson-Germer Prize (1974), presidency of the American Physical Society (1978–1979), IEEE Medal of Honor (1984), Rabi Prize (1985), Compton Medal (1986), Oersted Medal (1988), and National Medal of Science (1988). The 1989 Nobel Prize in Physics was shared by Ramsey, for the invention of the separated oscillatory fields method and its use in the hydrogen maser and other atomic clocks, and by Hans Dehmelt of the University of Washington and Wolfgang Paul of the University of Bonn, for the development of the ion trap technique.

Phoenix Roberts

Sources

Frangsmyr, Tore, ed. *The Nobel Prizes, 1989.* Stockholm, Sweden: Nobel Foundation, 1990.

Ramsey, Norman. *Molecular Beams.* Wotton-under-Edge, UK: Clarendon, 1956.

———. "Science as an Art: A Lecture." John Hamilton Fulton Memorial Lectureship in the Liberal Arts, Middlebury College, Middlebury, VT, 1969.

RELATIVITY

In physics, "relativity" refers to the possibility of variation in physical laws due to the observer's position, motion, or other variables. For most of human history, an object's position and velocity were considered absolute values measured with respect to the unvarying and stationary Earth.

This absolutist view of the universe was modified by Galileo Galilei in his 1632 treatise "Dialogue Concerning the Two Chief World Systems," in which he argued that physical laws of the universe are the same for all observers at rest, which is an inertial frame of reference. The popularity of "Galilean relativity" waxed and waned over the centuries, with such luminaries as Isaac Newton stating that, while physical laws made it impossible to identify, an absolutely stationary frame surely existed.

The Scottish physicist James Clerk Maxwell contributed to the discussions of absolute reference frames in 1873 when he wondered how the speed of light can be measured when it depends on the motion of the measurer. Maxwell assumed that there is a substance of unknown composition that pervades all of space, "ether," which allows for material objects to pass through it without resistance. Ether provided a stationary frame of reference.

Americans Albert Michelson and Edward Morley attempted to identify and measure the phenomenon of ether in an 1887 experiment. If the ether exists, they reasoned, Earth's motion around the sun should create an ether "wind," which could be detected with a properly manipulated light beam. The experiment detected no such effect, however, implying that Earth is always stationary with respect to the ether.

Albert Einstein, in his 1905 paper on the theory of special relativity, denied the existence of ether, arguing that the speed of light does not vary, but that light is the same for all observers, regardless of their velocity. Einstein's theory also states that lengths of objects contract in their direction of motion, that time intervals change with speed, and that objects are limited to speeds less than that of light in vacuum.

Special relativity deals only with inertial observers, excluding any observer experiencing a force; yet in nature, physical objects are always subject to forces such as gravity. Moreover, for gravity, the universal speed limit of special relativity posed a serious problem. Newtonian theory required gravity to act instantaneously over distances in order to conserve angular momentum. Similar problems in electromagnetics had been solved when Maxwell and others introduced the concepts of electric and magnetic fields. Presumably, a gravitational field theory was needed.

Einstein's general theory of relativity, formulated in 1916, provided an explanatory model to show the relationship of gravity to observation, speed, and time. Einstein imagined a windowless rocket ship in which the occupant is uncertain whether he is in the idealized force-free weightlessness of special relativity, or in a free fall near the surface of Earth. The two situations must be equivalent, Einstein reasoned, and if one is inertial, both must be inertial.

The easiest manner in which force-free acceleration can be produced is to suppose that it is a property of space itself: that is, to suppose that space is curved. Einstein was able to use Riemannian geometry, the theory of nonflat spaces, to formulate gravitational field theory. The result was a geometric theory in which matter and energy produce the effects of gravity by curving space and in which the laws of physics are the same for all observers.

General relativity has had great success in its predictions and has been supported by all experiments designed to test it. The global positioning satellite system (GPS) routinely makes use of general relativity to determine locations of objects on Earth to within a centimeter. General relativity has made modern cosmology possible, allowing for prediction and models about the formation of the universe and its development. Among the predictions and concepts stemming from general relativity are peculiarities in the orbit of Mercury, gravitational lensing, the expansion of the universe, the big bang theory of the formation of the universe, gravitational radiation, and black holes.

R. Dwayne Ramey

Sources

Einstein, Albert. *Relativity: The Special and General Theory.* Trans. Robert W. Lawson. New York: Routledge, 2001.

Feynman, Richard P., Robert B. Leighton, and Matthew Sands. *The Feynman Lectures on Physics.* Vol. 1. Reading, MA: Addison-Wesley, 1977.

Schutz, Bernard. *Gravity from the Ground Up: An Introductory Guide to Gravity and General Relativity.* Cambridge, UK: Cambridge University Press, 2003.

RITTENHOUSE, DAVID (1732–1796)

David Rittenhouse's many-faceted career centered on his activities as a mechanic and astronomer. He was not a rigorous, systematic scientist, but he did make significant contributions to scientific inquiry, especially as one of the foremost instrument makers in eighteenth-century America. Rittenhouse was born to Matthias and Elizabeth Williams Rittenhouse on April 8, 1732, at Paper Mill Run, near Germantown, Pennsylvania. He had little formal education, being mostly self-taught on his father's farm, about twenty miles north of Philadelphia.

It is not known what, if any, science books Rittenhouse read as a boy, but some sources say he owned an English translation of Newton's *Principia Mathematica.* We do know that he had a natural propensity for tinkering, which was put to use constructing models. As he got older, these projects became more advanced and more useful; they included clocks, barometers, thermometers, hygrometers, and surveying equipment, such as compasses, levels, and transits. Rittenhouse also built telescopes and was one of the first to use spider webs for crosshairs in the eyepiece.

Early on, his abilities caught the attention of patrons in colonial Pennsylvania, such as his friend Thomas Barton (who married Rittenhouse's sister in 1753), as well as provincial surveyor John Lukens and Richard Peters, who was the secretary to the governor of Pennsylvania. Rittenhouse worked as a surveyor and was involved with surveying many of Pennsylvania's borders, and to a lesser extent the borders of New York.

Rittenhouse's most important scientific devices were his "orreries." These detailed models of the solar system, named after an earlier European model constructed for the Earl of Orrery, had moving parts that represented the motion of

An eighteenth-century instrument maker, astronomer, and director of the U.S. Mint, David Rittenhouse of Pennsylvania is credited with building the first telescope in America, as well as working models of the solar system called "orreries." *(MPI/Hulton Archive/Getty Images)*

the planets around the sun. The orreries' primary use was in the lecture hall, where instructors could demonstrate to their students the operation of the Newtonian universe. Rittenhouse began to construct his first orrery in 1767; it was the most intricate of those constructed in the colonies. While not as polished as British models, such as those produced by Benjamin Martin and Thomas Wright, Rittenhouse's orreries were praised for their accuracy by American contemporaries, who also marveled at their beauty.

In 1767, before his first orrery was completed, Rittenhouse was elected a corresponding member of the American Society for Promoting and Propagating Useful Knowledge. In 1768, he was elected a member of the American Philosophical Society, reading their first scientific paper, his description of the orrery, which was subsequently published in the *Pennsylvania Gazette.* In 1771, the College of Philadelphia appointed him to perform experiments that demonstrated the lessons students had learned in their natural philosophy lectures. It was a post he did not enjoy and did not hold long.

John Witherspoon arranged, in 1771, to purchase Rittenhouse's first orrery for the College of New Jersey (now Princeton University). Another of his orreries went to the College of Philadelphia, where William Smith had arranged for Rittenhouse to be granted a master of arts degree. Thomas Jefferson did much for Rittenhouse's standing in America when he praised his orreries and encouraged Rittenhouse to build other models, for the College of William and Mary and for King Louis XVI of France; neither of those projects was ever completed.

Like many others of his day with scientific interests, Rittenhouse supplemented his income by providing astronomical calculations to almanacs, including the *Universal Almanack* and *Father Abraham's Almanack.* His most significant scientific ideas were disseminated in periodical publications, particularly the *Transactions* of the American Philosophical Society. By the early 1770s, Rittenhouse had gained an international reputation as an astronomer, largely through published papers on the transits of Mercury and Jupiter's satellites, and also for his observations of the transit of Venus in 1769—observations made from the observatory he had constructed at Norriton. After 1770, Rittenhouse lived in Philadelphia, where he continued to make astronomical observations. During the years of the American Revolution, he put his scientific knowledge to work in the production of cannon and saltpeter (potassium nitrate).

Rittenhouse was at various times secretary, curator, and librarian of the American Philosophical Society; in 1779, he was appointed its vice president. In 1782, he was elected a fellow of the American Academy of Arts and Sciences, in Boston, and the College of New Jersey granted him a doctor of laws degree in 1789. In 1791, following Benjamin Franklin's death, Rittenhouse was elected president of the American Philosophical Society, a post he held until his own death.

In the 1790s, Rittenhouse continued to publish papers in the *Transactions* of the American Philosophical Society; he contributed twenty-two papers in all. A number of these were on mathematical topics, including sines and logarithms. He also wrote about experiments he conducted on various topics, including the expansion of wood by heat, magnetism, and pendulums. Other papers, such as "Account of Several Houses in Philadelphia Struck With Lightning," dealt with electricity, a subject that was of great interest in the late eighteenth century.

By 1795, Rittenhouse's reputation was such that he was named a member of the Royal Society of London. He died on June 26, 1796, at his home in Philadelphia.

Mark G. Spencer

Sources

Ford, Edward. *David Rittenhouse: Astronomer-Patriot, 1732–1796.* Philadelphia: University of Pennsylvania Press, 1946.

Hindle, Brooke. *David Rittenhouse.* Princeton, NJ: Princeton University Press, 1964.

Rice, Howard C. *The Rittenhouse Orrery: Princeton's Eighteenth-Century Planetarium, 1767–1954.* Princeton, NJ: Princeton University Press, 1954.

SPECTROSCOPY

Spectroscopy, the study of the properties of matter as it interacts with light, is a branch of science that incorporates both visible and invisible (electromagnetic) sources of light. Research areas include ordinary light, radiation, radio waves, X-rays, sound waves, microwaves, and other sources. The fields of physical chemistry, astronomy, nuclear physics, and radiology all use spectroscopy for research and analysis.

Early work in spectroscopy was done by British physicist Isaac Newton, who proposed an explanation for the spectrum of visible light. In his publication *Opticks* (1704), Newton addressed the reflection and refraction of light, the production of spectra by prisms, the properties of colored light, the composition and dispersion of white light, and light as distinct particles with immutable refractive properties. In 1802, British chemist William Hyde Wollaston established the existence of dark lines (not visible) in the spectrum of the sun while trying to answer the question of how many primary colors exist in the solar spectrum. Twelve years later, German optician Joseph von Fraunhofer found 574 dark solar lines while measuring the dispersive powers of glass for light and different colors.

One of the most important contributors to the American science of spectroscopy was David

Alter, a physician and inventor from Freeport, Pennsylvania. Alter published *On Certain Physical Properties of Light Produced by the Combustion of Different Metals in an Electric Spark Refracted by a Prism* (1854), in which he examined the spectrum of twelve metals and six gases, including hydrogen. These findings propelled spectroscopy as a tool for chemical analysis, which was especially important to astronomers seeking to determine the chemical composition of distant stellar bodies.

The development of the laser led to significant advances in spectroscopy. Gordon Gould, a graduate student at Columbia University in New York, came up with the laser design concept in 1957, suggesting the use of concentrated light beams with amplified power devices and special optical lenses. Theodore Maiman was the first to succeed in demonstrating the use of a laser, at the Hughes Research Laboratories in Malibu, California, in 1960. A more advanced laser helped discover new light frequencies deep in the ultraviolet index, and helped narrow light to better detect the composition and structure of objects. Multiple laser sources (argon, carbon dioxide, ion, and krypton) were developed by the 1990s; producing varied wavelengths, these provided a valuable new tool for spectroscopy research.

Modern spectroscopy is generally divided into two main areas: 1) absorption, or the measurement of the absorption of light by a sample, and 2) emission, in which a sample radiates into light energy following a chemical reaction, irradiation, or molecular collisions at high temperatures. Common types of absorption spectroscopy include ultraviolet and infrared light, both of which provide data for molecular content and structural information. For example, it is possible to distinguish between the chemicals phenol and benzene using infrared spectrometry.

Normally, however, a variety of spectrometric analyses have to be performed to determine the precise structure and identity of a sample. X-ray spectroscopy, a type of emission spectroscopy, is useful in determining the structure of crystalline samples and in the elemental analysis of solid samples. In nuclear magnetic resonance (NMR) spectroscopy, each carbon-13 atom or nonequivalent proton gives rise to a distinct peak in the spectrum because of its unique molecular environment. Unlike other spectroscopic methods, mass spectrometry measures the weight of molecular fragments or ions given off by a sample as a high-energy electron beam destroys it. NMR and mass spectroscopy remain the most widely used techniques for structure determination in modern chemistry.

James Fargo Balliett and Sean Kelly

Sources

Chapman, Brian. *Glow Discharge Processes.* Hoboken, NJ: John Wiley and Sons, 1980.

Marcus, R. Kenneth, and José A.C. Broekaert, eds. *Glow Discharge Plasmas in Analytical Spectroscopy.* Chichester, UK: John Wiley and Sons, 2003.

McGucken, William. *Nineteenth-Century Spectroscopy: Development of the Understanding of Spectra, 1802–1897.* Baltimore: Johns Hopkins University Press, 1969.

SUPERCONDUCTIVITY

Superconductivity, the state in which a material loses all electrical resistance and expels all magnetic fields from its interior, is a growing research field that holds much promise for important technical applications of science in transportation, communications, and energy.

Superconductivity was discovered and named in 1911 by the Dutch physicist Heike Kamerlingh Onnes during his work with liquid helium and supercooling. Onnes found that as he reduced the temperature of certain substances to very low values, the resistance abruptly plunged to zero. Onnes was awarded the Nobel Prize for this discovery in 1913.

Continuing with this work, Germans Walther Meissner and Robert Ochsenfeld in 1933 determined that a substance passing into the superconducting state also expels magnetic fields from its interior. This expulsion is known today as the Meissner effect. Since Onnes's original experiments, many elements and compounds have been found to have a superconducting state at sufficiently low temperatures. These compounds are grouped into Type I and Type II superconductors, with an additional subcategory of the high-temperature superconductor.

Americans John Bardeen of the University of Illinois (already a winner of the Nobel Prize for his work in inventing the transistor), Leon Cooper of

Brown University, and John Robert Schrieffer of the University of Pennsylvania won the 1972 Nobel Prize for their creation of the BCS (Bardeen Cooper Schrieffer) theory, which provided the first subatomic explanation of superconduction. BCS theory postulates an attractive force between electrons that can cause them to bind together in pairs known as Cooper pairs. In a normal conductor, as single electrons move through a material, they interact with the substance and are jostled and bumped, which causes a slowing called resistance.

In a superconductor, the binding energy of the electron pairs forms an energy gap that must be surmounted before the pairs can interact with the material. At low temperatures, these interactions do not have enough energy to cross the gap, and resistance vanishes. However, as the temperature increases, these interactions become large enough to overcome the gap and produce resistance. Substances transition to a superconducting state at a temperature characteristic of the material known as the critical temperature. For most materials, the critical temperature lies below 30 kelvin. A special set of materials, known as high-temperature superconductors, has been found with critical temperatures as high as 125 kelvin. These materials are not well explained by BCS theory.

A large magnetic field also can destroy a superconducting state. If an external field reaches a value known as the critical field strength, which is material dependent, superconductivity is destroyed, regardless of temperature. Type I superconductors pass into a normally conducting state as the magnetic field surpasses the critical value. Type II materials initially pass into a mixed state at a lower critical field value and then fall completely to a normally conducting state at a higher critical field value.

Important applications of superconductors derive largely from their magnetic field properties, as in magnetic resonance imaging (MRI) and particle accelerators. The Meissner effect has been used in several prototype trains to levitate cars off their tracks and reduce friction. Superconducting quantum interface devices (SQUIDs) are used to measure magnetic fields with a high degree of precision. Electric companies have begun to experiment with superconductors in generating facilities and transmission lines.

Researchers have experimented with high-temperature superconductors, but their hardness and brittleness resist efforts to create wiring of commercially viable lengths. Research is ongoing in the manipulation of these compounds and creation of materials with higher critical temperatures.

R. Dwayne Ramey

Sources

Matricon, Jean, Georges Waysand, and Charles Glashausser. *The Cold Wars: A History of Superconductivity.* Piscataway, NJ: Rutgers University Press, 2003.
Tinkham, Michael. *Introduction to Superconductivity.* New York: McGraw-Hill, 1996.

TELLER, EDWARD (1908–2003)

A controversial figure in the American scientific community, the theoretical physicist Edward Teller advanced the study of nuclear fission and fusion. Teller's career included work on the Manhattan Project, which developed the first nuclear weapons. His research on spectroscopy increased scientific understanding of the properties of light particles.

Teller was born on January 15, 1908, in Budapest, Hungary, where his parents were prominent members of a thriving Jewish community. He finished an undergraduate degree in chemistry at the University of Karlsruhe in Germany but switched to physics for his doctoral work. He studied under the quantum theorist Werner Heisenberg, receiving his doctorate at the University of Leipzig in 1930.

With the rise of Nazism and anti-Semitism in Germany, Teller emigrated to the United States in 1934. He joined the faculty at George Washington University in Washington, D.C., as a professor of physics the following year. In 1941, during World War II, he became a naturalized U.S. citizen. By 1943, Teller had moved on to the University of Chicago, where he and colleague Enrico Fermi began discussing the concept of a thermonuclear fusion reaction triggered by an atomic explosion.

Weapons Design

Teller joined the early phases of the Manhattan Project at Los Alamos, New Mexico, where

The controversial Hungarian American physicist Edward Teller is known as the "father of the hydrogen bomb" and an advocate of other advanced weapons systems. *(Nat Farbman/Time & Life Pictures/Getty Images)*

work on fission led to construction of the atomic bomb. But Teller was thinking beyond the immediate task of developing a fission bomb and pushed his idea of a more potent thermonuclear fusion weapon. The resulting research led to development of the hydrogen bomb.

Pressure to develop the hydrogen bomb in the United States increased after the Soviet Union detonated its first atomic bomb in October 1949. Teller believed, as he wrote in his *Memoirs,* "the survival of peace depended on the ability of the United States to maintain the edge in nuclear weaponry." Although the U.S. armed forces, congressional committees, and key members of the scientific community all favored development, the General Advisory Committee of the Atomic Energy Commission (AEC)—chaired by the former Manhattan Project director J. Robert Oppenheimer—opposed it. Following a recommendation by the National Security Council, President Harry S. Truman in January 1950 ordered work to begin. Edward Teller was placed in charge, a design was completed by 1951, and the first test was carried out in November 1952.

To continue research on thermonuclear weapons, the federal government established the Lawrence Livermore National Laboratory near Berkeley, California, in 1952. Teller served as its first director, from 1958 to 1960. During the course of the Cold War, he was a staunch advocate of advanced weapons systems, including a nuclear-powered, space-based antimissile system. He also supported the Reagan Administration's Strategic Defense Initiative, popularly referred to as "Star Wars," the design of which was to use lasers to shoot down missiles.

Conflict and Professional Isolation

In the 1950s, Teller had experienced several personal and philosophical conflicts with co-workers and scientific colleagues, including Oppenheimer. A public controversy erupted in 1954 when, at a security-clearance hearing by the AEC, Teller suggested Oppenheimer might be a security risk.

Teller was also criticized in various circles for his opposition to the proposed nuclear test ban in the 1960s and his support for nonmilitary uses of nuclear explosives. One such early project, Operation Plowshare in 1958, proposed the use of nuclear explosives to dig a deepwater harbor near Hope Point, Alaska. The project was shelved due to predicted radioactive fallout and the fact that the waterway would be frozen for nine months of the year.

Teller was sometimes referred to in the popular press as "the real Dr. Strangelove," referring to the unstable presidential adviser in Stanley Kubrick's 1964 movie *Dr. Strangelove, or How I Learned to Stop Worrying and Love the Bomb.* After the Three Mile Island nuclear reactor accident in 1979, Teller testified before Congress in defense of atomic energy and to counter statements by actress Jane Fonda and consumer advocate Ralph Nader against the nuclear industry. The next day, Teller suffered a heart attack, which he later blamed on Fonda in a two-page *Wall Street Journal* ad.

A few weeks before his death on September 9, 2003, Teller was awarded the Presidential Medal of Freedom by President George W. Bush for his efforts to "protect our nation and bring about the end of the Cold War."

James Fargo Balliett and William M. Shields

Sources

Goodchild, Peter. *Edward Teller: The Real Dr. Strangelove.* Cambridge, MA: Harvard University Press, 2004.

Herken, Gregg. *Brotherhood of the Bomb: The Tangled Lives and Loyalties of Robert Oppenheimer, Ernest Lawrence, and Edward Teller.* New York: Henry Holt, 2002.

Mark, Hans, and Sidney Fernbach. *Properties of Matter Under Unusual Conditions (In Honor of Edward Teller's 60th Birthday).* New York: John Wiley and Sons, 1969.

Teller, Edward, with Judith L. Shoolery. *Memoirs: A Twentieth-Century Journey in Science and Politics.* Cambridge, MA: Perseus, 2001.

York, Herbert. *The Advisors: Oppenheimer, Teller, and the Superbomb.* Stanford, CA: Stanford University Press, 1976, 1989.

THERMODYNAMICS

The study of the nature of heat is called thermodynamics, derived from the Greek words for heat (*therme*) and power (*dynamis*). Thermodynamics is the field of physics focusing on the exchange of heat—and resulting temperature equilibrium—between hot and cold substances. Work can be extracted through this process of heat transfer. Europeans initially led the way in the study of heat, followed by Americans Benjamin Thompson (Count Rumford) and Josiah Gibbs.

During the seventeenth and eighteenth centuries, the European scientists Galileo Galilei, Robert Boyle, Robert Hooke, and Isaac Newton defined heat as the movement of tiny particles inside matter. During the eighteenth century, scientists focused on the concept of the flow of heat, paralleling the flow of fluids, and determining the heat conductivity of materials, especially in metals.

The late eighteenth-century French scientist Antoine-Laurent Lavoisier's empirical research resulted in a quantitative theory of heat. Benjamin Thompson of Massachusetts, who fled to Europe during the American Revolution and assumed the title of Count Rumford, revisited the kinetic theory of heat to expand on the principles of thermodynamics. His breakthrough was the realization that heat was not a fluid but a conversion of energy process.

Following on the heels of Thompson's qualitative observation, English physicist James Prescott Joule proved that heat was not only a form of energy but equivalent to mechanical energy. He showed that, in an isolated system, work is converted to heat in a one-to-one ratio. Joule's discovery became known as the first law of thermodynamics, often called the law of the conservation of energy.

Research by French scientist Nicolas Léonard Sadi Carnot in the early nineteenth century explored the flow of heat from hot to cold regions. His seminal work, a treatise on the motive power of heat published in 1824, later led to the formulation of the second law of thermodynamics, which involves the unidirectional movement of heat. A hot cup of cocoa, for example, cools because of heat transfer to the surroundings, but heat will not flow from the cooler surroundings to the hotter cup of cocoa. Such common observations are proof of the second law of thermodynamics. Since heat is an interaction and not a fluid flow, a heat engine or heat pump must interact with both a cold sink and a hot source for work to be produced without pause.

The American scientist Josiah Gibbs explored a new area of thermodynamics and provided a strong foundation for much of the field of physical chemistry. Gibbs pioneered the field of chemical thermodynamics in the 1870s. In several papers written during the late 1870s, collectively called *On the Equilibrium of Heterogeneous Substances*, Gibbs introduced the phase rule, which describes the possible number of degrees of freedom in a closed system at equilibrium, including equilibrium at conditions of fixed pressure and temperature. Gibbs's contributions broadened the field of thermodynamics to encompass all transformations between thermal, chemical, mechanical, and electrical energy.

Robert Karl Koslowsky

Sources

Fenn, John B. *Engines, Energy, and Entropy.* New York: W.H. Freeman, 1982.

Van Wylen, Gordon J., and Richard E. Sonntag. *Fundamentals of Classical Thermodynamics.* New York: John Wiley and Sons, 1976.

THOMPSON, BENJAMIN (COUNT RUMFORD; 1753–1814)

Benjamin Thompson was a physicist, inventor, Tory, and expatriate American who was among the great thinkers of the early nineteenth century on both sides of the Atlantic Ocean.

Born in Woburn, Massachusetts, on March 26, 1753, Thompson briefly studied under John Winthrop IV, the Harvard physicist, before moving to Rumford (Concord), New Hampshire. There, he married a rich widow and established himself in polite society.

Thompson became friends with Governor John Wentworth of New Hampshire, with whom he planned a mountaineering trip to the White Mountains in 1773, though at the last minute Thompson was unable to go. In 1776, Thompson's political sympathies forced him to abandon his home and family. He fled to England and eventually ended up in Bavaria, where he was granted the title of Count of the Holy Roman Empire.

Count Rumford, as he was henceforth known, engaged his mind in a variety of scientific interests, ranging from the best way to make coffee to the principles of heat and cold, the nature and uses of gunpowder, and the best means to heat a home. His *Essays, Political, Economical, and Philosophical,* published in 1796, contains discussions on all of these topics and more.

Rumford's studies enabled him to design the most efficient fireplace of his time. Unlike the typical rectangular model of today, Rumford's fireplace was tall and wide, with the sides and fire back tapering elegantly in. Smoke rose up a thin throat that featured a small shelf in the flue, which helped generate a draft to bring cool air in to heat. Rumford's fireplace had the added advantage of being quite smoke-free, hence much healthier for inhabitants. Long concerned with medicine and public health, Rumford hoped his fireplace would benefit Europe's poor.

Rumford was also engaged in the international scientific community. He was an elected member of the Royal Society of London, contributing many papers to its meetings and transactions. He contributed to the American Academy of Arts and Sciences, the Bavarian Academy of Arts and Sciences, which he founded, and an organization in London called the Royal Institution.

Count Rumford lived in several of the great capitals of Europe, including London, Munich, and Paris. He died in Paris on August 21, 1814, having recently been divorced from his second wife, Madame Lavoisier, the widow of the great chemist.

Russell Lawson

Benjamin Thompson, a Massachusetts physicist and inventor whose Tory sympathies led him to move to London at the start of the American Revolution, contributed the theory that heat energy is a by-product of mechanical motion and not a substance. *(Hulton Archive/Getty Images)*

Sources

Brown, Sanborn C. *Benjamin Thompson, Count Rumford.* Cambridge, MA: MIT Press, 1981.

Orton, Vrest. *The Forgotten Art of Building a Good Fireplace: The Story of Benjamin Thompson, Count Rumford, an American Genius, and His Principles of Fireplace Designs Which Have Remained Unchanged for 174 years.* Collingdale, PA: Diane Publishing, 1999.

Thompson, Benjamin (Count Rumford). *Collected Works of Count Rumford.* Cambridge, MA: Harvard University Press, 1968.

UNCERTAINTY PRINCIPLE

The uncertainty principle, formulated in 1927 by the German physicist Werner Karl Heisenberg, states that it is not possible to accurately determine both the position and momentum of a particle such as an electron. The theory is also known as Heisenberg's uncertainty principle and as the principle of indeterminism.

The uncertainty principle is not the only contribution that Heisenberg made to quantum theory. During the period 1924–1926, Heisenberg worked in Copenhagen with the great theoretical physicist Niels Bohr. At Copenhagen, Heisenberg came to the conclusion that it was pointless for physicists to conceptualize the atom in visual terms. All knowledge of the atom comes from observable phenomena, such as its emitted light, its frequency, and its intensity. Heisenberg formulated the equations that were necessary to predict these phenomena. This version of quantum theory became known as "matrix mechanics," and it was the work for which Heisenberg received the 1932 Nobel Prize in Physics. For the remainder of the 1920s, Heisenberg went on to investigate other aspects of quantum theory. In 1927, he formulated his well-known uncertainty principle.

Heisenberg's work on observable quantum phenomena led him to conclude that it is impossible to know for certain the whereabouts and speed of atomic particles. The principle maintains that to locate the exact position of a particle, the observer must subject it to rays of short wavelengths, such as gamma rays. In so doing, the observer will alter the particle's momentum in an unpredictable way. Rays with longer wavelengths would not upset the momentum of the particle as much, but would lack the precision of the shorter wavelengths in determining the particle's location.

The uncertainty principle implies that any description in quantum mechanics may consist only of the relative probability of a value rather than exact numbers. This has important consequences for those seeking a unified field theory that unites the four known interactions: weak nuclear forces, strong nuclear forces, electromagnetism, and gravity.

Albert Einstein made one of the first attempts at formulating a unified field theory, but he rejected quantum theory altogether. One of the most important developments thus far has been made by the American physicist Steven Weinberg and the Pakistani physicist Abdus Salam. Their work contributed to supersymmetry theories and other concepts that are proving to be useful in cosmological models such as the inflationary theory of the universe. American scientists such as Alan Guth and Paul Steinhardt conducted much of the work in this field.

Gordon Stienburg

Sources

Dirac, Paul. *The Principles of Quantum Mechanics.* New York: Oxford University Press, 1982.
Heisenberg, Werner. *The Physical Principles of Quantum Theory.* New York: Dover, 1949.

WHEELER, JOHN (1911–)

John Archibald Wheeler, a pioneering figure in theoretical physics, made contributions in areas as diverse as the structure of the atomic nucleus, relativity, nuclear fission and fusion, unified field theory, and black holes.

Born in Jacksonville, Florida, on July 9, 1911, to parents who were librarians, Wheeler received his Ph.D. in physics from Johns Hopkins University at the age of twenty-one. In 1938, he was hired as a professor at Princeton University, where he worked with Niels Bohr and Albert Einstein.

During World War II, Wheeler worked on the Manhattan Project and did extensive research on general relativity, seeking to develop a unified theory that would encompass both relativity and quantum mechanics. After the war, he worked with Edward Teller on nuclear fusion, and, in 1951, he launched a magnetic fusion research program at Princeton with the help of Lyman Spitzer, Jr., a professor of astronomy.

Working at the Princeton Plasma Physics Laboratory, Wheeler oversaw Project Matterhorn B (for "bombs,"), while Spitzer ran Project Matterhorn S (for the "stellerator," a magnetic fusion design). Matterhorn B was instrumental in helping develop calculations for the thermonuclear hydrogen bomb test "Mike" on November 1, 1952. In 1967, at a conference on supernovae held at the Goddard Institute of Space Studies in New York, Wheeler coined the term "black hole" for a gravitationally collapsed stellar object.

Wheeler retired from Princeton in 1976 and took a teaching position at the University of Texas at Austin, where he focused on research in quantum physics. He did pioneering work in the field of quantum gravity and, in collaboration with Bryce DeWitt, developed the Wheeler-DeWitt equation, also referred to as the "wave function of the universe."

At the same time that he helped advance emerging theories in quantum information physics, Wheeler spent much of his time doing what he loved most: teaching. He mentored a generation of physicists during the course of his career, including 1965 Nobel laureate Richard Feynman. After a prolific career at the University of Texas, Wheeler returned to New Jersey in 1986 to emeritus professor status at Princeton.

Todd A. Hanson

Sources

Taylor, Edwin F., and John Wheeler. *Spacetime Physics.* New York: W.H. Freeman, 1992.

Wheeler, John. *Geons, Black Holes, and Quantum Foam: A Life in Physics.* New York: W.W. Norton, 1998.

WIGNER, EUGENE
(1902–1995)

Nobelist Eugene Paul Wigner was a quantum physicist and contributor to the Manhattan Project.

He was born in Budapest, Hungary, on November 17, 1902. Wigner studied chemical engineering at the Technical University of Berlin and served as a research assistant at the University of Berlin. He was lifelong friends with the mathematician and physicist John von Neumann and also was acquainted with fellow Hungarian scientists Leo Szilard and Edward Teller. During the 1920s, Wigner developed his interests in quantum mechanics and atomic symmetry. He wrote *Group Theory and Its Application to the Quantum Mechanics of Atomic Spectra* in 1931, the same year that he emigrated to America.

Teaching at Princeton University, Wigner had a difficult time adjusting to American social customs and lifestyle, and he wished to return to Germany. But when Adolf Hitler came to power in Germany in 1933, Wigner, a Jew, knew he could not return. So he stayed in United States, becoming a U.S. citizen in 1937.

During the 1930s, Wigner's research focused on solid-state physics and the behavior of protons and neutrons in the fission chain reaction. He worked closely with Hungarian physicist Leo Szilard and Italian physicist Enrico Fermi in theoretical and experimental research into chain reactions. Wigner joined Fermi at the University of Chicago in the early 1940s, where his work on plutonium was used in developing the reactor at Hanford, Washington, that produced the fissionable plutonium used in the first successful test of a nuclear weapon in July 1945. However, he was one of a group of scientists who opposed using the weapon against a civilian population.

After the war, Wigner served as director of the Oak Ridge Laboratory in Tennessee, which focused on the development of fissionable uranium. In 1947, he returned to Princeton, where he spent several decades in research and teaching, focusing in particular on quantum physics. For his work on nucleons (neutrons and protons of atoms), their movements, forces, and symmetry, he shared the 1963 Nobel Prize in Physics with Maria Goeppert Mayer. Wigner died on January 1, 1995.

Russell Lawson

Sources

Seitz, Frederick, Erich Vogt, and Alvin M. Weinberg. "Eugene Paul Wigner." In *Biographical Memoirs,* National Academy of Sciences, vol. 74. Washington, DC: National Academies Press, 1998.

Wigner, Eugene P. *Symmetries and Reflections.* Woodbridge, CT: Ox Bow, 1979.

WINTHROP, JOHN, IV
(1714–1779)

The physicist and mathematician John Winthrop IV was a member of the famous New England family; his great-great-grandfather was the first governor of the Massachusetts Bay Colony, and his great-granduncle was a founding member of the Royal Society of London and governor of

Connecticut (1660–1676). Winthrop's reputation, though, was built on his talent and achievements more than on the family name.

Born in Boston on December 19, 1714, Winthrop attended Boston Latin School and Harvard College, graduating in 1732. He spent the next six years in self-study at the home of his father, a judge. When one of his Harvard professors, Isaac Greenwood, was fired in 1738, Winthrop was appointed to replace him as the Hollis Professor of Mathematics and Natural Philosophy.

He supervised the mathematics tutors who provided instruction in arithmetic, algebra, and geometry, and he delivered scientific lectures to upperclassmen. Under his guidance, Harvard students studied dynamics in matter and fluids and Isaac Newton's *Principia Mathematica* from at least 1751. Winthrop also raised funds and contracted with instrument makers to rebuild Harvard's collection of scientific instruments after most items were destroyed by fire in 1764.

While he remained Hollis Professor until his death, Winthrop's most notable work was as an astronomical observer. He collected data on sunspots in 1739; transits of Mercury in 1740, 1743, and 1763; Halley's comet in 1759; transits of Venus in 1761 and 1769; and several comet passages in 1769–1770. Winthrop performed computations with his 1740s Mercury data to ascertain the longitude between Cambridge and Greenwich, England. For the 1761 Venus transit, he led an expedition of enlightened amateurs and Harvard students to St. John's, Newfoundland. The experience was so profound that he wrote two poems about the trip in addition to a scientific report.

Winthrop also gave public lectures on science, including his prediction of the return of Halley's Comet and a report on a 1755 earthquake, in which he argued that disturbances in Earth's crust were waves caused by heat. Between 1742 and 1774, he published a total of twelve papers in *Philosophical Transactions of the Royal Society.*

A Baconian empiricist, Winthrop kept a weather journal for thirty-five years. He set up an experimental laboratory for exploring the physical sciences in 1746—the first in America—and Benjamin Thompson, who was later known as Count Rumford, attended his demonstrations of the instruments. He was known as a supporter of Benjamin Franklin's one-fluid theory of electricity. The two men were also of like mind on the question of American independence, although Winthrop continued to correspond with colleagues at the Royal Society and the Royal Observatory at Greenwich after the American Revolution began. Winthrop provided support and advice to George Washington and John Adams.

Winthrop was elected a Fellow of the Royal Society in 1766 and to the American Philosophical Society in 1769. The University of Edinburgh awarded him an honorary degree in 1771; Harvard followed suit in 1773 with its first-ever honorary degree. These honors recognized that, for forty years, Winthrop set the standard for scientific, intellectual, and political involvement among colonial professors.

Amy Ackerberg-Hastings

Sources

Shute, Michael, ed. *The Scientific Work of John Winthrop.* New York: Arno, 1980.

Winthrop, John, and John Adams. "Correspondence Between John Adams and John Winthrop." *Collections of the Massachusetts Historical Society,* ser. 5, 4 (1878): 289–313.

DOCUMENTS

Count Rumford's Experiments in Heat

Benjamin Thompson, Count Rumford, made some of his most lasting contributions in the study of heat and motion. The following excerpt is from a paper that he presented to the Royal Society of London in 1798.

Being engaged, lately, in superintending the boring of cannon, in the workshops of the military arsenal at Munich, I was struck with the very considerable degree of heat which a brass gun acquires, in a short time, in being bored; and with the still more intense heat (much greater than that of boiling water, as I found by experiment) of the metallic chips separated from it by the borer.

The more I meditated on these phenomena the more they appeared to me to be curious and interesting. A thorough investigating of them seemed even to bid fair to give a farther insight into the hidden nature of heat; and to enable us to form some reasonable conjectures respecting the existence, or non-existence, of an igneous fluid: a subject on which the opinions of philosophers have, in all ages, been much divided. . . .

From whence comes the heat actually produced in the mechanical operation above mentioned?

Is it furnished by the metallic chips which are separated by the borer from the solid mass of metal?

If this were the case, then, according to the modern doctrines of latent heat, and of caloric, the capacity for the heat of the parts of the metal, so reduced to chips, ought not only to be changed, but the change undergone by them should be sufficiently great to account for all the heat produced.

But no such change had taken place; for I found, upon taking equal quantities, by weight, of these chips, and of thin slips of the same block of metal separated by means of a fine saw, and putting them at the same temperature (that of boiling water) into equal quantities of cold water (that is to say, at the temperature of $59\frac{1}{2}°$ F), the

portion of the water into which the chips were put was not, to all appearance, heated either less or more than the other portion, in which the slips of metal were to put.

This experiment being repeated several times, the results were always so nearly the same that I could not determine whether any, or what change, had been produced in the metal, in regard to its capacity for heat, by being reduced to chips by the borer.

From hence it is evident that the heat produced could not possibly have been furnished at the expense of the latent heat of the metallic chips. But, not being willing to rest satisfied with these trials, however conclusive they appeared to me to be, I had resource to the following still more decisive experiment:

Taking a cannon (a brass six-pounder) cast solid, and rough as it came from the foundry, and fixing it (horizontally) in the machine used for boring, and at the same time finishing the outside of the cannon by turning, I caused its extremity to be cut off; and, by turning down the metal in that part, a solid cylinder was formed, $7\frac{3}{4}$ inches in diameter, and $9\frac{8}{10}$ inches long.

This short cylinder, which was supported in its horizontal position, and turned round its axis, by means of the neck by which it remained united to the cannon, was now bored with the horizontal borer used in boring cannon.

This cylinder being designed for the express purpose of generating heat by friction, by having a blunt borer forced against its solid bottom at the same time that it should be turned round its axis by the force of horses, in order that the heat accumulated in the cylinder might from time to time be measured, a small round hole, 0.37 of an inch only in diameter, and 4.2 inches in depth, for the purpose of introduction a small cylindrical mercurial thermometer, was made in it.

This experiment was made in order to ascertain how much heat was actually generated by friction, when a blunt steel borer being so forcibly shoved (by means of a strong screw) against the bottom of the bore of the cylinder that the pressure against it was equal to the weight of about

10,000 pounds avoirdupois, the cylinder was turned round on its axis (by the force of horses) at the rate of about thirty-two times in a minute. . . .

To prevent, as far as possible, the loss of any part the heat that was generated in the experiment, the cylinder was well covered up with a fit coating of thick and warm flannel, the cylinder was carefully wrapped round it, and defended it on every side from the cold air of the atmosphere.

At the beginning of the experiment the temperature of the air in the shade, as also that of the cylinder, was just 60° F.

At the end of thirty minutes, when the cylinder had made 960 revolutions about its axis, the horses being stopped, a cylindrical mercurial thermometer, whose bulb was $32/100$ of an inch in diameter, and $3\frac{1}{4}$ inches in length, was introduced into the hole made to receive it, in the side of the cylinder, when the mercury rose almost instantly to 130° F. . . .

Finding so much reason to conclude that the heat generated in these experiments, or excited, as I would rather choose to express it, was not furnished at the expense of the latent heat or combined caloric of the metal, I pushed my inquiries a step farther and endeavored to find out whether the air did, or did not, contribute anything in the generation of it. . . .

Everything being ready, I proceeded to make the experiment I had projected in the following manner:

The hollow cylinder having been previously cleaned out, and the inside of its bore wiped with a clean towel till it was quite dry, the square iron bar, with the blunt steel borer fixed to the end of it, it was put into its place; the mouth of the bore of the cylinder being closed at the same time, by means of the circular piston, through the center of which the iron bar passed.

This being done, the box was put in its place, and the joining of the iron rod, and of the neck of the cylinder, with the two ends of the box, having been with cold water (viz., at the temperature of 60° F) and the machine was put in motion.

The result of this beautiful experiment was very striking, and the pleasure it afforded me amply repaid me for all the trouble I had had in contriving and arranging the complicated machinery used in making it.

The cylinder, revolving at the rate of about thirty-two times in a minute, had been in motion but a short time when I perceived, by putting my hand into the water and touching the outside of the cylinder, that heat was generated; and it was not long before the water which surrounded the cylinder began to be sensibly warm.

At the end of one hour I found, by plunging a thermometer into the water in the box (the quantity of which fluid amounted to 18.77 pounds avoirdupois, or $2\frac{1}{4}$ wine gallons) that its temperature had been raised no less than 47 degrees; being now 107° of Fahrenheit's scale.

When thirty minutes more had elapsed, or one hour and thirty minutes after the machinery had been put in motion, the heat of the water in the box was 142 F.

At the end of two hours, reckoning from the beginning of the experiment, the temperature of the water was found to be raised to 178° F. At two hours twenty minutes it was 200° F; and at two hours thirty minutes it *actually boiled!* . . .

By meditating on the results of all these experiments we are naturally brought to that great question which has so often been the subject of speculation among philosophers; namely:

What is heat? Is there any such thing as an *igneous fluid?* Is there anything that can with propriety be called *caloric?*

We have seen that a very considerable quantity of heat may be excited in the friction of two metallic surfaces and given off in a constant stream or flux, *in all directions,* without iteration or intermission, and without any signs of diminution or exhaustion.

From whence came the heat which was continually given off in this manner, in the foregoing experiments? Was it furnished by the small particles of metal, detached from the larger solid masses, on their being rubbed together? This, as we have already seen, could not possibly have been the case.

Was it furnished by the air? This could not have been the case; for in there of the experiments, the machinery being kept immersed in water, the access of the air of the atmosphere was completely prevented.

Was it furnished by the water which surrounded the machinery? That this could not have been the case is evident: first, because this water was continually *receiving heat* from the machinery and could not, at the same time, be *giving to,* and *receiving heat from,* the same body; and secondly, because there was no chemical decomposition of any part of this water. . . .

It is in hardly necessary to add that anything which any *insulated* body, or system of bodies, can continue to furnish *without limitation* cannot possibly be a *material substance:* and it appears to me to be extremely difficult, if not quite impossible, to form any distinct idea of anything, capable of being excited and communicated, in the manner the heat was excited and communication in these, except it be MOTION.

Source: Benjamin Thompson (Count Rumford), "Heat Is a Form of Motion: An Experiment in Boring Cannon," *Transactions of the Royal Society of London* 88 (1798).

The Physics of Sound

Nineteenth-century American scientists, in the wake of inventions in technology that allowed for the electronic transmission of the human voice, explored the physical nature of the voice, as the following journal excerpt reveals.

Of all the branches of natural philosophy, there was not one for a long time, which was so much behind as acoustics—the science of sound. . . .

Among the investigations of [its perception by living beings] is that of the determination of the duration of the residual sensation. It may *a priori* be concluded that the ear acts in this respect toward sound as the eye does toward light, and that the nervous sensation lasts longer than the actual impression. It is well known that a rapidly moving spark makes on the eye the impression of a luminous line, hence the circle of fire seen when a spark is swung around, and that two or more sparks rapidly moving the same line cannot be distinguished from each other, but appear as one single luminous line.

What space is to light, time is to sound; and so in the arts based on light and sound, painting and music, the first ornaments space, the second ornaments time; therefore if rapidly moving luminous points coalesce in space, tones sounded in rapid succession will coalesce in time, and that this coalescence is greater with slow vibrations or low tones than with rapid vibrations or high tones, every attentive listener to music must have observed; the same musical phrase which sounds clear and distinct in the higher octaves will often

become muddled and indistinct when rendered in the lower octaves, and even the greatest composers have often disregarded this, when giving rapid passages to the contra-basso, which never can give satisfaction to such hearers as wish intellectually to understand the meaning.

Source: "New Researches in Sound," *Manufacturer and Builder* 9:6 (June 1877).

Nineteenth-Century Understanding of the Forces of Attraction and Caloric

This paper, read before the Engineers' Society of Western Pennsylvania on November 15, 1881, explored the current scientific understanding of physics.

It is now a well established fact that matter, *per se,* is inert, and that its energy is derived from the physical forces; therefore all chemical and physical phenomena observed in the universe are caused by and due to the operations of the physical forces. . . .

There are but two physical forces, *i.e.,* the force of attraction and the force of caloric. The force of attraction is inherent in the matter, and tends to draw the particles together and hold them in a state of rest. The force of caloric accompanies the matter and tends to push the particles outward into a state of activity.

The force of attraction being inherent, it abides in the matter continuously and can neither be increased nor diminished; it, however, is present in different elementary bodies in different degrees, and in compound bodies relative to the elements of which they are composed.

The force of caloric is mobile, and is capable of moving from one portion of matter to another; yet under certain conditions a portion of caloric is occluded in the matter by the force of attraction. . . .

The force of attraction . . . tends to draw the particles of matter together and hold them in a state of rest; but as this force is inherent, the degree of power thus exerted is in an inverse ratio to the distance of the particles from each other. The effective force so exerted is always balanced by an equivalent amount of the force of caloric, and that modicum of caloric so engaged in balancing the effective force of attraction is static, because occluded in that work.

In solid or fluid bodies, where the molecules are held in a local or near relation to each other, the amount of static caloric will be in direct proportion to the effective force of attraction, but in gaseous bodies the static caloric is in an inverse ratio to the effective force of attraction; hence the amount of static caloric present in solid and fluid bodies will be greatest when the molecules are nearest each other, and greatest in gaseous bodies when the molecules are furthest apart.

Caloric, whether static or dynamic, is not phenomenal; therefore the phenomena of light, temperature, incandescence, luminosity, heat, cold, and motion, as well as all other phenomena, are due to the movement of matter caused by the physical forces. Thus we find that temperature is a phenomenal measure of molecular velocity, as we consider weight to be the measure of matter.

An increase of temperature denotes an increased molecular velocity, and this in solid and liquid bodies unlocks a portion of the static caloric and converts it into dynamic caloric, while an increased temperature of gases occludes additional caloric, thus converting dynamic into static caloric; and a reduction of molecular activity reverses this action. From this we see that a change of temperature either converts static to dynamic or dynamic to static caloric.

Thus we find that the amount of static caloric which a body possesses is in direct relation to its temperature, but . . . temperature is a phenomenal indication of molecular velocity, and as increased velocity separates the molecules to a greater distance, which reduces the effective force of attraction and unlocks a portion of caloric, it will be seen that the separation of the molecules from any other cause will have the same effect.

Source: Jacob Reese, "Electricity: What It Is, and What May Be Expected of It," *Scientific American Supplement* 312 (December 24, 1881).

Section 11

CHEMISTRY

The American Chemist

The first American chemists were alchemists. John Winthrop, Jr., for example, practiced metallurgy partly out of a religious belief that there is a spiritual force inherent in nature that can be discovered by the alchemist, providing knowledge of the sum of all things.

By the Enlightenment of the eighteenth century, Americans, influenced by European chemists, had moved beyond alchemy to forge the beginnings of a rational, empirical science. Chemistry during the nineteenth and twentieth centuries increasingly became the province of the American inventor, who developed such versatile substances as nylon, vulcanized rubber, and plastic.

Alchemical Origins

Modern chemistry developed from ancient and medieval chemistry, which combined observations of the physical world with speculations regarding the nature of the spiritual world. The first American colonies were founded at the end of the European Renaissance, when chemistry and alchemy were synonymous. The Renaissance alchemist was still beholden to an animistic view that there exists a spiritual, living component to Earth that the alchemist, using white magic, can understand, even manipulate. The earliest American scientists relied heavily on such white magic, believing that nature contained spiritual qualities that are unleashed by correct incantations, formulas, and words. Scientist-magicians believed that there exists an elixir of life that prevents disease. The alchemist sought to turn lead into gold through the process of transmutation.

American colonists who had been trained in alchemical ways included George Starkey, Gershom Bulkeley, and John Winthrop, Jr. Starkey, who was educated at Harvard, wrote under the pseudonym Eirenaeus Philalethes; he influenced the likes of the English chemist Robert Boyle. Bulkeley, of Connecticut, was a minister turned alchemist and physician. And Winthrop was the governor of Connecticut, a leader in alchemical work in seventeenth-century America, and the first American member of the Royal Society of London.

American Enlightenment

American chemists were informed by Europeans such as the French scientist Jean Beguin, author of *Beginner's Chemistry* (1610), and British scientist Robert Boyle, who argued that the first-century B.C.E. Epicurean Lucretius was correct when he wrote, in *On the Nature of Things,* that the universe is composed of invisible particles, atoms, or corpuscles, in constant motion. Another influence was the seventeenth-century German chemist John Becher, who formulated the phlogiston theory, based on his belief that there are five elements—air, water, vitreous earth, fatty earth, and a volatile fluid—and that fatty earth is combustible, releasing a substance called phlogiston when it burns.

In the late eighteenth century, French scientist Antoine-Laurent Lavoisier was establishing chemistry as a science, while British chemist Joseph Priestley was discovering oxygen, and British physicist John Dalton was formulating his atomic theory.

Most American chemists at this time still were amateurs who tinkered with experiments and communicated with others about the results. Two notable exceptions were Priestley, after he emigrated to America in 1794, and John Winthrop IV, a Hollis Professor at Harvard.

In 1781, amateur scientist Ebenezer Hazard of Philadelphia sent his friend and fellow amateur Jeremy Belknap two small glass bubbles filled with water, explaining, "I cannot find that they are of any use but to startle people with a sudden smart explosion." Hazard instructed Belknap to

A gunpowder mill founded on Brandywine Creek near Wilmington, Delaware, by French émigré Eleuthère Irénée du Pont in 1802 was one of the earliest chemical production facilities in America and the birthplace of the E.I. DuPont Company. *(Hulton Archive/Getty Images)*

scrape the ashes in the fireplace to one side and "lay one of them before and pretty close to the fire; the heat will make the water evaporate, and burst the glass. If you put it *in* the ashes, or among the coals, it will make them fly about the house." Belknap responded that "the glass bubbles you sent me are the same that I remember to have seen used in Dr. Winthrop's course of Experimental Philosophy," which Belknap took in the 1760s when he was a student at Harvard. The glass was used "to evince the elasticity of the air. One of them was put on a lighted candle, and exploded with a report equal to a pocket pistol. There is another sort made by dropping melted glass into water, which I think he told us was a '*Nodus philosophorum,*' and could not be explained satisfactorily. On breaking the point, the whole mass falls into dust." Winthrop had used the experiment to teach his students that air is matter in motion, corporeal and elastic, fluid and transparent, able to be measured and compressed.

Modern American Chemists

The concern of the American Enlightenment scientist for practical knowledge continued into the nineteenth and twentieth centuries. In 1802, chemist Eleuthère Irénée du Pont established a factory to produce gunpowder. In 1831, physician Samuel Guthrie invented chloroform, which was eventually used as an anesthetic in surgical procedures. In 1839, Charles Goodyear invented the process of vulcanization, combining rubber latex, lead, and sulfur to form a durable and long-lasting rubber. Johns Hopkins University chemist Ira Remsen, in 1879, tasted a sweet substance accidentally produced in the laboratory and soon thereafter invented saccharin. In 1886, Charles Hall, one of the founders of Alcoa, discovered electrolysis as a means of separating aluminum from ore.

Among other practical discoveries, American inventors also made a variety of discoveries in plastics, which are among the most diverse substances in use today. Leo Baekeland in 1907 invented the synthetic polymer Bakelite, a durable and versatile plastic. Wallace Carothers, working for the DuPont company in 1930, invented nylon.

American chemists were also involved in the research and development of the atomic bomb. Harold Urey, for example, developed deuterium (heavy water) in 1934. The following year, Arthur Dempster discovered the uranium isotope U-235 that was used in the atomic bomb "Little Boy" that was dropped on Hiroshima on August 6, 1945. In 1941, Glenn Seaborg developed plutonium, based on the uranium isotope U-238, which was used in the atomic bomb "Fat Man" that was dropped on Nagasaki on August 9, 1945.

In recent decades, the American chemist has been involved in the more theoretical aspects of chemistry. For example, Nobel Prizes have been awarded to Linus Pauling in 1954 for his research on chemical bonding, Robert Mulliken in 1966 for his research on molecular structure, and Roald Hoffman in 1981 for his work on chemical reactions. American chemists have also spearheaded research into DNA and RNA; recent Nobelists in this field are Paul Berg (1980), Walter Gilbert (1980), Thomas Cech (1989), and Sidney Altman (1989). American chemists also have been preeminent in research on proteins, enzymes, and synthetic chemistry.

Russell Lawson

Sources

Lawson, Russell M. "Science and Medicine." In *American Eras: The Colonial Era, 1600–1754*, ed. Jessica Kross. Detroit: Gale Research, 1998.

Leicester, Henry M. *The Historical Background of Chemistry.* New York: Dover, 1971.

Stearns, Raymond Phineas. *Science in the British Colonies of America.* Urbana: University of Illinois Press, 1970.

Eighteenth-Century Chemistry in America

The modern study of chemistry emerged in America in the eighteenth century, but its roots can be found in the alchemy of the sixteenth and seventeenth centuries. Alchemy was more than the mystical mixing and heating of various elements in a vain attempt to transform them into gold; it provided the empirical and theoretical foundations of Enlightenment "chymistry," a quantitative analytical science.

Isaac Newton and Robert Boyle were both influenced by the Harvard-educated "chymist" George Starkey, a seventeenth-century American expatriate living in England. Starkey's primary interest was finding the "universal remedy," a potion that could transmute substances and cure disease. Starkey was one of Boyle's mentors, and Newton's interest in "chymistry" was first piqued in 1687 by Starkey's writings.

The first American who actively engaged in structured chemical study was John Winthrop, Jr., the son of the first governor of the Massachusetts Bay Colony. A seventeenth-century farmer, self-educated physician, naturalist, charter member of the Royal Society of London (founded 1660), and governor of Connecticut (1659–1676), Winthrop had brought books, chemicals, instruments, laboratory apparatuses, and a love of medicine and alchemy with him when he came to America from England in 1631. He corresponded frequently with the Royal Society on matters ranging from corn (maize) to cornbread-based beer brewing, and his paper "Of the Manner of Making Tar and Pitch in New England," which he read before the Royal Society in 1662, marked the first scholarly presentation of an American colonial to a European scientific society.

Winthrop set up the first chemical laboratory in the British colonies in America, as well as the first scientific library. With his use of herbal medicines and natural compounds as remedies, Winthrop pursued iatrochemistry—chemistry for the treatment of disease. There is no indication that he was aware of Giovanni Borelli's understanding of the relationship of blood and disease. It is clear, however, that he was aware of Boyle's corpuscular mechanical philosophy.

The American scientist and naturalist Cadwallader Colden of New Jersey also had an interest in using chemistry for medical treatments. He advocated controlling the velocity of blood through the body by controlling "fermentation." His methods included cooling the body, bleeding, and administering blood-thinning medications.

Although eighteenth-century chemistry was a full-fledged, independent discipline, the science texts of the period placed chemistry within the natural sciences or medicine. The science curricula at American colleges reflected this as well. It was not until the advent of American medical schools in the late eighteenth century that chemistry was considered a science that should be taught along with mathematics, physics, and astronomy.

John Morgan's initial lectures on chemistry (1765 and 1767), the first in the colonies, at the College of Philadelphia (now the University of Pennsylvania Medical School), provided the impetus for the appointment in 1768 of Benjamin Rush, a Philadelphia-born and University of Edinburgh–trained physician, to the first chemistry professorship in America. Rush also wrote the first American textbook on the subject, *A Syllabus of a Course of Lectures on Chemistry* (1770). As of 1800, only six American universities (Pennsylvania, William and Mary, Harvard Medical School, Dartmouth, Columbia Medical School, and Princeton) were offering chemistry as part of the science curriculum.

Apart from the medical schools, chemistry's role in the colonies was more practical and artisanal than theoretical and experimental. Chemistry was used in the manufacture of glass, dyes, waxes, salt, and other useful materials, as well as by apothecaries to produce various medicines. Even Benjamin Rush emphasized chemistry's practical applications, especially its use in the production of gunpowder during the Revolutionary War.

The latter half of the eighteenth century marked a transition in chemistry toward a quantitative analytical study emphasizing elements, elemental compositions, and their relationships and interactions. The French scientist Antoine-Laurent Lavoisier defined the chemical element and compound, stated the law of conservation of mass, introduced the system foundational to chemical nomenclature, devised the "balance sheet" technique of weighing initial ingredients and final products in a chemical reaction, and explained combustion, going beyond the phlogistic chemistry of the English-American Joseph Priestley.

Lavoisier's ideas found fertile ground in America. John de Normandie's "An Analysis of the Chalybeate Waters of Bristol in Pennsylvania," published in 1769 in the *Transactions of the American Philosophical Society*, was the first American paper to address chemical identity and composition. Samuel L. Mitchill of Columbia (editor of the journal *Medical Repository*) and James Woodhouse of the University of Pennsylvania (the founder in 1792 of the first society in the country devoted to the study of chemistry, the Chemical Society of Philadelphia) adopted and promoted Lavoisier's ideas, such as his explanation of combustion.

The study and science of chemistry in America lagged behind that in Europe until the mid-nineteenth century because of shortages in instruments, research facilities, and advanced education at the college level, as well as a lack of educated practitioners. Many more European universities actively taught chemistry and engaged in experimentation. In addition, America suffered from a relative lack of science libraries and theoretical scientific societies. These shortfalls were compounded by public disinterest in any discipline that did not have an immediate and practical application.

It was not until the late nineteenth century, with the rise of the university, that Americans developed the requisite scientific knowledge, the educated base, and the commitment and funding for the modern discipline of chemistry to flourish.

Richard M. Edwards

Sources

Bedini, Silvio A. *Thinkers and Tinkers: Early American Men of Science.* New York: Charles Scribner's Sons, 1975.

Friedenberg, Zachary B. *The Doctor in Colonial America.* Danbury, CT: Rutledge, 1998.

Hindle, Brooke, ed. *Early American Science.* New York: Science History, 1976.

———. *The Pursuit of Science in Revolutionary America, 1735–1789.* Chapel Hill: University of North Carolina Press, 1956.

Newman, William R. *Gehennical Fire: The Lives of George Starkey, an American Alchemist in the Scientific Revolution.* Chicago: University of Chicago Press, 2002.

Newman, William R., and Lawrence M. Principe. *Alchemy Tried in the Fire: Starkey, Boyle, and the Fate of Helmontian Chymistry.* Chicago: University of Chicago Press, 2002.

Stearns, Raymond Phineas. *Science in the British Colonies of America.* Urbana: University of Illinois Press, 1970.

The Plastics Revolution

Plastics are moldable synthetic materials made of polymers, which are molecules consisting of many repeating units and usually made of carbon, hydrogen, oxygen, and a number of other possible elements. The versatility of plastics makes them functional alternatives to such natural materials as stone, metal, wood, and leather. As a result, plastics represent one of the largest manufacturing industries in the United States today, even though the invention of plastics is a relatively modern technological development.

Early Breakthroughs

Only in the last hundred years have scientists developed the fully synthetic polymers that are the basis of plastics manufacturing. One of the first commercially successful semisynthetic plastics resulted when John Wesley Hyatt responded to a challenge in the 1860s to develop a material to replace ivory in billiard balls. Hyatt modified cellulose to create pyroxylin, or "celluloid."

The first truly synthetic polymer, phenolic resin, was developed by Leo Baekeland in 1907 and unveiled in 1909. Phenolic resins soon found widespread use in adhesives, laminating resins, and molding compounds, and Baekeland's trademarked name, Bakelite, became a synonym for any hard plastic. This marked the beginning of a manufacturing revolution. As materials engineers and product designers recognized the versatility and low cost of plastics, these synthetic substances came to be used in an ever widening range of products and packaging.

Synthetic polymer development and the plastics industry in general flourished in the United States in the 1900s as a result of the activities of both individual researchers and corporate laboratories. Phenolics such as Bakelite are thermosets, or plastics that, once formed and cured, cannot be remelted without degradation. Next came the development of aminoplastics: thioureas (1928) could be more brightly colored than the phenolics, and melamines (1937) offered better heat and moisture resistance. These materials found decorative as well as industrial uses, from jewelry, dinnerware, and radio cabinets to connectors, handles, and switchgear.

In the 1930s, chemists began to formulate what some in the field call "true plastics," or polymeric materials derived from a hydrocarbon base. These are known as thermoplastics and, in contrast to thermosets, can be repeatedly melted and formed into useful shapes. The majority of thermoplastics were developed between 1930 and 1965, many in the United States. Waldo Semon, a chemist at the B.F. Goodrich tire and rubber company, developed polyvinylchloride (PVC) in 1933. Later that year, Dow Chemical researcher Ralph Wiley discovered polyvinylidene chloride (PVDC, Saran). In 1938, DuPont chemist Roy Plunkett discovered polytetrafluoroethylene (PTFE, Teflon). Among these early plastics were nylon, first developed by the DuPont Corporation and introduced to the public at the 1939 New York World's Fair. Most of the new plastics of the 1930s, however, from acrylic to polyurethane, were invented by scientists in Germany, which has a long tradition in chemical research.

Two well-known thermosets were introduced in the United States during World War II. Epoxy resins were initially used as industrial paints and linings. Polyester thermosets, which were readily molded at low pressures, were used for the encapsulation of electrical and electronic parts as well as in boat hulls, hardhats, and other reinforced plastics. Some of these plastics were put to use in the war effort—especially synthetic rubber, a critical defense material, since the sources of the natural substance in Southeast Asia were controlled by the Japanese. The war postponed the full implementation of these new plastics in consumer products.

The Postwar Explosion

The global postwar boom pushed plastics into everyday life. From the end of World War II through the 1960s, U.S. companies such as DuPont and General Electric were developing new techniques and machinery to mold and extrude plastics on a mass scale. The various

The strength of new plastic sheeting is demonstrated at a loading dock in the early 1950s. Lightweight, durable, versatile, and inexpensive, plastics have caused a global industrial revolution that continues to expand—and pose a growing environmental danger. *(Andreas Feininger/Time & Life Pictures/Getty Images)*

types of polymers and the techniques for producing them allowed for the introduction of plastics with specialized consistencies and uses.

Some of the new plastics in the polyethylene terephthalate (PET) family came in the form of synthetic fibers such as Dacron®, Lycra®, and polyester, which could be woven into "wash-and-wear" garments that needed no ironing. Another PET fiber introduced in the early 1950s was Mylar®, which came to be used in recording tape and plastic covering for glass (for shading or reducing shattering). But the most widespread use of inexpensive PET plastics was in packaging. The first plastic soda bottles were introduced in the 1950s. Such use of plastics became increasingly popular in a host of products in the 1960s and 1970s due to the material's light weight and resistance to breakage.

Teflon®, which was invented by the DuPont chemist Roy Plunkett in 1938, was derived from a different type of polymer—polytetrafluoroethylene (PTFE)—and it came into widespread use in the 1950s and 1960s as a nonstick coating for cookware. Derivatives of PTFE found other uses in the 1970s in everything from dental floss to surgical implants and synthetics such as Gore-Tex®. Unlike most other synthetic fibers, Gore-Tex had the ability to "breathe," or allow air and water vapor to pass between its interconnected fibers, while keeping out larger water molecules, making it ideal for foul-weather clothing. As outdoor sports activities gained in popularity during the last decades of the century, Gore-Tex, and similar permeable membranes, became increasingly associated with sports enthusiasts and those who wanted to appear athletic.

The versatility of plastics is witnessed by the fact that, while Gore-Tex was popular for its breathability, other plastics were purchased because of their ability to keep air out. Perhaps no line of items better acquainted Americans with the potential benefits of plastic than Tupperware®, a line of polyethylene storage containers with the patented "burp" seal that provides airtight protection of food.

Earl Tupper came up with the idea for a polyethylene-based airtight container while working as a chemist at DuPont in the 1930s. In 1938, he quit DuPont to form the Tupperware Company. He began marketing the product in stores in the late 1940s, but to middling success. In the early 1950s, Tupper tried a new approach. Discontinuing store sales, he launched a direct-marketing campaign in which homemakers invited their neighbors for demonstrations of the product in their living rooms. These Tupperware parties made the brand a household name, and the products became a fixture in the American kitchen.

Aiding the spread of of Tupperware, and new food-preserving plastics such as Saran Wrap®, was the increase in supermarkets. Instead of buying a day's worth of provisions at the local grocery store, consumers shopped at supermarkets and bought a week's worth of groceries—much of it requiring longer-term storage.

New materials and fibers, such as synthetic fleece (a wool substitute), Thinsulate™ (a replacement for bird down insulation), and Kevlar® (an impact- and tear-resistant fiber used in bulletproofing and other materials) have continued to come out since the golden age of plastics in the 1950s and 1960s. Today, thermoplastics are used in a wide range of products, including pipe (PVC), barrier films (Saran), rope and other twisted fibers (nylon and polypropylene), nonstick coatings

(Teflon), medical implants (polyethylene and silicones), safety lenses (polycarbonate), electrical insulation (polyimide), and automotive components (polyphenylene oxide or PPO). The plastics industry is now one of the largest in the United States, with revenues estimated in the hundreds of billions of dollars annually.

Maureen T.F. Reitman and James Ciment

Sources

Brydson, J.A. *Plastics Materials.* 7th ed. Burlington, VT: Butterworth-Heinemann, 1999.

DuBois, J. Harry. *Plastics History U.S.A.* Boston: Cahners, 1972.

Fenichell, Stephen. *Plastic: The Making of a Synthetic Century.* New York: HarperBusiness, 1996.

Meikle, Jeffrey L. *American Plastic: A Cultural History.* New Brunswick, NJ: Rutgers University Press, 1995.

Plastics Historical Society. http://www.plastiquarian.com.

CALVIN, MELVIN
(1911–1997)

Nobel laureate and chemist Melvin Calvin was born on April 8, 1911, in St. Paul, Minnesota. He received his bachelor's degree in science from the Michigan College of Mining and Technology in 1931. Four years later, he took a doctorate in chemistry at the University of Minnesota.

Following two years of postdoctoral study at the University of Manchester in England, where he became interested in phthalocyanines, Calvin became a chemistry instructor at the University of California, Berkeley. He rose to the rank of professor in 1947 and became a university professor of chemistry in 1971. From 1963 to 1980, Calvin also served as professor of molecular biology and director of the Laboratory of Chemical Biodynamics. From 1967 to 1980, he was associate director of the Lawrence Berkeley Laboratory, and, from 1981 to 1985, he served on the Energy Research Advisory Board of the U.S. Department of Energy.

Of the numerous academic honors Calvin received during the course of his career, none was greater than his 1961 Nobel Prize in Chemistry. At Berkeley, Calvin's initial work with Gilbert Lewis on the photochemistry of colored porphyrin analogs and the theoretical aspects of molecular structure and behavior of organic compounds led him to the problem of photosynthesis. By the late 1940s, the process of carbon dioxide assimilation was known to involve two interdependent processes: the light and dark reactions. Using the radioactive isotope carbon 14 as a tracer, Calvin's lab studied the dark reaction in the single-celled alga *Chlorella pyrerloidosa*. The researchers exposed cultures to radioactive carbon dioxide for varying lengths of time and then killed the algae. The intermediary products in the conversion of carbon dioxide and water to carbohydrates could thus be isolated and then identified via paper chromatography. Ultimately, Calvin and his associates mapped the complex cycle of reactions that have come to be known as the Calvin cycle, by which atmospheric carbon dioxide is converted into carbohydrates and other organic compounds.

Throughout his career, Calvin demonstrated a diverse range of research interests, which grew broader as he got older. He worked in hot atom chemistry, carcinogenesis, chemical evolution and the origin of life, organic geochemistry, immunochemistry, petroleum production from plants, farming, extraterrestrial geology and biology, and the feasibility of artificial photosynthetic systems. Among his more prominent publications are *The Path of Carbon in Photosynthesis* (1957), which served as a kind of bible for the first group of researchers that worked with radioactive carbon; *Chemical Evolution: Molecular Evolution Towards the Origins of Living Systems on Earth and Elsewhere* (1969); and an autobiography, *Following the Trail of Light: A Scientific Odyssey* (1992).

Calvin maintained a research group at Berkeley until 1996. He died of a heart attack on January 8, 1997.

Sean Kelly

Sources

Calvin, Melvin. *Following the Trail of Light: A Scientific Odyssey.* Oxford, UK: Oxford University Press, 1998.
Nobel Lectures. *Chemistry 1942–1962.* Amsterdam: Elsevier, 1964.

CAROTHERS, WALLACE
(1896–1937)

Wallace Hume Carothers, scientist, teacher, and researcher, was responsible for the basic research that developed neoprene (synthetic rubber) and nylon.

He was born in Burlington, Iowa, on April 27, 1896, the eldest of four children of Ira Hume Carothers and Mary Evelina McMullin. After completing his secondary education in Des Moines in 1914, he studied accounting at Capital City Com-

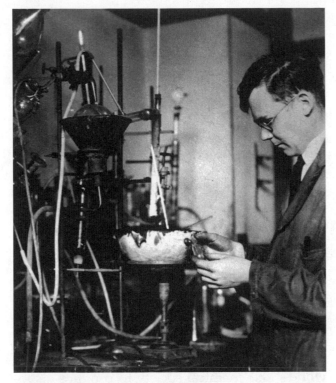

Organic chemist Wallace Carothers led the polymer research team at a DuPont laboratory that developed neoprene (a synthetic rubber) in 1930 and nylon (a synthetic fiber) in 1935. *(Hulton Archive/Getty Images)*

mercial College and earned a bachelor's degree in chemistry at Tarkio College. He earned a master's degree from the University of Illinois in 1921 and went on to teach chemistry at the University of South Dakota. While there, he began exploring the ideas of American chemist Irving Langmuir on double bonding.

Returning to the University of Illinois, he earned his Ph.D. in chemistry in 1924, writing a dissertation on hydrogenations with modified platinum-oxide-platinum-black catalysts. After remaining at Illinois for another two years as an instructor, Carothers began teaching chemistry at Harvard in 1926 and took up experiments with the chemical structure of long-chain polymers.

In 1928, E.I. du Pont de Nemours and Company hired Carothers to head a basic research team at its central laboratory in Wilmington, Delaware. With his own lab and team of scientists, Carothers began studying the acetylene chemical family. Building on work begun by Notre Dame chemist Julius Arthur Nieuwland, the team generated a number of patents and original research papers. Turning their attention

to the synthesis of vinylacetylene with chlorine, they created synthetic rubber (neoprene) in 1930.

When worsening political conditions in Asia during the 1930s threatened the silk trade with China, the DuPont corporation turned its attention to the possibility of producing artificial silk. Company experiments produced a variety of polyesters and polyethers. In 1935, a polyamide was formed by combining hexamethylene diamine and adipic acid, which led to the invention of nylon.

In 1936, Carothers became the first organic chemist associated with private industry to be elected to the National Academy of Sciences. The following January, his sister Isobel, a successful radio musician (as Lu in the musical trio Clara, Lu, and Em), died unexpectedly. Carothers fell into a depression that led to suicide on April 29, 1937.

Andrew J. Waskey

Sources

Hermes, Matthew E., ed. *Enough for One Lifetime: Wallace Carothers, Inventor of Nylon.* Washington, DC: American Chemical Society, 1996.

Mark, H., and G.S. Whitby, eds. *Collected Papers of Wallace Hume Carothers on High Polymeric Substances.* New York: Interscience, 1940.

CELLULOID

Celluloid is the common name for a type of synthetic plastic developed in the mid-nineteenth century, using nitrocellulose, camphor, and other materials. Believed to be the original thermoplastic, celluloid is easily shaped into a number of industrial and commercial applications, including waterproofing for clothing, billiard balls, pen bodies, toys, photographic paper, film for movies, and other products.

Due to the instability and flammability of its original formula, the original celluloid was replaced in the late 1920s by more advanced materials, including the family of polyethylene plastics. For some seventy years, however, celluloid was the object of ongoing technological development and innovative design with myriad applications in everyday American life.

An early form of celluloid was invented in 1856

by Alexander Parkes, a metallurgist in Birming-ham, England, who did not succeed in marketing the product. In the 1860s, John Wesley Hyatt, an American printer and inventor, picked up where Parkes had left off. Hyatt was after a $10,000 prize offered by the Phelan and Collander company, a maker of billiard balls, for anyone who could de-velop a practical substitute for ivory. Hyatt had already developed expertise in hot-compression molding, fabricating dominoes from mixtures of cellulose (the chief component of the cell walls in wood, cotton, hemp, and other plants, and the most abundant of all naturally occurring organic compounds), and shellac. In 1870, Hyatt applied the same techniques to create celluloid, a solu-tion of nitrocellulose and camphor, the latter added as the plasticizing agent.

Hyatt and his brother, Isaiah Smith Hyatt, reg-istered the trade name "celluloid" in 1871 and es-tablished the Celluloid Manufacturing Company in Albany, New York; the following year, they re-located the company to Newark, New Jersey. Among the many unique products that Hyatt developed were composite billiard balls and molded composites of ivory dust and cellulose nitrate. He prepared his best compositions for celluloid by evaporating aqueous mixtures of cellulose nitrate and camphor, then molding the residue with heat and pressure.

Applications for celluloid during the early 1900s included knife handles, toothbrushes, combs, corset stays, shirt collars, and privacy side curtains for buggies. Later applications included privacy curtains for automobiles, piano keys, and the product for which the material is best known—photographic film. Celluloid film be-came an essential component of the handheld Kodak camera, which popularized photography as a hobby and art form in the 1890s. Moving-picture cameras likewise employed celluloid film and gave rise to the movie industry in the twen-tieth century. Other products made from cellu-loid included eyeglass frames, pens, toilet seats, and safety glass.

In part due to the absence of competitive plas-tics, the celluloid industry continued to grow until about 1925. But because celluloid was highly flammable and required high-priced and toxic camphor, which was imported from Asia, it was soon replaced by cheaper, more stable syn-thetic polymers. Hyatt's Newark plant closed in 1949, and celluloid is no longer manufactured on a large scale in the United States. Ping-pong balls are among the last products still made of celluloid.

James Fargo Balliett and George B. Kauffman

Sources

Friedel, Robert. *Pioneer Plastic: The Making and Selling of Cel-luloid.* Madison: University of Wisconsin Press, 1983.

Meikle, Jeffery L. *American Plastic: A Cultural History.* New Brunswick, NJ: Rutgers University Press, 1995.

Seymour, Raymond B., and George B. Kauffman. "The Rise and Fall of Celluloid." *Journal of Chemical Education* 69:4 (1992): 311–14.

CHEMICAL SOCIETY OF PHILADELPHIA

The Chemical Society of Philadelphia, the world's first chemical society to produce a sub-stantial record of writing and publication, was founded in 1792 by James Woodhouse, who re-mained its senior president throughout its exis-tence. Woodhouse, a physician like many early chemists, was appointed professor of chemistry at the University of Pennsylvania in 1795. An-other early leader of the society with a medical background and connections to the university was the physician and medical historian John Redman Coxe. An early champion of vaccina-tion, Coxe succeeded Woodhouse as professor of chemistry after the latter's death in 1809.

Despite the medical backgrounds of many so-ciety members, its chemical interests were not restricted to pharmacology. Inspired by a utili-tarian approach to chemistry, one of its missions was encouraging the exploitation of America's rich mineral resources. In pursuit of that end, the society maintained a laboratory and offered free analyses of mineral samples sent in by the pub-lic. Another mission was to advance the new, anti-phlogistonic chemistry of Antoine-Laurent Lavoisier in America. The society published pamphlets on analysis, as well as some of the an-nual orations delivered before it. Thomas Peters Smith's published 1798 oration, *A Sketch of the Revolutions in Chemistry*, was one of the earliest attempts at a written history of chemistry.

One of the most important papers presented

to the society was Robert Hare's "Memoir of the Supply and Application of the Blow-Pipe: containing an account of a new method of supplying the blow-pipe either with common air, or oxygen gas, and also of the effects of the intense heat produced by the combustion of the hydrogen and oxygen gases," read on December 10, 1801, and published by the society the following year. Hare's oxyhydrogen blow-pipe, capable of producing intense heat, was the ancestor of the modern welding torch.

The society had about seventy members and one corresponding member, the English experimental chemist Elizabeth Fulhame (who, as a woman, was excluded from scientific societies in her own country). The Chemical Society of Philadelphia disbanded sometime in the first decade of the nineteenth century, and it was succeeded by another short-lived group, the Columbian Chemical Society of Philadelphia.

William E. Burns

Sources

Greene, John C. "The Development of Mineralogy in Philadelphia, 1780–1820." *Proceedings of the American Philosophical Society* 113:4 (August 15, 1969): 283–95.

Miles, Wyndham D. "John Redman Coxe and the Founding of the Chemical Society of Philadelphia in 1792." *Bulletin of the History of Medicine* 30 (1956): 469–72.

CONANT, JAMES B. (1893–1978)

James Bryant Conant was a preeminent chemist, college administrator, adviser to presidents, and contributor to the Manhattan Project.

Conant was born on March 26, 1893, in Dorchester, Massachusetts. He attended Harvard University, earning a Ph.D. in chemistry in 1916. During World War I, he worked in chemical weapons research, specifically on lewisite, a chemical agent similar to mustard gas that blisters and burns the skin and lungs.

At the end of the war, Conant returned to Harvard, where he served as a professor of organic chemistry until 1933. He wrote more than 100 scientific papers highlighting his research on chlorophyll, polymers, radiocarbon, and organic chemical reactions. Conant became chair of the Chemistry Department and, in 1933, president of the university.

In 1940, Conant became involved in government research in preparation for possible war, serving as head of the National Defense Research Committee. During World War II, he was an adviser to President Franklin D. Roosevelt on the Manhattan Project to build an atomic bomb as well as in chemical weapons research and development. Conant recruited American scientists to research chemical and nuclear weapons, synthetic rubber, and radar technology. He was a chief adviser to President Harry S. Truman regarding implementation of the atomic bomb and, in May 1945, served on the Interim Committee that recommended to Truman how, where, and when to use the bomb against Japan.

At the end of the war, Conant became a member of the Atomic Energy Commission. He took a tough stance against the Soviet Union throughout the Cold War and served as ambassador to West Germany in the 1950s. Conant was awarded the Presidential Medal of Freedom by President John F. Kennedy in 1963.

Over the years, Conant wrote numerous books. One of his research interests was the English chemist Robert Boyle's "Experiments in Pneumatics," but most of his publications involved more contemporary issues regarding science and public policy. He published a discussion of the nature of science, *On Understanding Science: An Historical Approach*, in 1947; this work was expanded in 1951, in *Science and Common Sense*.

In the following passage from *On Understanding Science*, Conant anticipated science historian Thomas Kuhn's concept of paradigms in the process of scientific revolutions:

> We can put it down as one of the principles learned from the history of science that a theory is only overthrown by a better theory, never merely by contradictory facts. Attempts are first made to reconcile the contradictory facts to the existing conceptual scheme by some modification of the concept. Only the combination of a new concept with facts contradictory to the old ideas finally brings about a scientific revolution. And when once this has taken place, then in a few short years discovery follows upon discovery and the branch of science in question progresses by leaps and bounds.

Conant also was involved in education and social policy, and he wrote an important book, *Slums and Suburbs* (1961), examining the differences between inner-city and suburban schools in America. Toward the end of his life, he penned an autobiography, *My Several Lives: Memoirs of a Social Inventor* (1970). Conant died in Hanover, New Hampshire, on February 11, 1978.

Russell Lawson

Sources

Conant, James B. *On Understanding Science: An Historical Approach.* New York: Mentor, 1951.

————. *Science and Common Sense.* New Haven, CT: Yale University Press, 1964.

————. *Slums and Suburbs.* New York: McGraw-Hill, 1961.

Hershberg, James B. *James B. Conant: Harvard to Hiroshima and the Making of the Nuclear Age.* New York: Alfred A. Knopf, 1993.

DOW CHEMICAL

The Dow Chemical Company, a multinational corporation headquartered in Midland, Michigan, was founded by Herbert Henry Dow in 1897 to extract chlorides and bromides from brine deposits under Midland. Dow is the world's largest chemical company, producing calcium chloride, ethylene oxide, acrylates, surfactants, ethylcellulose resins, and agricultural chemicals such as the pesticide Lorsban. It is also the largest producer of plastics, such as polystyrene, polyurethanes, polyethylene terephthalate, polypropylene, and synthetic rubbers. Styrene, Saran Wrap, Ziploc bags, Handi-Wrap, Scrubbing Bubbles, Styrofoam, and Silly Putty are among Dow's well-known consumer products.

Herbert Dow initially envisioned the vertical integration of the production and sale of bleach. The company expanded during the first decades of the twentieth century into the production of agricultural chemicals, elemental chlorine, phenol, magnesium metal, and other dyestuffs. It became involved in thermoplastics in the 1930s and eventually in various consumer products. Dow also grew by means of acquisitions, subordinate mergers, the development of international operations, and cooperative agreements with Midland Chemical in 1900, Dow Chemical of Canada in 1942, Asahi-Dow in 1952, the pharmaceutical company Merrill Dow in 1980, Hoechst South Africa in 1999, and Union Carbide in 2001.

Dow's core chemical and plastics industries, along with its production of magnesium extracted from seawater, used in the fabrication of lightweight airplane parts, made the company strategically important to the United States during World War II. Dow's expertise in the production of synthetic rubber, its burgeoning partnership with Corning in the production of silicone (1943), and its ability to safely produce incendiary napalm in high volumes made Dow essential to the war effort and established defense products as a core business that continued into the twenty-first century.

Although many of Dow's products are essential to industry, agriculture, and the military, a number have faced ethical and legal challenges, including Agent Orange, an herbicide produced for the Vietnam conflict; dioxin, a toxic manufacturing by-product; and Dow Corning's silicone breast implants. To ensure the safety and propriety of its products and manufacturing processes, Dow created a number of protocols: a formal pollution control program (1936); Global Pollution Control Guidelines (1972); International Business Principles (1975); Global Ethics and Compliance (1998); and Code of Business Conduct (1999).

Richard M. Edwards

Sources

Brandt, E. Ned. *Growth Company: Dow Chemical's First Century.* Lansing: Michigan State University Press, 2003.

Doyle, Jack. *Trespass Against Us: Dow Chemical and the Toxic Century.* Monroe, ME: Common Courage, 2004.

Whitehead, Don. *Dow Story: The History of the Dow Chemical Company.* New York: McGraw-Hill, 1968.

GIAUQUE, WILLIAM (1895–1982)

A physical chemist and educator, William Francis Giauque received the 1949 Nobel Prize in Chemistry "for his contributions in the field of classical thermodynamics, particularly concerning the behavior of substances at extremely low temperatures."

The son of railroad worker William Tecumseh Sherman Giauque and Isabella Jane Duncan, he was born in Niagara Falls, Ontario, on May 12, 1895. The family lived in Michigan, where Giauque attended school until age thirteen, when his father died and the family returned to Niagara Falls.

Two years of work at the Hooker Electrochemical Company led Giauque to study chemical engineering. In 1916, he entered the University of California, Berkeley, majoring in chemistry. After graduating with a B.S. in 1920, he earned a Ph.D. in chemistry with a minor in physics in 1922. Giauque spent the rest of his career at Berkeley, serving as an instructor (1922–1927), assistant professor (1927–1930), associate professor (1930–1934), and professor (1934–1977). In 1932, he married Muriel Frances Ashley, a physicist and botanist.

Giauque's earliest research focused on entropy (a measure of disorder) at low temperatures and the third law of thermodynamics, the validity of which he was the first to demonstrate. He studied the properties of matter at the lowest attainable temperatures throughout his career. In 1920, he and his mentor, George Ernest Gibson, stated the third law as currently accepted: the entropy of a perfect crystal is zero at absolute zero (0 K, –273° C, –459° F). He calculated entropies and other thermodynamic properties of substances at low temperatures with ten times the accuracy of earlier researchers. These entropies, together with heats of formation, permit the calculation of free energies (the work put into a system during certain reversible processes), which can be used to predict the direction and extent of chemical reactions.

Because many of his low-temperature studies were of long duration and required constant attention, he and his assistants worked as many as sixty hours straight. During the 1920s and early 1930s, Giauque and his students continued third-law confirmations with a series of studies on diatomic gases (those whose molecules consist of two atoms), experiments that verified the use of quantum statistics (the statistical description of particles or systems of particles the behavior of which must be described by quantum mechanics rather than classical mechanics), and studies on the partition function (the sum over allowed states of an exponential quantity) in calculating entropy.

From 1928 to 1932, Giauque worked on the entropy of gases, showing that some crystallize with residual entropy as the temperature approaches absolute zero. His results supported physicist Werner Heisenberg's 1927 proposal that hydrogen and other diatomic molecules of elements could exist in two forms: ortho (antisymmetrical), in which the spins of the two nuclei are parallel, and para (symmetrical), in which the spins of the two nuclei are antiparallel.

Giauque's most significant contribution, the adiabotic (involving no gain or loss of heat) magnetic cooling technique, resulted from his third-law studies. It allowed scientists to obtain temperatures close to absolute zero, to understand better the principles and mechanisms of electrical and thermal conductivity, to determine heat capacities, and to study the behavior of superconductors at very low temperatures. At these temperatures, the usual constant-volume gas thermometer was useless, so Giauque invented a highly sensitive carbon (lampblack) resistance thermometer for temperatures below 1 kelvin.

In 1929, while studying the entropy of oxygen, Giauque ascribed the presence of faint lines in its spectra to small amounts of two hitherto unknown isotopes (forms of an element with the same atomic numbers but different atomic weights): oxygen-17, and oxygen-18. His discovery of oxygen-18 provided an isotopic tracer for studying the mechanisms of photosynthesis and respiration, and it led to Harold Urey's discovery of deuterium (hydrogen-2) in 1931.

Giauque published 183 articles and trained fifty-one graduate students. A cryogenics laboratory on the Berkeley campus named in his honor opened in 1966. He died on March 28, 1982, in Oakland, California.

George B. Kauffman

Sources

Giauque, William F. *Low Temperature, Chemical, and Magneto Thermodynamics: The Scientific Papers of William F. Giauque.* New York: Dover, 1969.

Jeffers, William A., Jr. "William Francis Giauque (1895–1982)." In *Nobel Laureates in Chemistry 1901–1992,* ed. Laylin K. James. Washington, DC: American Chemical Society, 1993.

Nobel Lectures. *Chemistry 1942–1962.* Amsterdam, The Netherlands: Elsevier, 1964.

HALL, LLOYD AUGUSTUS (1894–1971)

Lloyd Augustus Hall was one of the foremost scientists to have worked in the field of food chemistry.

He was born on June 20, 1894, in Elgin, Illinois, and raised by his parents, Augustus and Isabel Hall, in the nearby town of Aurora. His father was a Baptist minister and the son of the pastor for the first African-American church in Chicago. His mother was the daughter of a woman who had escaped enslavement on the Underground Railroad. The lessons of perseverance and resolve were passed from his grandparents to his parents and on to him, enabling Hall to weather the trials that awaited him during his career as a scientist.

Hall's curiosity in science was piqued when he took a chemistry class at East Aid High School, where he was an honors student. By the end of high school, Hall had earned scholarships to four different colleges; he chose Northwestern University. Majoring in chemistry, Hall graduated in 1916. He landed a job with the Chicago Department of Health, rising to the post of senior chemist in just one year.

During World War I, Hall was commissioned as a lieutenant and served as assistant chief inspector of explosives for the U.S. Ordnance Department. After the war, he married Myrrhene Newsome, and the couple moved to Chicago, where Hall took a job with Boyer Chemical Laboratory.

At Boyer, Hall began working in food chemistry. Specifically, he researched meat preservation, exploring how different chemical compounds made from salts would keep food fresh the longest. In the course of his research, Hall discovered the process of "flash drying," a preservation technique that revolutionized the meat-curing industry.

In 1925, Hall became chief chemist and director of research at Griffith Laboratories in Illinois. From 1925 to 1959, he developed and refined processes for the sterilization and preservation of food that are still in use today. Among his innovations were methods of sterilizing spices and cereals; the development of meat-curing products, seasonings, and emulsions; and the development of yeast foods. Over the course of his forty-three-year career, he obtained more than 100 patents for his work.

After retiring, Hall served as a consultant to the United Nations Food and Agriculture Organization and sat on the American Food for Peace Council. He died in Pasadena, California, on January 2, 1971.

Paul T. Miller

Source

Carwell, Hattie. *Blacks in Science: Astrophysicist to Zoologist.* New York: Exposition, 1977.

INORGANIC CHEMISTRY

Although the science of chemistry was of little interest to most early colonists, the history of American chemistry can be traced back to the early seventeenth century. Not long after his arrival in Boston in 1631, John Winthrop, Jr., using equipment, reagents, and literature imported from England, established the first chemical laboratory within the North American British colonies. He became the first colonial American to present a scholarly paper to a scientific organization when he read his "Of the Manner of Making Tar and Pitch in New England" before the Royal Society in 1662.

For much of the ensuing century, those colonials who were interested in chemistry tended to study with apothecaries or in medical schools. It was only in the mid-eighteenth century that chemistry began to penetrate medical curricula.

In the latter half of the eighteenth century, French chemist Antoine Lavoisier revolutionized the field. In addition to the introduction of a new nomenclature, which named chemicals according to their composition, Lavoisier differentiated organic from inorganic chemistry. Today, inorganic chemistry refers to the chemistry of all non-carbon-based compounds.

Nineteenth-century chemistry within the United States came to focus primarily on identifying and analyzing the country's untapped mineral wealth, though academic debates were not entirely absent. The founder of the Chemical Society of Philadelphia, James Woodhouse, for example, spent close to a decade in the 1790s debating (in the pages of the *Medical Repository*) Joseph Priestley over the latter's phlogiston theory of combustion. Thomas Jefferson, in particular, encouraged scientists to focus on useful arts and sciences to further the country's prestige, power, and interests. Trained chemists, however, remained scarce.

At the turn of the nineteenth century, only six schools—Pennsylvania, William and Mary, Harvard Medical School, Dartmouth, Columbia, and Princeton—offered separate courses in chemistry. Although teachers, textbooks, and laboratory supplies remained in short supply, the number of schools that offered instruction in chemistry grew tremendously. By the late nineteenth century, almost a third of the nation's 101 major chemists, those who were distinguished enough to appear in the *Dictionary of American Biography* (New York, 1928–1958), worked in either industrial or agricultural chemistry. Among the more prominent nineteenth-century American chemists was Charles Eliot, who as president of Harvard from 1869 to 1909 directed the school's transformation from a divinity school into a modern research university. Schools across the country reformed themselves along German lines between 1870 and 1910. Such changes paved the way for the tremendous achievement of American chemists in the latter half of the twentieth century.

In 1914, becoming only the second American to receive a Nobel Prize for scientific research, Theodore W. Richards was honored for accurately determining the atomic weights of twenty-five elements. Among the other American chemists to win the Nobel Prize were Irving Langmuir in 1932 for his work in surface chemistry and Harold C. Urey in 1934 for the discovery of heavy hydrogen.

All Nobelists are obviously illustrious, but Linus Pauling is widely considered the foremost chemist of the twentieth century, and he is the only person to receive two unshared Nobel Prizes. He was recognized first in 1954 for his work on the nature of the chemical bond and then again in 1962 with the Nobel Peace Prize for his tireless campaign on the dangers of nuclear weapons.

Sean Kelly

Sources

Bruce, Robert V. *The Launching of Modern American Science, 1846–1876.* New York: Alfred A. Knopf, 1987.

Greene, John. *American Science in the Age of Jefferson.* Ames: Iowa State University Press, 1984.

Hindle, Brooke. *The Pursuit of Science in Revolutionary America, 1735–1789.* Chapel Hill: University of North Carolina Press, 1956.

Ihde, Aaron J. *The Development of Modern Chemistry.* New York: Harper and Row, 1964.

LIBBY, WILLARD F. (1908–1980)

Willard Frank Libby, the discoverer of radiocarbon (carbon-14) dating, was born on December 17, 1908, to Eva May and Ora Edward Libby on a farm in Grand Valley, Colorado. He had schooling in Sebastopol, California, from 1913 to 1926. He received his B.S. and Ph.D. from the University of California, Berkeley, where he became an instructor in 1933 and, later, an associate professor.

Libby's Guggenheim Memorial Foundation Fellowship at Princeton University was interrupted from 1941 to 1945, when he worked at Columbia University on the Manhattan Project for making the atomic bomb. His research on separating uranium isotopes contributed to the preparation of U-238 for the atomic bomb dropped on Hiroshima in August 1945. Libby also discovered that after evaporation, water remains in the atmosphere for nine days and subsequently precipitates as rain.

Libby joined the Department of Chemistry and Institute for Nuclear Studies at the University of Chicago, where he was professor of chemistry from 1945 to 1954. While working on radioactive substances, he discovered the unstable radioactive isotope carbon-14. With a half-life of about 5,730 years, radiocarbon decays at a constant rate after the death of an organism. The age of an organism can be known by measuring the remnant C-14. Libby and his students thus developed the carbon-14 dating technique.

In the first application of this carbon-dating technique, Libby calibrated the date of acacia wooden items found in the tomb of the Egyptian pharaoh Zoser. He also took specimens of wood and mud from the American and European continents and calculated the period of glaciation.

Libby's technique came to be used in archeology, geology, geophysics, hydrology, and oceanography. He wrote about his discovery in *Radiocarbon Dating* (1952). For his work in this field, Libby was awarded the Nobel Prize in Chemistry in 1960.

Libby was appointed a member of the U.S. Atomic Energy Commission on October 1, 1954, and he served in that capacity for five years. He then joined the University of California as a professor of chemistry and was named director of the Institute of Geophysics and Planetary Physics in 1962.

Libby's other honors include the Research Corporation Award in 1951, the Chandler Medal of Columbia University in 1954, and the Day Medal of the Geological Society of America in 1961. He died on September 8, 1980, in Los Angeles.

Patit Paban Mishra

Sources

Aitken, Martin J. *Science-Based Dating in Archaeology.* London: Longman, 1990.

Libby, Willard F. *Radiocarbon Dating.* Chicago: University of Chicago Press, 1952.

Nobel Lectures. *Chemistry 1942–1962.* Amsterdam, The Netherlands: Elsevier, 1964.

MACLEAN, JOHN
(1771–1814)

The Scottish immigrant John Maclean was a founder of the discipline of chemistry and chemical education in the early American republic. The son of a surgeon, he was a student of the great Scottish chemist Joseph Black at the University of Edinburgh, which he entered at the age of thirteen.

Like many early chemists, Maclean was a medical student who became a physician. He studied in London and revolutionary Paris, where he was exposed to the new chemical ideas of Antoine-Laurent Lavoisier and his followers. Maclean became a convert to the Lavoisier school and its denial of the existence of "phlogiston," a substance believed to be given off during burning, in favor of the theory of combustion where oxygen is absorbed by substances. Sympathetic to American republicanism and hoping that America would offer a wider field for professional advancement, Maclean migrated in 1795. On the advice of Benjamin Rush, he established himself as a physician in Princeton, New Jersey, offering a course of lectures on chemistry. At this time, few Americans were aware of Lavoisier's chemistry, and Maclean's lectures were popular. On the strength of them, the College of New Jersey (later Princeton) hired Maclean as a professor of chemistry and natural history beginning with the fall term in 1795. Maclean's was the first appointment of a chemistry professor outside a medical school at an American university. (He later added mathematics and natural philosophy to his portfolio.)

As an instructor, Maclean pioneered the use of laboratory work in chemical teaching, demanding that his pupils perform experiments rather than simply observe the teacher's demonstrations. He taught applied as well as theoretical chemistry. Devoting himself full-time to teaching, he abandoned his medical practice.

Maclean engaged in a controversy with America's leading chemist, the recent immigrant from England, Joseph Priestley. Priestley was virtually the last leading chemist to advocate the phlogiston theory, a position he defended in *Considerations on the Doctrine of Phlogiston, and the Decomposition of Water* (1796). Maclean's *Two Lectures on Combustion: Supplementary to a Course of Lectures on Chemistry* (1797) was a forthright attack on Priestley's position. Its two main subjects, the composition of metals and the decomposition of water, are the same as those of Priestley's work. The resulting dispute between Priestley and Maclean was carried on in the pages of the *Medical Repository.*

Maclean left the College of New Jersey in 1812 for the College of William and Mary in Virginia, but the breakdown of his health forced him to retire from teaching and return to Princeton. He died two years later at age forty-three.

William E. Burns

Source

Maclean, John. *Considerations on the Doctrine of Phlogiston, and the Decomposition of Water, by Joseph Priestley; and Two Lectures on Combustion and an Examination of Doctor Priestley's Considerations on the Doctrine of Phlogiston, by John Maclean; edited, with a sketch of the life and letters of Doctor Maclean.* Ed. William Foster. Princeton, NJ: Princeton University Press, 1929.

NYLON

Nylon is a synthetic fabric developed by Wallace H. Carothers and his team at the DuPont corporation in the 1930s. Nylon resulted from the company's decision to invest in a program of research into mechanisms of polymerization. By 1930, some of that work had suggested the possibility of a fiber that could compete with natural silk and its principal artificial rival, the cellulose-based rayon. In the summer of 1931, DuPont filed a patent application, establishing priority of claim.

After DuPont had solved a host of production and quality problems, it publicly announced its development of nylon in late 1938 and displayed the material at the New York World's Fair the following year. When nylon stockings finally went on sale nationally in May 1940, they were greeted with unrestrained enthusiasm. Nylon soon became an essential wartime material—used, for example, for parachutes. Many members of the DuPont nylon team went on to the Manhattan Project. The resumption of civilian production in 1945 created "nylon riots" when the stockings, with which "nylons" soon became synonymous, again became available in stores.

Chemically, nylon is a family of synthetic, linear polyamides. The Federal Trade Commission defines the substance as a "manufactured fiber in which the fiber-forming substance is a long-chain synthetic polyamide in which less than 85 percent of the amide linkages are attached directly to two aromatic rings." Nylon is a generic rather than trade name.

A specific form called nylon 6,6 (due to the two stretches of six carbon atoms on the backbone chain of the polymer) is covered by DuPont's nylon patent. Nylon 6, created first by chemist Paul Schlak at the I.G. Farben company in Germany in 1938, while chemically distinct, has essentially the same properties and now accounts for over 60 percent of the world's nylon. In production, polyamide flakes are melted and then forced through the small holes of a spinneret, somewhat similar to the way water comes through a showerhead. As the material cools, it solidifies into fibers.

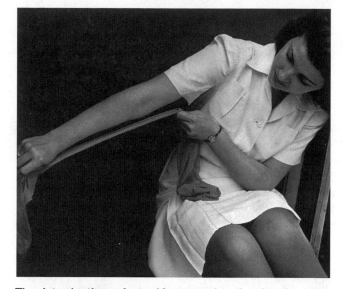

The introduction of stockings made of nylon in 1940 created a public sensation. The new synthetic material found a myriad of industrial, engineering, and commercial applications during World War II and the consumer products boom of the postwar era. *(Walter Sanders/Time & Life Pictures/Getty Images)*

Nylon's characteristics include its strength, elasticity, luster, excellent abrasion resistance, and ease of maintenance. It also can be combined with a wide range of other fibers and included in composite materials. In consumer use, nylon is most familiar in a variety of apparel, including hosiery and outdoor wear, as well as in carpets. The material also has industrial applications, ranging from tire cord to parachutes to fishing line.

Although many synthetics have been developed to compete with nylon, worldwide production and consumption of the material continues to grow. Production currently is more than 4 million metric tons, increasing by about 2 percent per year. While U.S. production has remained constant in recent years, the percentage manufactured by new Asian producers such as Taiwan and Korea has grown. DuPont remains the largest single producer of nylon.

James Hull

Sources

Handley, Susannah. *Nylon: The Story of a Fashion Revolution.* Baltimore: Johns Hopkins University Press, 1999.

Hounshell, David A., and John Kenly Smith, Jr. *Science and Corporate Strategy: DuPont R & D, 1902—1980.* New York: Cambridge University Press, 1988.

ONSAGER, LARS
(1903–1976)

A Norwegian American theoretical chemist and physicist, Lars Onsager carried out extensive research on the properties of liquids and solids, thermodynamics, and low-temperature physics. He was awarded the 1968 Nobel Prize in Chemistry for his theoretical studies of the thermodynamics of irreversible processes. In the field of low-temperature physics, he predicted and explained the occurrence of vortexes in superfluid helium.

The son of Norwegian supreme court barrister Erling Onsager and teacher Ingrid Onsager (née Kirkeby), Onsager was born on November 27, 1903, in Kristiania (now Oslo), Norway. In 1920, he became a chemical engineering student at the Norwegian Institute of Technology (now the Norwegian University of Science and Technology), where he studied Brownian motion, the random movement of small particles in liquids or gases. In 1925, the year he received his chemical engineering degree, Onsager challenged the Dutch chemist Peter Debye on the Debye-Hückel theory of electrolytic solutions. Debye was so impressed that he offered Onsager a position in Zurich as his assistant. Onsager worked with Debye for two years, studying the behaviors of electrolytes. His revision of the Debye-Hückel theory laid the groundwork for a comprehensive theory of ion movement in electrified solutions.

After immigrating to the United States in 1928, Onsanger became a teaching associate in chemistry at Johns Hopkins University in Baltimore, Maryland. His students complained that he was a poor instructor, however, and he was dismissed after one semester. In the fall of 1929, he became a research instructor at Brown University in Providence, Rhode Island, where he taught statistical mechanics. Onsager's students called his class "Advanced Norwegian I" or "Sadistical Mechanics," because of his accent and obscure lecturing style.

In 1929, Onsager presented his ideas on reciprocal relations at a meeting of the Scandinavian Physical Society at Copenhagen. The fully formed theory was published in two parts in 1931 as "Reciprocal Relations in Irreversible Pro-cesses." While it received little attention from other scientists until after World War II, it was this work that ultimately earned Onsager the 1968 Nobel Prize in Chemistry. His theory introduced a new law to the study of thermodynamics: the law of reciprocal relations. A general mathematical formula regarding the behavior of substances in fluids, it represented a major breakthrough in theoretical chemistry.

Onsager received a Sterling postdoctoral fellowship in 1933 at Yale University in New Haven, Connecticut. To their embarrassment, the chemistry faculty realized upon his arrival that Onsager did not possess a doctorate. Waiving the course requirements, the Yale faculty awarded him a Ph.D. in 1935 on the basis of a recent paper he had written on deviations from Ohm's law in weak electrolytes. The subject was so complex that the chemistry and physics faculty had to seek help from the mathematics department to understand it; the mathematicians openly applauded the work. In total, Onsager spent thirty-nine years at Yale, holding the J. Willard Gibbs professorship of theoretical chemistry, from 1945 to 1972.

After his retirement from Yale, he served as a distinguished university professor at the University of Miami's Center for Theoretical Studies. He was found dead of an aneurysm at the age of seventy-two in his home in Coral Gables, Florida, on October 5, 1976.

James Fargo Balliett and George B. Kauffman

Sources

Kirkwood, John G. "The Scientific Work of Lars Onsager." *Proceedings of the American Academy of Arts and Sciences* 82 (1953): 298–300.

Longuet-Higgins, H. Christopher, and Michael E. Fisher. "Lars Onsager, 27 November 1903–5 October 1976." *Biographical Memoirs of Fellows of the Royal Society* 24 (1978): 442–71.

Nobel Lectures. *Chemistry 1963–1970.* Amsterdam, The Netherlands: Elsevier, 1972.

Onsager, Lars. *The Collected Work of Lars Onsager.* Hackensack, NJ: World Scientific, 1996.

ORGANIC CHEMISTRY

Although the French chemists Claude Berthollet and Antoine Lavoisier undertook the first systematic study of organic composition in the last decades of the eighteenth century, a tentative definition of organic chemistry as the chemistry of carbon-based compounds was not proposed until the mid-nineteenth century. Until this point, chemistry had been primarily a science of discovery and analysis.

The first American paper that addressed chemical identity and composition was published in 1771 by Dr. John de Normandie. The subject of the paper was a mineral spring near Bristol, Pennsylvania. Throughout the nineteenth century, American chemists focused largely on industrial and agricultural chemistry. This focus was directed by the belief, held by Thomas Jefferson and others, that science had to be practical, as well as by the demands of an increasingly industrialized economy.

By the time a workable theory of chemical structure emerged, which paved the way for the golden age of organic synthesis in the latter half of the nineteenth century, American chemists had fallen way behind their German and British counterparts. Ira Remsen, who earned his Ph.D. at the University of Göttingen in 1870 and subsequently became one of the original faculty members at Johns Hopkins (where he and a colleague in 1878 discovered the artificial sweetener saccharin), believed most of his American-trained colleagues did not know enough even to appreciate their own ignorance.

At the time, only three American schools offered lab instruction in organic chemistry. Many schools still saw their task as one of imparting culture, not practical knowledge. Furthermore, the concerns of private capital, which funded most scientific activity in late nineteenth- and early twentieth-century America, also contributed to keeping the focus on applied chemistry.

The number of Americans studying chemical engineering rose almost tenfold between 1900 and 1918, and American universities continued to reform themselves along German lines. Americans did not, however, copy the German industrial research lab, so that before 1914 a domestic chemical industry had not developed in the United States.

In the aftermath of World War I, universities and companies such as DuPont were able to lure large numbers of German scientists to the United States. The number of companies with in-house research labs and the number of industrial researchers tripled during the 1920s. Simultaneously, the expansion of graduate education, which had begun in the latter half of the nineteenth century, finally began to produce enough chemistry Ph.D.s to staff American universities.

The results of these processes proved remarkable. Lured away from Harvard in 1928, for example, Wallace Carothers at DuPont stumbled onto the world's first synthetic fiber, neoprene rubber, in 1930. Four years later, his lab synthesized the first polyamide, later named nylon.

Within the academic world, American Nobelists tended to be inorganic chemists before. Melvin Calvin won the Nobel Prize in Chemistry in 1961 for his work on carbon dioxide assimilation in plants. Calvin's research was typical of the increasing convergence of organic chemistry and biochemistry in the second half of the twentieth century.

Sean Kelly

Sources

Bruce, Robert V. *The Launching of Modern American Science, 1846–1876.* New York: Alfred A. Knopf, 1987.

Hounshell, David. "DuPont and the Management of Large-Scale Research and Development." In *Big Science: The Growth of Large-Scale Research,* ed. Peter Galison and Bruce Hevly. Stanford, CA: Stanford University Press, 1992.

Hugill, Peter, and Veit Bachmann. "The Route to the Techno-Industrial World Economy and the Transfer of German Organic Chemistry to America Before, During, and Immediately After World War I." *Comparative Technology Transfer and Society* 3:2 (August 2005): 159–86.

Ihde, Aaron J. *The Development of Modern Chemistry.* New York: Harper and Row, 1964.

PATENT MEDICINE

Patent or proprietary medicine refers to a class of unregulated elixirs, nostrums, and curatives of dubious quality produced and marketed directly to the public by private companies and individuals of questionable reputation. While virtually all contemporary prescription and over-the-counter medications are patented by their developers or manufacturers, the term "patent medicine" is used almost exclusively as a synonym for quack cures, few of which have actually ever been patented.

The heyday of patent medicine in North America, the second half of the nineteenth century, mirrored both the rise of mass-media advertising and the regulation and legitimization of the medical profession. The popularity of patent medicines reflected both the desire for a quick, simple cure and the lingering suspicion of modern medical practice.

Patent medicines flourished in an age when the line between quackery and legitimate medicine was still difficult to discern. The causes, and therefore cures, of most diseases were poorly understood prior to the germ theory and bacteriology of the nineteenth century. Such medical practices as bleeding, cupping, and blistering were commonly used until well into the nineteenth century, and regulated medical education was a rarity in the

United States prior to the 1890s. Most doctors served only brief apprenticeships before going into practice for themselves. While the nascent medical profession waged a constant war of words with the purveyors of questionable tonics and practices, to the average patient with a complaint, there often seemed little difference between the two.

Prior to the passing of the Pure Food and Drug Act of 1906, there were few restraints on the contents of patent medicines, which routinely contained alcohol, opium, and chloroform, among other ingredients. Many a parent soothed their child's teething pain with patent medications containing morphine. Other patent medicines were little more than sugar water.

Though some were created and sold by sincere individuals believing they were passing along valuable cures, most patent medicines were concocted for the sole purpose of making money. A few of what Oliver Wendell Holmes referred to as the "Toadstool Millionaires" became wealthy and famous, building not only manufacturing plants for their medicines, but also glassworks to make the bottles they were sold in and shops to build the boxes in which they were shipped. Some developed nationwide distribution networks, employing dozens of salespeople.

The popularity and success of patent medicines was the result, at least in part, of the shrewd use of modern advertising. Without regulation, manufacturers made outlandish claims for their products and pioneered such advertising tools as color printing, billboards, mass-produced fliers, and sworn testimonials. Many claimed that their goods were "secret" cures from "exotic" cultures such as the Chinese, Egyptian, and, most often, Native American.

During their golden age in the nineteenth century, patent medicines were a part of a larger pattern of alternative and quack treatments such as phrenology and homeopathy. These served as precursors to later fads such as electrical treatments, magnetism, and crystal healing.

John P. Hundley

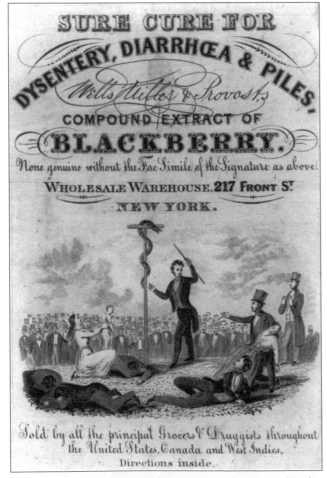

Patent medicines—unregulated, secret formulas for the treatment of everyday maladies and serious illnesses—thrived in the latter half of the nineteenth century on the strength of advertising, packaging, and the public desire for simple cures. *(Library of Congress, LC-USZ62–47350)*

Sources

Hecthlinger, Adelaide. *The Great Patent Medicine Era, or Without Benefit of Doctor.* New York: Galahad, 1970.

Young, James Harvey. *The Toadstool Millionaires: A Social History of Patent Medicines in America Before Federal Regulation.* Princeton, NJ: Princeton University Press, 1961.

PAULING, LINUS
(1901–1994)

A Nobel laureate and a leading quantum chemist and biochemist of the twentieth century, Linus Pauling applied concepts of quantum mechanics to chemistry. His research helped found a new discipline in the field of biology— molecular biology—and his studies on crystal and protein structures within cells contributed significantly to the discovery of DNA.

Linus Carl Pauling was born on February 28, 1901, in Portland, Oregon, to Lucy Darling and Herman Pauling, a pharmacist whose lack of success in business forced the family to move to several different Western cities between 1903 and 1909. Shortly after returning to Portland in 1910, Herman Pauling died of a perforated ulcer, leaving his wife to care for their three children.

In grammar school, Linus Pauling often visited a friend who had a chemistry set, which triggered his interest in the field and his desire to become a researcher. By his sophomore year in high school, Pauling had assembled a classroom laboratory from parts he had collected at an abandoned steel company where his grandfather worked as a night watchman. Before finishing his senior year, however, Pauling dropped out of school to protest courses he thought were useless; he did not receive a high school diploma until 1962.

Theoretical Chemistry

In 1917, Pauling was accepted at the Oregon Agricultural College in Corvallis. To support himself and his family, he taught quantitative analysis beginning in his second year. In 1922, he was awarded his bachelor's degree in chemical engineering and was accepted for graduate study at the California Institute of Technology in Pasadena.

Pauling's research—under the direction of Roscoe Dickinson, a chemist known for his work on X-ray crystallography, and Richard Tolman, a physicist and chemist who worked in statistical mechanics—focused on ways to use X-ray technology to determine the crystal structure of minerals. After three years, which included the

Linus Pauling won the 1954 Nobel Prize in Chemistry for his research on the nature of chemical bonds and the 1962 Nobel Peace Prize for his campaign against nuclear testing. He gained further attention in his later years for advocating large doses of Vitamin C to fight colds. *(Pictorial Parade/Hulton Archive/Getty Images)*

publication of seven papers, Pauling received his doctorate in physical chemistry and mathematical physics.

In the mid-1920s, advanced research in the new field of quantum mechanics was being performed in Europe. Through a Guggenheim Fellowship, Pauling joined those efforts in 1926–1927, working under a number of researchers. Among them was the Austrian physicist Erwin Schrödinger, in Zurich, who worked on the space and time dependence of quantum mechanical equations. Another was the German physicist Arnold Sommerfeld, in Munich, who pioneered atomic research. Working with these scientists in the emerging field of quantum mechanics and atomic structure, Pauling discovered a theoretical approach to molecular structure and chemical properties and functions that would serve him well in his career. Returning to the United States in 1927, he accepted a position at the California Institute of Technology (Cal Tech) as a professor of theoretical chemistry.

Pauling became known as a versatile and driven researcher. In his first five years at Cal

Tech, he wrote fifty papers on X-ray crystal studies and articulated five new scientific principles, known as Pauling's Rules, on the molecular and chemical structure of complex crystals. In 1932, he introduced the concept of electronegativity—the ability of an atom or molecule to attract electrons to itself—and the Pauling Electronegativity Scale to measure it. His 1939 textbook *The Nature of the Chemical Bond and the Structures of Molecules and Crystals* changed the way scientists thought about chemistry. In the book, Pauling explained chemical processes as a function of quantum mechanics operating on the chemical bond at the molecular level. He detailed the concept of orbital hybridization, the relationship between ionic and covalent bonding, and the structure of aromatic hydrocarbons.

Opposition to Nuclear Weapons and War

In 1942, theoretical physicist J. Robert Oppenheimer invited Pauling to head the chemistry division of the Manhattan Project, working to develop the atomic bomb. Pauling refused. During the course of World War II, however, he contributed to the U.S. war effort by working on rocket propellants, explosives, oxygen meters in submarines, and a synthetic form of blood plasma for battlefield medical treatment. In 1948, he was awarded the Presidential Medal for Merit by President Harry S. Truman for his accomplishments during the war.

After World War II, Pauling became active in political and peace issues. He joined the Emergency Committee of Atomic Scientists (chaired by Albert Einstein) in 1946 to raise awareness of the threats associated with nuclear weapons. He also gave a number of public lectures with other committee scientists on the dangers of nuclear war.

In 1949, Pauling joined with civil rights activist W.E.B. Du Bois, actress Uta Hagen, and others in organizing the American Continental Congress for Peace, which met in Mexico City. His participation in that event, the American Peace Crusade, and other left-wing organizations and activities drew the attention of government officials during the Cold War. Senator Joseph McCarthy accused him of being a Communist, and, in 1952, the State Department denied Pauling a passport to travel to London for a peace conference. The Soviet Union, meanwhile, had announced that Pauling's chemistry work was "pseudo-scientific" and "hostile to the Marxist view."

In 1954, Pauling was awarded the Nobel Prize in Chemistry for his research on the nature of chemical bonds, the energy that gives atoms the ability to form molecules that become the foundation for physical matter. Four years later, still active in the campaign against nuclear testing and the dangers of radioactive fallout, he organized a petition of 11,000 "concerned scientists" and presented the signed document to the United Nations. In 1962, at the age of 61, he became the only person ever to be awarded a second undivided Nobel Prize, this time for peace.

During the 1950s, Pauling had conducted research on diverse topics pertaining to abnormal cell molecules, including the causes of schizophrenia, hereditary faults in body chemistry, and the role of nucleic acids in cell development. His later career focused on medical issues. He discovered that sickle-cell anemia is a hereditary molecular disease, the first association of a specific protein with human disease. And he pioneered the field of megavitamin therapy, generating heavy publicity—and debate in the scientific community—by advocating large doses of vitamin C to combat the common cold and diseases such as cancer. Pauling died of prostate cancer at the age of ninety-three in California on August 19, 1994.

James Fargo Balliett

Sources

Goertzel, Ted, and Ben Goertzel. *Linus Pauling: A Life in Science and Politics.* New York: Basic Books, 1995.

Hager, Thomas. *Linus Pauling and the Chemistry of Life.* New York: Oxford University Press, 1998.

Marinacci, Barbara, ed. *Linus Pauling in His Own Words.* New York: Touchstone, 1995.

Pauling, Linus. *General Chemistry.* North Chelmsford, MA: Courier Dover, 1988.

———. *Selected Scientific Papers.* New York: World Scientific, 1999.

PHARMACEUTICAL INDUSTRY

The American pharmaceutical industry was born during the Revolutionary War era and became a major business in the years after the

American Civil War. Following World War I, the industry was a significant scientific innovator, equaling and sometimes surpassing its European counterparts in the development of new drugs. By the years after World War II, American pharmaceutical companies dominated the global marketplace.

America's first pharmaceutical manufacturing establishment was founded in Carlisle, Pennsylvania, in 1778, under the Continental Army's apothecary general, Andrew Craigie, to supply medicines to the troops. The first U.S. commercial operation—the firm of Christopher and Charles Marshall—was founded eight years later in Philadelphia, producing muriate of ammonia, an expectorant used in the treatment of bronchitis, and the laxative Glauber's salt, a form of sodium sulfate. During the early part of the nineteenth century, Philadelphia became the center of the embryonic American pharmaceutical industry and included such firms as John Farr, which was founded in 1818. Another Philadelphia firm, Rosengarten and Sons, founded in 1822, was the first American manufacturer to extract quinine sulfate, an early medicine for the treatment of malaria, then a prevalent disease in the United States.

As the nation expanded to the west, so did the pharmaceutical industry. One of the key firms was the company founded by Frederick Stearns in Detroit, Michigan, in 1855. A harsh critic of the widespread medical quackery of his day, Stearns invented the term "ethical pharmaceuticals" to describe products, such as his own, that had undergone the most up-to-date testing for purity and efficacy.

The American Civil War, with its rampant epidemics and battlefield carnage, provided a major impetus for the domestic pharmaceutical industry. Quinine was the "miracle drug" of the war, used to treat not only malaria but also venereal disease and diarrhea. The most lasting innovation produced during the war years was the drug tablet, first produced by the Philadelphia druggist John Dunton in 1863, whereby the active ingredients of a medicine were compressed into a portable and long-lasting pill form that could be easily swallowed. The last half of the nineteenth century saw dramatic growth in the drug industry, with the establishment of numerous pharmaceutical companies.

Some, such as Parke Davis (established in 1871), Smith Kline (1875), and Eli Lilly (1876), still dominate the American and global trade.

While many of these firms practiced "ethical" pharmacology, useless nostrums continued to be peddled by untrained individuals who claimed their drugs could cure everything from headaches to cancer. Even drugs considered efficacious, such as opium, used in the treatment of pain, were often adulterated with ingredients that rendered them ineffective and even dangerous. To stop this practice, the federal government passed the Food and Drug Law of 1906, "for preventing the manufacture, sale, or transportation of adulterated or misbranded or poisonous or deleterious foods, drugs, medicines, and liquors." This was followed in 1914 by the Harrison Narcotic Act, restricting commerce in opiates and coca products, then commonly found in all kinds of patent medicines, and the Food, Drug, and Cosmetics Act of 1938, which required manufacturers to submit new drugs to the Food and Drug Administration for approval.

With these regulatory mechanisms in place, and with the astounding pharmacological advances of the twentieth century, American firms advanced to the forefront of the world's pharmaceutical industry, with more than $200 billion in annual revenues by the beginning of the twenty-first century. In recent years, the industry also has undergone rapid consolidation.

The modern-day pharmaceutical industry has been criticized for maintaining high prices on medications. It also is under scrutiny for its emphasis on creating profitable "lifestyle drugs" for the developed world (such as the erectile dysfunction medication Viagra), rather than medications to treat diseases rampant in the developing world.

James Ciment

Sources

Chandler, Alfred D., Jr. *Shaping the Industrial Century: The Remarkable Story of the Evolution of the Modern Chemical and Pharmaceutical Industries.* Cambridge, MA: Harvard University Press, 2005.

Liebenau, Jonathan. *Medical Science and Medical Industry: The Formation of the American Pharmaceutical Industry.* Basingstoke, UK: Macmillan, 1987.

PHARMACOLOGY

Modern pharmacology is the science of determining the therapeutic effects of synthesized compounds on the body. Originally, the term referred to the study of all aspects of drugs, from their origins to their chemical properties to their physiological effects, but the meaning gradually changed over the course of the nineteenth century.

Pharmacology is nearly as old as human civilization itself. Records from ancient Greece and China, among other civilizations, discuss experimentation with the effect of natural compounds on the body. Modern pharmacology, however, only arose with the discovery that organic compounds could be synthesized artificially, a breakthrough first achieved by German scientist Friedrich Wohler in the early nineteenth century. Until that time, it was believed that organic substances could be created only through what was called the "vital force," a form of energy said to exist only in living plants and animals.

In the United States, the science got its start at the University of Michigan, where the first chair in pharmacology was established in 1891 under John Jacob Abel, a chemist who had studied at the University of Strasbourg under Oswald Schmiedeberg, widely considered the founder of modern pharmacology. In 1893, Abel was hired as a professor of pharmacology by the just-opened Johns Hopkins University medical school. Until his retirement from that institution in 1933, Abel trained many of the leading American pharmacologists of the twentieth century. Abel was also a highly accomplished researcher. His pioneering work in isolating epinephrine, a hormone produced by the adrenal gland, and histamine, an ammonia-based compound created in the pituitary gland, rank among the first great American achievements in pharmacology.

By the early 1900s, many of the nation's major medical schools had established pharmacology departments. Some of the finest were those at Western Reserve University in Ohio, the University of Pennsylvania, and Columbia University's College of Physicians and Surgeons in New York. At the same time, various government agencies at the federal level, including the Department of Agriculture and the Public Health Service, had hired pharmacologists and established laboratories to do research on new and existing drugs. The private sector was also active, with Parke-Davis among the first pharmaceutical companies to hire professionally trained pharmacologists to discover new medicines and test the efficacy of existing ones.

Professionalization of the field was proceeding as well. In 1908, Abel helped organize the American Society for Pharmacology and Experimental Therapeutics, the nation's first professional organization for pharmacologists. The *Journal of Pharmacology and Experimental Therapeutics*, America's first professional journal in the field, was established the following year.

Pharmacology's acceptance as a serious field of research science—rather than a purely technical field or applied science—became evident in the first half of the twentieth century, as pharmacology departments were shifted from medical schools to academic health science programs. As a result, an increasing number of pharmacologists began to earn Ph.D.s rather than medical degrees. New subdisciplines, such as toxicology and molecular pharmacology, emerged after World War II. By the early 2000s, more than 200 American universities offered Ph.D. programs in pharmacology, up from fewer than a dozen in 1930.

James Ciment

Sources

Chen, Ko Kuei, ed. *The American Society for Pharmacology and Experimental Therapeutics, Incorporated: The First Sixty Years.* Bethesda, MD: American Society for Pharmacology and Experimental Therapeutics, 1969.

Parascandola, John. *The Development of American Pharmacology: John J. Abel and the Shaping of a Discipline.* Baltimore: Johns Hopkins University Press, 1992.

PHLOGISTON

"Phlogiston," a word coined in the early eighteenth century by the German physician and chemist Georg Ernst Stahl from the Greek *phlogistos*, meaning "burned," refers to a hypothetical substance believed to be present in all combustible matter. It was thought that the liberation of phlogiston caused burning. When hydrogen was discovered, it was at first believed to be pure phlogiston because of its low weight and flammability.

Johann Joachim Becher had postulated in 1669 that substances are composed of three kinds of earth—vitrifiable, mercurial, and combustible—and that burning liberated the combustible earth. Stahl's theory involved "phlogisticated" substances, that is, substances containing phlogiston, which Stahl asserted was colorless, odorless, tasteless, and weightless. On burning, phlogisticated substances were said to be "dephlogisticated."

The ash or residue of the burned material was believed to be the essential substance, a theory that explained burning, oxidation, calcination (post-combustion metal residue called calx), and breathing. Support for the hypothesis was drawn by observing the burning of charcoal, heating of metals, and suffocation of mice, but the total residual products of these processes were never quantified.

When the hypothesis was tested quantitatively, however, questions arose. Why did dephlogisticated organic substances (ash) appear to weigh less than the original phlogisticated substances, while some metals, such as magnesium, gained weight when burned? If phlogiston was given off when a metal formed a calx, why did the calx weigh more than the phlogisticated metal? Theoretically it should weigh less, since phlogiston had been liberated. Stahl responded with the unverifiable postulation that air was entering the metal and filling the vacuum left by the liberated phlogiston. He eventually retreated to the argument that phlogiston was not an actual substance but an immaterial principle.

The French chemist Antoine Lavoisier and his followers (antiphlogistians) discredited the phlogiston theory in a series of experiments from the 1770s to 1790s. They studied the weight gain or loss of the oxidation (combustion) or deoxidation (reduction) of lead, tin, sulfur, iron, phosphorus, and other substances. These laboratory experiments allowed the quantification of all residual substances, including the gases produced. Unlike earlier nonlaboratory experiments that did not quantify all residual substances, these experiments demonstrated that the total weight of all the residual products of oxidation and deoxidation always exceeded the original weight, even though the phlogiston theory projected a loss of weight or no change at all. When iron rusted completely, the rust weighed more than the original iron, and when charcoal burned completely, the carbon dioxide produced weighed more than the original charcoal.

The antiphlogistians demonstrated through these experiments that oxygen, recently discovered by Joseph Priestley in 1774, was part of the chemical process. Lavoisier's oxygen combustion theory was generally accepted by 1800, with the notable exception of Priestley.

When Priestley discovered oxygen by heating red oxide of mercury, he initially named it dephlogiscated air. He also discovered nitrous oxide in 1776 and, after emigrating to America, discovered carbon monoxide in 1779. Ironically, Priestley never abandoned the phlogiston theory. With every new question or anomaly, he adjusted the theory to account for the question or anomaly. At the time of his death, Priestley was the sole defender of the phlogiston theory.

Richard M. Edwards

Sources

Gribbin, John. *The Scientists: A History of Science Told Through the Lives of Its Greatest Inventors.* New York: Random House, 2003.

Partington, James Riddick. *Historical Studies on the Phlogiston Theory.* New York: Arno, 1981.

PINKHAM, LYDIA (1819–1883)

Lydia Estes Pinkham's name is associated with one of America's most successful patent medicine companies in the nineteenth century.

She was born on February 9, 1819, in Lynn, Massachusetts, to a Quaker family that was dedicated to reform movements, including abolition, temperance, and woman suffrage. After marrying Isaac Pinkham in 1843, she spent the next thirty years of her life tending to her husband, daughter, and three sons, until an economic downturn threatened the family with poverty. To make ends meet, Pinkham began what turned out to be a highly lucrative career, manufacturing and marketing her trademarked remedy, Pinkham's Vegetable Compound.

It is unclear why Pinkham's tincture—touted as the "Greatest Cure in the World" for "female complaints"—was so effective. Studies by the American Medical Association in the early twentieth century concluded that the remedy, composed of various herbs and roots suspended in a solution of 18 percent alcohol, had little pharmaceutical value and owed its efficacy to what today might be called the "placebo effect." In the mid-twentieth century, however, scientists found that the compound contained estrogens that could have actually helped relieve the symptoms of menopause.

Initially, the family business—which relied heavily on the business acumen of Pinkham's sons—began selling the Vegetable Compound locally around Lynn, until they registered the label at the Patent Office and began paying for advertising space in large metropolitan newspapers. Advertising proved to be a wise investment, as Pinkham's cure quickly became a national commodity.

The company's advertising relied heavily on testimonials sent in from satisfied customers, a grandmotherly photo of Lydia E. Pinkham herself, and inflated claims as to what the Vegetable Compound could cure, including prolapsed uteruses, leucorrhoea (vaginal infections), irregular and painful menstruation, kidney diseases, neurasthenia, and discomforts associated with menopause. Pinkham's marketing, which played off the hesitation of many women to discuss private issues with male physicians, celebrated the remedy as "A medicine for women. Invented by a woman. Prepared by a woman." She supported this claim—and helped develop a market niche—by personally responding to thousands of letters sent by women asking for medical advice.

By the time of Pinkham's death in 1883, the product earned the family $300,000 a year. The company continued to grow, despite turn-of-the-century progressive reforms aimed at checking the sales of patent medicine. Muckraking journalists, seeking to uncover fraud, reported in 1905 that the company misled customers into thinking that Pinkham was still alive and answering letters, when, in fact, she had been dead for decades. The American Medical Association also ran a series of reports aimed at discrediting Pinkham's Vegetable Compound and other patent medicines.

Legislation supported by the Food and Drug Administration in 1906 forced medicine manufactures to print ingredients on labels, which exposed the tincture's high alcohol content and shocked many customers dedicated to temperance. This may have actually been a boon for the company, however, as profits reached $3 million in 1925, when Prohibition had made most other forms of alcohol illegal. The family-owned business finally sold the rights to Pinkham's Vegetable Compound to Cooper Laboratories in 1968.

David G. Schuster

Sources

Burton, Jean. *Lydia Pinkham Is Her Name.* New York: Farrar, Straus, 1949.

Stage, Sarah. *Female Complaints: Lydia Pinkham and the Business of Women's Medicine.* New York: W.W. Norton, 1979.

PRIESTLEY, JOSEPH (1733–1804)

Renowned as a Unitarian minister, scientist, theologian, and educator, Joseph Priestley was a founder of modern chemistry, best remembered for his discovery of oxygen and carbon monoxide. He was greatly respected for his views on science, educational philosophy, political theory, and Christian theology. His published works on these and other topics fill twenty-six volumes.

Born in Fieldhead, England, on March 13, 1733, Priestley demonstrated a prodigious intellect and was set on a course of study for the Presbyterian ministry. At age nineteen, he enrolled at the liberal, nonconformist Daventry Academy, where he ultimately rejected the Calvinist orthodoxy of his youth in favor of Arianism, an early Christian heresy which denied that Christ was "of the same substance" as God the father.

At Daventry, he was profoundly influenced by David Hartley's *Observations on Man* (1740), a work that became Priestley's philosophical foundation. A pioneer in the field of psychology, Hartley, in the empiricist tradition of John Locke, introduced the doctrine of Associationism, wherein ideas are derived from sense experience and all mental phenomena are governed by physical laws. For Hartley, this meant that education was of crucial importance and could lead to unlimited progress, a central tenet of the Enlightenment. Under Hartley's influence, Priestley resolved that he could at once be a materialist, a necessarian (or determinist), and a Christian, despite his unorthodox theology.

Aligned with nonconformist churches (diverse Protestant congregations that did not conform to the Church of England), Priestley became a minister and schoolmaster in Suffolk, England. There, he began to nourish an insatiable interest in science, and he procured for the school such instruments as an air pump and a static generator for electrical demonstrations.

Priestley's teaching career advanced in 1761, when he was hired as a tutor of languages for the Nonconformist Academy at Warrington, Lancashire. He taught courses in oratory, criticism, grammar, history, and law; wrote *Rudiments of English Grammar* (1761) and *A Chart of Biography* (1765); earned an honorary doctorate from the University of Edinburgh; and was elected to the Royal Society of London. During the 1760s, he met and befriended Benjamin Franklin, who inspired Priestley to write *History and Present State of Electricity* (1767), a book that would evolve beyond a mere historical work to include notes on his own electrical experiments.

Among his discoveries, Priestley demonstrated that charcoal was an effective conductor of electricity, disproving the established belief that only water and metals could conduct electricity. He also contributed new insight into the relationship between electrical and chemical change. Describing one of his chief contributions, Priestley stated that "the attraction of electricity is subject to the same laws as that of gravitation," a discovery that helped establish electrical theory as an exact science.

Convinced that ministry was his highest calling, Priestley took the pastorate at the church of Mill Hill, Leeds, from 1767 to 1772. The congregation was congenial to his unorthodox theology, and his light ministerial duties afforded him leisure time to pursue science. Reflecting on this time, Priestley noted, "nothing engaged my attention while at Leeds so much as the

prosecution of my experiments relating to electricity, and especially the doctrine of air." His experiments with air were aided by the good fortune of living next to a brewery from which he acquired generous quantities of carbon dioxide.

In 1772, he discovered soda water (then called windy water) by dissolving in water the carbon dioxide produced by fermentation; the discovery earned him the Copley Medal of the Royal Society. Also in 1772, he read his paper "On Different Kinds of Air" to the Royal Society and was established as a leading chemist. He had discovered four previously unknown gases during his years at Leeds and would later bring five more to light.

In 1772, William Fitzmaurice Petty, Second Earl of Shelburne, hired Priestley as his librarian and literary companion, and tutor to his children. (Shelburne would later negotiate the Treaty of Paris that ended the American Revolution.) Priestley set up a laboratory at Shelburne's estate and on August 1, 1774, he discovered that by heating red mercuric oxide he liberated a gas (oxygen), which he called "dephlogisticated air" in adherence to the then-popular but errant phlogiston theory. (Nearly three years before, the Swedish scientist Carl Scheele independently discovered oxygen and also called it dephlogisticated air.) An invitation to Paris by the distinguished chemist Antoine-Laurent Lavoisier allowed Priestley to present his findings on gases before eminent French scientists. Lavoisier began his own experiments on Priestley's dephlogisticated air, eventually proving it to be an element and calling it oxygen.

Priestley's public support for the French Revolution inflamed the passions of a royalist mob that in 1791 burned down his house and laboratory in Birmingham. Forced to flee England in 1794, he migrated to Pennsylvania. He refused the offer of a chair in chemistry at the University of Pennsylvania and chose instead to settle in the small town of Northumberland, Pennsylvania, where he built a home and set up the first scientifically equipped laboratory in the United States. It was there that he made his major contribution to science during his American years: the identification of carbon monoxide as a distinctive gas, which he called "heavy inflammable air."

Priestley's scientific contributions also included a description of photosynthesis, the invention of the gum eraser, the use of compressed gases to produce refrigeration, and an understanding of oscillatory discharge that would later prove crucial to the development of radio and television. In addition to his scientific work, Priestley established a Unitarian Society in Philadelphia and wrote a *History of the Christian Church* in six volumes.

He maintained close ties with Franklin, John Adams, and Thomas Jefferson; his correspondence with Jefferson influenced the philosophical design of the American liberal arts curriculum. Priestley died of yellow fever in Northumberland, Pennsylvania, on February 6, 1804.

Stephen Peterson

Sources

Brown, Ira, ed. *Joseph Priestley: Selections from His Writings.* University Park: Pennsylvania State University Press, 1962.

Davis, Kenneth S. *The Cautionary Scientists: Priestley, Lavoisier, and the Founding of Modern Chemistry.* New York: G.P. Putnam's Sons, 1966.

Gibbs, F.W. *Joseph Priestley: Revolutions of the Eighteenth Century.* Garden City, NY: Doubleday, 1967.

Passmore, John A., ed. *Priestley's Writings on Philosophy, Science, and Politics.* New York: Collier, 1965.

RADIOCARBON DATING

The development of radiocarbon dating represented a major breakthrough in the establishment of absolute age determination as opposed to relative age scales. It remains a widely used technique, but it is only applicable to organic materials such as peat, plant remains, wood, bone, and teeth that are less than 60,000 years old.

The element carbon is present in the environment as three isotopes: C-12, C-13, and C-14; only C-14 is radioactive. It is produced in the upper atmosphere when the nuclei of nitrogen atoms are bombarded by cosmic radiation. Once formed, C-14 enters the biogeochemical cycle of carbon and behaves like the nonradioactive carbon isotopes. Thus, C-14 is oxidized to carbon dioxide in the atmosphere, is absorbed by plants in photosynthesis, and enters animals through their links with plants in the food chain. Upon death, organisms no longer exchange C-14, because they no longer feed or respire; the amount of C-14 in the tissues therefore begins to decline.

In 1949, University of Chicago scientist Willard F. Libby discovered the rate at which C-14 decays, notably the time required for decay to half the original C-14 present. This is the half-life, established by Libby as $5,568 \pm 30$ years (since corrected to $5,730 \pm 40$ years). To determine the age of a sample, the C-14 content is measured and compared with that of a modern sample; the difference represents the passage of time since death. The chemical analysis is reliable up to approximately 60,000 years; if the fossil is older, the method does not work. Older objects have limited C-14 residue, and the background radiation affects it considerably.

C-14 can be measured in two ways. Conventional measurement involves the counting of beta particles emitted as the C-14 decays. They are usually measured when C-14 is released from a sample as a gas, such as carbon dioxide or methane. Innovations in recent decades have resulted in accelerator mass spectrometry (AMS), which allows the direct measurement of C-14 atoms by passing charged particles at high speeds through a magnetic field. The major advantage over beta particle measurements is that AMS requires only small samples of a few milligrams.

Radiocarbon dating has revolutionized historical, archeological, and scientific understandings of past cultures and the temporal framework of cultural evolution. The technique has had a major impact on such disciplines as geology, geophysics, hydrology, atmospheric science, oceanography, and biomedicine.

Patit Paban Mishra and A.M. Mannion

Sources

Aitken, Martin J. *Science-Based Dating in Archaeology.* London: Longman, 1990.

Geyh, Mebus A., and Helmut Schleicher. *Absolute Age Determination: Physical and Chemical Dating Methods and Their Application.* New York: Springer-Verlag, 1990.

Greene, Kevin. *Archaeology: An Introduction.* London: Routledge, 2002.

Libby, Willard F. *Radiocarbon Dating.* Chicago: University of Chicago Press, 1952.

Taylor, R.E. and Martin J. Aitken, eds. *Chronometric Dating in Archaeology: Advances in Archaeological and Museum Science.* Vol. 2. Oxford, UK: Oxford University Press, 1997.

REMSEN, IRA
(1846–1927)

Ira Remsen, a professor of chemistry at Johns Hopkins University, was instrumental in establishing the academic discipline of chemistry in the United States.

He was born in New York City on February 10, 1846, the only child of James Vanderbilt Remsen, a merchant, and Rosanna Secor. At the age of fourteen, he entered the Free Academy (later the City College of New York), but instead of finishing the four-year course, he was apprenticed at his father's urging to a physician. Dissatisfied with the apprenticeship, he enrolled in the College of Physicians and Surgeons of Columbia University and received his M.D. in 1867.

Attracted to the study of chemistry, Remsen attended the University of Göttingen in Germany from 1867 to 1870, earning a Ph.D. in 1870 for his research, under Rudolph Fittig's supervision, on the structure of piperic and piperonylic acids. Remsen accompanied Fittig to the University of Tübingen, where he remained until 1872 as Fittig's lecture and laboratory assistant.

Upon his return to the United States, Remsen became a professor of chemistry and physics at Williams College in Williamstown, Massachusetts. There, despite an atmosphere indifferent to chemistry, he pursued his research and developed a simple, lucid lecturing style.

In 1876, he wrote *The Principles of Theoretical Chemistry,* an influential text that provided a consistent scale of atomic and molecular weights. That same year, Remsen became professor of chemistry at the newly established Johns Hopkins University in Baltimore. He served as head of the chemistry laboratory until 1908 and as president of the university from 1901 to 1913. He helped to build the institution on a continental European model, making it a place for discovery rather than merely transmission of knowledge. He introduced many of the teaching methods of Germany, which influenced chemistry instruction throughout the United States, especially at the graduate level. By the end of the century, more than half of the leading American chemists had been educated at Johns Hopkins.

Remsen founded the *American Chemical Journal* in 1879, the first continuing periodical devoted to American chemical research. He served as its chief editor until 1911, when it was incorporated into the *Journal of the American Chemical Society.* In 1907, President Theodore Roosevelt appointed Remsen head of the advisory commission established by the Food and Drug Act of 1906. In addition to serving as the twenty-third president (1902) of the American Chemical Society, he was president of the American Association for the Advancement of Science (AAAS), the Society of Chemical Industry, and the National Academy of Sciences.

Although Remsen and his students published more than 170 articles, he is best known as a teacher, mentor of students, textbook writer (his seven texts went through twenty-eight editions and fifteen translations), and builder of one of America's most distinguished universities. In 1923, Remsen became the first recipient of the Priestley Medal, the American Chemical Society's highest honor. He died on March 4, 1927, in Carmel, California.

George B. Kauffman

Sources

Getman, Frederick H. *The Life of Ira Remsen.* Easton, PA: *Journal of Chemical Education,* 1940.

Harrow, Benjamin. *Eminent Chemists of Our Time.* New York: D. Van Nostrand, 1927.

Hawthorne, R.M., Jr. "Ira Remsen." In *American Chemists and Chemical Engineers,* ed. Wyndham D. Miles. Washington, DC: American Chemical Society, 1976.

SEABORG, GLENN T.
(1912–1999)

American nuclear chemist Glenn Theodore Seaborg is best known for his work in isolating and identifying most of the transuranium elements—those heavier than uranium, with an atomic weight of 93 and higher. His work resulted in a significant expansion of the periodic table and earned him the 1951 Nobel Prize in Chemistry, which he shared with Edwin Mattison McMillan. Seaborg was also a member of the Manhattan Project, an adviser to ten U.S. presidents on nuclear and science issues, and chair of

Chemist Glenn Seaborg contributed to the discovery of most transuranium elements, one of which is named for him (seaborgium). He was also a professor, chancellor of the University of California at Berkeley, and chairman of the Atomic Energy Commission. *(Fritz Goro/Time & Life Pictures/Getty Images)*

the U.S. Atomic Energy Commission under three presidents.

The son of Selma Ericksson and Herman Theodore Seaborg, poor Swedish immigrants, he was born on April 19, 1912, in Ishpeming, Michigan, a small iron mining town. When he was ten, the family moved to Home Gardens (now South Gate), California, a suburb of Los Angeles.

An avid reader who kept a journal from 1927 to 1998, Seaborg was not a particularly motivated student until his junior year in high school, when he was inspired by his chemistry and physics teacher. He attended the University of California, Los Angeles, where he received his undergraduate degree in chemistry in 1934, and the University of California, Berkeley, where he was awarded his Ph.D. in 1937. One event of his student days that left a strong impression was a brief meeting with Albert Einstein.

As a graduate student and research associate at Berkeley from 1936 to 1939, Seaborg focused on the isolation of radioisotopes. Hired as a member of the faculty in 1939, he rose through the ranks and became a full professor of chemistry in 1945. He served as chancellor of the university from 1958 to 1961.

On February 23, 1941, with colleagues Arthur C. Wahl and Joseph W. Kennedy, Seaborg pro-

duced and identified plutonium, element 94 on the periodic table, a discovery that would change the course of science. By 1955, the "radioisotope hunter," as he was called, had identified ten new elements: numbers 94–102 and 106 on the periodic table. The last, seaborgium, was named in his honor, making him the only person for whom a chemical element was named during his lifetime. Seaborg spent World War II as "chemistry chief" at the University of Chicago Metallurgical Laboratory, a cover name for one of the most important sections of the Manhattan Project. At the "Met Lab" as it was known, he was responsible for isolating plutonium and extracting ultramicroscopic amounts for potential use in an atomic bomb.

In 1945, Seaborg went against his colleagues' advice by proposing the most significant change to the periodic table since its conception by Russian chemist Dmitry Mendeleyev in 1869. In Mendeleyev's periodic table, the elements are arranged in vertical rows (groups) and horizontal columns (periods). Each element generally resembles the element directly above it in the same group. Thus, the elements thorium (90) through lawrencium (103) would be expected to resemble the elements hafnium (72) through astatine (85). Seaborg proposed that the fourteen closely related elements heavier than actinium (89), rather than resembling the elements immediately above them, belong to a separate family in the table—the actinides. These, he argued, are analogous to the fourteen elements heavier than lanthanum (57)—cerium (58) through lutetium (71)—called the lanthanides, or rare earths. Elements 90 through 103 were henceforth called the actinide series. His addition was accepted by the scientific community as an important clarification and restructuring.

As Seaborg recalled in his *Memoirs,* "I showed my new table to the two leading inorganic chemists in the world before publishing it. The idea went over like a lead balloon. 'Don't do it, Glenn,' they warned me, 'it will ruin your scientific reputation.' It was just so hard to conceive that the periodic table had been this wrong. I didn't have any scientific reputation, so I published it anyway." According to Seaborg, it was "the key to the subsequent discovery of a number of transuranium elements."

In 1961, Seaborg moved to Washington, D.C., at the request of President John F. Kennedy, who appointed him chair of the Atomic Energy

Commission, a federal agency established in 1948 to oversee atomic policy and development. Seaborg was the first scientist to serve as chair of the commission, holding the position until 1971. He was also a leader in the movement to improve scientific education in the United States and was a member of federal advisory committees that revamped high school and college chemistry curricula.

As an adviser to American presidents from Franklin D. Roosevelt to George H.W. Bush, Seaborg visited sixty-three countries to promote international scientific cooperation and nuclear arms control. He considered the control of nuclear weapons the most critical problem of the times and made a number of substantive contributions to the Nuclear Non-Proliferation Treaty of 1968, eventually signed by 188 countries. In 1971, he returned to the University of California, Berkeley, where he served as university professor, associate director at large of the Lawrence Berkeley Laboratory, and chair of the Lawrence Hall of Science.

A prolific writer, Seaborg co-authored approximately 500 scientific papers and nearly fifty books. He held more than forty U.S. patents, most of them for his discoveries of chemical elements, and he was awarded more than fifty honorary degrees. He died on February 25, 1999, in Lafayette, California, of complications from a stroke.

James Fargo Balliett and George B. Kauffman

Sources

Kauffman, George B. "Beyond Uranium." *Chemical and Engineering News,* November 19, 1990, 18–23, 26–29.

———. "In Memoriam Glenn T. Seaborg (1912–1999)." *Chemical Educator* 4:2 (1999): 1–6.

———. "Transuranium Pioneer: Glenn T. Seaborg." *Today's Chemist* 4:3 (1991): 18–20, 23, 24, 32.

Seaborg, Glenn T. *A Chemist in the White House: From the Manhattan Project to the End of the Cold War.* Washington, DC: American Chemical Society, 1998.

———. *The Transuranium People: The Inside Story.* Singapore: World Scientific, 1999.

SILLIMAN, BENJAMIN
(1779–1864)

Benjamin Silliman, an early American chemist and founder of the *American Journal of Science and Arts,* was born in Trumbull, Connecticut, on August 8, 1779, of parents who traced their lineage to seventeenth-century Puritans. Like his father and paternal grandfather, Silliman expected to follow the law as a profession. He entered Yale College at age thirteen; after his four years there, he studied law privately and was admitted to the bar in 1802. In the same year, he was offered a professorship of chemistry and natural history at Yale, where he had been serving as a tutor.

For the next two years, he spent most of his time learning what he was supposed to teach. This took him to Philadelphia and Princeton and, after his earliest lectures on chemistry in 1804, to Great Britain and Holland. Upon his return from Europe, Silliman added lectures on mineralogy and geology to his teaching repertoire. He continued to teach at Yale until his retirement in 1853.

Silliman's scientific discoveries were less important than his advocacy activities. He offered public lectures on science in New Haven, inviting women to attend, and he continued to widen his outreach to other audiences eager to learn about natural science. After he gave enormously popular lectures on chemistry and geology at Boston's Lowell Institute, the demand for his talents took him as far west as St. Louis and as far south as New Orleans. He was instrumental in founding Yale Medical School in 1801, where he lectured on chemistry and pharmacy. He also edited textbooks on chemistry and geology that were used at Yale and elsewhere.

Perhaps Silliman's greatest contribution to the advancement of American science was the journal he founded in 1818. The *American Journal of Science and Arts* was not the first periodical in the country devoted entirely to science, but, from the beginning, it embraced the widest areas of interest. It has also proved to be the longest lasting, though its editorial content today is restricted to the earth sciences. Under Silliman's editorship, about a third of the journal's pages were devoted to geology and mineralogy, often written by the editor himself. The remainder of the contents depended on what he could persuade his wide circle of acquaintances to produce, including articles on botany, zoology, chemistry, mathematics, and natural philosophy.

Silliman was the sole editor of the journal for two decades and, during its early years, when readers were delinquent in paying for their subscriptions, he helped pay for the printing out of his own pocket. So identified with the publication did he become that it was familiarly referred to as "Silliman's Journal." His triumph was to see it become the best-known and most respected American periodical devoted to science on either side of the Atlantic.

Silliman died in New Haven on November 24, 1864. His son, Benjamin Silliman, Jr., carried on his work, eventually bringing about the establishment of the Sheffield Scientific School at Yale.

Charles Boewe

Sources

Brown, Chandos Michael. *Benjamin Silliman: A Life in the Young Republic.* Princeton, NJ: Princeton University Press, 1989.

Dana, Edward Salisbury. "The American Journal of Science from 1818 to 1918." In *A Century of Science in America with Special Reference to the American Journal of Science 1818–1918*, ed. Edward Salisbury Dana. New Haven, CT: Yale University Press, 1918.

Wilson, Leonard G., ed. *Benjamin Silliman and His Circle: Studies on the Influence of Benjamin Silliman on Science in America.* New York: Science History, 1979.

SQUIBB, EDWARD R. (1819–1900)

The physician, pharmacist, scientist, inventor, author, and entrepreneur Edward Robinson Squibb was one of the earliest and most influential nineteenth-century voices on behalf of drug purity in the United States. His death on October 25, 1900, came six years before passage of the Federal Food and Drug Act, which brought his aspirations and efforts to fruition.

Born on July 5, 1819, in Wilmington, Delaware, Squibb graduated in 1845 from Philadelphia's prestigious Jefferson Medical College. Contrary to the antiwar doctrine of his Quaker upbringing, Squibb served until 1857 as an assistant surgeon in the U.S. Navy after receiving his M.D. degree. During the latter part of his navy years, he established the Brooklyn Naval Laboratory, initially devoted to research that led to the purification and standardization of ether. One of his early publications describing the apparatus for preparing ether appeared in the September 1856 issue of the *American Journal of Pharmacy*.

After leaving the navy, Squibb became involved in a drug manufacturing partnership in Louisville, Kentucky. He was soon motivated to start a business of his own, building a small laboratory in Brooklyn for the purpose of manufacturing pure drug products for sale to physicians and pharmacists.

As a member of the Committee of Revision for the 1860 United States Pharmacopoeia, Squibb was able to wield substantial influence in his lifelong crusade for drug-product purity backed by sound scientific research. His precise and direct manner of attacking problems gained him widespread support and prestige. As a prominent figure in American medicine, pharmacy, and chemistry, he provided important leadership in favor of high drug-quality standards and research throughout his life. He was always ready to share research results and patentable information, as it was his belief that such knowledge belonged to the world.

The outbreak of the Civil War in 1860 provided substantial incentive for Squibb to expand his manufacturing facility to meet Union Army demands for anesthetics and bandaging supplies. Like many Americans of the day, Squibb, with friends residing in the South, had divided emotions about the war. Nevertheless, his support for the Union and scrupulous adherence to its laws helped his business grow during those turbulent years.

Despite time spent attending to a waning business and family illness in postbellum Brooklyn, Squibb continued his scientific interests in drug purity, publishing frequently and attending relevant local and national meetings. Improvement in the general business climate and the reputation of his company led to greater financial success.

The name of the company was changed from Edward R. Squibb, M. D., to E.R. Squibb and Sons in 1892, in recognition of his successors, Edward Hamilton Squibb and Charles Fellows Squibb. They continued to manage the business after their father's death in 1900. In 1989, the company became part of Bristol-Meyers Squibb as a result of a major international merger.

Carl Buckner

Sources

Blochman, Lawrence G. *Doctor Squibb: The Life and Times of a Rugged Idealist.* New York: Simon and Schuster, 1958.

Florey, Klaus, ed. *The Collected Papers of Edward Robinson Squibb, M.D. (1819–1900).* Princeton, NJ: Squibb Institute, 1988.

STARKEY, GEORGE
(1627–1665)

The seventeenth-century alchemist George Starkey, son of a Scottish minister, was born in Bermuda and educated at Harvard. There, he learned a version of Aristotelian natural philosophy that emphasized "corpuscles," the smallest particles into which matter could be divided. He later criticized the Harvard curriculum as given to empty scholastic dispute rather than the teaching of true philosophy, although many of his own later writings were organized in an academic question-and-response form.

Starkey began the independent study of chemistry under the tutelage of the Charlestown, Massachusetts, physician Richard Palgrave in 1644. He also was associated with the circle around John Winthrop, Jr., as well as with Winthrop's efforts to establish an ironworks in New England. Another associate was the physician and metallurgist Robert Child.

Although Starkey lacked a medical degree, in physician-poor New England he was able to practice medicine successfully upon his graduation with a B.A. in 1646. Difficulties obtaining good laboratory equipment in New England caused him to move to England in 1650.

Starkey wrote several alchemical tracts, some published during his lifetime and others posthumously. Some appeared under his own name, and others were attributed to a pseudonym, Eirenaeus Philalethes, Latin for "peaceful lover of truth." Starkey claimed that Philalethes was an alchemist living in America who performed wonders such as restoring an old woman's hair and teeth. Philalethes assumed an existence independent of his creator—the English natural philosopher Kenelm Digby claimed to have met him, and he was believed to have been alive as late as the mid-eighteenth century.

Starkey's alchemy, principally based on the ideas of the Belgian Johannes Baptista van Hel-mont, was expressed in the traditionally obscure alchemical style. He sought to achieve both traditional goals of alchemy—to make gold and to cure diseases. As an alchemist, he emphasized the use of mercury to prepare the Philosopher's Stone, unlike many other seventeenth-century alchemists who emphasized salts.

Starkey got along well with England's Puritan rulers in the 1650s, but, upon the restoration of Charles II in 1660, he attempted to ingratiate himself with the Royalists by publishing monarchical tracts. His hopes of patronage were disappointed, and the last years of his life were spent in desperate poverty.

Starkey joined with others in forming the Society of Chemical Physicians in London to advance Helmontian medicine, but he died while tending the sick during the great London plague of 1665. His corpuscular alchemy was a major influence on the chemistry of Robert Boyle, Isaac Newton, and the German physician Georg Stahl, the originator of the phlogiston theory. Starkey's works were reprinted into the eighteenth century.

William E. Burns

Sources

Newman, William R. *Gehennical Fire: The Lives of George Starkey, an American Alchemist in the Scientific Revolution.* Cambridge, MA: Harvard University Press, 1994.

Newman, William R., and Lawrence M. Principe. *Alchemy Tried in the Fire: Starkey, Boyle, and the Fate of Helmontian Chymistry.* Chicago: University of Chicago Press, 2002.

UREY, HAROLD CLAYTON
(1893–1981)

The chemist and 1934 Nobel laureate Harold Clayton Urey was born on April 29, 1893, in Walkerton, Indiana, one of the three children of Reverend Samuel Clayton Urey, a schoolteacher and lay minister, who died when Harold was six, and Cora Rebecca Reinsehl. After graduation from high school in 1911, Urey taught for three years in country schools in Indiana and then Montana, where his family had moved. He entered the University of Montana in 1914, and in 1917 he received his B.S. degree in zoology with a minor in chemistry.

Although Urey intended to be a biologist, the U.S. entry into World War I led him to spend two years as an industrial research chemist before returning to Montana in 1919 as an instructor. In 1921, he entered the University of California, Berkeley, to work under Gilbert N. Lewis on calculating thermodynamic properties from molecular spectra and the distribution of electrons among the orbits of excited hydrogen atoms. He received his Ph.D. in chemistry in 1923.

Urey spent the next year in Copenhagen at Niels Bohr's Institute for Theoretical Physics as an American-Scandinavian Foundation fellow and then spent five years as an associate in chemistry at Johns Hopkins University. In 1926, he married Frieda Daum, with whom he had three daughters and one son.

He became an associate professor of chemistry at Columbia University in 1929 and a professor in 1934. During this time, he co-authored the book *Atoms, Molecules, and Quanta* (1930), describing re-

Harold Urey won the 1934 Nobel Prize in Chemistry for his discovery of deuterium (heavy hydrogen). He later headed a division of the Manhattan Project and worked in such diverse fields as geochemistry, planetary evolution, and the origin of life. *(George Karger/Pix, Inc./Time & Life Pictures/Getty Images)*

cent advances in quantum mechanics. He also was the founding editor of the *Journal of Chemical Physics,* serving in that capacity from 1933 to 1940.

In the 1930s, Urey developed a method for concentrating heavy hydrogen isotopes (forms of an element with the same atomic numbers but different atomic weights) by fractionally distilling liquid hydrogen, which led to the discovery of deuterium (hydrogen with atomic weight 2, in contrast to common hydrogen with atomic weight 1). With Edward W. Washburn, he devised an electrolytic method for the separation of hydrogen isotopes and carried out detailed investigations of their properties, especially the vapor pressure of hydrogen and deuterium and the equilibrium constants of exchange reactions. In 1934, Urey received the Nobel Prize in Chemistry "for his discovery of heavy hydrogen."

Urey remained at Columbia through World War II. In 1945, he became distinguished service professor of chemistry at the Institute for Nuclear Studies, University of Chicago, and then the Martin A. Ryerson Professor in 1952. In 1956–1957, he was the George Eastman Visiting Professor at Oxford, and, in 1958, he became a professor-at-large of the University of California, San Diego.

Urey had worked on the separation of uranium isotopes for the nuclear bomb, but believing that the U.S. government intended to produce nuclear weapons beyond those needed for the war, he began to work for the control of nuclear energy. Other, later interests included the measurement of temperatures of ancient oceans, the origin of the planets, and the chemical problems of the origin of Earth. His book *The Planets: Their Origin and Development* (1952) was the first to systematically apply chemical principles to the origin of the solar system.

Urey's research showed a strongly quantitative approach, a consideration of the entire problem rather than only a part of it, and a willingness to follow his conclusions into areas beyond his initial expertise. He received numerous honors, awards, and honorary degrees.

Urey retired in 1970. He died at age eighty-seven on January 5, 1981, in La Jolla, California. In 1984, the Harold C. Urey Prize in Planetary Science was established by the Division of Planetary Sciences of the American Astronomical Society.

George B. Kauffman

Sources

Cohen, K.P., S.K. Runcorn, H.E. Suess, and H.G. Thode. "Harold Clayton Urey." *Biographical Memoirs of Fellows of the Royal Society* 29 (1983): 623–59.

Nobel Lectures. *Chemistry 1922–1941.* Amsterdam, The Netherlands: Elsevier, 1966.

Ruark, Arthur Edward, and Harold Clayton Urey. *Atoms, Molecules, and Quanta.* New York: McGraw-Hill, 1930.

Urey, Harold Clayton. *The Planets: Their Origin and Development.* New Haven, CT: Yale University Press, 1952.

Urey, Harold Clayton, Ferdinand G. Brickwedde, and George M. Murphy. "Hydrogen Isotope of Mass 2 and Its Concentration." *Physical Review* 40 (1932): 1–15.

VULCANIZATION

Vulcanization, also called "curing," is a process used in the production of commercial rubber to improve its quality. The term "rubber" can refer to the finished product as well as to the coagulated natural gum (*latex*) from the sap of more than 200 species of plants popularly called rubber trees. The name is derived from the substance's use as an eraser for "rubbing out" pencil marks.

Vulcanization uses pressure and heat in conjunction with a curative agent, most commonly sulfur (though peroxide, gamma radiation, and other organic additives such as aniline are also used), to irreversibly link rubber molecules. This process creates a stronger, more flexible, and more durable rubber that is less affected by temperature variations.

The rubber molecule has a number of "cure sites" to which sulfur atoms attach and from which chains of two to ten sulfur atoms form molecular bridges that join the rubber molecules (monomers) into chains creating larger molecules (polymers) with elastic properties (elastomers). As the ratio of sulfur atoms to rubber molecules increases, the number of these bridges increases, resulting in a harder rubber. A lower sulfur atom to rubber molecule ratio produces a softer rubber. This variation in consistency permits vulcanized rubber to be formulated for applications ranging from surgical gloves to rubber mallets.

Rubber was usable only within a narrow temperature gradient until Charles Goodyear accidentally discovered the vulcanization process in 1839, when he dropped an experimental rubber and sulfur mixture on a hot stove. The use of thermoset (vulcanized) rubber greatly improved the efficiency, durability, and performance of machines and engines. Previously, when heated engine and machine parts expanded and separated, gases and lubricants escaped, resulting in both a diminished compression that decreased power output and increased friction that more rapidly degraded the components. Leather soaked in oil had been used to stem the leakage by filling the potential gaps, but the more tightly the oiled leather was compacted, the greater the friction and the more rapid the degradation of the seal. Vulcanized rubber could maintain an elasticity range within temperature tolerances, and it could be molded and conformed into gaskets, washers, and other parts that filled the gaps that needed to be sealed. This not only increased the efficiency, durability, and performance of existing engines and machines but also made possible the development of higher-performance engines and machines capable of sustained operations over longer periods at substantially wider temperature and pressure ranges.

Goodyear applied for a patent in 1844, once he felt he had perfected his process. By then, however, several others, most notably Horace H. Day, also claimed the discovery. The U.S. Circuit Court in Trenton, New Jersey, declared Goodyear the sole inventor of vulcanization when the court adjudicated his patent infringement case against Day in 1852. A cold vulcanization process using a sulfur bath was later developed by Alexander Parkes (1846), but the end product was less moldable than the product created by Goodyear's hot vulcanization.

Vulcanized rubber has multitudinous uses in industries ranging from health care to space exploration. Today, it is found commonly in products such as shoes, engines, vehicle tires, and much more.

Richard M. Edwards

Sources

Alliger, Glen, and Irvin Julian Sjothun. *Vulcanization of Elastomers: Principles and Practice of Vulcanization of Commercial Rubbers.* Melbourne, FL: Krieger, 1978.

Korman, Richard. *The Goodyear Story: An Inventor's Obsession and the Struggle for a Rubber Monopoly.* New York: Encounter, 2002.

Peirce, Bradford. *Trials of an Inventor: Life and Discoveries of Charles Goodyear.* Seattle, WA: University Press of the Pacific, 2003.

Slack, Charles. *Noble Obsession: Charles Goodyear, Thomas Hancock, and the Race to Unlock the Greatest Industrial Secret of the Nineteenth Century.* New York: Hyperion, 2003.

WINTHROP, JOHN, JR. (1605–1676)

John Winthrop, Jr., had wide-ranging interests in multiple branches of the natural sciences, applied his scientific expertise to the development of industry and commerce, practiced medicine, knew several languages, had one of the largest libraries in the New World, and communicated regularly with major European scholars of the day. Winthrop was an alchemist associated with Richard Starkey. Some scholars have argued that Winthrop was Eirenaeus Philalethes ("Peaceful Lover of Truth"), the pen name for a mid-seventeenth century alchemical writer, but Starkey has the better claim. Alchemy was considered an honorable and scientific pursuit in Winthrop's time, practiced not only by colonials such as Winthrop and Starkey but also by English savants such as Isaac Newton.

Winthrop (also known as John Winthrop the Younger) was born in 1605 in Groton, England. He entered Trinity College in Dublin, then studied law in London and became a barrister. He soon gave up a career in law, however. Having gained an appointment as secretary to a navy captain in 1627, Winthrop sailed on a disastrous mission to support the Huguenot garrison at La Rochelle, France. After this debacle, he left the navy and traveled across Europe for more than a year.

Upon his return to England in 1629, his father, John Winthrop, Sr., who was to become the first governor of Massachusetts, decided to emigrate to America. The elder Winthrop left for the colonies in 1630. John Winthrop, Jr., followed with his wife in 1631 and helped found the town of Ipswich, Massachusetts. He remained there until the death of his wife and infant daughter in 1634 and then returned to England.

In 1635, Winthrop received a one-year commission as governor of a new plantation on the Connecticut River, and he returned to America with his second wife. He again settled in Ipswich and spent part of 1636 (until his commission expired) overseeing construction of the plantation at Saybrook. Winthrop believed that establishing an industrial base was necessary for the colony's survival, and he returned to England in 1641 with the goal of attracting capital and skilled workers for an ironworks and other industrial initiatives.

In 1644, he established two furnaces in Massachusetts and founded a settlement in Connecticut for the same purpose. He moved his family to what is now New London, and later to the colony of New Haven. After being elected to several local government offices, he was elected governor of Connecticut in 1659, moved to Hartford, and served in that capacity until his death in 1676.

Although raised as a Puritan, Winthrop was not devoutly religious and as governor was tolerant of many who were harshly treated in Massachusetts. His commercial endeavors—including iron, lead, and salt works as well as various other ventures—were largely unsuccessful. Concerned about his financial status, he attempted to resign as governor three times, but each time he was refused.

Winthrop's most important act as governor was likely his return to England from 1661 to 1663 to gain a charter for Connecticut. He received a broad and liberal charter incorporating the former colony of New Haven within Connecticut's borders.

During this visit to England, Winthrop was elected to the Royal Society (in 1662), becoming the first member from America. He presented papers to the society on tar and pitch making, shipbuilding, and Indian corn (maize). He returned to Connecticut in 1663, acting as the Royal Society's correspondent for North America. Over the next fourteen years, Winthrop shipped a wide range of natural specimens back to the Royal Society, including an unusual species of starfish, horseshoe crabs, hummingbirds' nests, barnacles, milkweed fibers, and a sealed box of poison ivy. He also engaged in extensive communications with society members on topics including tides, comets, and agriculture.

Winthrop was interested in a wide range of applied scientific issues. He studied various agricultural diseases and pests, including wheat blights and tent caterpillars, and discussed possible modifications of Indian corn. He found that cornstalks yielded a "syrup sweet as sugar," known today as corn syrup. Among the industrial processes of interest to him were metallurgy, charcoal production, and mining.

Winthrop was also a self-trained physician, and he carried out an extensive medical practice. For his time, Winthrop can be considered a "modern" physician, in that he abandoned the concept described by Galen of bodily humors (blood, phlegm, yellow bile, black bile), and used chemicals and herbs in his treatments. His favorite treatment was "rubila," a mixture of nitre and antimony, which he prescribed for many illnesses.

An avid astronomer, Winthrop imported the first telescope to the colonies and tentatively reported viewing Jupiter's fifth satellite. However, once this satellite's existence was confirmed more than 200 years later, his telescope was demonstrated not to have been powerful enough for such a sighting; Winthrop likely mistook a star for this satellite. He presented his telescope to Harvard in the winter of 1671–1672.

At times Winthrop appeared to regret being "so greatly separated from happy Europe," and he appreciated his extensive correspondence with scientists and other notables in England, including Robert Boyle, Robert Hooke, Oliver Cromwell, Charles II, and Milton. Many of his contacts sent him books, and Winthrop's library was perhaps the largest in the colonies, described as having well over a thousand books in Latin, French, Italian, English, German, and Dutch. Winthrop died in 1676 in Boston.

Michael T. Halpern

Sources

Benton, Robert M. "The John Winthrops and Developing Scientific Thought in New England." *Early American Literature* 7:3 (1973): 272–90.

Black, Robert C. *The Younger John Winthrop.* New York: Columbia University Press, 1968.

DOCUMENTS

Poison Gas in World War I

German scientists developed the first chemical weapons of mass destruction during World War I, as this excerpt from the official U.S. history of the Great War describes.

The first use of asphyxiating gas was by the Germans during the first battle of Ypres. There the deadly compound was mixed in huge reservoirs back of the German lines. From these extended a system of pipes with vents pointed toward the British and Canadian lines. Waiting until air currents were moving steadily westward, the Germans opened the stop-cocks shortly after midnight and the poisonous fumes swept slowly, relentlessly forward in a greenish cloud that moved close to the earth. The result of that fiendish and cowardly act was that thousands of men died in horrible agony without a chance for their lives.

Besides that first asphyxiating gas, there soon developed others even more deadly. The base of most of these was chlorine. Then came the lachrymatory of "tear-compelling" gases, calculated to produce temporary or permanent blindness. Another German "triumph" was mustard gas. This is spread in gas shells, as are all the modern gases. The Germans abandoned the cumbersome gas-distributing system after the invention of the gas shell. These make a peculiar gobbling sound as they rush overhead. They explode with a very slight noise and scatter their contents broadcast. The liquids carried by them are usually of the sort that decompose rapidly when exposed to the air and give off the acrid gases dreaded by the soldiers. They are directed against the artillery as well as against intrenched troops. Every command, no matter how small, has its warning signal in the shape of a gong or a siren warning of approaching gas.

Gas masks were speedily discovered to offset the dangers of poison gases of all kinds. These were worn not only by troops in the field, but by artillery horses, pack mules, liaison dogs, and by the civilian inhabitants in back of the battle lines. Where used quickly and in accordance with instructions, these masks were a complete protection against attacks by gas.

The perfected gas masks used by both sides contained a chamber filled with a specially prepared charcoal. Peach pits were collected by the millions in all the belligerent countries to make this charcoal, and other vegetable substances of similar density were also used. Anti-gas chemicals were mixed with the charcoal. The wearer of the mask breathed entirely through the mouth, gripping a rubber mouthpiece while his nose was pinched shut by a clamp attached to the mask.

In training, soldiers were required to hold their breath for six seconds while the mask was being adjusted. It was explained to them that four breaths of the deadly chlorine gas was sufficient to kill; the first breath produced a spasm of the glottis; the second brought mental confusion and delirium; the third produced unconsciousness; and the fourth, death. The bag containing the gas mask and respirator was carried always by the soldier.

Source: Francis A. March, in collaboration with Richard J. Beamish. *History of the World War: An Authentic Narrative of the World's Greatest War* (Philadelphia: United Publishers of the United States and Canada, 1919).

Joseph Priestley's Observations on the Theory of Oxygen

Joseph Priestley, the British chemist, wrote this treatise after emigrating to America in 1794. He stubbornly refused to accept the new theory of oxygen advocated by Antoine-Laurent Lavoisier and his protégés, whom he labeled "Antiphlogistians," or opponents of the theory of phlogiston.

There have been few, if any, revolutions in science so great, so sudden, and so general, as the prevalence of what is now usually termed the *new system* of chemistry, or that of the *Antiphlogistians*. ... Though there had been some who occasionally expressed doubts of the existence of such a principle as that of *phlogiston*, nothing had been advanced that could have laid the foundation of another system before the labours of Mr. Lavoisier and his friends, from whom this new system is often called that of the *French*. ...

It is no doubt *time*, and of course opportunity of examination and discussion, that gives stability to any principles. But this new theory has not only kept its ground, but has been constantly and uniformly advancing in reputation, more than *ten years*, which, as the attention of so many persons, the best judges of everything relating to the subject has been unremittingly given to it, is no inconsiderable period. Every year of the last twenty or thirty has been of more importance to science, and especially to chemistry, than any ten in the preceding century. So firmly established has this new theory been considered, that a *new nomenclature*, entirely founded upon it, has been invented, and is now almost in universal use; so that, whether we adopt the new system or not, we are under the necessity of learning the new language, if we would understand some of the most valuable of modern publications.

In this state of things, an advocate for the old system has but little prospect of obtaining a patient hearing. And yet, not having seen sufficient reason to change my opinion, and knowing that free discussion must always be favourable to the cause of truth, I wish to make one appeal more to the philosophical world on the subject, though I have nothing materially new to advance. For I cannot help thinking that what I have observed in several of my publications has not been duly attended to, or well understood. I shall therefore endeavour to bring into one view what appears to me of the greatest weight, avoiding all extraneous and unimportant matter; and perhaps it may be the means of bringing out something more decisive in point of *fact*, or of *argument*, than has hitherto appeared.

No person acquainted with my philosophical publications can say that I appear to have been particularly attached to any hypothesis, as I have frequently avowed a change of opinion, and have more than once expressed an inclination for the new theory, especially that very important

part of it the *decomposition of water*, for which I was an advocate when I published the sixth volume of my experiments; though farther reflection on the subject has led me to revert to the creed of the school in which I was educated, if in this respect I can be said to have been educated in any school. However, whether this new theory shall appear to be well founded or not, the advancing of it will always be considered as having been of great importance in chemistry, from the attention which it has excited, and the many new experiments which it has occasioned, owing to the just celebrity of its patrons and admirers.

Source: Joseph Priestley, *Considerations on the Doctrine of Phlogiston and the Decomposition of Water* (Philadelphia: Thomas Dobson, 1796).

The Home Chemist

Henry Hartshorne, a professor at the University of Pennsylvania, published The Household Cyclopedia of General Information, *a do-it-yourself handbook for the practical American, in 1881. The following selections provide guidelines for cooling and dyeing materials. (For historical interest only—do not attempt.)*

Artificial Cold

When a solid body becomes liquid, a liquid vapor, or, when a gas or vapor expands, heat is abstracted from neighboring bodies, and the phenomena or sensation of cold is produced.

Evaporation produces cold, as is seen familiarly in the chilliness caused by a draught of air blowing on the moist skin. Water may be cooled to 30°, in warm climates, by keeping it in jars of porous earthenware; a flower-pot, moistened and kept in a draught of air, will keep butter, placed beneath it, hard in warm weather. In India water is exposed at night in shallow pans, placed on straw in trenches, and freezes even when the thermometer does not fall below 40°. Water may be frozen by its own evaporation under the receiver of an air-pump over sulphuric acid; the process is a delicate one, and not adapted for use on the large scale.

Compressed Air. Air, when compressed, gives out heat which is reabsorbed when it is allowed to expand. By forcing the air into a strong receiver and carrying off the heat developed by a stream of water, it may, on expanding, re-absorb enough to reduce the temperature below 32°. It is thus used in the paraffine works in England, and would be an excellent method of at once ventilating and cooling large buildings.

Freezing Mixtures. Depend upon the conversion of solid bodies into liquids. There are two classes, those used without ice and those in which it is employed. Where extreme cold is required, the body to be frozen should be first cooled as much as possible by one portion of the mixture, and then by a succeeding one.

Without Ice.—Four oz. each of nitre and sal ammoniac in 8 of water will reduce the temperature from 50° to 10°.

Equal parts of nitrate of ammonia and water, from 50° to 4°. The salt may be recovered by evaporation and used over again.

Equal parts of water, crystallized nitrate of ammonia, carbonate of soda, crystallized and in powder, from 50° to 7°.

Five parts of commercial muriatic acid and 8 of Glauber's salt in powder, from 50° to 0°.

With Ice.—Snow is always preferable. Ice is best powdered by shaving with a plane like a carpenter's, or it may be put into a canvas bag and beaten fine with a wooden mallet.

Equal parts of snow and common salt will produce a temperature of –4°, which may be maintained for hours. This is the best mixture for ordinary use.

Three parts of crystallized chloride of calcium and 2 of snow will produce a cold sufficient to freeze mercury, and to reduce a spirit thermometer from 32° to –50°. The chloride may be recovered by evaporation. There are many other freezing mixtures given in the books, but none are so cheap and efficient as the above.

Dyeing

The art of dyeing has for its object the fixing permanently of a color of a definite shade upon stuffs. The stuffs are animal, as silk wool, and feathers, or vegetable, as cotton and linen. The former take the colors much more readily, and they are more brilliant.

In some cases, as in dyeing silk and wool with coaltar colors, the color at once unites with the fiber; generally, however, a process of preparation is necessary. In certain other cases, as in dyeing silk and wool yellow by nitric acid, the color is due to a change in the stuff, and is not properly dyeing.

Insoluble colors are managed by taking advantage of known chemical changes; thus chromate of lead (chrome yellow) is precipitated by dipping the stuff into solutions, first of acetate of lead, and then of bichromate of potassa.

Mordants (bindermittle, middle binder of the Germans) are bodies which, by their attraction for organic matter, adhere to the fibre of the stuff, and also to the coloring matter. They are applied first, but in domestic dyeing they are often mixed with the dye-stuff. By the use of a mordant, a dye which would wash out is rendered permanent.

Some mordants modify the color; thus alum brightens madder, giving a light-red, while iron darkens it, giving a purple.

Mordants. The principal mordants are alum, cubic-alum, acetate of alumina, protochloride of tin, bichloride of tin, sulphate of iron, acetate of iron, tannin, stannate of soda.

Dye-Stuffs. The materials used in dyeing are numerous; the following are the most important: Madder, indigo, logwood, quercitron, or oak-bark, Brazil wood, sumach, galls, weld, annato, turmeric, alkanet, red launders, litmus or archil, cudbear, cochineal, lac; and the following mineral substances: ferrocyanide of potassium, bichromate of potash, cream of tartar, lime-water, and verdigris.

Source: Henry Hartshorne, *The Household Cyclopedia of General Information* (New York: Thomas Kelly, 1881).

Section 12

MATHEMATICS AND COMPUTER SCIENCE

ESSAYS

Euclidean and Non-Euclidean Geometry

After years of being overshadowed by Europeans, American mathematicians of the twentieth century made important contributions to the field of non-Euclidean geometry. Euclidean geometry, commonly taught in U.S. secondary schools and institutions of higher education as a system of logic, was developed by the Greek mathematician Euclid during the fourth century B.C.E. Euclid's *Elements* provides a set of self-evident ideas and definitions as well as self-evident assumptions: axioms or postulates. One example of a postulate is the statement that a straight line may be drawn between any two given points. Postulates and definitions, along with previously proven theorems, are then used to construct proofs of additional theorems, creating a system of logic.

During the subsequent two millennia, Euclidean geometry served as an accurate description of the nature of the space. Nineteenth-century mathematicians, however, challenged the role of Euclidean geometry, developing a system known collectively as non-Euclidean geometry. The key to understanding these new geometries is Euclid's fifth postulate—the parallel postulate—which describes a flat space on which, given a line and a point not on that line, there is only one possible line that can be drawn parallel to the given line through the given point. For mathematicians such as John Playfair of Scotland in the late eighteenth century, who were seeking to make Euclidean geometry more rigorous, the parallel postulate was not self-evident.

Over the course of centuries, various mathematicians had tried unsuccessfully to prove the postulate as a theorem. If Euclid's fifth postulate could not be proved as a theorem, then perhaps it is not an absolute, and additional spaces could be defined by a different parallel postulate. By the early 1800s, some mathematicians took the intellectual leap and began to study what would result if there were either an infinite number of parallel lines or no parallel lines at all. This endeavor marked the beginning of non-Euclidean geometry.

There are two general types of non-Euclidean geometry. The first kind explored a curved space in which there are an infinite number of possible parallel lines. This is called hyperbolic geometry. Work in this area is credited to the nineteenth-century Russian Nikolai Ivanovich Lobachevsky. The second kind is called elliptical or spherical geometry. Developed by the German mathematician G.F. Bernhard Riemann in the 1850s, it describes a curved space where there are no parallel lines.

After Riemann, non-Euclidean geometry as an acceptable field of study expanded and the old work of Lobachevsky and others was recovered and translated. George B. Halsted, a professor at the University of Texas, translated the Hungarian Janos Bolyai's paper "The Science of Absolute Space" (1832) and Lobachevsky's "Theory of Parallels" (1840). Thus Halsted is credited with introducing the field of non-Euclidean geometry to the United States.

By the early twentieth century, new research in physics showed that the world of non-Euclidean geometry was very real. Given Riemann's insight, Euclidean geometry was shown to be only an approximation of the true space, locally accurate, but still an approximation. Albert Einstein's general theory of relativity demonstrated that space was actually curved and thus could be accurately described only by non-Euclidean geometry. In explaining his general theory, Einstein drew heavily on geometry, especially the insights of Riemann. Ultimately, Einstein's work suggested space was both Euclidean and non-Euclidean. As mass increases in an area of space, it curves space, creating gravity, making the spatial geometry less Euclidean and more non-Euclidean. Einstein brought geometry back to its beginning as a description—albeit a much more complicated description—of the world around us.

Research into non-Euclidean geometry during the first four decades of the twentieth century moved American mathematics beyond translations and commentaries to seminal contributions. The primary American contributions to geometry came in the area of differential geometry and a related branch of mathematics called topology. Differential geometry is a subfield of geometry that studies the nature of curved surfaces (such as the curved space-time described by Einstein) by using the tools of calculus. Princeton University was the center of this research in the United States. Besides Einstein, Princeton mathematicians such as Luther Eisenhart and Oswald Veblen developed work in areas of geometry that furthered the mathematical understanding of Einstein's general theory of relativity. Veblen contributed to various forms of geometry, eventually publishing work that first defined the differentiable manifold, a concept in geometry where a surface seems flat locally but is, in reality, curved (like our experience of Earth).

Mathematics, like many fields of science in the United States, tended to focus on applications rather than theory. But during and after World War II, Chinese mathematicians who emigrated to the United States energized the study of mathematics, especially theory. Shying-shen Chern, for example, joined the faculty at Princeton's Institute for Advanced Study in 1943. In 1949, he moved to the University of Chicago, where he began the first large-scale training of graduate students in geometry. In 1960, he established another center of mathematics research at the University of California, Berkeley. While contributing to various areas of geometry, including differential geometry and manifold theory, Chern also trained a new generation of geometers who helped integrate various fields of geometry, topology, and differential equations.

Other important centers of geometry developed during the 1960s and 1970s. These include the State University of New York at Stony Brook and the University of Pennsylvania. In 1980, the first Fields Medal awarded for work in differential geometry was awarded to Shing Tung Yau and William Thurston, both graduates of Berkeley. With this prestigious award, American contributions clearly rose to world-class levels.

Geometric science in the twenty-first century is returning to its roots as a tool for better understanding the world around us. Developments in fields such as computational geometry marry the methods of geometers to the needs of geographers, molecular biologists, astrophysicists, and product designers; geometry is used for constructing and understanding digital images in tomography and digital mapping. Thus, geometry, a tool as old the Neolithic Age, has become vital in hundreds of ways in the information age.

Paul Buckingham

Sources

Bonola, Roberto. *Non-Euclidean Geometry: A Critical and Historical Study of Its Developments.* Trans. H.S. Carslaw. New York: Dover, 1955.

Duren, Peter, Richard A. Askey, and Uta C. Merzbach, eds. *A Century of Mathematics in America.* 3 vols. Providence, RI: American Mathematical Society, 1988–1989.

Gray, Jeremy. *Ideas of Space: Euclidean, Non-Euclidean, and Relativistic.* 2nd ed. Oxford, UK: Clarendon, 1989.

Greenberg, Marvin J. *Euclidean and Non-Euclidean Geometry: Development and History.* New York: W.H. Freeman, 1993.

Halsted, George. "Biography: Lobachevsky." *American Mathematical Monthly* 2:5 (1895): 136–9.

Heilbron, J.L. *Geometry Civilized: History, Culture, and Technique.* New York: Oxford University Press, 2000.

Honsberger, Ross. *Episodes in Nineteenth and Twentieth Century Euclidean Geometry.* Washington, DC: Mathematical Association of America, 1996.

Rosenfeld, Boris A. *A History of Non-Euclidean Geometry: Evolution of the Concept of a Geometric Space.* Trans. Hardy Grant, with Abe Shenitzer. New York: Springer-Verlag, 1988.

Ryan, Patrick J. *Euclidean and Non-Euclidean Geometry.* New York: Cambridge University Press, 1986.

Tarwater, Dalton, ed. *The Bicentennial Tribute to American Mathematics, 1776–1976.* Papers presented at the fifty-ninth annual meeting of the Mathematical Association of America commemorating the nation's bicentennial, Buffalo, New York, 1977.

The Computer Revolution

The computer is a programmable electronic machine that uses transistor technology ("hardware") to perform a number of mathematical tasks ("software"). The precursors to modern computers date back thousands of years, developing from simple devices such as the abacus to more technically advanced calculators. In America, the foundations of modern computer technology were established in the late 1800s with the work of such mathematicians as Herman Hollerith, who invented calculating machines.

World War II and the subsequent Cold War encouraged the rapid development of electronic computers for military purposes; systems and complex software programs were devised to complete enormous mathematical tasks in a fraction of the time humans would take to perform the same functions. Computers remained large and expensive until the 1970s, when American inventors introduced microcomputers, the memory chip, and speedy computer processors. The computer science field, which has expanded greatly in the last two decades, involves the study of numerical and theoretical information, as it relates to computation, including human calculation or computation by electronic computer systems.

Prototypes

Nineteenth-century British inventor Charles Babbage was first to discover how to construct a programmable computer. After years spent studying mathematical calculation tables that often produced erroneous results, he envisioned a fixed device, which he called a "difference engine," that would deliver accurate results. In 1822, Babbage devised a machine with large metal, steam-driven gears that required roughly 25,000 parts; however, he never completed his machines due to the complexity involved.

He designed his last model shortly before his death in 1871. The bulky machine, which he called an "analytical engine," measured 60 feet by 30 feet. It was able to hold 1,000 numbers of fifty digits each, and it could make a difficult calculation using a series of programmed punch cards as the programming input. Although Babbage died before he could complete the analytical engine, his ideas would inspire the creation of the first electronic computers a century later.

William Bundy, a clockmaker and jeweler, and Herman Hollerith, a statistician, combined ideas and inventions to fuel the development of mechanical computers. Bundy invented a device in 1888 that was both a clock and a tool to measure employee hours; it later became the punch clock. Bundy's company began mass-producing these clocks in 1889, along with other measurement machines. Hollerith invented the Electric Tabulating Machine, which proved highly successful in calculating 1890 U.S. Census results. Hollerith founded the Tabulation Machine Company, which merged with Bundy's company to form the Computing Tabulating Recording (CTR) Corporation in 1911 in Endicott, New York. The company produced a range of mechanical devices, including voting punch-card readers, typing machines, scales for weight measurement, and employee punch-card systems for payroll.

When CTR expanded into Canadian markets in 1917, it became International Business Machines (IBM). Over the next few decades, the company designed engine parts, tabulating equipment, and high-technology weapons components. In the 1930s, IBM scientists researched an advanced electronic machine to do more than perform simple mathematical calculations. Working with Harvard's Howard Aiken, in 1944, the company constructed the Mark I. Standing 6 feet tall by 5 feet wide, the Mark I could process limited programs to calculate logarithms and trigonometric functions. It was slow, requiring five seconds for simple multiplication, but once programmed, it could complete long computations. Aiken ultimately built a series of these machines, each surpassing the previous model in its capabilities.

In 1940, George Stibitz of Bell Labs developed the Complex Number Calculator, which could be operated remotely via the telephone. Two University of Pennsylvania researchers, John Mauchly and J. Presper Eckert, built a massive computer in 1945 called the Electrical Numerical Integrator and

Computer (ENIAC); it took up nearly 1,800 square feet and used more than 18,000 vacuum tubes and multiple electronic units that sent or routed computations with a series of switches and settable tables. ENIAC was the first "wire your own" technology, as it was able to handle various mathematical formulas, versus the common "fixed design" machine that handled only one or a few fixed tasks. The ENIAC used ten decimal digits, punch-card input and output, one multiplier machine, a divider-square rooter, and twenty adding machines. It could store limited information, and it could count at a speed of 0.0002 seconds per number.

Software

John von Neumann, a leading mathematician who worked at the Institute for Advanced Study in Princeton, New Jersey, conceptualized a computer that had "variable hardware components" capable of executing thousands of software configurations. In 1945, he designed computer concepts with programs that could be stored and accessed, so that the information was not in sequence. Von Neumann's contributions allowed the next generation of computer builders to put data and instruction programs together in the same place and have these elements function cooperatively. By 1947, computer designs included the first random access memory program, which von Neumann designed.

By this time, computers were down to the size of a grand piano and had fewer than 2,500 vacuum tubes, but these machines still required extensive maintenance and upkeep. Many colleges and universities began to open computer science programs and departments, investing in new faculty and research programs to develop new machines and software.

One of the first commercial software programs was the Beginner's All-Purpose Symbolic Instruction Code (BASIC), written by Thomas Kurtz and John Kemeny in 1963 at the Dartmouth Mathematics Lab in Hanover, New Hampshire. Kurtz and Kemeny created an understandable software language for simple applications, such as a code written to make a message appear repeatedly on a monitor screen. The applications were limitless but required individual programming.

Around the same time, other software languages were written. IBM produced FORTRAN for mathematical calculations. A collective of computer programmers in Zurich, Germany, developed C, which became the standard language for business computing in the 1960s.

The Home Computer

In 1968, Bob Noyce and Gordon Moore, two computer design engineers, formed a small company called Intel in Santa Clara, California, to design and manufacture computer components. Intel raised over $2.5 million in venture capital and, in 1971, released its first computer brain, or microprocessor. This microprocessor was an integrated electronic circuit board that could process four bits of data at a time. Previously, only the military, government agencies, and universities could afford to design and manage a computer system. But now Intel was designing "consumer" components such as processors and memory chips that could process and save data.

Another company, Micro Instrumentation and Telemetry Systems (MITS), used the Intel processor in a new computer called the Altair 8800. Released in 1975, and the first machine to be referred to as a "personal computer" (PC), the Altair 8800 sold for $397. It came with no software; the users had to write their own software, using a series of switches on the machine.

During the 1980s, the personal computer became widespread because of developments in microprocessors, silicon data chips, keyboards, monitors, and a handheld control device called a "mouse." IBM released the IBM PC in 1981, which had a 16,000-character memory and cost $1,265. IBM also began installing the first version of "servers" in businesses, which included multiple computers sharing information.

Apple, a computer company formed by Steve Jobs and others in 1976, took computing to the next consumer level in 1984 when the company released the Macintosh line of desktops. This computer looked like a small television set; the unit had all the components built in, with plugs for a mouse, keyboard, and printer.

Microsoft, founded in 1975 by Bill Gates and Paul Allen, devised the MS-DOS operating system. This system dominated the software market in the 1980s. DOS was eventually replaced by Microsoft's Windows operating system, which, by 1990, had captured almost 90 percent of the PC market.

Advances in the 1990s included Microsoft's Windows 95 operating system, which was able to run multiple word processing, data analysis, communication, and other programs at the same time. On the hardware side, Intel experienced parallel success, growing into the largest chip manufacturer, producing a range of semiconductor products, including computer processor motherboard chips, network cards, memory chips, graphic chips, and other communications devices. Intel released the Pentium chip in 1993, which was made up of over 4 million individual transistors and allowed users to perform tasks more than two times faster than previous chips.

By the late 1990s, the platform for processor chips was advanced from 32-bit to 64-bit, providing a substantial increase in a computer's capacity and ability to handle simultaneous mathematical calculations. In 2000, Intel released the Pentium 4 chip, featuring 42 million transistors and running at a speed of 1.5 gigahertz. The Intel Pentium 4 processor was able to run multiple computer applications, communicate via video conferencing, deliver television programs and movies, and be connected to the Internet—all at the same time.

Modern computer technology—especially the personal computer, the Internet, and the World Wide Web—has spurred the development of a global economy, changed the speed and efficiency by which humans communicate, and become an essential part of academic, government, and consumer communications and data management. Computers have become the backbone of modern technology. The speed by which information can be stored, analyzed, and transmitted has resulted in a worldwide society, able to quickly share information. Especially in America, people have come to expect technological change and to anticipate future advances with increasing eagerness.

James Fargo Balliett

Sources

Allan, Roy. *A History of the Personal Computer.* London, Ontario, Canada: Allan, 2001.

Burks, Alice. *Who Invented the Computer?* New York: Prometheus, 2003.

Campbell-Kelly, Martin, and William Aspray. *Computer: A History of the Information Machine.* New York: Basic Books, 1996.

Cringely, Robert. *Accidental Empires.* New York: Perseus, 1992.

Kaplan, David. *The Silicon Boys and Their Valley of Dreams.* New York: HarperCollins, 2000.

The Internet

The Internet, which links millions of computers around the world, is arguably the most powerful medium of communication ever devised. Like the telegraph, the telephone, and the radio, the Internet has dramatically reduced the significance of geographical distance, while facilitating commerce, the sharing of resources, and the exchange of information.

Some elements of the Internet—most notably uniform resource locators (URLs) and hypertext markup language (HTML)—were developed by European computer scientists. But most of the theoretical underpinnings and practical applications that made the Internet possible were developed in the United States.

LANs, WANs, and ARPANET

The roots of the Internet date to the 1940s, when the first computer network was constructed by George Stibitz and a team of Bell Laboratories scientists in New Jersey. The Stibitz network was composed of a small handful of computers linked by telephone lines and housed entirely in the Bell offices. Small networks like this would soon come to be known as local area networks (LANs). They remain the most common type of computer network in use today.

Stibitz's team laid the groundwork for the Internet only in a general sense, because the Internet is not a LAN but a wide area network (WAN).

WANs differ from LANs in two critical ways. First, LANs tend to be highly centralized, organized around one or more powerful central computers called servers. WANs, on the other hand, tend to be decentralized. Second, and more important, LANs are much smaller than WANs. LANs are confined to a limited geographical area, usually a single building; their reach is sometimes defined as a maximum of one square kilometer. WANs are spread across a much broader geographical area—a city, a country, even the entire world.

The first major WAN, and the direct predecessor of the Internet, was the U.S. Department of Defense's Advanced Research Projects Agency Network (ARPANET). Node 1 of ARPANET went online at the University of California at Los Angeles in 1969. It was soon followed by nodes at Stanford University, the University of California at Santa Barbara, and the University of Utah. By the end of the year, twenty universities were part of the network; three years later, another forty-six universities and research institutions were added.

In the 1960s and 1970s, computing resources were scarce and very valuable. ARPANET allowed researchers working on Department of Defense (DOD) projects to share these resources. It was not long, however, before the members of the network discovered other uses for the interconnectivity of ARPANET. In 1971, a Cambridge, Massachusetts, computer engineer named Ray Tomlinson developed a system for exchanging electronic mail messages between different nodes of the network. E-mail soon became a significant part of ARPANET traffic.

Shortly thereafter, development began on a system of posting messages to topical "news groups." In 1979, this system became formalized with the establishment of Usenet. Early Usenet groups soon strayed from news to discussions of topics such as sex, drugs, and rock and roll.

Sharing information about sex, drugs, and rock and roll was not what the Department of Defense had envisioned as the use of its computer network. So by the mid-1980s, with the original purpose of ARPANET largely lost and e-mail and newsgroup postings making up nearly all of the traffic on the network, the DOD began to distance itself from the project. Control

of ARPANET passed to the federal National Science Foundation (NSF). At roughly the same time, a number of other WANs were established. Some of these, such as BITNET and CSNET, were based at universities. Others, such as PSINet, Commercial Internet Exchange, Portal, and Netcom, were privately owned and accessible to the general public, which was increasingly wired, thanks to the personal computer revolution. The Internet was born when the NSF allowed these networks to be connected to ARPANET in the late 1980s.

By 1990, it was clear that ARPANET's use was evolving in different directions from the DOD's original plan. This led to a major change in the federal government's approach to the Internet. ARPANET was taken off-line, and the Defense Department shifted its resources to a different WAN called MILNET. The rest of ARPANET's traffic was moved to a new and more robust backbone called NSFNet.

The High Performance Computing Act of 1991, sponsored by Senator Al Gore of Tennessee, extended the Internet's reach into community colleges, high schools, and elementary schools. No longer was the Internet envisioned as a means of sharing computer power; instead, it had become a means of sharing information.

World Wide Web

The government's reorganization of the Internet came just as a new phase in the Internet's development was getting under way. As popular as e-mail and Usenet are, their significance pales in comparison to that of the World Wide Web, which came into being in the early 1990s.

The Web's existence is dependent on three basic innovations. The first was hypertext, credited to American computer engineers Ted Nelson and Doug Engelbart, who worked independently of one another in the 1960s. The basic concept of hypertext is that information in other texts and on other computers can be instantly accessed by clicking on a link in the original text or on the original computer, a particularly effective and dynamic way of organizing content.

The second and third innovations that made the Internet possible were both developed in 1989 by researcher Tim Berners-Lee at the European

Particle Physics Laboratory in Geneva, Switzerland. His uniform resource locators (URLs) allow computer files to be uniquely identified, and his hypertext markup language (HTML) provide a means for combining text and hyperlinks into a single document readable by any computer.

Even with the creation of URLs and HTML, however, the World Wide Web was not quite fully realized. For the first several years after Berners-Lee's groundbreaking work, the Web was made up entirely of text. While HTML has the capacity to accommodate pictures and graphics, the software necessary to display those elements took additional time to develop.

In 1992, two graphical Web browsers, Viola and Mosaic, were made available to the public. Viola, the work of the University of California, Berkeley, was not successful, but Mosaic, developed by the University of Illinois at Urbana-Champaign, soon found its way onto millions of computers.

Mosaic project leader Marc Andreesen went into business for himself, forming the Netscape Corporation. Netscape's Navigator software, released in 1994, was the first successful commercial browser, followed a year later by Microsoft's Internet Explorer. The importance of graphical browsers cannot be overstated. When Mosaic was first released, the World Wide Web comprised only fifty sites. Three years later, when Microsoft entered the fray, the number of sites had jumped to 25,000.

With such rapid growth, it is no surprise that the Internet quickly drew the attention of corporations such as Microsoft. Perhaps the most salient Internet trend of the mid-1990s was the increased presence of businesses in cyberspace. The most obvious manifestation of this trend was the rise of Web-based commerce, but there were other important ways in which corporate interests asserted themselves.

Before 1995, the software that drove the Internet was generally created by public entities. The University of Minnesota's Gopher, Dartmouth University's Fetch, and the University of Illinois at Urbana-Champaign's Telnet and Mosaic are but a few examples. These programs, and their source codes, were given away freely. But beginning with Netscape, the software driving the Web was created largely by private businesses.

This gave corporations a great deal of control over the direction of the Internet—sometimes too much. In 2001, for example, the Microsoft Corporation faced a major antitrust lawsuit because of its manipulation of the Web browser market.

The infrastructure that makes up the Internet also became privatized. ARPANET was under the control of the federal government, as was NSFNet. Shortly after the NSFNet backbone was created, however, it began to be broken up and sold off to private telecommunications companies. By 1994, what had been NSFNet was entirely in the hands of corporations, including MCI, Sprint, AT&T, and Netcom. Connections to these privately held networks were provided mostly by large companies such as Microsoft and America Online (AOL). By the year 1996, AOL controlled some 55 percent of the market. Meanwhile, the withdrawal of NSF left management of the Internet in the hands of public entities such as the Internet Engineering Task Force and private companies such as Network Solutions.

Broadband

High-speed, or broadband, access to the Internet was becoming increasingly crucial as Web sites became more elaborate and more data heavy, and thus slower to download by conventional dial-up telephone line access. American scientists and corporations pioneered the way.

The earliest work on one form of broadband—digital subscriber lines (DSL)—was conducted in the late 1980s by computer technologist Joe Lechleider at Bell Communications Research, the research consortium of the regional Bell telephone companies. Around the same time, scientists at the American-owned corporation Motorola were helping to perfect the other main type of wired broadband access, the cable modem. While DSL works over telephone lines, the cable modem provides a connection via the same wiring that brings consumers their cable television.

However, the future of broadband access, most computer experts say, is wireless. The precursor to current high-speed wireless Internet access, or WiFi, was invented in 1991 by scientists working for Lucent, the manufacturing wing of AT&T.

Christopher Bates

Sources

Bronson, Po. *The Nudist on the Late Shift: And Other True Tales of Silicon Valley.* New York: Random House, 1999.

Cassidy, John. *Dot.con: The Greatest Story Ever Sold.* New York: HarperCollins, 2002.

Hafner, Katie, and Matthew Lyon. *Where Wizards Stay Up Late: The Origins of the Internet.* New York: Touchstone, 1998.

Kaplan, Philip J. *F'd Companies: Spectacular Dot-Com Flame-outs.* New York: Simon and Schuster, 2002.

Kuo, J. David. *dot.bomb: My Days and Nights at an Internet Goliath.* New York: Little, Brown, 2001.

Stoll, Clifford. *Cuckoo's Egg: Tracking a Spy Through the Maze of Computer Espionage.* New York: Pocket Books, 2000.

———. *Silicon Snake Oil: Second Thoughts on the Information Highway.* New York: Anchor, 1996.

AIKEN, HOWARD
(1900–1973)

Electrical engineer Howard Aiken designed the Mark I, the first large-scale automatic computer capable of performing thousands of complex mathematical calculations. The Mark I was built for Harvard University by International Business Machines (IBM), and Aiken went on to design several more advanced computers, all housed at Harvard.

Howard Hathaway Aiken was born on March 8, 1900, in Hoboken, New Jersey. Aiken was an only child, and he and his parents moved to Indianapolis, Indiana, when he was a teenager. After his father deserted the family, Aiken dropped out of school and got a job installing telephones. He later completed correspondent courses and graduated from Arsenal Technical High School in Indianapolis in 1919. The family then moved to Madison, Wisconsin, where Aiken took a job with a utility company and attended the University of Wisconsin. After graduating with a bachelor's degree in electrical engineering in 1923, he held a string of jobs with various electric companies over the next decade.

At the age of thirty-three, and with more experience than most graduate students, Aiken was accepted into the graduate physics program at Harvard University, where he did research on the engineering of antennas and the thermionic emissions of electrons. His thesis was on the conductivity of vacuum tubes, which used power to amplify a signal by controlling the movement of electrons. After earning his Ph.D. in electrical engineering in 1939, he was hired by Harvard as an assistant professor. He would teach there for the next twenty-two years.

One of Aiken's goals was to build a computer. He had been fascinated with Charles Babbage's 1822 concept of a mechanical "difference engine" to perform complex mathematical calculations. In his thesis research, Aiken had performed extensive differential equations in cylindrical coordinates, which had taken large amounts of time and energy. He had completed his design for a computing device as a graduate student but did not have the funding or facilities to begin building it.

Although the physics department at Harvard did not really see the need for what it regarded as a large calculator, it eventually agreed to fund the "computing machine." Aiken picked IBM to build it. Construction began in 1939 at the IBM facilities in Endicott, New York. The project would take five years to complete, at a cost of $250,000.

When Aiken was called to active duty in the U.S. Navy in the spring of 1941, he appointed Harvard graduate student Robert Campbell to oversee the project in his absence. The five-ton device was finished in early 1944 and shipped in boxes to Harvard for reassembly. The navy released Aiken to work on the final assembly and initial programming at Harvard.

Originally called the Automatic Sequence Controlled Calculator (ASCC), the machine was renamed the Mark I. It measured 51 feet (15.5 meters) long, 8 feet (2.4 meters) high, and 3 feet (9 meters) deep. It used 530 miles (850 kilometers) of copper wire and contained 760,000 electrical parts. Although it was the property of the U.S. Navy, the computer was housed at Harvard with a large support staff.

Aiken was closely involved in the first tasks of the Mark I, which ran twenty-four hours a day, seven days a week. The initial problem-solving tasks pertained to military matters, such as protecting ships from mines with magnetic fields and the operation of radar technology. Aiken also brought the Mark I to bear on problems encountered by scientists working on the atomic bomb at Los Alamos, New Mexico. At Harvard, a new academic department (called "computer science") was created and Aiken developed a series of courses based on the Mark I. He created the Harvard Computation Laboratory in 1947 and served as its director until his retirement in 1961. The Mark I remained in active use for fourteen years, until 1959.

Using the experience from his first machine, Aiken designed the Mark II—commissioned by the navy—with more advanced electromagnetic relays and built-in hardware for a diversity of trigonometric, reciprocal, square root, and exponential functions. During an early test, a moth was found trapped between two relay points—this "bug in the system" led to the now common term for any computer problem.

Aiken's next computer design, for the Mark III, pioneered the use of magnetic drum memory to store commands and data. The Mark IV, developed for the U.S. Air Force, employed a fully magnetic core memory that was completely electronic. Completed in 1952, the Mark IV was the first computer to integrate metal circuitry (ferrite) to improve the core memory.

Aiken's computer science program at Harvard flourished in the 1950s, turning out an all-star roster of computer experts for the federal government and cutting-edge private companies such as IBM, Intel, Texas Instruments, and Hewlett Packard. He continued teaching until his retirement in 1961.

In 1970, Aiken received the Institute of Electrical and Electronics Engineers (IEEE) Edison Medal for his "pioneering contributions to the development and application of large-scale digital computers and important contributions to education in the digital computer field." He died in his sleep on March 14, 1973.

James Fargo Balliett

Sources

Aiken, Howard. *Synthesis of Electronic Computing and Control Circuits.* Cambridge, MA: Harvard University Press, 1952.

Campbell-Kelly, Martin, and William Aspray. *Computer: A History of the Information Machine.* New York: Basic Books, 1996.

Cohen, Bernard I. *Howard Aiken: Portrait of a Computer Pioneer.* Cambridge, MA: MIT Press, 2000.

———. *Makin' Numbers: Howard Aiken and the Computer.* Cambridge, MA: MIT Press, 1999.

AMERICAN MATHEMATICAL SOCIETY

The American Mathematical Society (AMS) was founded in New York City on November 24, 1888, for the purpose of promoting cooperation between mathematicians and advancing mathematical scholarship and research. The AMS continues to accomplish these goals today through regular meetings and publications. It has nearly 30,000 members (about 7,300 outside the United States) and maintains offices in Providence and Pawtucket, Rhode Island; Washington, D.C.; and Ann Arbor, Michigan.

The AMS was envisioned by Thomas S. Fiske, Harold Jacoby, and Edward L. Stabler of Columbia College (Columbia University from 1896). They proposed an organized society for American mathematicians and announced their intention through the publication of a notice of an impending meeting in November 1888, on the morning of Thanksgiving Day. In attendance were Fiske, Jacoby, Stabler, J.H. Van Amringe, John K. Rees, and James Maclay. During the second meeting, on November 29, a constitution was adopted and the New York Mathematical Society (NYMS) was formally established. Van Amringe was elected president, and Fiske was designated secretary. Constitutional amendments passed on July 1, 1894, transformed the NYMS into the AMS. A further reorganization on May 3, 1923, incorporated the AMS under the laws of the District of Columbia. Membership in the AMS grew progressively. By the end of 1889, there were sixteen chartered members, and, at the time of incorporation, membership had grown to 1,250. The expansion of the membership network coincided with the publication of the first journal of the AMS.

During a meeting on December 5, 1890, Van Amringe, retiring as president, proposed that the AMS should publish a society bulletin. The realization of his vision hinged on adequate increases in membership subscriptions, but the increases depended on the publication of a journal. Several publishing houses, notably Macmillan and John Wiley and Sons, helped breach the impasse by providing lists of people with an interest in mathematics. Fiske sent them prospectuses of the bulletin along with an invitation to join the AMS. Having expanded its membership and subscription base, the society began publication of the AMS *Bulletin* in 1891 with Fiske as editor-in-chief.

By 2005, the AMS *Bulletin* and *Notices* circulated to more mathematicians throughout the world than any other mathematical journal. The AMS publishes more than 100 books and nine research journals annually, containing more than

1,000 articles. It has more than 3,000 research monographs, collected works, proceedings, and textbooks in print; it also maintains the Mathematical Reviews database, facilitating online searching of more than 1.7 million reviews and citations. More than 200 people are employed by the AMS in publishing, setting fiscal and scientific policies, organizing the profession, advancing education, establishing prizes and awards, and arranging national and international meetings and conferences.

Santi S. Chanthaphavong

Sources

American Mathematical Society. http://www.ams.org.

Archibald, R.C. *A Semicentennial History of the American Mathematical Society, 1888–1938.* New York: American Mathematical Society, 1938.

Pitcher, Everett. *A History of the Second Fifty Years, American Mathematical Society, 1939–1988.* Providence, RI: American Mathematical Society, 1988.

APPLE COMPUTERS

Steven Jobs and Stephen Wozniak were high school friends, both of whom dropped out of college and began working for firms in Silicon Valley (on the peninsula south of San Francisco), then emerging as the design center of the computer industry. The two men had ambitions of starting a computer firm that would transform the personal computer into the dominant platform within the industry. In the mid-1970s, Wozniak developed a personal computer that Jobs would eventually help him market as the Apple 1. On April 1, 1976, Apple Computers was formally incorporated.

The company did not become profitable until 1977, when the Apple II, the first PC to feature color graphics and to be contained in a plastic casing, was introduced. It gained a firm footing the following year, after the release of the Apple Disk II, the first PC to feature an affordable and readily accessible floppy drive.

By 1980, when the Apple III was introduced, the company had thousands of employees and a formal corporate structure. Thereafter, the company faced several crises. Seriously injured in a plane crash, Wozniak took a leave of absence, and Jobs assumed the chairmanship. Intensifying competition in the industry combined with a national economic recession to cause the first decline in Apple sales and the first employee layoffs in the company's history.

In the late 1970s and early 1980s, Apple's major competitor was IBM, which, in 1981, introduced its PC, using Microsoft's MS-DOS operating system. Jobs oversaw the development of the Macintosh PC, which featured a graphical user interface (GUI) that greatly simplified the process of initiating program commands. To market the Macintosh, Jobs recruited John Sculley, the president of Pepsi-Cola, to become president and CEO of Apple. Although the Macintosh had promising initial sales, several deficiencies eroded its appeal. Jobs and Sculley blamed each other for the declining sales. In

Computer engineer and entrepreneur Steve Jobs cofounded Apple Computer with Steve Wozniak in the mid-1970s. The Apple II, introduced in 1977, was the company's first popular microcomputer and the most successful of all early personal desktops. *(Ralph Morse/Time & Life Pictures/Getty Images)*

May 1985, after dramatic behind-the-scenes corporate scheming, Sculley managed to oust Jobs from the corporate hierarchy.

Under Sculley's leadership, Apple experienced a dramatic change in fortune. The company became involved in a protracted legal battle with Microsoft over the latter's alleged incorporation of the Macintosh GUI in the Windows operating system. The dispute shook consumer and industry confidence in the company and led to declines in both its stock price and sales. In 1985, Apple was forced to lay off about 20 percent of its workforce, or more than 1,200 employees. In the midst of this crisis, however, Apple designers developed PageMaker, the first desktop publishing program, and Laser-Writer, the first relatively inexpensive laser printer, for use with the company's new PC, the Mac II. In the late 1980s, the company's profits regularly exceeded Wall Street expectations.

Apple's success was short-lived, because Microsoft had begun marketing its Windows 3.0 operating system to the increasing number of companies marketing clones of the IBM PC. In attempting to maintain proprietary control of its hardware and software, Apple had become an anachronism in the industry. In 1991, the company successfully introduced the PowerBooks line of lightweight laptop computers, but the Newton Personal Digital Assistant (PDA), introduced in 1993, had disappointing sales. Sculley was forced out by the board of directors in June 1993.

Under Sculley's successor, Michael Spindler, Apple developed the PowerMac, using the PowerPC processor. When the company could not meet the market demand for PowerMacs, orders began to level off. In addition, the company successfully marketed the inexpensive Performa PC, but the profit margin was so narrow that the sales success did little for Apple's overall bottom line.

In January 1996, Spindler was removed and replaced by Gil Amelio. The company was reporting quarterly losses of three-quarters of a billion dollars, but Amelio forced through a dramatic restructuring that restored modest profitability. Apple acquired NeXT, planning to develop NeXTstep into a new Mac operating system to be called Rhapsody. Steven Jobs was brought back to oversee this project and, when

Amelio resigned under pressure in July 1997, Jobs and Fred Anderson, the company's CFO, jointly managed the company.

As interim chair of Apple, Jobs revamped the corporate structure and replaced most of the directors on the board. He dramatically announced a cooperative agreement with Apple's longtime nemesis, Microsoft, and introduced a series of new PowerMacs and the tremendously successful iMac. The company again became highly profitable, its stock price rose throughout the late 1990s, and Jobs was able to drop "interim" from his title.

Apple's next major innovation, the Cube, was as disappointing as the iMac had been successful. The company subsequently concentrated on marketing proprietary ancillary devices and services, such as the iPod and iTunes.

Martin Kich

Sources

Brackett, Virginia. *Steve Jobs: Computer Genius of Apple.* Berkeley Heights, NJ: Enslow, 2003.

Butcher, Lee. *Accidental Millionaire: The Rise and Fall of Steve Jobs at Apple Computers.* New York: Paragon, 1988.

Kendall, Martha E. *Steve Wozniak: Inventor of Apple Computer.* New York: Walker, 1994.

Linzmayer, Owen W. *Apple Confidential: The Real Story of Apple Computer, Inc.* San Francisco: Publishers Group West, 1999.

Moritz, Michael. *The Little Kingdom: The Private Story of Apple Computer.* New York: William Morrow, 1984.

Rose, Frank. *West of Eden: The End of Innocence at Apple Computer.* New York: Viking, 1989.

Applied Mathematics

American achievements transformed applied mathematics from an embryonic discipline in the nineteenth century to a discipline that proved crucial to the Allied victory in World War II. In the postwar period, this discipline became of central importance to industry.

Applied mathematicians begin with a consideration of real-world problems, envision separate elements of the problem under consideration, and abstract those elements into mathematical representations and structures. The nineteenth-century Yale professor Josiah Willard Gibbs was the first important American applied mathematician, working especially in statistical mechanics and thermodynamics.

During the early to mid-twentieth century, American mathematicians developed mathematical theories in statistics, physics, quantum mechanics, chemistry, and artificial intelligence. Harvard mathematician G.D. Birkhoff developed the ergodic theorem and worked in statistical physics and quantum mechanics. The ergodic theorem was also important in developing a kinetic theory of gases. MIT professor Norbert Wiener was a pioneer in stochastic processes, particularly Brownian motion, which explains the random nature of subatomic particles.

In 1935, Richard Courant founded the Courant Institute for Mathematical Sciences, focusing on applied mathematics, at New York University. In the spring of 1937, Samuel S. Wilks taught mathematical statistics to undergraduates at Princeton. By 1940, serious work in mathematical statistics was being done by Harold Hotelling at Columbia and Jerzy Neyman at Berkeley. In 1941, William Prager and R.G.D. Richardson established a Program of Advanced Instruction and Research in Applied Mechanics at Brown University. Despite these significant advances, applied mathematics remained a discipline dominated by physicists and engineers and practiced by few professional mathematicians.

Applied mathematics rose to prominence when America mobilized its ablest mathematicians in support of the nation's military effort in World War II. The emphasis on applied mathematics derived from a recognition that the outcome of the war depended on obtaining solutions to mathematical problems in submarine warfare, radar, electronic countermeasures, explosives, rocketry, operations research, and cryptanalysis. The Applied Mathematics Panel (AMP) was established in 1942 to provide leadership in applied mathematics during the war. The AMP promoted programs in applied mathematics at Princeton, Columbia, New York, UC Berkeley, Brown, Harvard, and Northwestern, and it involved such important American mathematicians as Vannevar Bush, Warren Weaver, Richard Courant, and Oswald Veblen.

Mathematicians who contributed to the development of computer science during the war included John von Neumann, who was also instrumental to the success of the Manhattan Project, and George Stibitz of Bell Telephone Laboratories (a war contractor), who exhibited a machine for computing complex numbers with telephone relays. In 1946 the Moore School of the University of Pennsylvania operated ENIAC, the first electronic computer. The U.S. leadership in computer technology became possible through the establishment of the National Applied Mathematics Laboratories of the National Bureau of Standards and the Institute for Numerical Analysis at UCLA.

The growing importance of applied mathematics in science and technology led to the establishment of the Society for Industrial and Applied Mathematics (SIAM) in 1951 in Philadelphia. Mathematicians who had worked in cryptography also supported industrial efforts to commercialize computers between 1946 and 1953.

During the latter half of the twentieth century, applied mathematics involved game theory, control theory, and operations research. Linear, dynamic, and integer programming has become an indispensable tool of economics, business, and finance.

Santi S. Chanthaphavong

Source

Duren, Peter, Richard A. Askey, and Uta C. Merzbach, eds. *A Century of Mathematics in America.* 3 vols. Providence, RI: American Mathematical Society, 1988–1989.

BANNEKER, BENJAMIN (1731–1806)

Benjamin Banneker, an eighteenth-century independent scholar from Maryland, was a surveyor, astronomer, and almanac publisher. His intellectual accomplishments were impressive in and of themselves, especially in light of his status as an African American in a slave state, where free African Americans such as Banneker were subjected to prejudice and persecution.

Banneker's maternal grandmother, Molly Welsh, a dairymaid in Devon, England, was accused of stealing milk from her employer in 1682. She was sentenced to seven years of indentured servitude in Maryland, where, after completing her sentence, she was given fifty acres of arable

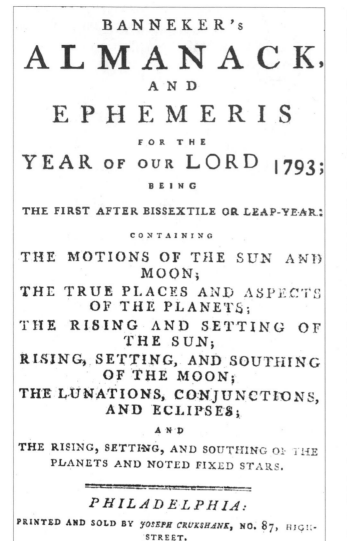

BANNEKER's

ALMANACK,

AND

EPHEMERIS

FOR THE

YEAR OF OUR LORD 1793;

BEING

THE FIRST AFTER BISSEXTILE OR LEAP-YEAR:

CONTAINING

THE MOTIONS OF THE SUN AND MOON;

THE TRUE PLACES AND ASPECTS OF THE PLANETS;

THE RISING AND SETTING OF THE SUN;

RISING, SETTING, AND SOUTHING OF THE MOON;

THE LUNATIONS, CONJUNCTIONS, AND ECLIPSES;

AND

THE RISING, SETTING, AND SOUTHING OF THE PLANETS AND NOTED FIXED STARS.

PHILADELPHIA:

PRINTED AND SOLD BY *JOSEPH CRUKSHANK*, NO. 87, HIGH-STREET.

A free black tobacco farmer, clockmaker, surveyor, and self-taught astronomer from Maryland, Benjamin Banneker published the widely respected *Banneker's Almanack* from 1792 through 1797. *(MPI/Hulton Archive/Getty Images)*

farmland as required by law and managed to buy two African American slaves. One of the slaves was known as "Banneka," and Welsh eventually freed and married him. Their grandson, Benjamin Banneker, was born near Ellicott's Lower Mills, Maryland, on November 9, 1731.

Many details of Banneker's life are unknown or unclear. As a boy, he was befriended by a Quaker schoolmaster named Peter Heinrich, who encouraged him in his studies. Heinrich lent him books, conversed with him, and perhaps enrolled Benjamin in his school. Regardless of his education, Banneker was extremely well read. He demonstrated extraordinary intelligence and intellectual curiosity throughout his life.

As a young adult, Banneker supported himself by farming, pursuing intellectual endeavors such as astronomy in his free time. In 1753, he amazed his community with a wooden clock he had designed and built on his own; in that era, timepieces were rare.

Later, he developed a close friendship with George Ellicott, a fellow intellectual who recognized Banneker's genius and appreciated the knowledge and wisdom that Banneker imparted. Since Ellicott was white, however, the relationship was awkward. Nevertheless, in 1790, when the U.S. Congress began plans to create a permanent national capital, Ellicot's uncle, Andrew Ellicott, was appointed to the position of surveyor general. He, in turn, appointed Banneker as one of his personal assistants, risking his reputation by selecting an African American. For several months, Banneker proudly worked alongside the white men, though racism made life difficult from day to day.

Probably Banneker's greatest claim to fame was his *Almanac,* which contained, in his words, "the motions of the sun and moon, the true places and aspects of the planets, the rising and setting of the sun, rising, setting, and southing of the moon, the lunations, conjunctions, and eclipses, and the rising, setting, and southing of the planets and noted fixed stars." The first edition was dated 1792, and Banneker published new editions each year until 1797.

These almanacs were well regarded and widely respected. For some buyers, the author's African American heritage was a selling point; many wished to support the work of an African American, and others considered the book an amusing novelty. Banneker began to receive a multitude of letters from admiring readers, as well as a steady stream of visitors to his home.

Just before the publication of his first almanac, Banneker took a bold step. He sent a handwritten advance copy to Thomas Jefferson, then secretary of state, together with a lengthy personal letter. Generally respectful in tone, the letter nevertheless included a sharp and direct criticism of Jefferson's slaveholding. Jefferson replied with a polite but evasive note praising the almanac.

Banneker died on October 9, 1806. On the day of the funeral, his home was burned to the ground in an apparent act of arson. Many valuable records were lost, including a mass of personal writings and the famous wooden clock. Historians have had limited success in documenting his life and work, but he is still widely regarded as an inspirational figure.

Andrew Perry

Sources

Beddini, Silvio A. *The Life of Benjamin Banneker.* Rancho Cordova, CA: Landmark Enterprises, 1984.

Cerami, Charles. *Benjamin Banneker.* New York: John Wiley and Sons, 2002.

BROWNIAN MOTION

Brownian motion is the random movement of microscopic particles suspended in a gas or liquid that is caused by the particles' collisions with molecules in the surrounding liquid or gas. Observations of this movement in the nineteenth century provided the first visible evidence for the existence of molecules. By the start of the twenty-first century, Brownian motion had become the basis for a variety of mathematical models used in science, manufacturing, and economics.

Brownian motion is named after Robert Brown, the Scottish botanist who first described the phenomenon in 1827, after observing the random motions of evening primrose pollen as it floated in water on a microscope slide. After conducting more research, Brown theorized that the motion of the tiny particles was not due to the pollen being alive, but to the action of invisible molecules in constant collisions with the tiny particles. In 1905, Albert Einstein published a paper that connected the Brownian motion phenomenon to the molecular-kinetic theory of heat and provided the first mathematical theory explaining the concept. Subsequent studies of Brownian motion by the French physicist Jean-Baptiste Perrin verified Einstein's explanation in 1908 and ultimately confirmed the atomic nature of all matter. For his work, Perrin received the 1926 Nobel Prize in Physics.

Research into the nature and applications of Brownian motion by American scientists began in earnest in the 1920s when MIT mathematics professor Norbert Wiener, who is perhaps more widely known for his work in cybernetics, began relating Brownian motion to stochastic process mathematical theories based on random variables. Wiener's work was so influential that Brownian motion is now often called the Wiener process.

By the 1930s, physicist George Uhlenbeck was doing pioneering work leading to a new theory of macroscopic Brownian motion as a process. Throughout the 1940s, Brownian motion emerged as a fairly well-known and well-explicated physical phenomenon, yet it was not until the late 1950s that the concept became more widely known outside the fields of physics and mathematics.

In 1959, M.F.M. Osborne, a physicist at the U.S. Naval Research Laboratory, applied the concept of Brownian motion to his studies of the apparent randomness of the stock market. Drawing heavily on the theories of the late nineteenth-century French mathematician Louis Bachelier, Osborne developed "econophysics," a model combining physics and financial economies. His research would eventually challenge several of the long-established macroeconomic and neoclassical models in economics.

By the 1970s, notions of Brownian motion had influenced the work of IBM researcher Benoit Mandelbrot on fractal geometry, which he defined as "a rough or fragmented geometric shape that can be subdivided in parts, each of which is . . . a reduced-size copy of the whole." Mandelbrot's fractal research revolutionized the field of digital graphics in everything from medical imaging to computer gaming.

Today, the concept of Brownian motion is at the heart of such diverse fields as statistical physics, complex systems analysis, and manufacturing operations research.

Todd A. Hanson

Sources

Mandelbrot, Benoit B. *The Fractal Geometry of Nature.* New York: W.H. Freeman, 1982.

Mazo, Robert. *Brownian Motion: Fluctuations, Dynamics, and Applications.* New York: Oxford University Press, 2002.

BUSH, VANNEVAR
(1890–1974)

Professor, engineer, and a leading supporter of scientific and technological development, Vannevar Bush helped to shape some of the emerging trends in twentieth-century America. Some of his most notable achievements were in the development of the computer and the increased involvement of the federal government in providing funding and setting national policy in science and technology.

Born on March 11, 1890, Bush grew up in Chelsea, Massachusetts. He attended Tufts University and in 1916 earned a doctorate in engineering from Harvard University and the Massachusetts Institute of Technology (MIT). He taught and directed research in engineering at MIT from 1919 to 1938, serving the last six years as dean of the engineering school.

While teaching at MIT, Bush developed an interest in the construction of computers that could solve the complex mathematical equations commonly used by engineers. In the 1920s, Bush and his students developed a machine that could find the area under a curve where the curve is represented by the product of two functions. This device, called a network analyzer, was completed in 1927.

In the following years, Bush and his students produced several other computing devices, including the differential analyzer (in 1930), which could solve the complex equations used in creating electrical power networks. A second, more advanced differential analyzer, completed in 1942, cut the programming time (over a day) of the original down to under five minutes. These computers, as well as others designed or envisioned by Bush, were analog devices that would be replaced by speedier and more advanced digital computers following World War II.

Besides his contributions as a teacher and engineer, Bush was also a key figure in the growing involvement of the federal government in supporting science. As president of the Carnegie Institution starting in 1939, he directed grants to support a variety of scientific pursuits.

As Bush became involved in national scientific policy, he pushed for the creation of an organization that could coordinate civilian and military research needs. He believed such an

Computer design pioneer Vannevar Bush (far left) works with colleagues at MIT in the 1930s on his differential analyzer, a predecessor to the analog computer. Bush later helped direct U.S. science policy during World War II. *(General Photographic Agency/Hulton Archive/Getty Images)*

organization was vital, given the failure of civilian–military cooperation during World War I and the growing tensions in Europe. President Franklin D. Roosevelt took his advice on this issue, and in 1940, Bush was appointed chair of the newly created National Defense Research Committee (NDRC). The committee, which reported to the president as an advisory body, had representatives from both the U.S. Army and U.S. Navy. When the NDRC was reorganized in 1941 and absorbed into the new Office of Scientific Research and Development (OSRD), Bush became director of the OSRD. He held this position until 1946.

The OSRD oversaw many federally funded science projects during World War II, including the government laboratories that developed radar, sonar, and the atomic bomb. Bush, at the request of President Roosevelt, also made recommendations on ways to continue the interaction between the federal government and the scientific community after the war. He recommended expanded federal assistance to restart basic research projects that had been set aside because of the war and to deal with the increasing cost of maintaining and purchasing equipment. Bush urged the federal government to take on a new peacetime role as a financial supporter of basic scientific research, not necessarily through government-run labs, but by funding private research. His hopes and plans in this area were laid out in his 1945 publication *Science: The Endless Frontier*. Some, but by no means all, of his recommendations resulted in the 1951 creation of the National Science Foundation.

When the Red Scare swept across the nation in the early 1950s, Bush publicly defended the loyalty of Robert Oppenheimer. In 1955, Bush retired as president of the Carnegie Institution and left Washington and policy-making circles, discouraged at the new emphasis on national security at the expense of innovation.

In his later years, Bush spent his time raising turkeys in New Hampshire and completing a few engineering projects. He sometimes spoke out on the nuclear arms race, the importance of entrepreneurship for engineers and scientists, and the growing commercialization of American life. He died on June 30, 1974.

Paul Buckingham

Sources

Bush, Vannevar. *Operational Circuit Analysis.* New York: John Wiley and Sons, 1929.
———. *Science, the Endless Frontier.* A Report to the President by Vannevar Bush, Director of Office of Scientific Research and Development, July 1945. Washington, DC: U.S. Government Printing Office, 1945.
Nyce, James M., and Paul Kahn, eds. *From Memex to Hypertext: Vannevar Bush and the Mind's Machine.* Boston, MA: Academic Press, 1992.
Zachary, G. Pascal. *Endless Frontier: Vannevar Bush, Engineer of the American Century.* New York: Free Press, 1997.

CALCULUS

Calculus is a set of general methods for solving mathematical problems involving rates of change and the accumulation of quantities, especially problems involving infinite or infinitesimal quantities. Its discovery in the seventeenth century was a milestone in the history of mathematics and contributed significantly to the development of the physical sciences.

Before calculus, mathematics and the associated physical sciences had been hampered by a lack of general methods and notation for solving certain problems. Mathematicians usually addressed problems one at a time, often by developing a different method for each problem. Though parts of what became calculus can be found in the work of many mathematicians, there was no recognition of the existence of a general method prior to the late seventeenth century.

Calculus is composed of two general methods: integral and differential calculus. The two types are related, as solutions to problems found by methods of integration can be verified with the tools of differential calculus, and those solved via differentiation can be verified through integration. This concept is called the fundamental theorem of calculus, and its expression by Gottfried Leibniz and Isaac Newton mark them as the co-discoverers of calculus. Newton probably developed his ideas first, but Leibniz, who developed his method independently, was the first to publish, in 1684. This sparked an epic conflict over who developed this key mathematical concept first. Leibniz's notation is regarded as more flexible than that of Newton and is the one generally used today for both differential and integral calculus.

Each type of calculus is used for different categories of problems. Differential calculus is a method of solving problems involving rates of change over time, such as in the acceleration of a solid body or the growth of principle at a given interest rate. Integral calculus is used for solving problems involving the accumulation of quantities, such as determining forces exerted by liquids or measuring the volume and surface area of solids.

The core method of calculus was in place by the eighteenth century, and the system was further fleshed out in the nineteenth century, but Americans were not yet contributing significant advances in the theoretical branches of mathematics. The mathematics community in nineteenth-century America was largely centered around the teaching of basic mathematics for practical purposes, such as mapmaking, navigation, and engineering. Scholarship was largely confined to translations, commentaries, and edited editions of the works of prominent, but not necessarily contemporary, European mathematicians.

Only in the late nineteenth century were there scholarly communities in the United States that treated mathematics as more than a tool and began making important contributions in fields associated with calculus. American mathematicians have contributed in such fields as the calculus of variations, differential equations, vector calculus, complex analysis, and differential geometry.

Paul Buckingham

Sources

Berlinski, David. *A Tour of the Calculus.* New York: Pantheon, 1995.

Duren, Peter, Richard A. Askey, and Uta C. Merzbach. *A Century of Mathematics in America.* 3 vols. Providence, RI: American Mathematical Society, 1988–1989.

Hall, A. Rupert. *Philosophers at War: The Quarrel Between Newton and Leibniz.* Cambridge, UK: Cambridge University Press, 1980.

Tarwater, Dalton, ed. *The Bicentennial Tribute to American Mathematics, 1776–1976.* Papers presented at the fifty-ninth annual meeting of the Mathematical Association of America commemorating the nation's bicentennial, Buffalo, New York, 1977.

Chaos Theory

Chaos theory deals with the irregular, erratic side of nature, such as the swirling of smoke and the Earth's weather patterns. It was first developed in 1961 by Edward Lorenz, a meteorologist at the Massachusetts Institute of Technology (MIT).

Lorenz was using a computer program to study the possibility of predicting long-term weather patterns, and he wanted to examine one weather sequence at greater length. To begin the simulation in the middle, he typed in a data sequence carried to three decimal places. The computer's memory retained six decimal places, however, creating a difference in the initial data that was equivalent to only one part in a thousand. That small difference eventually created a wild divergence from the previous weather pattern and showed how an extremely small initial change in a sequence can lead to much larger changes over time.

This phenomenon, "sensitive dependence on initial conditions," is sometimes referred to as "the butterfly effect." The name comes from the idea that a butterfly flapping its wings in New York can create a change in the atmosphere that, while slight, might eventually lead to large-scale changes that produce hurricanes in Hong Kong. In other words, insignificant changes at the beginning of a process or system may become increasingly significant and yield larger and larger changes as time passes.

Lorenz's discovery led to the hope of finding mathematical models to explain erratic systems that do not obey the laws of Newtonian physics. These systems, such as the unpredictability of a gambling game, have been problematic, because they do not seem to conform to any consistent mathematical explanations. Chaos theory posits that these systems are not actually random but are deterministic and can be predicted using mathematical models.

Chaos theory is closely related to the study of fractals, geometric shapes that are highly complex and infinitely detailed. Fractals contain copies of themselves—a property called "self-similarity"—with each section containing as much detail as the whole fractal. The more the fractal is enlarged, the more detail can be seen—a property that goes on infinitely. Fractals are related to chaos theory, because they are extremely complex, seemingly chaotic structures that exhibit definite properties. By studying fractals, scientists may be able to discover mathematical models that can be applied to larger systems.

The study of chaos theory has been greatly enhanced—and in many ways even made possible—through the use of digital technology and computer simulations. Prior to the advent of computers, it was impossible to perform the number of computations necessary to create the long-term simulations that provide insight into the workings of chaotic systems. The science of chaos is sometimes regarded as the third great revolution in science, after relativity and quantum mechanics.

Beth A. Kattelman

Sources

Gleick, James. *Chaos: Making a New Science.* New York: Viking, 1987.

Grebogi, Celso, and James A. Yorke, eds. *The Impact of Chaos on Science and Society.* New York: United Nations University Press, 1997.

COMPUTER APPLICATIONS

Computer applications, also known as computer programs or end-user programs, are sets of instructions that direct a computer to perform a desired sequence of operations, thus enabling it to carry out a specific task.

The earliest computers had fixed programs that allowed the computer to carry out only a specific type of task. In order to change the task, or "program," of these computers, they would have to be physically rewired, a time-consuming and tedious process. In 1945, however, mathematician John von Neumann put forth the idea that a computer could have a fixed physical structure and yet be directed to execute various functions and computations through the use of a programmed control.

In 1947, scientists produced the first generation of computers that could take advantage of von Neumann's programming ideas. This development enabled programmers to write a series of instructions in code and change the computer's operation without having to physically rewire the machine.

These computers also included the important innovation of random access memory (RAM), a memory structure designed to give constant access to any particular piece of infor-

mation. This memory structure facilitated the use of subroutines—small bits of prewritten, repetitive code—and opened up new possibilities in the creation of computer applications.

Throughout the 1960s, the surge in development of computer hardware continued to spur the creation of computer software. By the end of the 1960s, computer applications were in such demand that companies began to sell prepackaged software to businesses.

With the advent of the personal computer (PC) in the early 1980s, the market for packaged computer applications grew rapidly and software companies began to create programs designed to run on particular PC operating systems. The two most popular operating systems in this early PC market were Apple Corporation's graphical user interface (GUI) and Microsoft's Windows operating system.

Today, the creation of computer applications has become a fairly standard process. All applications are written in one of a number of particular programming language. The programmer uses the selected language to create instructions for the computer, also known as the source code. Of the numerous programming languages that can be used to create computer source code, common examples are Basic, C++, Java, and Perl.

Once the application source code is written, it must be run through a compiler and an assembler program, in order to translate it into a set of instructions that the computer can understand. This final format is known as machine language. When this format is reached, the computer can understand and execute the instructions contained in the application.

Computer applications can be grouped in several major categories, depending on what type of actions they are created to perform. Some of the most common categories are word processing, database applications, spreadsheets, drawing and other image interfaces, slide and graphics presentations, desktop publishing, media players, and Internet browsers.

Beth A. Kattelman

Sources

Campbell-Kelly, Martin, and William Aspray. *Computer: A History of the Information Machine.* New York: Basic Books, 1996.

Glassborow, Francis. *You Can Do It!: A Beginners Introduction to Computer Programming.* Hoboken, NJ: John Wiley and Sons, 2004.

Ifrah, Georges. *The Universal History of Computing: From the Abacus to the Quantum Computer.* Trans. E.F. Harding. New York: John Wiley and Sons, 2001.

Williams, Brian K., and Stacey C. Sawyer. *Using Information Technology: A Practical Introduction to Computers and Communications.* 5th ed. New York: McGraw-Hill, 2003.

CYBERNETICS

Cybernetics is the study of communication and control in living organisms and in machines. It emphasizes the similarities that exist between animals and machines and recognizes that, even though they may be constructed of very different materials, their operation is essentially the same. The field also posits that, because of the similarity, scientific methods can be used to study both.

Feedback—especially negative—is an important element in cybernetics. When part of the output of a system is fed back into the system, it affects the system and its subsequent output. This creates a closed loop in which the system output continually affects its own value, thus directing future operation of the system. Negative feedback helps the system determine what not to do and thereby acts as a stabilizing factor. The system attempts to establish equilibrium and becomes self-regulating. The ability of a complex system to regulate itself—also known as homeostasis—can be applied to numerous structures, including complex machines, biological entities, and social systems. One of the most important contributions of cybernetics is that it offers a single vocabulary and a single set of concepts for representing the most diverse types of systems.

Cybernetics is an interdisciplinary field that cuts across various natural and social sciences. Its origin is usually traced to the publication of an influential paper in 1943, "Behavior, Purpose and Teleology," by American mathematician Norbert Wiener, American electrical engineer Julian Bigelow, and Mexican neurophysiologist Arturo Rosenbleuth. In the paper, they established a clear link between animate behavior and that of feedback-control systems.

The term "cybernetics" was applied to the field by Wiener in his 1948 book *Cybernetics, or Control and Communication in the Animal and Machine.* The word derives from the Greek *kybernetes,* meaning "one who steers." After the publication of the book, the term was picked up by others studying similar principles, and it became widely applied to many studies of communication and control.

Wiener expanded the influence of cybernetics by exploring the social implications of its principles in his book *The Human Use of Human Beings* (1950), now considered the seminal work in the field. The book notes the application of cybernetic principles to social institutions and draws analogies between automated machines and human institutions.

The study of cybernetics continued to grow, and the American Society for Cybernetics was founded in 1964. Cybernetics discoveries have been influential in a wide range of scientific fields, including artificial intelligence, computing, sociology, and medicine.

Beth A. Kattelman

Sources

Ashby, W. Ross. *An Introduction to Cybernetics.* London: Chapman and Hall, 1956.

Cordeschi, Roberto. *The Discovery of the Artificial: Behavior, Mind, and Machines Before and Beyond Cybernetics.* Boston: Kluwer Academic, 2002.

Rose, John. *The Cybernetic Revolution.* New York: Barnes and Noble, 1974.

ENIAC

The Electronic Numerical Integrator and Computer, or ENIAC, was the world's first digital electronic computer. This significant advance in computer design is credited to John Adam Presper Eckert, Jr., and John William Mauchly. Eckert, an electrical engineer, and Mauchly, a physicist interested in meteorology, met in 1941 and turned a shared interest in electronic counting circuits into a U.S. Army contract to develop a new electronic computer.

The U.S. Army needed artillery fire tables for quick battlefield computation of shell trajectories, but none of the contemporary mechanical and electronic computers were fast enough to meet its requirements. Construction on ENIAC began at the University of Pennsylvania's Moore School of Electrical Engineering in July 1943. The 30 ton, 1,800 square foot computer was ready for operation in fall 1945, but it was not unveiled publicly until early 1946.

To meet the needs of the U.S. Army, Eckert and Mauchly incorporated two major breakthroughs in their ENIAC design: digital circuits and vacuum tube relays. Analog electronic machines stored numbers by reading electronic pulses. Ten pulses equaled the number ten. The digital circuits of ENIAC counted the number ten with one pulse, a pulse in the tens-digit circuit. Seventy-two would not be seventy-two pulses, as in the analog systems, but seven pulses in the tens-digit circuit and two pulses in the ones-digit circuit. ENIAC could store twenty ten-digit numbers with accompanying positive/negative signs. The machine could handle basic arithmetic operations and, by plugging in different circuit panels, could process data using various mathematical functions.

Advances in design gave ENIAC a computation speed estimated at 500 times that of contemporary analog computers. It was able to process 5,000 addition operations per paired circuit per second. Multiplication and other operations took a bit longer. Further increases in speed were derived from the use of 17,468 vacuum tubes instead of the standard mechanical relay switches. Because programming ENIAC for a specific problem involved setting as many as 3,000 switches, it could take as long as two months to program the system.

The computer itself was never used for its intended purpose, as the war ended before it was fully functional. It was used, however, in atomic weapons research, including some associated ballistics work, until it was removed from service in 1955.

Eckert and Mauchly did not profit significantly from their key role in the design of ENIAC. In fact, their contributions were variously usurped, challenged, or simply forgotten over the years. They did market the ENIAC design and developed other more advanced computers, such as BINAC and UNIVAC, but they did not succeed as entrepreneurs.

Patent rights for some of ENIAC's design advances were assigned in 1947 to the Eckert-Mauchly Computer Company, which proved

ENIAC—the world's first operational, large-scale, electronic digital computer, unveiled in 1946—occupied 1,800 square feet (167 square meters), weighed thirty tons, and contained nearly 18,000 vacuum tubes. *(Time & Life Pictures/Getty Images)*

unsuccessful. In 1955, the patents came into the possession of Sperry Corporation, prompting a series of court battles between Sperry and other computer corporations who were using equipment based on the ENIAC design. A federal court decision in *Honeywell v. Sperry* (1973) invalidated the Eckert-Mauchly patents to avoid creating a monopoly in the increasingly important computer industry. These legal and patent conflicts further undermined the credit owed to Eckert and Mauchly for their design.

Paul Buckingham

Sources

McCartney, Scott. *ENIAC: The Triumphs and Tragedies of the World's First Computer.* New York: Berkley, 2001.

Metropolis, Nicholas, Jack Howlett, and Gian-Carlo Rota, eds. *A History of Computing in the Twentieth Century: A Collection of Essays.* New York: Academic Press, 1980.

Norberg, Arthur L. *Computers and Commerce: A Study of Technology and Management at Eckert-Mauchly Computer Company, Engineering Research Associates, and Remington Rand, 1946–1857.* Cambridge, MA: MIT Press, 2005.

FARRAR, JOHN (1779–1853)

The Harvard professor of mathematics and natural philosophy John Farrar was born in Lincoln, Massachusetts, on July 1, 1779. He attended Phillips Academy on scholarship before entering Harvard College in 1799. He distinguished himself at both ends of the classroom, teaching in common schools during winter breaks and winning John Bonnycastle's *An Introduction to Astronomy* (1786) in an academic contest in 1802.

After Farrar graduated from Harvard in 1803, his financial patron, Mrs. Samuel Phillips, paid his tuition to Andover Theological Seminary, but the Harvard Overseers interrupted his divinity career by hiring him as a tutor of Greek in 1805. Two years later, they promoted him to Hollis Professor of Mathematics and Natural Philosophy.

Farrar spent the rest of his career in this professorship. As was standard practice, he supervised the mathematics tutors and delivered lectures on natural philosophy to the juniors and seniors. He was considered a caring and engaging instructor who displayed a flair for showmanship with the physical and electrical experiments he performed during class. He restored the prestige the chair had known under John Winthrop IV by collecting meteorological and astronomical data, purchasing scientific instruments, and researching the possibility of constructing an observatory in Cambridge. He associated professionally and socially with liberal Unitarians, including the Transcendentalist circle.

Farrar's most significant contribution was in refocusing American collegiate mathematics and science education. When he became Hollis Professor, Harvard students learned all of their mathematics by memorizing and reciting portions of Webber's *Mathematics* (1801), which was derived from outdated English textbooks, and Euclid's *Elements of Geometry* (1482). While Yale's Jeremiah Day included French mathematics textbooks among his sources in the series he prepared in the 1810s, Farrar recommended that students learn directly from the treatises written by Silvestre-François Lacroix and Adrien-Marie Legendre in the 1790s. These books appeared to be superior, because they were analytical in style, emphasizing algebraic techniques and a "natural" mode of laying out mathematical principles in historical order, while incorporating recent discoveries in mathematics. In general, Farrar argued, these works were more successful at challenging the student mind to develop disciplined patterns of thinking.

Farrar published seven volumes in the Cambridge Series of Mathematics between 1818 and 1824; various Harvard tutors seem to have done most of the literal translation work. Farrar followed this series with the five-volume Cambridge Series of Natural Philosophy (1825–1827), which was mainly a translation of several works by Jean-Baptiste Biot.

Farrar's mathematics textbooks were quickly adopted throughout the United States at institutions ranging from the U.S. Military Academy to Bowdoin College and the University of Virginia. They helped create an increased emphasis on science and mathematics in the nineteenth-century college curriculum and led to the establishment of mathematics prerequisites for college admission. They also inspired other professors to prepare mathematics textbooks that melded British and French influences into a uniquely

American style. Textbook series such as those compiled by Charles Davies surpassed Farrar's series in popularity during the 1830s.

The physical problems and nervous ailments Farrar suffered throughout his life forced him to take a leave of absence from Harvard in 1831–1832 and to retire completely in 1836. He traveled to Europe for his health, visiting scientists in France, England, and Scotland. He was married twice and had no children. After 1840, he lived in seclusion in Cambridge, where he died on May 8, 1853.

Although his scientific and mathematical achievements were overshadowed by younger, professional mathematicians and scientists, such as Benjamin Peirce, Farrar prepared the intellectual and cultural setting in which the next generation worked.

Amy Ackerberg-Hastings

Source

Palfrey, John Gorham. "Professor Farrar." *Christian Examiner* 55 (1853): 121–36.

FORRESTER, JAY WRIGHT (1918–)

The American electrical engineer Jay Wright Forrester, a pioneer in computer engineering and management, developed random access magnetic core memory (the predecessor of random access memory, or RAM, used in modern computers) as well as system dynamics, which deals with the simulation of interactions between objects in dynamic systems. He is the author of multiple books and papers on system principles and dynamics, and he is currently Germeshausen Professor Emeritus and Senior Lecturer at the Massachusetts Institute of Technology (MIT) Sloan School of Management.

Born in Nebraska on July 14, 1918, Forrester grew up on a cattle ranch. While still in high school, he built a wind-driven, 12-volt electric plant from old automobile parts that provided the ranch's first electricity. He graduated from the University of Nebraska with a B.S. in electrical engineering, then attended MIT. Under the mentorship of Gordon S. Brown, a pioneer in feedback control systems, he studied electric and hydraulic servomechanisms (control systems) for radar antennas and gun mounts.

In 1945, Forrester began to lead the development of a new aircraft stability and control analyzer for the U.S. Navy. The navy wanted the ability to test the effects of aircraft design changes by simulating aircraft behavior, including a plane's real-time response to pilot actions. The project initially called for development of an analog computer, but to meet real-time performance demands, Forrester constructed an experimental digital computer, inspired by the U.S. Army's ENIAC computer.

Known as Whirlwind, the new digital computer occupied 25,000 square feet and contained thousands of vacuum tubes, each with a life expectancy of approximately 500 hours. Forrester discovered that he could prolong tube life to 500,000 hours through the use of silicon-free cathode material. He further enhanced reliability by installing a system to automatically check for components showing early indications of failure, thus allowing operators to repair or replace a component before an error occurred. Despite these improvements, memory was still a problem, so Forrester began work on an eventual replacement for the slow one-dimensional mercury delay line and the faster, but much less dependable, two-dimensional Williams tube.

In 1949, as the U.S. military was losing interest and confidence in the Whirlwind project, the Soviets developed an atomic weapon. This created an urgent demand for an air-defense system that could identify and intercept incoming aircraft or missiles. Forrester was put in charge of implementing Whirlwind as the heart of a new defense system called SAGE (Semi-Automatic Ground Environment). By 1953, SAGE had the capability of monitoring forty-eight aircraft simultaneously. That year, Whirlwind's memory was replaced with Forrester's new three-dimensional memory, which was made of a magnetic ferrite core combined with a wire grid. This innovation doubled Whirlwind's speed while improving reliability and lowering costs. Forrester left the project in 1956, two years before SAGE began operation. It was used as a U.S. air-defense system until the 1980s.

As a professor at MIT's Sloan School of Management, Forrester continued to break new ground by applying his engineering view of electrical systems to the field of human systems. Focusing on case studies of organizational policy, he used computer simulations to analyze social systems and predict the implications and outcomes of different models. This methodology, called system dynamics, has been applied to a wide range of problems, including urbanization, industrialization, energy and the environment, and biological and medical modeling.

Glenda Turner

Sources

Forrester, J.W. *Industrial Dynamics.* Cambridge, MA: Productivity, 1961.
System Dynamics Society. http://www.systemdynamics.org.

GIBBS, JOSIAH WILLARD (1839–1903)

In addition to his major discoveries in chemistry and physics, Josiah Willard Gibbs is known for his work in mathematics, where he is credited with the invention of vector analysis, a form of calculus that explains the movement of objects through space. Gibbs was born on February 11, 1839, in New Haven, Connecticut. In 1854, he entered Yale College (now Yale University), where his father was a professor of sacred literature. Nine years later, Gibbs graduated with the first engineering doctorate—one of the first doctorates of any kind—awarded by an American institution of higher learning. From 1866 to 1869, he did postdoctoral work at various institutions in Paris and Berlin, ending up at the renowned University of Heidelberg in Germany, then a leading center for the study of chemistry and physics.

Upon his return to the United States, Gibbs began teaching at Yale and was appointed professor of mathematical physics in 1871. Gibbs's appointment was unorthodox in that the young scientist, then thirty-two, had never published a scholarly paper. Over the next seven years, however, he published a series of papers cumulatively titled *On the Equilibrium of Heterogeneous Substances*; this work is regarded by historians of science as one of the most important in nineteenth-century physics and mathematics. In these papers, Gibbs applied the principles of thermodynamics—the branch of physics that studies the effect of temperature, pressure, and volume on physical systems—to the understanding of chemical reactions.

In the 1880s, Gibbs began his work on vector analysis, also known as vector calculus. (In mathematics, a vector is a number combined with a direction.) His findings were recorded largely in the form of unpublished notes that he used for teaching his students. Not until 1901 did one of his students compile the notes into a scholarly article that was published. Even as Gibbs was pioneering vector analysis, he was also writing on optics, the branch of physics that describes the property of behavior of light, and statistical mechanics, which applies probability theory to the study of particles subjected to force. In the 1890s, Gibbs conducted research into crystallography, the study of the atomic arrangement of solids. He also applied his theories of vector analysis to astronomy, using it to more precisely determine the orbits of planets and comets.

For much of his life, Gibbs's work in all of these fields went largely unrecognized. This was partly because European scientists, then at the forefront of physics and mathematics, considered America a scientific backwater, and partly because Gibbs published his work in the *Transactions of the Connecticut Academy of Sciences,* an obscure journal barely read in the United States and virtually unknown in Europe. By the 1890s, his writings were being translated and published in French and German scientific journals, but by then several European scientists had duplicated some of his work.

Gibbs never married. He lived with his sister and her husband in New Haven for most of his life. He died on April 28, 1903.

James Ciment

Sources

Rukeyser, Muriel. *Willard Gibbs.* Garden City, NY: Doubleday, Doran, 1942.
Wheeler, Lynde Phelps. *Josiah Willard Gibbs: The History of a Great Mind.* New Haven, CT: Yale University Press, 1952.

GORENSTEIN, DANIEL (1923–1992)

Daniel Gorenstein worked in algebraic geometry, which uses abstract algebraic equations to solve problems in the geometric mathematics of points, lines, curves, and surfaces. Known for his ability to render complex mathematics into accessible prose, he authored numerous works on algebraic geometry, including *Finite Groups* (1967) and *Finite Simple Groups: An Introduction to Their Classification* (1982).

Gorenstein was born in Boston on January 1, 1923. From an early age, he showed an aptitude for mathematics, teaching himself the basics of calculus by the time he was twelve. His formal education took place at the prestigious Boston Latin School, the oldest secondary school in America, and at Harvard University, where he earned a bachelor's degree in 1943.

During World War II, Gorenstein accepted a position at Harvard teaching mathematics to army personnel. After the war, he stayed at Harvard as a graduate student. Working under Oscar Zariski, a pioneer in the development of algebraic geometry, Gorenstein earned a Ph.D. in mathematics in 1950.

Over the course of four decades, Gorenstein taught at a number of institutions of higher learning, including Clark University (1951–1964, with a year's hiatus to teach at Cornell in 1958–1959), Northeastern University (1964–1969), and Rutgers University (1969–1992), where he was chair of the mathematics department from 1975 to 1981. From 1989, he also served as director of the National Science Foundation Science Technology Center in Discrete Mathematics and Theoretical Computer Science, a joint project of Rutgers and Bell Laboratories.

Gorenstein's first major contribution to the field of algebraic geometry came with his doctoral dissertation, in which he introduced the concept of the Gorenstein ring, a key element in commutative ring theory. But he is best remembered for his work in the classification of finite simple groups.

In 1960 and 1961, he attended the University of Chicago's Group Theory Year, a symposium of scientists dedicated to resolving fundamental problems in the group theory subfield of algebraic geometry. Gorenstein provided the overall leadership and intellectual guidance to this symposium of scientists from around the world, an extraordinary achievement in that the final proof offered by the conference ran to some 10,000 pages in about 500 journal articles by more than 100 scholars. The Group Theory Year is regarded by most scholars in the field as one of the crowning achievements of twentieth-century mathematics.

Over the years, Gorenstein earned many honors. He was a Guggenheim Fellow, a Fulbright Scholar, and a member of the National Academy of Sciences and the American Academy of Arts and Sciences, and he won the American Mathematical Society's Steel Prize for mathematical exposition in 1989. He died on August 26, 1992.

James Ciment

Sources

Christensen, Lars Winther. *Gorenstein Dimensions.* New York: Springer, 2000.
Gorenstein, Daniel. *Finite Groups.* New York: Harper and Row, 1967.

HOLLERITH, HERMAN (1860–1929)

An innovative American statistician, engineer, and early computer scientist, Herman Hollerith pioneered modern statistical analysis and information processing with his automatic tabulating machines. He revolutionized the process by which the U.S. Census was tabulated by implementing a system of punch cards and mechanical readers to analyze millions of pieces of data that previously had to be hand calculated. Hollerith also founded the Tabulating Machine Company, which eventually became the International Business Machine Corporation (IBM).

Herman Hollerith was born on February 29, 1860, in Buffalo, New York, to Franciska Brunn and Johann Hollerith, who had emigrated from Germany in 1848. He was a bright child but struggled to learn spelling; a determined teacher

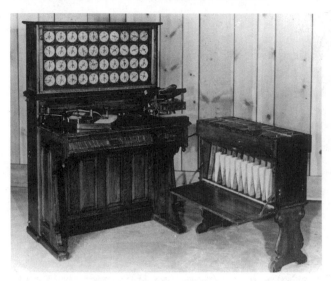

Herman Hollerith invented a mechanical tabulating machine, used in the 1890 U.S. Census, based on punch cards for recording data. Punch cards became a basic input mechanism for later digital computers. His firm eventually merged with two others to form IBM. *(Hulton Archive/Getty Images)*

made his life miserable with constant efforts to improve his spelling, causing him to skip school whenever possible. As a result of these problems in school, he was taught at home by the family's Lutheran minister, an environment in which he excelled. Hollerith entered New York City College in 1875 at the age of fifteen and the School of Mines of Columbia University in New York two years later.

After graduating with a degree in mining engineering in 1879, he went to work as an assistant to his professor, William Trowbridge, at Columbia. Trowbridge subsequently was appointed chief special agent for the U.S. Census Bureau in Washington, D.C., heading up the data analysis section, and he asked Hollerith to join him as a statistician.

At the Census Bureau, Hollerith became acquainted with John Shaw Billings, the director of vital statistics for the 1880 census. The nation's population boom and the addition of new questions to the census that year posed enormous data collection and analysis problems. Billings and Hollerith discussed the possibility of using machines and punch cards to automate the processing of census data—an idea that Hollerith enthusiastically pursued.

In the fall of 1882, Hollerith became a lecturer on mechanical engineering at the Massachusetts Institute of Technology (MIT) in Cambridge, devoting his nonteaching time to designing a mechanical system to more efficiently sort and compile data. One of his first prototypes employed a roll of paper tape with holes in predetermined locations. The paper would roll across a mechanical spool with needles. When the needles came to a hole, an electric circuit would be completed and a number automatically registered.

Returning to Washington after a year, Hollerith became an examiner at the U.S. Patent Office, but he left that position after building his first working computing device. On September 23, 1884, he filed a patent for a machine that tabulated data using paper tape and a roller with pins. An electrical current was created when the pins passed through the holes and into mercury located underneath; the electrical charge activated a mechanical counter. Later, Hollerith replaced the paper tape with punch cards as a means to record census data. Each punch card represented one person, and each hole on the card indicated data such as birth date, age, sex, occupation, and race. Presorted punch cards allowed for a logical sequence of computations.

On June 8, 1887, Hollerith took out a patent for a machine-driven punch-card sorter. To prove that the system worked, he offered to organize the voluminous and disarrayed health records of the city of Baltimore, Maryland. Although he had to punch the data manually onto the cards, the Baltimore trial succeeded. When the results were repeated in the compilation of statistics for the health departments of the state of New Jersey and the city of New York, Hollerith drew the attention of the newspapers.

Hollerith modified the punch cards to accommodate complex data. On December 9, 1888, he convinced the U.S. surgeon general's office to use his machines for their data management. His system was unveiled at the Paris Exposition of 1889, and it won the gold medal; it also won the bronze medal at the Chicago world's fair of 1893.

The 1880 U.S. Census, which was manually processed, was not completed until 1888, by which time the statistics had become obsolete. The problems of a larger population and more extensive questionnaire were compounded by the difficulty of collecting data from the new, far-flung Western states. To address these issue, the

Census Bureau held a competition for a system that would reduce the effort in collecting and processing the population and demographic data. Hollerith submitted his system and won the contract for the 1890 census.

The Hollerith Electric Tabulating System, based on the reading and sorting of punch cards, completed all data processing for the 1890 census in just six weeks and saved the federal government $5 million over the 1880 effort. For his invention, Hollerith received the Elliot Cresson Medal, awarded for outstanding achievements in science and technology, from the Franklin Institute of Philadelphia in 1890. That same year, he was awarded a doctorate in Engineering from Columbia University in New York.

Hollerith's Tabulating Machine Company was incorporated in New Jersey on December 3, 1896. With successive tabulators, his invention became a commercial success, especially with foreign governments, railroad, and life insurance companies. Hollerith also invented the first key-punch device, allowing manual entry into a data machine and an automatic card mechanism that fed a stack of cards into a reader.

Hollerith did the data tabulation for the 1900 census as well, but the high cost of his machines led the federal government to build its own system for the 1910 census. Hollerith sued the government for patent infringement, but he ultimately lost the seven-year litigation.

In 1911, the Tabulating Machine Company merged with Computing Scale and Bundy Manufacturing to form the Computing Tabulating Recording (CTR) Corporation. When Hollerith sold his shares and retired in 1921, he was a millionaire. With his wife, Lucia Talcott, he raised cattle near the Chesapeake Bay, in Maryland. In February 1924, CTR changed its name to International Business Machines. Hollerith died of a heart attack on November 17, 1929.

James Fargo Balliett and Santi S. Chanthaphavong

Sources

Austrian, Geoffrey D. *Herman Hollerith: Forgotten Giant of Information Processing.* New York: Columbia University Press, 1982.

Kidwell, Peggy A., and Paul E. Ceruzzi. *Landmarks in Digital Computing: A Smithsonian Pictorial History.* Washington, DC: Smithsonian Institution, 1994.

Shurkin, Joel N. *Engines of the Mind: A History of the Computer.* New York: W.W. Norton, 1984.

LEVINSON, NORMAN (1912–1975)

Norman Levinson was a mathematician who worked in the fields of number theory, nonlinear differential equations, and complex analysis.

He was born on August 11, 1912, in Lynn, Massachusetts. By 1934, he had completed both his B.S. and M.S. degrees in electrical engineering from the Massachusetts Institute of Technology (MIT). In his studies, he especially excelled in mathematics, and his abilities in this field, along with the support and encouragement of his mentor and mathematics professor, Norbert Wiener, led him to apply to the MIT doctoral program in mathematics.

Because of his exceptional work to this point, Levinson earned a traveling fellowship from the MIT mathematics department that allowed him to study at Cambridge University in England. Based on the papers he produced at Cambridge, MIT granted Levinson a Doctor of Science degree in mathematics in 1935. He then earned a National Research Council Fellowship, which he used to study at Princeton's Institute for Advanced Study for nearly two years. The fellowship ended early when Levinson was appointed as an instructor in mathematics at MIT in 1937, beginning a thirty-eight-year teaching career.

Levinson published *Gap and Density Theorems* in 1940 as part of a prestigious American Mathematical Society Colloquium series usually reserved for senior scholars. His 1955 *Theory of Ordinary Differential Equations,* written with Earl Coddington, was a widely used course text. Levinson also worked in geophysics and signal processing, a key field dedicated to studying the amplification and interpretation of radio and other electromagnetic signals.

Besides his contributions to mathematics, Levinson was also embroiled in the politics of his time. As a Jew in the New England academic community, he faced resistance and, at times, anti-Semitism. His personal feelings opposing racial discrimination and anti-Semitism influenced his decision to join the American Communist Party in 1931. He left the party after eleven years, because he wished to focus on social equality, not Stalinism.

His membership in the Communist Party came back to haunt him years later during the Red Scare of the 1950s when federal investigators began to look into the affairs of high-profile college faculty, especially those involved in research in sensitive areas such as signal processing. During Levinson's testimony before congressional committees investigating the Army Signal Corps, he was forthright about his involvement in the Communist Party and managed to avoid lasting harm to his career. He continued during the next two decades, as one colleague put it, as the heart of the mathematics community at MIT.

Levinson's awards included a Guggenheim Fellowship in 1948, the 1953 Bocher Prize for his work in differential equations, and the 1971 Chauvenet Prize for a paper on the prime number theorem. He was in the midst of an important set of projects related to the Riemann hypothesis, one of the great unsolved problems in mathematics, when he died on October 10, 1975.

Paul Buckingham

Source

Levinson, Norman. *Selected Papers of Norman Levinson.* 2 vols. Ed. John A. Nohel and David H. Sattinger. Boston, MA: Birkhäuser, 1998.

MARK I

The Harvard Mark I computer was the first widely known, fully automatic (program-controlled) electromechanical calculating device. Also known as the IBM-Harvard Automatic Sequence Controlled Calculator, the Mark I was commissioned in 1944. The computer was the brainchild of Howard Hathaway Aiken, a graduate student in theoretical physics at Harvard University. Aiken's completed machine, nicknamed "Bessie" by its operators, cost his sponsors—International Business Machines (IBM) and the U.S. Navy—nearly half a million dollars.

Aiken proposed the development of a digital mechanical calculating device to solve certain complex nonlinear differential equations—called Bessel functions—in his 1939 thesis on vacuum tube design. Analog computers then available, such as Vannevar Bush's Differential Analyzer, had great difficulty calculating such functions, which had no exact solutions. Aiken envisioned "a switchboard on which are mounted various pieces of calculating machine apparatus. Each panel of the switchboard is given over to definite mathematical operations." Aiken later claimed that his device was inspired by a chance encounter with a remnant of Charles Babbage's famous nineteenth-century mechanical computer, the Difference Engine.

The completed Mark I was the largest electromechanical computer ever built. It stood 8 feet high, 51 feet long, and 2 feet deep. It weighed five tons, including its 750,000 individual parts and 500 miles of wiring. According to one observer, the computer made as much noise as "a roomful of ladies knitting." The Mark I could add, subtract, multiply, and divide. Each addition or subtraction operation took one-third of a second, each multiplication took six seconds, and each division took fifteen seconds. Mathematical operations could be calculated to twenty-three significant figures.

Critics downplay the Mark I as stillborn technology (almost immediately obsolete), but it provided a valuable architecture for study and practice. One of the machine's chief programmers was naval officer Grace Murray Hopper, who created her first program on the Mark I. That program calculated coefficients for the interpolation of the arc tangent for Harvard's Bureau of Ships Computation Project. Hopper wrote the definitive *Manual of Operation for the Automatic Sequence Controlled Calculator* (1946) and was the chief developer of Flow-matic, an early English-like higher-level programming language. Several of the basic commands introduced in Flow-matic, including "add," "execute," and "stop," were later incorporated into the popular computer language COBOL.

Hopper also overcame one of the chief limitations of the computer, the lack of conditional branches, by reusing loops of punched paper tape encoded with instructions. Still, because the Mark I had no memory by which to store programs internally, Hopper complained that she was forced to start nearly from scratch each day.

The Mark I computer was also used by the Russian American economist Wassily Leontief in developing the input-output method of economic analysis, which earned him a Nobel Prize in 1973. Astronomer James Baker used the machine to design telephoto lenses for the U.S. Army Air Corps for reconnaissance purposes. John von Neumann, who developed today's most common computer architecture, visited the Mark I—his first brush with digital computing—to feed it implosion calculations in the development of the first atomic bomb. And the U.S. Navy's Bureau of Ordnance Computation Project used the machine to compute ballistics tables for gunnery range-finding and fire control.

Aiken dismantled the Mark I computer after fifteen years of service to Harvard, but he also developed several successors—Mark II through IV—each increasing in reliability as solid-state components replaced older electromechanical relay and vacuum tube technology. The Mark computers and their Harvard architecture did not fare well with the advent of the von Neumann architecture, which made reading, writing, and storage of programs a feasible process, and they were defunct by the 1960s.

Philip Frana

Sources

Bashe, Charles. "Constructing the IBM ASCC (Harvard Mark I)." In *Makin' Numbers: Howard Aiken and the Computer,* ed. I. Bernard Cohen and Gregory W. Welch. Cambridge, MA: MIT Press, 1999.

Cohen, I. Bernard. *Howard Aiken: Portrait of a Computer Pioneer.* Cambridge, MA: MIT Press, 1999.

Oettinger, Anthony G. "Howard Aiken." *Communications of the ACM* 5:6 (1962): 298–99.

MICROSOFT

The world's largest computer software company, the Microsoft Corporation was founded in Albuquerque, New Mexico, on April 4, 1975, by William "Bill" Gates, Jr., and Paul Gardner Allen, two young computer technicians and entrepreneurs from Seattle, Washington. Four years later, the company relocated to the Seattle suburb of Bellevue, and in 1986 moved to its current headquarters in Redmond.

While technology historians generally do not consider Microsoft a major software innovator, the company has nevertheless been expert at developing and marketing inexpensive and adaptable computer technologies, most notably MS-DOS,

The Windows operating system made Microsoft the world's dominant software provider and Bill Gates, its co-founder and chairman, the world's richest man. *(Carol Halebian/Getty Images)*

the Windows operating system, and the set of office software products that includes the MS Word word processing program, the Excel spreadsheet program, and the PowerPoint visual presentation system. Over the years, Microsoft expanded into other areas, such as computer gaming and television broadcasting.

Microsoft got its start adapting an early computer programming language known as BASIC (Beginner's All-purpose Symbolic Instruction Code) for use in the Altair 8800, one of the first personal computers. But the company's major breakthrough came with the creation of MS-DOS in 1981, which was designed for use in the first personal computer built by International Business Machines (IBM), then the world's largest manufacturer of mainframes. As with BASIC and Altair, Microsoft adapted an existing software program, a version of QDOS (Quick and Dirty Operating System) from another small software company and converted it for use in the IBM personal computer.

From their experience with mainframes, IBM executives believed that the largest share of revenues in the new personal computer market would come from hardware; thus, it did not bother buying exclusive rights to the use of what it called PC-DOS. This proved to be a mistake with enormous consequences. Other companies soon built cheaper versions of the IBM PC, but they had to install software that was compatible with the IBM PC. The result was rapid growth for Microsoft, as MS-DOS, the name it gave its operating system, became the standard for most of the exploding personal computer market.

At the same time, a major early competitor, Apple Computers, was developing its own proprietary operating system. While MS-DOS required users to learn code to operate their PCs, Apple's software used a graphic interface: Users could point and click at icons on the screen to perform simple tasks. In 1985, Microsoft introduced its Windows operating system, which employed a graphic interface similar to Apple's. With its greater financial resources and marketing savvy, Microsoft succeeded in marginalizing Apple to a small share of the personal computer market.

In subsequent years, Microsoft continued to piggyback on existing technologies, using its enormous financial resources and market muscle to imitate a competitor's product and then situate its own as the industry standard. In the late 1980s, Excel and MS Word sidelined Novell's Lotus spreadsheet program and Corel's WordPerfect word processing program, respectively. Explorer browsing software overwhelmed Netscape in the mid-1990s.

Microsoft's aggressive marketing, as well as the ubiquity of its operating systems, eventually led to antitrust action by the U.S. Department of Justice and the European Union. Microsoft was forced to take remedial action and had to pay substantial damages to competitors as a result of court rulings, but the corporation has not been broken up or forced to divest itself of major assets. Many computer experts, however, point to one new threat to Microsoft's future—the ability of computer users to avoid buying operating systems altogether and use those increasingly available over the Internet.

In the early 2000s, the company had revenues of nearly $50 billion annually and employed in excess of 75,000 employees in more than 100 countries. Arguably, no technology firm has done more to make the personal computer (PC) a ubiquitous part of everyday life.

James Ciment

Sources

Manes, Stephen, and Paul Andrews. *Gates: How Microsoft's Mogul Reinvented an Industry and Made Himself the Richest Man in America.* New York: Simon and Schuster, 1994.
Wallace, James, and Jim Erickson. *Hard Drive: Bill Gates and the Making of the Microsoft Empire.* New York: HarperBusiness, 1993.

NUMBER THEORY

Number theory is a branch of mathematics that deals with the properties of numbers and the intrinsic powers of integers (whole numbers and zero). It draws on a range of mathematical fields—from algebra and statistics to advanced calculus and quadratic equations—depending on the type of problem being solved. One of the largest branches of pure mathematics, it has proven useful to problem-solving in computing, physics, chemistry, biology, and astronomy.

Mathematicians sometimes use the terms "number theory," "arithmetic," and "higher arithmetic" interchangeably, but number theory is different from simple arithmetic. Number theory has a rich history, challenging mathematicians with complex problems that have gone unanswered for 2,000 years and more. It has grown in prominence and importance in recent years, especially with the development of cryptography and statistical mechanics.

In nineteenth-century America, a handful of mathematicians took on segments of number theory in their work. Harvard professor Benjamin Peirce became a major figure in the history of number theory by proving that there is no odd perfect number with fewer than four distinct prime factors. Peirce's introductory textbooks in algebra became standards, and his more advanced *A System of Analytical Mechanics* (1855) was considered the most important work in mathematics published in the United States to that time.

At the University of Chicago, mathematics professor Eliakim Moore did groundbreaking work in abstract algebra, the foundations of geometry, and integral equations. In 1893, while studying algebraic structures and groups, he proved that every finite field is a Galois field (named after French mathematician Évariste Galois)—that is, it contains a finite number of elements. In the early 1900s, Moore reformulated Hilbert's axioms, a set of twenty geometry assumptions devised by German mathematician David Hilbert in 1899, to turn the original undefined lines, points, and planes into undefined points only. Moore's *On the Projective Axioms of Geometry* (1902) revealed that Hilbert's axioms contained redundancies.

Moore's students included such leading American mathematicians of the next generation as George David Birkhoff, Anna Wheeler, Oswald Veblen, and Leonard Dickinson. Birkhoff, whose work encompassed many areas of mathematics, is best remembered for his contribution to ergodic theory, a foundation of contemporary chaos theory pertaining to statistical physics.

Anna Wheeler, who had also worked with David Hilbert, was known for her work on integral equations with an emphasis on infinite dimensional linear spaces. Wheeler was the first woman to give the Colloquium Lectures at the American Mathematical Society meetings, in 1927. She was one of the first mathematicians to work in the area known as "functional analysis."

Oswald Veblen was one of the founders and the first professor of the Institute for Advanced Study at Princeton University (1930), where he played a key role in recruiting such luminaries as Albert Einstein and John von Neumann. Veblen was also accomplished in the field of algebraic topology and worked in relativity as well.

Leonard Dickson, who became a mathematics instructor at the University of Chicago in 1900, did important work in abstract algebra and spent much of his career on number theory. His three-volume *History of the Theory of Numbers* (1919–1923) covers every significant number theory from the dawn of mathematics to the 1920s, including those pertaining to divisibility, prime numbers, Diophantine analysis, and quadratic equations.

Julia Robinson, a mathematician at the University of California, Berkley, contributed significant advances in the 1940s that led to the solution in 1970 of the number theory problem known as Hilbert's Tenth Problem—a Diophantine equation (polynomial equation with integral coefficients and only integral solutions) written by David Hilbert in 1900. Robinson was the first woman elected to the mathematical section of the National Academy of Science, in 1975,

and she was the second woman, after Wheeler, to give the Colloquium Lectures at the American Mathematical Society meetings, in 1980.

In 1995, Princeton mathematician Andrew Wiles and Harvard professor Richard Taylor solved one of the longest-standing problems in mathematics by providing a proof of Fermat's Last Theorem. The theory, devised by the seventeenth-century French government official and amateur mathematician Pierre de Fermat, states: "It is impossible to separate any power higher than the second into two like powers." In a notebook, Fermat had written that he had devised a proof of this theory, but that it was too large to fit in the margin. For 350 years, various number theorists tried to work out a proof for the theorem. Wiles thought he found a solution in 1993, and he gave a series of lectures about his discovery, but then he identified an error and spent two more years collaborating with former student Richard Taylor to complete the proof using algebraic geometry.

Dan Goldstein, working at the Institute of Mathematics in Palo Alto, California, in 2003 achieved a breakthrough on another long-standing problem in number theory—proving the "twin prime conjecture." Proposed by the ancient Greek mathematician Euclid about 300 B.C.E., the theory states that there are an infinite number of "twin primes," or prime numbers that differ by two. (Prime numbers cannot be divided by any number smaller than themselves, other than 1, without leaving a remainder. Those that differ by two include 3, 5, 7, 11, and so on.) Goldstein's findings brought the math world a few steps closer to understanding the frequency and location of prime number patterns, in particular advancing the knowledge of how prime numbers, especially larger ones, are distributed. Meanwhile, several advanced computing projects were under way to find the largest twin prime number; as of early 2007, the largest on record was 58,711 digits long.

Number theory, long an esoteric field even in academic circles, holds a central position among American mathematicians and theoretical scientists in the twenty-first century, often involving intense competition, international collaboration, and the help of advanced computing systems. Researchers often work in secret for years to solve problems that have frustrated number theorists for centuries, though practical applications have abounded. The attributes of prime numbers and factors are of particular importance in the development of security codes for access to and transmission of sensitive database information.

James Fargo Balliett

Sources

Aczel, A.D. *Fermat's Last Theorem: Unlocking the Secret of an Ancient Mathematical Problem.* New York: Penguin, 1996.

Artemiadis, Nikolaos K. *History of Mathematics: From a Mathematician's Vantage Point.* Providence, RI: American Mathematical Society, 2004.

Yan, Song Y. *Number Theory for Computing.* Berlin: Springer-Verlag, 2000.

PEIRCE, BENJAMIN (1809–1880)

The mathematician, astronomer, and educator Benjamin Peirce—the father of logician and philosopher Charles Sanders Peirce—was born on April 4, 1809, into a prominent Massachusetts family. His father served in the state legislature and later worked as the librarian at Harvard College.

Peirce entered Harvard in 1825 and graduated with an A.B. degree in 1829. He befriended the son of mathematician and astronomer Nathaniel Bowditch and assisted Bowditch with the translation and editing of Pierre-Simon Laplace's *Mécanique Céleste* (*Celestial Mechanics*, 1829–1839). He taught at the Round Hill School in Northampton, Massachusetts, from 1829 to 1831, before being appointed tutor of mathematics at Harvard. He received a master's degree there in 1833.

By this time, Hollis Professor of Mathematics and Natural Philosophy John Farrar was increasingly unable to fulfill his duties. In 1833, Peirce was appointed to the new and unendowed position of university professor of mathematics and natural philosophy. He became known as an enthusiastic teacher who was difficult to understand; similarly, only the most capable students could follow the mathematics textbook series he prepared in the 1830s.

In 1842, Peirce was appointed the first Perkins Professor of Astronomy and Mathematics, an endowed position that he held until his death. His interest in tracking students so that only those truly interested in mathematics would take advanced courses helped lead to the founding of the Lawrence Scientific School, founded at Harvard University in 1847.

Peirce conducted research in mathematics and astronomy. He published lectures on the 1843 comet, edited the mathematical sections of the *American Almanac and Repository of Useful Knowledge,* and was appointed director of longitude determination of the U.S. Coast Survey in 1852. He was consulting astronomer for the *American Ephemeris and Nautical Almanac* from 1849 to 1867, and he developed the first statistical significance test for outliers, which was published in the *Astronomical Journal* in 1852. Peirce engaged in a priority dispute with George Phillips Bond over the structure of Saturn's rings in the early 1850s, and he wrote *A System of Analytical Mechanics* in 1855. In 1867, he was promoted to superintendent of the Coast Survey. After he retired from that post in 1874, he served as the consulting geometer.

Peirce's most notable publication was *Linear Associative Algebra* (1870), a relatively brief treatise in which he provided a method for classifying algebras according to their defining properties. He synthesized earlier work by George Peacock and William Rowan Hamilton. His justification for the book revealed his desire to study mathematics for theological edification, but Peirce also identified a connection between mathematics and deductive thought that became commonly accepted in the twentieth century. Although this work was only lithographed and distributed in small numbers at first, the work was republished in 1881 in the *American Journal of Mathematics,* the first American research periodical in mathematics, which Peirce had edited in the 1870s.

Peirce was elected to Phi Beta Kappa in 1829, the American Philosophical Society in 1842, the Royal Astronomical Society of London in 1850, and the Royal Society of London in 1852. He served on the founding committees of the Smithsonian Institution (1847), the Dudley Observatory at Albany, New York (1855–1858), and the National Academy of Sciences (1863). He was also one of the mid-nineteenth-century American advocates for mathematical and scientific research who described themselves as the Scientific Lazzaroni (scientific wanderers).

Although Peirce did not directly train any of the first-generation of American research mathematicians, he was a force on behalf of professionalization in at least three ways: he advocated research-oriented higher education, he interested American mathematicians in new algebras and in statistics, and he helped to create numerous institutions devoted to research. In addition, he actively participated in the concerns of nineteenth-century American astronomy, mathematics, and surveying.

Amy Ackerberg-Hastings

Sources

Cohen, I. Bernard, ed. *Benjamin Peirce: "Father of Pure Mathematics" in America.* New York: Arno, 1980.

Grattan-Guinness, Ivor. "Benjamin Peirce's 'Linear Associative Algebra' (1870): New Light on Its Preparation and Publication." *Annals of Science* 54 (1997): 597–606.

Hogan, Edward. "'A Proper Spirit Is Abroad': Peirce, Sylvester, Ward, and American Mathematics, 1829–1843." *Historia Mathematica* 18 (1991): 158–72.

Lenzen, Victor F. *Benjamin Peirce and the United States Coast Survey.* San Francisco: San Francisco Press, 1968.

PEIRCE, CHARLES S. (1839–1914)

Regarded by many as the greatest logician of his time, Charles S. Peirce was the originator of the philosophy of pragmatism (testing the meaning of something by its consequences) and semiotics (the study of signs). Peirce left an impressive body of original work and profoundly influenced such philosophical giants as William James and John Dewey. Ironically, he held only one brief academic appointment and lived on the charity of others for much of his life.

The son of one of the nation's foremost mathematicians, Peirce was born in Cambridge, Massachusetts, on September 10, 1839. In 1861, he began working for the U.S. Coast and Geodetic Survey. He remained there for thirty years, while pursuing his philosophical studies.

In a series of articles including "The Fixation of Belief" (1877) and "How to Make Our Ideas Clear" (1878), Peirce argued that thought and action are linked and that the meaning of an idea is tested by its practical consequences. Although he originally called his philosophy "pragmatism," he eventually began calling it "pragmaticism," in order to differentiate it from the work of James, which he found too subjective and insufficiently rigorous, and that of Dewey, which he found too sociological in nature. Peirce quipped that the term "pragmaticism" was "ugly enough to keep it safe from kidnappers."

In 1879, Peirce became a lecturer in logic at Johns Hopkins University, but school authorities dismissed him in 1884 when they learned that he had lived with his wife before their wedding. Peirce never obtained another university appointment, and his private life was often troubled. Suffering from physical ailments and the side effects of drugs he took to treat them, he

A founder of the pragmatic movement in American philosophy and an innovator in mathematics, physics, and astronomy, Charles Sanders Peirce regarded himself first and foremost as a logician. Some regard him as America's greatest. *(National Oceanic and Atmospheric Administration/Department of Commerce)*

had two nervous breakdowns. After being forced to resign from the U.S. Coast and Geodetic Survey in 1891, he had to rely on friends for support, chiefly William James.

Lacking a permanent academic position as a forum for his ideas, Peirce was often under appreciated both during his lifetime and for many years after his death on April 19, 1914. The publication of his complete works, an ongoing project, and renewed scholarly interest in his ideas have confirmed his status as one of America's most important philosophers.

Fred Nielsen

Sources

Apel, Karl-Otto. *Charles S. Peirce: From Pragmatism to Pragmaticism.* Amherst: University of Massachusetts Press, 1981.

Brent, Joseph. *Charles Sanders Peirce: A Life.* 1993. Reprint, Bloomington: Indiana University Press, 1998.

Menand, Louis. *The Metaphysical Club.* New York: Farrar, Straus and Giroux, 2002.

Peirce, Charles. *Writings of Charles S. Peirce: A Chronological Edition.* Bloomington: Indiana University Press, 1982.

SAGE

SAGE, an acronym for Semi-Automatic Ground Environment, is the predecessor of most real-time computer systems, command-and-control military systems, and air defense systems in operation today. Devised during the Cold War, SAGE laid the foundation for the Federal Aviation Administration (FAA) national air-traffic control system in the United States, as well as the first commercial airline reservation system (SABRE). The SAGE system, now dismantled, is also notable as one of the largest software projects of the 1950s and early 1960s.

Work on SAGE began at MIT's Lincoln Laboratory in 1954, led by engineers George Valley and Jay Forrester. The most important function of the system as they conceived it was to coordinate radar information received from operators at remote air-defense direction centers—especially the echo signatures of Soviet intercontinental missiles—and return instructions to local anti-aircraft batteries or fighter interceptors. Sector control officers combined data received from the network of individual SAGE installations to create

unique geographic air-situation displays on 19 inch cathode-ray tubes. The SAGE system was capable of refreshing its displays with new information 200 times every 2.5 seconds. Operators targeted individual aircraft on the display with a unique light gun. A mockup of a SAGE display unit was prominently featured in the doomsday film *Fail-Safe* (1964).

At the heart of SAGE was AN/FSQ-7 (Army-Navy Fixed Special Equipment), an early operating system developed by the U.S. Air Force. The SAGE AN/FSQ-7 system got its name, in turn, from the FSQ-7 Whirlwind II, an early random-access core memory computer and direct successor to the first real-time control computer, known as Whirlwind.

The AN/FSQ-7 system required one of the largest programming efforts of its time, employing hundreds of programmers well before the formal establishment of computer science as a discipline. Some 100,000 instructions, or roughly 1,250 per computer program, were written for the system by 1956. Ten years later, the AN/FSQ-7 comprised more than 500,000 instructions. Because of the operating system's size, AN/FSQ-7 programmers introduced novel software for checking and compiling code, as well as new computer communications capabilities. Among them were the Lincoln Utility System, designed to assist inexperienced programmers in structuring, documenting, debugging, and communicating their work.

The complete AN/FSQ-7 system—twenty-four hardware installations in all—processed system status data, buffer storage tables, and basic system programs by dividing them into blocks, or pieces of code between 25 and 4,000 words long. The computer transferred these blocks into and out of core memory as needed. SAGE systems were closely coordinated with a sequence-control program, which ran each air-defense monitoring program repeatedly—some every few seconds and others every few minutes. The system's reliability and redundancies meant that the SAGE system experienced less than ten hours of downtime each year.

International Business Machines (IBM)—which worked closely with the Lincoln Lab on computer projects—developed its own commercial real-time airline reservation system, the Semi-Automatic Business-Research Environment

(SABRE), by reusing software and expertise developed in completing the massive SAGE network. IBM implemented the first SABRE system for American Airlines between 1962 and 1964. The system included a half-million lines of code and reportedly cost $40 million. IBM sold variants of SABRE to Pan American and Delta Airlines in 1965. The company later developed an off-the-shelf version of its system called the Programmed Airline Reservations System (PARS).

Philip Frana

Sources

Astrahan, Morton M., and John F. Jacobs. "History of the Design of the SAGE Computer: The AN/FSQ-7." *Annals of the History of Computing* 5 (October 1983): 340–49.

Everett, Robert R., et al. "SAGE: A Data Processing System for Air Defense." In *Proceedings of the Eastern Joint Computer Conference.* Washington, DC: IRE-ACM-AIEE, 1957.

Jacobs, John F. "SAGE Overview." *Annals of the History of Computing* 5 (October 1983): 323–29.

STATISTICS

Statistics is a branch of mathematics that entails collecting and analyzing numerical data and making inferences or predictions based on the analysis. Subdisciplines are descriptive statistics, inferential statistics, and probability theory.

Descriptive statistics involves the collection of data by means of scientific observation and experiment, and the subsequent analysis of data by means of simple mathematical calculations. Analysis can range from tables of percentages to graphs of frequency distributions and more sophisticated measures of ranking data. Inferential statistics involves the use of analyzed data to make inferences about tendencies or patterns. Analysis of large collections of data often yields patterns that allow scientists to assert the probability of a given situation or result to recur repeatedly.

Modern statistics largely developed in the twentieth century in response to the increasing demand for analysis of large amounts of data collected by social and natural scientists. The English philosopher and mathematician Karl Pearson, for example, contended in 1910 that any empirical argument should be backed with "statistics on the table." Pearson was especially known for developing the Pearson correlation, a means of finding correlation among disparate variables.

Early leaders in the field included Ronald Aylmer Fisher, who published *Statistical Methods for Research Workers* (1925), and Harald Cramer, who published *Mathematical Methods of Statistics* (1945). L.H.C. Tippett studied the problem of the distribution of extremes and coined the phrase "random number." Emil J. Bumbel examined extremes in distributions in *Statistics of Extremes* (1958). The foundations of probability theory were examined by a number of mathematicians, in particular F.N. David, who wrote *Games, Gods, and Gambling* (1962). The theories of these European statisticians would find a home in the practical needs of a growing America.

Formal statistical inquiry in the United States began in the nineteenth century. The American Statistical Association (ASA) was founded in Boston in 1839 to "collect, preserve, and diffuse statistical information in the different departments of human knowledge." In 1888, the American Statistical Association began publishing the *Journal of the American Statistical Association.* Many of its members worked for the U.S. Census Bureau, reflecting the need to track the nation's booming population. Herman Hollerith, for example, a member of the American Statistical Association, worked for the Census Bureau and devised the idea of calculating machines. Others worked for the Bureau of Labor Statistics, reflecting the growing productivity of the American industrial economy.

During and after World War II, American statistician William Edwards Deming applied statistical analysis to manufacturing processes for the purpose of quality control. Deming's ideas were adopted by the Japanese in their industrial development after World War II. In Japan, the Deming Prize for the advancement of statistical quality control is given each year to an individual and a company.

Throughout the twentieth century, American statisticians employed European statistical models in the biological, physical, and social sciences. Johns Hopkins University opened a biostatistics program in 1918. The American Statistical Association began publishing a biometrics journal in 1945. In 1959, the ASA published *Technometrics,* statistical methods in the physical sciences.

During the 1960s and 1970s, interest in statistical methodology spread to the social sciences, causing a revolution in social scientific inquiry and methods. SPSS (Statistical Package for the Social Sciences) became an essential tool for social scientists seeking quantitative data to support theory and inference.

Andrew J. Waskey and Russell Lawson

Sources

Agresti, Alan, and Barbara Finlay Agresti. *Statistical Methods for the Social Sciences.* San Francisco: Dellen, 1979.

American Statistical Association. http://www.asstat.org.

Salsburg, David. *The Lady Tasting Tea: How Statistics Revolutionized Science in the Twentieth Century.* New York: Henry Holt, 2001.

Stigler, Stephen M. *Statistics on the Table: The History of Statistical Concepts and Methods.* Cambridge, MA: Harvard University Press, 1999.

UNIVAC

On the last day of March 1951, the Philadelphia branch of the United States Census Bureau ushered in what historian Paul Ceruzzi has called "the true beginning of the computer age" when it took delivery of the Universal Automatic Computer, or UNIVAC I, the first commercially available electronic digital computer in the country. Though capable of the then blazing speed of 1,000 calculations per second, the device looked anything but nimble. It weighed in at 16,000 pounds (7,300 kilograms), contained 5,000 vacuum tubes, and required 1,000 cubic feet (28 cubic meters) of floor space.

Bulk aside, the UNIVAC offered exactly what organizations such as the Census Bureau needed: the ability to perform repetitive calculations on large volumes of numerical data quickly and reliably. Indeed, UNIVAC proved so useful in this regard that the Census Bureau ordered two more of the behemoths in the early 1950s, with the U.S. Air Force and General Electric (GE) following suit with single purchases. GE used its UNIVAC to become the first company in the world to institute a computerized payroll system, thus firmly establishing the computer in the technological arsenal of modern corporate America.

This was exactly the outcome anticipated by UNIVAC's designers, physicist J. William Mauchly and engineer J. Presper Eckert, Jr., when they began work on its predecessor—the Electrical Numerical Integrator and Calculator (ENIAC)—at the University of Pennsylvania's Moore School of Engineering in 1943. Like so many other research and development endeavors of the time, ENIAC, the world's first operational large-scale electronic digital computer, was intended to aid the war effort, specifically to calculate complex ballistics tables for the U.S. Army Ordnance Department. But Mauchly had long harbored the belief that computers eventually would have broad applications in the scientific and commercial fields. His desire to patent ENIAC and promote the machine for commercial use—despite its having been developed under the semiofficial auspices of the university—led to a permanent rift between the physicist and the parent institution in 1946.

Soon thereafter, Mauchly and Eckert transferred their operations to an old textile mill in Philadelphia. They landed a preliminary contract with the Census Bureau, formed an official business partnership, the Electronic Control Company (ECC), and began development of UNIVAC. Despite naysayers—most famously computer pioneer Howard Aiken of Harvard, who argued that there would be no need and no demand for more than one or two of these machines in the entire nation—Mauchly's marketing skills soon landed several additional contracts for the as yet unfinished UNIVAC.

Due to the exciting nature of this pioneering venture, ECC had little trouble attracting enthusiastic and talented programmers and engineers from high-caliber institutions such as Harvard, the Massachusetts Institute of Technology, and the original Moore School team. Veterans of the UNIVAC project later recalled a cooperative and egalitarian work environment where even comparatively minor employees could make significant contributions to the emerging field. Moreover, at a time when women were particularly rare in the fields of engineering and mathematics, the UNIVAC team boasted several talented female members in both areas, including most famously Grace Murray Hopper, a trailblazing programmer on other early computer projects, such as Harvard's Mark I program.

UNIVAC's innovative design team produced equally innovative solutions for this first generation of commercial-application computers, including the first magnetic tape storage system, which provided a much faster medium for data input and output than the then dominant punch-card systems from International Business Machines (IBM). Moreover, UNIVAC's ability to read both alphabetical as well as numerical symbols and to store programs in its memory as well as data represented significant breakthroughs in making the computer a flexible, all-purpose tool for business.

In 1947, ECC incorporated as the Eckert-Mauchly Computer Company (EMCC), but unfortunately, the business acumen of the company's founders did not match the design and development skills of its production team. Like many early computer concerns, EMCC had evolved from a government-financed, university-based project team that lacked significant experience in manufacturing or business management. Mauchly and Eckert consistently underestimated their development and production costs and contracted to deliver UNIVACs at optimistically low, fixed prices, rather than the more typical cost-plus format that had been employed for much war-production work in World War II. Chronically strapped for cash, Mauchly and Eckert sold EMCC to Remington Rand in 1950. They continued to develop UNIVACs as a division of that corporation until their UNIVAC model was officially retired in 1957.

Although only six UNIVACs were ever produced (one was sent to the Smithsonian Institution in 1957), the Eckert-Mauchly machine proved that computers could play a vital role in efficiently processing the vast streams of data generated by the increasingly bureaucratized world of modern corporations and government agencies. The UNIVAC also managed to capture the public's imagination when one of the Census Bureau's machines accurately predicted the outcome of the Eisenhower–Stevenson presidential contest during nationally televised coverage of the 1952 election.

Jacob Jones

Sources

Ceruzzi, Paul. *A History of Modern Computing.* Cambridge, MA: MIT Press, 1998.

Kidwell, Peggy A., and Paul E. Ceruzzi. *Landmarks in Digital Computing: A Smithsonian Pictorial History.* Washington, DC: Smithsonian Institution, 1994.

Lukoff, Herman. *From Dits to Bits: A Personal History of the Electronic Computer.* Portland, OR: Robotics, 1979.

Stern, Nancy. "The Eckert-Mauchly Computers: Conceptual Triumphs, Commercial Tribulations." *Technology and Culture* 23:4 (1982): 569–82.

———. *From ENIAC to UNIVAC: An Appraisal of the Eckert-Mauchly Computers.* Bedford, MA: Digital, 1981.

VON NEUMANN, JOHN (1903–1957)

The Hungarian American mathematician John von Neumann made significant contributions to such diverse fields as pure mathematics, logic, game theory, quantum physics, computer science, and meteorology.

He was born Janos Neumann on December 28, 1903, in Budapest, Hungary, and emigrated to the United States in the face of rising anti-Semitism in Nazi Germany during the 1930s. He was an outspoken critic of communism and did much work for the U.S. armed forces, including helping to develop the atomic bomb. While still in gymnasium (high school), von Neumann did original research in mathematics. A paper he wrote in high school, "Zur Einführung der transfiniten Ordnungszählen" (Toward the Introduction of Transfinite Ordinal Numbers), published in 1923, expanded upon the work of German mathematician George Cantor in ordinal numbers.

Von Neumann passed the entrance exam for the chemical engineering program at Zurich's prestigious Eidgennoissische Technische Hochschule—a test Albert Einstein had failed on his first try—and enrolled simultaneously in a doctoral program in mathematics at Budapest University, splitting time between the two cities. He did his doctoral research on the axiomization of set theory, and, in 1926, at the age of twenty-two, he earned his Ph.D. (with highest honors) in chemical engineering. That same year, he took a position at Göttingen University in Germany, where he worked with David Hilbert and other mathematicians. He was a prolific author, publishing thirty-two papers on physics, mathematics, and economics by 1929.

In the early 1930s, German society became increasingly difficult for Jews. Fortunately, von Neumann received an attractive job offer in

1933: a lifetime professorship at the newly formed Institute for Advanced Study in Princeton, New Jersey. The position included a handsome salary and no teaching duties. Von Neumann quickly accepted the offer, joining Einstein and other luminaries at the prestigious institute. He became an American citizen in 1937.

During World War II, von Neumann turned to applied mathematics in an effort to help his new country's military efforts, working for the National Defense Research Council and other agencies on a wide range of practical problems. These included studies of the efficiency of different geometric shapes in designing explosives, the effectiveness of aircraft bombing patterns, explosive blast waves, and other topics. Toward the end of the war, von Neumann worked with the Manhattan Project at Los Alamos, New Mexico, helping to design the atomic bomb and joining discussions on where best to deploy it. He served on the U.S. Atomic Energy Commission from 1955 until his death.

Von Neumann was also a major pioneer in the developing field of game theory, a mathematically oriented subfield of economics. Along with Oskar Morganstern, he published the classic *Theory of Games and Economic Behavior* in 1946. According to von Neumann, economic problems "are often stated in such vague terms as to make mathematical treatment a priori hopeless." In the book, von Neumann restated the issues with clarity and precision before proceeding with his own insightful analysis.

Near the end of his life, he became increasingly interested in the development of the first computers. At the Institute for Advanced Study he directed the Computer Project, building a computer that proved particularly useful to meteorologists. In 1958, he published *The Computer and the Brain*. The Computer Project shut down soon after his death from cancer on February 8, 1957, but von Neumann's work in economics and computer science continues to be influential.

Andrew Perry

Sources

Macrae, Norman. *John von Neumann.* New York: Pantheon, 1992.

Morganstern, Oskar, and John von Neumann. *The Theory of Games and Economic Behavior.* Princeton, NJ: Princeton University Press, 1944.

WHIRLWIND

Upon its completion in 1951, after five years of work and $5 million in government investment, Whirlwind was "the fastest real-time digital computer in the world," according to historian Bruce Wheaton. Project Whirlwind also made substantial contributions to the nascent field of computer engineering, including magnetic-core random access memory (RAM) and block diagram system designs.

Like several other computer projects of the mid-twentieth century, Whirlwind grew out of a World War II military contract. In this case, the Special Devices Division of the U.S. Navy's Bureau of Aeronautics wanted a training machine that could analyze pilot responses while simulating the flight characteristics of a broad range of aircraft types. Since the Electrical Engineering Department at the Massachusetts Institute of Technology (MIT) was already working on related feedback systems, it was awarded the contract for the flight simulator in December 1944. The lead engineer on the project, a young MIT graduate student named Jay Forrester, soon realized that the analogue computers used in such work were too slow to produce the sort of rapid feedback needed for complex, real-time flight simulation. By the following year, Forrester had determined that only some version of the digital computers then being developed, such as in the ENIAC project at the University of Pennsylvania, would be capable of achieving the necessary speeds. Responding to Forrester's arguments, the navy agreed to expand the original contract to accommodate the development of a digital computer. Thus, Project Whirlwind was officially launched in March 1946.

Forrester recruited his friend and fellow MIT graduate student Robert R. Everett to manage Whirlwind's system design, or block diagram section. Block diagrams—visualizations of the interaction of computer components and operation sequences—were one of the major contributions made by the Whirlwind project to early computer design. As stated by Whirlwind's chief historian, Thomas Smith, block diagrams "bridged the gulf between the abstract logical concepts underlying the digital computer and the engineering concepts transforming the logic of the computer into corresponding engineering-design problems susceptible of solution."

Forrester, meanwhile, was making his own major contribution to computer design history by fashioning Whirlwind's internal magnetic-core storage unit, which gave the computer the sort of RAM capacity necessary to perform real-time functions. Magnetic-core storage would be the industry standard, until it was superseded by transistors in the 1960s.

Despite its major advances in general computer design, however, as the Whirlwind project neared the end of its funding cycle in 1948, the team still had not finished the flight simulator they were originally contracted to produce. As a result, Whirlwind's new supervisory agency—the Office of Naval Research (ONR)—threatened to phase out funding for the project. ONR also balked at Whirlwind's expense. Forrester's group had, after all, been consuming over $1 million a year when comparable programs such as UNIVAC were getting by on less than half that amount. Forrester responded that the military needed a fast, general-purpose computer that could be used for any number of real-time command and control applications, such as coordinated fire control for artillery and missile systems. As for the cost, Forrester argued that mathematically precise component engineering and system design were necessary for the level of reliability required for real-time operations, and thorough design work at the beginning would make the machine more easily replicable by private industry when the latter took over production.

Then, in August 1949, the Soviet Union successfully tested a nuclear weapon. The end of the U.S. nuclear monopoly meant that the Soviets eventually could launch a nuclear attack on the United States using long-range bombers flying over the North Pole. To counter this threat, the U.S. Air Force looked to build a nationwide network of radar posts to provide early warning of such an attack. Coordinating the information from these radar sites and providing the air force with real-time information for fighter intercepts would require the kind of computing capacity provided by Whirlwind. Thus, the U.S. Air Force became the new funding agency for the Whirlwind project.

In April 1951, the computer passed its first real-world test with flying colors, successfully coordinating information of a mock attack from several radar sites on Cape Cod, while also providing trajectories for U.S. Air Force counterattack. Whirlwind would become the prototype for the FSQ-7 computer used by the early warning air defense system known as SAGE (Semi-Automatic Ground Environment) developed by MIT and the MITRE Corporation, headed by Whirlwind veteran Everett. The FSQ-7 also would provide the foundation for future International Business Machines (IBM) dominance of the digital computer industry, as IBM was tapped to replicate the computer for both the defense establishment and the larger commercial market.

Jacob Jones

Sources

Redmond, Kent C., and Thomas M. Smith. *From Whirlwind to MITRE: The R & D Story of the SAGE Air Defense Computer.* Cambridge, MA: MIT Press, 2000.
———. *Project Whirlwind: The History of a Pioneer Computer.* Bedford, MA: Digital, 1980.
Smith, Thomas M. "Project Whirlwind: An Unorthodox Development Project." *Technology and Culture* 17:3 (July 1976): 447–64.

WIENER, NORBERT (1894–1964)

The American mathematician Norbert Wiener, who taught at the Massachusetts Institute of Technology (MIT) for more than forty years, developed an interdisciplinary approach to the study of communication and control in living organisms and machines—a field he dubbed cybernetics.

Born on November 26, 1894, Wiener was a child prodigy, a role thrust upon him by his father, Leo Wiener, a Slavic languages specialist at Harvard University. Enrolled at Tufts University at the age of eleven, Wiener finished his doctoral work in mathematical logic at Harvard at age eighteen.

As an instructor at MIT beginning in 1919, he worked out mathematical solutions by talking to anyone who would listen on his rambles around campus. He was known for his absentmindedness and quirks, many revolving around the paradox of his extreme egocentric behavior and personal generosity. His classes could be chaotic. He had a propensity for punctuating lectures with unrelated commentaries, self-reflections, and problems worked out on the fly.

Cybernetics involves the study of communication and control in living organisms and in machines, particularly in respect to the principles of biological feedback, how information is transferred, and inner organization and self-containment (of both humans and machines). Today, cybernetic thought permeates multidisciplinary activity in computer science, engineering, biology, and the social sciences, although the term itself is no longer widely used in the West.

Wiener, who derived the word cybernetics from a Greek word meaning "to steer," inspired many others to adopt the science as a unifying force binding together game theory, operations research, automata theory, logic, and information theory. He fashioned cybernetics in the context of World War II and the development of advanced weaponry in the struggle against the Axis powers. Wiener envisioned sophisticated fire-control systems for shooting down enemy planes. These weapons systems would be patterned after the feedback mechanisms, or self-adjusting physiological abilities, of the human body.

At the end of the war, Wiener emphasized the peaceful uses of cybernetics and computer automation. His cybernetics had a profound effect on the development of theories of human and machine intelligence after 1950. He argued that humans and machines should be considered together in fundamentally interchangeable terms. As he wrote in a 1948 article for *Scientific American*, "Cybernetics attempts to find the common elements in the functioning of automatic machines and of the human nervous system, and to develop a theory which will cover the entire field of control and communication in machines and in living organisms."

The debates of the Macy Conferences on Cybernetics in the 1940s and 1950s, which featured

Longtime **MIT** professor and mathematician Norbert Wiener is known as the founder of cybernetics—the interdisciplinary study of the feedback process in machines, living organisms, and social organizations. *(AIP Emilio Segre Visual Archives)*

many leading scientists from all disciplines, nurtured and further solidified the ideas of Wiener and others who wanted to use cybernetics to create a new machine-based artificial intelligence. One of the Macy Conferences' most active participants, the mathematical biophysicist Walter Pitts, appreciated Wiener's direct comparison of neural tissue to vacuum tube technology. Pitts considered the electrical devices ideal representatives of the most fundamental units of human thought.

Wiener was a professor at MIT from 1919 to 1960, during which time he worked on Brownian motion, the random action of subatomic particles; probability theory, which influenced stochastic processes; potential theory, which is related to electromagnetism; and harmonic analysis, which relates to irregular functions in mathematics. He wrote two autobiographies, *Ex Prodigy: My Childhood and Youth* (1953) and *I Am a Mathematician* (1956). Wiener died in Stockholm, Sweden, on March 18, 1964.

Philip Frana

Sources

Galison, Peter. "The Ontology of the Enemy: Norbert Wiener and the Cybernetic Vision." *Social Studies of Science* 23 (1993): 107–27.

Mahoney, Michael S. "Cybernetics and Information Technology." In *Companion to the History of Modern Science,* ed. R.C. Olby, et al. New York: Routledge, 1990.

Wiener, Norbert. "Cybernetics." *Scientific American* 179 (1948): 14–19.

———. *Cybernetics, or Control and Communication in the Animal and the Machine.* Cambridge, MA: MIT Press, 1948.

DOCUMENTS

A Nineteenth-Century Calculating Machine

The mechanical and electronic calculator has a long history. American scientists contributed to the development of calculators through pragmatic inventions that sought to martial information more quickly and efficiently. The following document, from an 1888 issue of Scientific American, *describes the invention of "An Improved Calculating Machine" by a Chicago inventor.*

There has lately been invented by Mr. Dorr E. Felt, of Chicago, a calculating machine which he has named the comptometer. It is a practical machine operated by keys for the computation of numbers and the solution of mathematical problems. The rapidity and accuracy with which computations are made on the comptometer when in the hands of a skillful operator are calculated to meet the approval and win the admiration of all.

In the construction of the comptometer all the operating parts are made of the finest hardened steel, thus insuring the greatest degree of durability. The accuracy and durability of the machine have been thoroughly tested in the actuary's department of the United States Treasury at Washington, where one is in constant use. It will add, subtract, multiply, and divide, from which it is evident that all arithmetical problems can be solved on it. Particular attention is called to its availability in computing interest, discount, percentage, and exchange. It is a neat, compact machine, fourteen and one-quarter inches long, seven and one-quarter inches wide, and five inches high, weighing eight and a half pounds.

By referring to the cut, it will be seen that each key has two numbers on its top, one large and the other small, but for the present leave the small one out of consideration, and understand every reference to be to the large one only. It will be seen that the keys resolve themselves into rows running from right to left and rows running from the operator. For convenience in explaining, the rows running from right to left will be called rows, and those running from the operator will be called series. It will be further noticed that every key in the first row has the figure 1 on its top, those in the second the figure 2, those in the third the figure 3, etc. The figures on the tops of the keys in the series run from one to nine inclusive. The first series represents units, the second tens, and the third hundreds, etc. To add, it is merely necessary to touch on the machine the numbers to be added; if we have 5,673 will be shown on the register; we next touch 9 in the third series, 3 in the second, and 2 in the first, when the sum of the two numbers, 6,605, will be shown by the register. This operation can be continued until the limit of the machine is reached, which in the standard size is 999,999,999.

Subtraction, multiplication, and division can each be as rapidly and as easily performed.

By again referring to the cut, it will be seen that at the front of the machine is a plate in which are a number of square openings, which is called the register plate. At these openings are shown all results by numeral wheels, which are below the plate and which stand side by side on the same shaft, and each of these numeral wheels is acted upon by its keys direct and also by the carrying part of the numeral wheel next lower in order, something that has never been practically accomplished before in any mathematical calculator operated by keys. The carrying mechanism in this machine is entirely independent of the keys struck, and the power required for carrying is gradually accumulated and automatically released at the proper moment, therefore requiring no additional effort to depress the key when, through the operation of the carrying device, the next numeral wheel in order above has to be moved, than when such is not the case; therefore, when a succession of nines occur on the register, and a key is struck in one of the lower orders, it is impossible to discover that any more power is required than when one nine only appears on the register. In this machine two positive stops are employed for each numeral wheel, one to prevent over-rotation of the numeral wheel under the impulse of the key stroke, and the other to prevent over-rotation of the numeral wheel when actuated by the carrying mechanism. As there is no frictional device employed to prevent over-rotation, the machine always responds to a light touch on the keys; and as each numeral wheel is always in positive engagement with its controlling devices, absolute accuracy is insured at all times. It having been stated that the carrying device is independent, it will be at once seen that when a key of one of the higher orders is struck, the carrying device of the next lower order is at once released, allowing the numeral wheel on which the key struck acts to move independently of all numeral wheels lower in order. The result of any operation being obtained, the machine is returned to naught by depressing the lever which appears on the right and turning the knob above it until the figures seven appear on the register, when release the lever and continue turning the knob, and the machine will stop at the ciphers.

Source: "An Improved Calculating Machine," *Scientific American* 59 (August 1888).

Herman Hollerith's Electric Tabulating System

Herman Hollerith, the founder of IBM, revolutionized the statistical manipulation of data with his Hollerith Electric Tabulating System, which was used by the U.S. Census Bureau in 1890. The following is a description of the machine published at the time.

[O]ne of the most important elements of the Hollerith electric tabulating system [is] the machine which is used for transferring the individual records to cards by punching holes, the relative position of these holes in the cards determining the characteristics of the individual. As used in the census work, it should be understood, that, in the compilation of the statistics of population, a card must be punched for each individual reported, so that for this class of observations alone no less than 65,000,000 cards will have to be punched. It is to facilitate this work that the key-board punch . . . has been devised.

The machine consists of a base, to which is secured a card-holder, in which individual cards can readily be inserted and removed. In front of this is a key-board, formed with a number of holes, each lettered and numbered, the letters and numbers corresponding to the designations given to the different statistical items to be recorded. Moving on this base is a swinging arm, pivoted at the back, and capable of being swung from right to left, and forward and backward. Secured to this swinging arm is a punch, which is connected with, and operated by, a lever terminating in a knob and pin directly over the key-board. The holes in the key-board are arranged in such curves, that, by moving the pin along them, the punch will move in straight lines parallel to the edge of the card. The pin, punch and key-board are in such relative positions that the punch will not cut the paper until the pin is depressed in one of the holes, thus securing the punching of the holes in proper position in the card.

To transcribe a record, a card is first put in the holder, and then, beginning at the upper left-hand corner of the key-board, the items are successively recorded by punching according to the letters and numbers on the key-board. The movement of the punch, or knob, being from the upper left-hand corner to the right, and then back along the front edge of the key-board to the left, leaves the punch in convenient position for removing the punched card and inserting a new one. The general arrangement of the key-board conforms with the arrangement of the items on the schedule which are to be transcribed. Simple items, such as sex, race, conjugal condition, etc., are designated on the key-board directly by abbreviations, suitably arranged in groups by heavy lines. Thus, for example, sex would be included in a small space, one hole being designated M, and one F. The operator becoming familiar with the position in which these items are found on the key-board, little time is lost in transcribing such records. In the case of occupations and birthplaces, and other items where a large amount of detail is required to be recorded in a comparatively small space of the card, combinations are used. Thus, a group of occupations, such, for instance, as the agricultural occupations, would be indicated by A; then, again, professional occupations may be indicated by B, another group of occupations by C, etc. . . .

The cards used for the purpose of transcribing the individual records of the census, will be made of think manilla stock, 3¼ inches wide and 6⅝ inches long. They will be perfectly blank, except a printed number, which serves to identify the given card, if necessary. Besides this number, the cards will have a number of holes punched by means of these key-board punches. The position of each of these punch marks upon each card designates, in accordance with a pre-arranged scheme, some distinguishing characteristic of the individual, such, for example, as the race, . . . the sex; the age; the conjugal condition, in case of females, whether a mother, and if so, how many children, and how many of these living; the place of birth; the place of birth of father; the place of birth of mother; if foreign born, the number of years in the United States; whether naturalized or not; the profession, trade or occupation; the number of months unemployed during the census year; whether the person attended school or not during the census year; whether able to read, able to write, able to speak English, if not, what language spoken.

Source: "Further Details of the Hollerith Electric Tabulating System," *Manufacturer and Builder* 22:5 (May 1890).

Electronic Tabulation of the 1890 Census

The following account describes the efficiency of the Hollerith Electric Tabulating System in the manipulation of population data for the 1890 U.S. Census.

The chiefs of the Population Division of the Census Office celebrated the completion of the count of the population of the United States. . . . The Hollerith electric tabulating system has been in use by the Census Office for the tabulation of the schedules of the population taken under the eleventh census. Superintendent Porter . . . spoke as follows: "For the first time in the history of the world, the count of the population of a great nation has been made by the aid of electricity. The number named on every one of fifteen million schedules has been registered twice by the nimble and expert fingers of the counters. . . . In June, these blanks were distributed throughout the country. In July and August they find themselves back in the Census Office, counted twice, and ready for the next statistical treatment. . . . We have actually counted 128,000,000 in six weeks, or the entire population of 64,000,000 twice in that period. . . . [W]e could, with these electrical machines, count the entire population of the United States in ten days of seven working hours each."

Source: "The Hollerith System in the Census Work," *Manufacturer and Builder* 22:9 (September 1890).

Section 13

APPLIED SCIENCE

The American Inventor

America has been a place for the jack of all trades, the entrepreneur, the tinker, the inventor. America was the *new world,* and as the eighteenth-century writer Hector St. John de Crevecoeur exclaimed in *Letters from an American Farmer* (1782), it was a place where "new men" could develop new ideas and a fresh perspective on life, unencumbered by the traditions, philosophies, and institutions of Europe. Americans were not metaphysicians, but rather practical thinkers who applied the great ideas of European scientists and philosophers to the particular problems of the American environment.

Over the course of American history, American inventors took European scientific theories and made useful items. Europeans worked out the theory and structure of electricity, but Thomas Alva Edison invented the light bulb. European physicists discovered the nature of the atom and the theories of fission and nuclear chain reaction, but Americans built the first atomic bomb. Americans used European mathematical foundations to build the first electronic computers. In agriculture, industry, communication, transportation, and military technology, American inventors created, developed, manufactured, multiplied, utilized, built, and succeeded in reshaping their environment.

The American innovative mindset was present at the beginning of colonial settlement, when the mere provision of the simple necessities of life, such as food and shelter, demanded a practical and empirical approach to problem solving. The first permanent English colony at Jamestown, for example, succeeded because Captain John Smith recognized that the initial mindset of the Jamestown colonists—that a successful colony required the discovery of mines of gold and silver—had to be replaced by the mindset of self-sufficiency in the new land. Smith realized that the wealth of North America lay not in precious metals but in the fertility of the soil and the plenty of the forests, rivers, and sea.

John Smith grew up in Elizabethan England at the same time as another innovative thinker, Francis Bacon. Bacon firmly believed that the problems of everyday life as well as the problems of science were best solved by means of an empirical mindset. In his book *Novum Organum* (*New Method*), published in 1620, Bacon argued that once problems are identified, the thinker must form a hypothesis of the solution and then test the hypothesis by experiment; trial and error is often necessary to discover the solution to the problem. Knowledge and behavior must change in accordance with the solution. Francis Bacon was a synthesizer: He realized that the empirical method was being used on a daily basis by English men and women to solve everyday problems, and he formulated a method to make systematic what previously had been sporadic. English emigrants to America brought Bacon's method to bear on the manifold problems they confronted in building communities in the wilderness.

Bacon's essay *New Atlantis* (1627) hypothesized a society based on the empirical method, where the goal of the populace was the increase and application of knowledge. The New Atlantis, which Bacon imagined to be off the coast of America somewhere in the Pacific, was said to be a land reflecting the manifold accomplishments of applied knowledge. The hero of this society was the inventor.

Likewise in America, especially during the eighteenth and nineteenth centuries, the inventor was eulogized. Benjamin Franklin, the most well-known and respected man of Revolutionary America, was the leader of the American effort to promote useful knowledge. Franklin invented practical items such as the bifocal lens, Franklin stove, catheter, and lightning rod, and he began useful institutions such as the public library, fire department, and American Philosophical Society.

Independence from England and the demands of creating a productive, successful society inspired in Americans a willingness to embrace technology to encourage improvements in production, transportation, and communication. The Industrial Revolution thrived in America in the late eighteenth and early nineteenth centuries because of the American personality, which focused on building and doing.

The stage for industrialization was set by the tremendous productivity of the American farmer, which resulted in available capital and food surplus to support a growing population, creating a demand for more products. Farm productivity was the result of individual and collective ingenuity. John Deere, for example, addressed the needs of Midwestern farmers in the 1830s by inventing a durable steel plow that could furrow the thick soil and scour the ploughshare. He mass-produced this steel plow at his factory in Moline, Illinois, meeting the demand of the increasing numbers of American farmers migrating west.

Also in great demand because of the growing American population was clothing. Do-it-yourself Americans sought the means to make their own inexpensive, durable clothing. In the 1840s and 1850s, inventors Elias Howe and Isaac Singer responded to American demand with the sewing machine, which allowed for both individual and collective production of clothing for personal and consumer use. The sewing machine helped break down class barriers by standardizing clothing, so that the factory laborer could buy similar clothes, and dress in the same fashion, as the factory owner.

Science and technology encouraged the standardization and democratization of invention, production, and consumption. Henry Ford, for example, was an American hero, because he integrated the elements of manufacture into an efficient assembly-line process. The result was a product without variety—the ownership of a Ford automobile was not distinctive and its use was not unique—and thus Ford drastically lowered the cost of the automobile, making it available to all Americans. This process would extend to numerous other consumer products.

During the late nineteenth and twentieth centuries, natural limitations on human movement yielded to American inventiveness. Communication across vast distances was enabled by Samuel F.B. Morse's telegraph (first demonstrated in 1838) and Alexander Graham Bell's telephone (1876). As the limitations of space and time were circumvented by communication and transportation devices, likewise the most fundamental of natural laws limiting human movement, gravity, was overcome by American inventors. In December 1903, bicycle makers Orville and Wilbur Wright flew a heavier-than-air machine over the sands of Kitty Hawk, North Carolina. In 1940, Russian immigrant Igor Sikorsky successfully flew the first helicopter.

One of the most outstanding examples of American inventiveness is the perfection of the rocket. Yankee ingenuity is illustrated in the life and accomplishments of Robert Goddard, a native of Worcester, Massachusetts, and graduate of Worcester Polytechnic Institute. A reclusive physics professor at Clark University, his pioneering genius gave birth to jet propulsion and liquid-propellant rocketry, and he secured an astonishing number of patents through his work. His suggestive paper, "A Method of Reaching Extreme Altitudes," published in 1919, was followed by the successful launch of a liquid-fueled rocket in 1926.

Goddard left New England in 1930 for the quiet and climatic constancy of Roswell, New Mexico, where he continued his experiments with liquid fuel, particularly liquid oxygen. In April 1932, he launched an 11 foot (3.35 meter) rocket to an altitude of 2,000 feet (620 meters) at 500 miles (800 kilometers) per hour. At Roswell, Goddard worked to ensure vertical stability by means of gyroscopes and pendulums; optimum power and acceleration with the most efficient weight; remote control; and reusable combustion chambers. During World War II, he continued his work on liquid-fueled rockets for the U.S. Navy.

Other American scientists and engineers continued the work in rocketry, producing the most astonishing examples of American ingenuity: rockets capable of reaching the moon and, in 1969, landing a human on its surface.

Russell Lawson

Sources

Boorstin, Daniel. *The Americans.* 3 vols. New York: Vintage Books, 1964–1973.

Cohen, I. Bernard. *Benjamin Franklin's Science.* Cambridge, MA: Harvard University Press, 1990.

Lawson, Russell M. "Science." In *Encyclopedia of New England.* New Haven, CT: Yale University Press, 2005.

———. "Science and Medicine." In *American Eras: The Colonial Era, 1600–1754,* ed. Jessica Kross. Detroit: Gale Research, 1998.

The Bounty of North America

The first European explorers of North America were struck by the incredible, seemingly unlimited bounty of the sea, rivers, forests, mountains, and plains. Learning the farming techniques of the American Indians, using trial and error, and bringing from Europe assumptions and methods of science, colonial Americans learned how to forage for food, medicine, and building materials; grow crops for consumption and trade; dig iron from bogs and smelt it into wrought iron and pig iron; build mills next to rivers and streams; and apply knowledge and technology in many other ways to make the land productive and their lives secure.

Fruit of the Forest

The forests and plains provided plentiful materials and wholesome food for colonists and the native peoples alike. They used basswood to make ropes and mats; flowering dogwood, black walnut, and black oak for dye; arrowwood branches for the shafts of arrows; the inner bark of the slippery elm as a glue; red mulberry fibers for clothing; and wax of the honeybee for candles.

Trees in which the hunter could find beehives filled with honey and wax included the basswood and southern bayberry. Fruit trees yielded their bounty in summer and fall. The persimmon, a sweet, pungent fruit, was dried by Indians and used in bread. Natives and settlers alike enjoyed the pawpaw. The fruit of the black cherry tree could be made into jelly and wine. The berries of the blackhaw made a good jam.

The inner bark of the sweet gum tree provided chewing gum. Indians ate the sprouts of the smooth sumac. Chewing sumac fruit helped ward off thirst. Winged sumac berries provided winter food. Maples produced sugar, as did the box elder and showy orchid. Many trees produced edible flowers in the spring; chief among these were the redbud, the flowers of which were used in salads. The common blue violet also was a salad material. The jack-in-the-pulpit was boiled and eaten.

The American Indians and settlers also gathered the nuts of the forest—the chestnut, pecan, black walnut, and butternut. The nut of the shagbark hickory could produce, upon boiling, an oily substance called *pawcohicorn* that Indians used as an ingredient in various foods. Bitternut hickory nuts produced oil for lamps.

Rural Economy

Agriculture was the foundation of colonial American society, culture, and economy. All thirteen of the colonies were agriculturally based, some more than others. New England colonies such as New Hampshire and Maine also were heavily dependent on fishing, shipbuilding, trade, and lumber. The Northern and Middle Colonies raised sheep for wool. Breeding of sheep, cattle, and horses was carried on actively.

The heavy, rocky terrain of New England required draft animals and sturdy plows to penetrate the soil. The plow was an essential tool, used with more frequency after 1650. During the colonial period and into the nineteenth century, plows became more sophisticated, with iron then steel plowshares (used to cut the soil) and wooden then iron plated moldboards (used to turn the soil). Plows developed during the colonial period included the Carey Plow, which used a wrought-iron plowshare, and the shovel-plow, which used a shovel-like plowshare to cut the soil with a furrowing action.

An abundance of lumber, the bounty of the sea, and the importance of maritime trade to the colonial economy made shipbuilding a vital industry in early America. *(MPI/Hulton Archive/Getty Images)*

Weavers and fullers produced cloth for domestic consumption from wool, flax, and cotton. New Englanders established fulling mills soon after settlement in the early to mid-1600s. Every colonial family had spinning wheels and hand-looms for making thread and fabric for home-spun clothes. Flax, a versatile plant that can be used for fibers and food, was cultivated in colonial America primarily for a thread that was strong and sturdy.

Southern colonies produced cash crops such as tobacco, sugar, indigo, and rice. Rice was introduced in the Carolinas during the mid-seventeenth century. Tobacco was introduced in Virginia much earlier, shortly after the founding of Jamestown.

Eliza Lucas of South Carolina was famous for her experiments on and production of indigo, a plant that forms a blue dye, which was increasingly in demand during the eighteenth century, particularly in England among cloth manufacturers. The best dye is produced through the leaves, but sometimes the whole plant is used. Lucas experimented for several years, trying different planting times and different soils to see which crop would be best. In 1741, samples of her indigo harvest were sent to England and declared the finest yet seen, even better than those produced by the French in their Caribbean colonies. Lucas shared seed with other South Carolina farmers, and the indigo harvest grew yearly, becoming an important part of the colony's economy.

Colonial Industries

Mills of various types were a necessity in burgeoning towns of early America. Mills were used to grind grain (gristmills), cut wood (sawmills), improve cloth (fulling mills), and make paper (paper mills). A moving stream of water pushed a waterwheel that powered a system of gears that rotated a saw blade or moved a millstone or other devices. Undershot and overshot mills were erected perpendicular to a stream: the waterwheel of an undershot mill moved clockwise; in an overshot mill, the waterwheel moved counterclockwise. Horizontal mills lay parallel to a stream.

North America had plentiful bogs that provided iron ore for those who knew how to smelt and hammer it into useful implements. Lacking a good iron source, England encouraged colonial production of pig iron, formed into bars or ingots that could be exported to the mother country. The English were less enthusiastic about colonial iron production that resulted in wrought iron for use in farm and building materials.

One of the first ironworks in the colonies was located at Saugus, Massachusetts, where, in the 1640s, Joseph Jenks built a furnace and forge to heat the iron with charcoal and limestone to produce pig iron. At its peak in the late 1640s, the Saugus Ironworks produced one ton of pig iron per day.

Fishing

When John Smith cruised the New England coast in 1614, he learned from the native Algonquin that the number of fish in the sea, rivers, and streams was comparable to the hairs on the head—

uncountable. Smith became a proponent of the colonial fishing industry, which came to fruition shortly after the founding of initial settlements at Plymouth, Massachusetts Bay, and Strawbery Banke (New Hampshire). Smith envisioned the center of the fishing industry at the Isles of Shoals off the coast of New Hampshire and Maine.

Indeed, during the 1600s and 1700s, the Isles of Shoals, particularly the town of Gosport, became the center of the New England fishing industry. Here fishermen and fishwives spent their days fishing, cleaning the catch, and drying the cleaned fish on stakes, wooden rafters that allowed the fish to be dried by the sun and the wind. Salt-fish was produced in huge quantities, as was dumb-fish, a fish (particularly cod) allowed to mellow, sometimes by being buried in the ground for several days, before it was boiled.

Russell Lawson

Sources

Lawson, Russell M. "Science and Medicine." In *American Eras: The Colonial Era, 1600–1754*, ed. Jessica Kross. Detroit: Gale Research, 1998.

Russell, Howard. *Indian New England before the Mayflower.* Hanover, NH: University Press of New England, 1980.

Smith, John. *The Complete Works of Captain John Smith.* Ed. Philip Barbour. 3 vols. Chapel Hill: University of North Carolina Press, 1986.

Science and the Industrial Revolution

Historians conceptualize the development of science and industry as a process that occurred over time through successive periods of advancement: the agricultural revolution of the pre-industrial sixteenth and seventeenth centuries, the Industrial Revolution of the mid-eighteenth and mid-nineteenth centuries, and a second industrial revolution during the late nineteenth and early twentieth centuries. While others were early leaders in developing theories of science, Americans were skilled at applying scientific theory, yielding technologies that spurred industrial and economic growth.

Agricultural Revolution

The social and technological changes in sixteenth- and seventeenth-century Europe and America known as the agricultural revolution produced a new economy that gave impetus to the even greater technological innovation of the Industrial Revolution. With the demise of serfdom, and with more efficient food production reducing the need for agricultural workers, many people migrated to the larger population centers in search of employment, which provided the ready pool of workers that made the Industrial Revolution possible.

The migration of peasants from the landed estates of Europe contributed a considerable number of settlers to the New World. These settlers introduced the new techniques of the agricultural revolution to the primarily agrarian society of the Americas, which was eager to employ any technology that would stimulate production. The American agricultural economy was unencumbered by outmoded land-use laws, which existed in Europe, and Americans had already demonstrated a willingness to develop and market new crops such a tobacco and corn.

American landowners sought to decrease the number of workers needed and the total cost of production by increasing yield per worker through new farming technology. Innovations that contributed to this greater efficiency included new crops, new techniques of land use, and mechanization. New crops such as potatoes and clover improved the crop yield per acre in two ways: These new root and fodder crops produced higher yields of animal feedstuffs than the grasslands they replaced, and the mix of crops intended for human consumption changed to higher-yielding crops. A secondary effect of these new crops was to shift the human diet away from cereals as a staple toward meat and other animal products. In addition, increased crop yields produced new marketable goods for a rapidly increasing population in America.

From the end of the eighteenth century to the beginning of the nineteenth century, the agricultural revolution was characterized by changes in land use, larger farms, regional specialization, and the introduction of new machinery and improved fertilizers, livestock feedstuffs, and drainage. In land-rich America, new lands were being cleared, and crop rotation, selective breeding, and herd management were introduced.

Two particular land reclamation innovations, both technologically based, were clearing and drainage, and fertilization and restoration. Clearing and drainage not only increased the quantity of available arable land, but it also allowed the land to be worked by mechanized farming techniques, using new plows and new plowing methods. Chemical fertilizers and revitalizing cover crops restored fertility to depleted lands. For example, chemical nitrogen fertilization and crop-based nitrogen fertilization using leguminous crops, such as clover, returned some overworked lands to production and improved production on in-use farm lands by as much as 60 percent.

The unenclosed, open field system of the Middle Ages was based on a three-year crop rotation, with different crops planted in each of two fields and a third field lying fallow. But the Dutch introduced a four-field crop rotation system by planting turnips and clover in fields that would have remained fallow, thus increasing overall production. Clover also proved to be an excellent fodder crop, used for livestock feedstuffs. The new system of crop rotation increased the area of arable land; the increased grain and cereal crops increased livestock production, which in turn yielded more manure for fertilizer.

The agricultural revolution brought improvements to old technologies and introduced new ones, primarily mechanization. For example, Jethro Tull's seed drill more evenly distributed seeds than was possible before its use, and Disney Stanyforth's and Joseph Foljambe's swing plough cultivated soil that traditional ploughs could not, making more land arable. Though small improvements to basic farm implements increased their efficiency, the introduction of mechanized harvesting—using thrashing or threshing machines, reapers, steam engines, and internal combustion engines—exponentially increased productivity.

With more production per acre, landowners needed fewer workers. Agrarian wealth became concentrated in the hands of fewer people. Unskilled workers sought jobs in industry and commerce rather than on farms, providing urban centers with factory workers and consumers. Between 1790 and the end of the nineteenth century, the number of people employed in agriculture dropped by 40 percent.

Industrial Revolution

Application-oriented science, often based on trial and error, was employed in the Industrial Revolution launched in the mid-eighteenth century. For example, the developing American textile industry relied on American and European inventiveness.

Of the former was Eli Whitney, who invented the cotton gin, allowing for the efficient production of cotton for New England factories, and the concept of interchangeable parts—subsequently called the American system of manufacturing—that he developed as a consequence of mass producing muskets for the U.S. military. Americans also turned European ideas and inventions into the efficient production of goods. English immigrant Samuel Slater, for example, brought Richard Arkwright's idea of the spinning frame to America, introducing its use at Slater's Mill in Pawtucket, Rhode Island.

During the second half of the nineteenth century, industrialization was driven by innovations in chemicals, petroleum refining and distribution, machinery, food and consumer production and distribution, electrical industries, and engine-based industries such as the automotive and farm machinery industries. At the same time, there was a shift in scientific, industrial, and technological leadership away from Great Britain to the United States. The primary technological changes in America included new basic materials, new energy sources, new machines and inventions, new organizational structures, new forms of transportation and communication, and new applications of science to industry and technology.

From the end of the nineteenth century to the first half of the twentieth century, science and industry were characterized by advances in and the predominance of steel, chemicals, the internal combustion engine, and electrical equipment and motors. It was in the last half of the nineteenth century that the distinction between engineering, applied science, and the pure sciences was first drawn, and it was at this time that science and technology transitioned from technology-driven science to science-driven technology.

For example, with advances in the science of metallurgy, American steel manufacturers were able to improve their processes and create products to the varying specifications of individual industries. In a similar vein, Nikola Tesla's knowledge of electromagnetism, specifically his rotating magnetic field principle in 1882, improved Thomas Edison's direct-current dynamos. And it was Tesla's polyphase system of alternating-current dynamos, transformers, and motors that allowed the efficient transmission of electricity over great distances.

The United States took the lead in the second industrial revolution by virtue of its abundant natural resources, assembly-line manufacturing techniques, advances in transportation and communication, and the application of science to technology. With its substantial raw materials and energy resources that were made accessible by an expanding railroad system, the United States began to exceed the industrial production of Western Europe in the late nineteenth century. By the first quarter of the twentieth century, the United States was well ahead of Western Europe in terms of industrial output.

New forms of transportation, such as the steam locomotive, steamship, automobile, and airplane, reduced travel time between various parts of the country, and made new areas accessible. New and improved forms of communication, such as the steam-powered rotary printing press, telegraph, telephone, and radio, enhanced the flow of information and allowed individuals, families, and businesses to more easily maintain contact across distances.

In business, improvements in transportation and communication allowed companies to link factories with off-site corporate management as well as outside suppliers. In one example, increased communication allowed General Motors to decentralize its management structure and give its operating divisions substantial autonomy within the parameters of the overall business plan. Levi Strauss was able to leave his brothers with the family dry goods business in New York and strike out for California to establish a long-distance partnership. Additionally,

businesses no longer had to move to be near their customer base; they simply dispatched a sales force and shipped their goods as needed.

The new forms of transportation allowed for increased distribution of goods throughout the world and allowed for the vertical integration of an industry without having to place a factory at the production site of components or the source of raw materials. Coke and iron could be brought to the steel mill, and parts for the production of automobiles could be shipped to the assembly plant. Carnegie Steel used railroads and ships to move the raw materials for steelmaking to its mills.

Henry Bessemer's pneumatic steelmaking process made bulk steel production possible, and Karl Wilhelm Siemens's open-hearth process allowed the efficient production of quality steel; these processes made possible the increase in industrial production that epitomized the second industrial revolution. The rapid introduction of the Bessemer process in the United States by Andrew Carnegie's J. Edgar Thomson Steel Works (later Carnegie Steel Company) during the 1870s catapulted American steel production beyond that of Great Britain.

The ready availability of steel for durable rails helped expand the growing railroad industry. The expansion of American railroads opened vast portions of the nation to settlement, manufacturing, and increased utilization of natural resources, thereby creating correlative growth in the demand for consumer goods.

Evolving industries relied on the accessibility and transportability of natural resources such as coal, iron, and petroleum. Moreover, the increasing availability of petroleum products at an affordable price powered the development of the internal combustion engine, giving rise to the American automotive and farm machinery industries. By 1929, the United States accounted for 85 percent of the world's output of automobiles, and it was the world leader in gasoline-powered farm machinery, construction equipment, and trucks.

Henry Ford adapted Eli Whitney's American system for the first assembly-line-manufactured automobile. Ford's innovation was to make the system a continuous process that enabled nonstop manufacturing with less worker intervention. In continuous-process manufacturing, machines are idled only for repairs or scheduled maintenance, requiring a workforce to be present twenty-four hours a day, seven days a week. The process was initially introduced in the mass production of foodstuffs in the United States. By the mid-twentieth century, it was used to manufacture everything from automobiles to tools to pharmaceuticals.

In the early twentieth century, industries that were dependent on the basic sciences, such as the chemical and electrical industries, began to recruit engineers, physicists, chemists, and other scientists. Thus, they were able to provide a competitive edge in the development and refinement of manufacturing processes and products.

Such companies as AT&T, B.F. Goodrich, Westinghouse, General Electric, and Bayer led their respective industries into the late twentieth century because of their long-standing commitment to basic research and the application of scientific knowledge. Companies that did not make this commitment, such as International Harvester, General Motors, and U.S. Steel, eventually lost market shares to companies that did.

Some historians believe that the twentieth century was characterized by a third industrial revolution, identified by various inventions and communications technologies. These include Lee de Forest's invention of the three-electrode vacuum tube in 1907, the invention of the transistor at Bell Laboratories in 1948, and the first patent for a silicon chip, by Jack Kilby of Texas Instruments, in 1959. Some historians also point to the development of the first mainframe, the desktop computer, the Windows operating system, and the browsable Internet. For other historians, the major transition now under way is essentially an electronic or information revolution.

Richard M. Edwards

Sources

Ackoff, Russell L. *The Second Industrial Revolution.* Washington, DC: Alban Institute, 1975.

Chandler, Alfred D., Jr. *The Visible Hand: The Managerial Revolution in American Business.* Cambridge, MA: Belknap Press, 1977.

Felman, Lewis. *Second Industrial Revolution.* Upper Saddle River, NJ: Prentice Hall, 1985.

Finkelstein, Joseph. *Windows on a New World: The Third Industrial Revolution.* Westport, CT: Greenwood, 1989.

Koning, Niek. *The Failure of Agrarian Capitalism: Agrarian Politics in the UK, Germany, the Netherlands, and the USA, 1846–1919.* New York: Routledge, 1994.

Overton, Mark. *Agricultural Revolution in England: The Transformation of the Agrarian Economy 1500–1850.* Cambridge Studies in Historical Geography. Cambridge, UK: Cambridge University Press, 1996.

Rutledge, John, and Deborah Allen. *Rust to Riches: The Coming of the Second Industrial Revolution.* New York: HarperCollins, 1989.

Stearns, Peter N. *The Industrial Revolution in World History.* 2nd ed. Boulder, CO: Westview, 1998.

Albert Einstein and Atomic Power

In the years immediately after World War II, a common perception throughout the world was that Albert Einstein, the physicist who developed the theory of relativity, was also directly involved in the development of the atomic bomb used against Japan in August 1945. Later that year, however, Einstein remarked, "I do not consider myself the father of the release of atomic energy. My part in it was quite indirect. I did not, in fact, foresee that it would be released in my time. I believed only that it was theoretically possible."

In a quirk of fate, the German physicist Otto Hahn accidentally split the uranium atom a year before Hitler's armies overran Poland and began World War II in Europe. Atomic energy was irreversibly linked to the war. By 1940, other experiments by other physicists had shown how fission could pass from one atom to another, a chain reaction of splitting atoms that released tremendous energy. American scientists took the lead in research and development early in the war.

Einstein's involvement consisted of a single, timely letter to President Franklin Roosevelt in August 1939 that helped inaugurate the Manhattan Project, America's official involvement in developing an atomic weapon. When the United States tested a plutonium bomb in July 1945, and, a few weeks later, used a uranium bomb against Hiroshima and a plutonium bomb against Nagasaki, Einstein agreed that such destruction was justified from the standards of wartime morality. The end of war was achieved. But what would the peace be like?

From 1945 until his death ten years later, Einstein threw himself into the political, scientific, and moral debates taking place throughout the world respecting the development and implementation of atomic power. The issue was not the validity of science or its applications. Atomic energy would no doubt be a useful part of human existence. To be sure, he stated in an address to scientists in 1948, uncontrolled technology may lead to destruction, but there is both good and bad in everything; the good must be coaxed, cultivated like a weak plant. "Penetrating research and keen scientific work have often had tragic implications for mankind, producing, on the one hand, inventions which liberated man from exhausting physical labor, making his life easier and richer; but on the other hand, introducing a grave restlessness into his life, making him a slave to his technological environment, and—most catastrophic of all—creating the means for his own mass destruction."

Einstein believed that the marriage of humanity and technology is filled with love and hate, but divorce is unthinkable. Rather, if the match threatens evil, humans must alter the conditions under which the match exists. According to Einstein in a 1946 radio broadcast, "the development of technology and of the implements of war has brought about something akin to a shrinking of our planet. Economic interlinking has made the destinies of nations interdependent to a degree far greater than in previous years. The available weapons of destruction are of a kind such that no place on earth is safeguarded against total destruction. The only hope for protection lies in the securing of peace in a supranational way. A world government must be created which is able to solve conflicts between nations by judicial decision."

He went on, "Moral authority alone is an inadequate means of securing the peace." Scientists, trained to examine problems objectively and to arrive at disinterested solutions, should take the lead in formulating a world society based on the peaceful control of nuclear weapons.

Einstein knew from personal experience that reason must be inspired by a sense of the beauty of the universe, a childlike wonder of the mysteries of existence that all scientists share. Science involves creativity. So, too, does the solution of peace in our time. The passionate logic of the world's scientists was exactly what the world needed in the nuclear age. Einstein was eccentric and naive enough to believe that the creative impulse, the human willingness to surrender to beauty, would overwhelm the forces of militarism, ideology, and inhumanity. In a world of icons and isms, life becomes so abstract as to be meaningless.

"The release of atomic energy," Einstein declared in 1946, "has not created a new problem. It has merely made more urgent the necessity of solving an existing one. One could say that it has affected us quantitatively, not qualitatively. As long as there are sovereign nations possessing great power, war is inevitable." Unrestrained technology in a world where the state is an "idol"

```
                              Albert Einstein
                              Old Grove Rd.
                              Nassau Point
                              Peconic, Long Island

                              August 2nd, 1939

F.D. Roosevelt,
President of the United States,
White House
Washington, D.C.

Sir:

        Some recent work by E.Fermi and L. Szilard, which has been com-
municated to me in manuscript, leads me to expect that the element uran-
ium may be turned into a new and important source of energy in the im-
mediate future. Certain aspects of the situation which has arisen seem
to call for watchfulness and, if necessary, quick action on the part
of the Administration. I believe therefore that it is my duty to bring
to your attention the following facts and recommendations:

        In the course of the last four months it has been made probable -
through the work of Joliot in France as well as Fermi and Szilard in
America - that it may become possible to set up a nuclear chain reaction
in a large mass of uranium,by which vast amounts of power and large quant-
ities of new radium-like elements would be generated. Now it appears
almost certain that this could be achieved in the immediate future.

        This new phenomenon would also lead to the construction of bombs,
and it is conceivable - though much less certain - that extremely power-
ful bombs of a new type may thus be constructed. A single bomb of this
type, carried by boat and exploded in a port, might very well destroy
the whole port together with some of the surrounding territory. However,
such bombs might very well prove to be too heavy for transportation by
air.
```

Albert Einstein's fateful letter to President Franklin D. Roosevelt in August 1939 warned of the possibility that Germany might build an atomic bomb and helped persuade FDR to fund the Manhattan Project. After the war, Einstein favored nuclear disarmament. *(MPI/Hulton Archive/Getty Images)* ensures that conflict will continue. Yet, Einstein maintained, "that was true before the atomic bomb was made. What has been changed is the destructiveness of war." Humans remain the same: they continue to be volatile, jealous, passionate, and warring. The ends of war never alter, but the means do. Technological changes demand comparative changes in human society, politics, institutions, beliefs, and even the assumptions on which science is based.

In 1951, commenting on Johannes Kepler (1570–1631), Einstein praised the German mathematician and astronomer for his ability to go beyond established preconceptions, to question and search for new answers, and to break from paradigms that shackle the human mind to scientific constructs of the past. Such, Einstein thought, was the dilemma of his own time. Humans still adhered to ideas untenable in an age of nuclear destruction. New ideas must be developed, new assumptions conceived. Einstein believed that scientists who, like Kepler, train themselves to go beyond the past and seek new truths ultimately will have the tools necessary to lead the world to peace.

Russell Lawson

Sources

Clark, Ronald W. *Einstein: The Life and Times.* New York: Avon, 1999.

Einstein, Albert. *Essays in Humanism.* New York: Philosophical Library, 1950.

AGRICULTURAL ENGINEERING

Traditionally, agricultural engineers have applied their knowledge and training to the design and upkeep of farm machinery, the architecture and arrangement of farm structures, power generation and use, and the conservation of resources. More recently, the field has expanded to include aquaculture, information technology, and renewable energy production.

The American Society of Agricultural Engineers (ASAE), founded in Madison, Wisconsin, in 1907, recently changed its name to the American Society of Agricultural and Biological Engineers (ASABE), reflecting vast changes in the methods of food production and processing and the advent of computerized farming and biotechnology. Still, agricultural engineers today focus mainly on increasing the efficiency of farm operations—usually by substituting capital for labor and thereby increasing production per worker—while setting the standards for farm machinery and processes most likely to achieve those efficiency gains.

Long before agricultural engineering became a recognized profession, American farmers and inventors (often one and the same) were developing machinery and techniques to make farm operations less burdensome and more profitable. As early as the late eighteenth century, seed drills had slowly begun to replace broadcast sowing, the cradle was making headway against the sickle for grain harvesting, and numerous tinkerers developed improved plow designs, including Thomas Jefferson who invented a better-scouring moldboard plow. Nonetheless, many farmers saw little need to change their traditional practices, particularly if self-sufficiency rather than the production of a large marketable surplus was their primary goal, as indeed was the case for thousands of small farmers in America.

Several factors converged in the nineteenth century, however, that permanently altered the nation's farming regimen. A steadily rising urban population increased demand for the products of American farms, while the expansion of a national transportation network facilitated the marketing of those crops and simultaneously reordered the geographic distribution of production and specialization. The center of national wheat growing, for example, shifted steadily westward over the course of the nineteenth century, as farmers transformed the fertile prairie lands of the Midwest into grain fields.

With their matted layers of grass and sometimes sparse supplies of water, the prairies proved a daunting obstacle for farmers, as well as an opportunity for the empirical, trial-and-error variety of invention and engineering that predominated in the 1900s. Steel plows, reapers and binders, windmills, and barbed wire eventually emerged as dominant systems in prairie farming, but only after literally thousands of variations had been tried and discarded. In fact, there were still dozens of different standards for many of these devices on the eve of the ASAE's organization. At the same time, the closing of the western frontier in the late nineteenth century heralded the beginning of a new era of intensive farming, in which agriculturists would have to learn how to get more production out of existing acreage rather than simply move west and start over again whenever the soil became exhausted.

The ASAE's message of standardization and efficiency promised scientific solutions to the manifold problems of modern farming, substituting systematic professionalism for the empiricism of previous decades. Agricultural engineers, for instance, argued that the confusing network of early twentieth-century equipment standards was inherently inefficient. Why should there be dozens of design types for ubiquitous farming equipment such as plows or harrows when soil conditions did not warrant such variety?

This approach was not without its critics. Farmers and their organizations were sometimes suspicious of the close relationship between agricultural engineers and farm implement companies, and some farmers simply resented advice from "college" types. Meanwhile, civil, mechanical, and later electrical engineers wondered why they should define themselves under the nebulous title of "agricultural" engineer when their own professional organizations already provided the necessary identity and standards.

Yet as farming became increasingly mechanized, irrigated, and electrified in the first half of the twentieth century, and as farm operations grew even more consolidated and specialized, the unique engineering requirements of agriculture became more obvious, helping to define the field in the process. So, too, did the establishment of numerous degree-granting land grant university programs in agricultural engineering.

In recent decades, the increasing mechanical and biological complexity of U.S. farm operations, particularly with the advent of new environmental concerns, as well as advances in biotechnology and information technology, have ensured that the profession of agricultural engineering is here to stay. Moreover, as these same concerns have spread around the world, the profession has become increasingly international. The ASABE now claims 9,000 members in more than 100 countries.

Jacob Jones

Sources

American Society of Agricultural and Biological Engineers. http://www.asabe.org.

Fitzgerald, Deborah. *Every Farm a Factory: The Industrial Ideal in American Agriculture.* New Haven, CT: Yale University Press, 2003.

Gates, Paul W. *The Economic History of the United States.* Vol. 3, *The Farmer's Age: Agriculture, 1815–1860.* Armonk, NY: M.E. Sharpe, 1960.

Kohlmeyer, Fred W., and Floyd L. Herum. "Science and Engineering in Agriculture: A Historical Perspective." *Technology and Culture* 2:4 (1961): 368–80.

Stewart, Robert E. *Seven Decades That Changed America: A History of the American Society of Agricultural Engineers, 1907–1977.* St. Joseph, MI: American Society of Agricultural Engineers, 1970.

AGRICULTURAL EXPERIMENT STATIONS

Agricultural experiment stations were first established in the United States in the 1870s. Affiliated with state universities, these laboratories conduct agricultural research and experiments to improve American agriculture.

American farmers relied on European farming methods until the nineteenth century, when expanding market demand for agricultural products led them to become more innovative. By 1825, courses in agriculture and sciences were being offered at schools and colleges, resulting in nationwide research and agricultural experimentation. To encourage such agricultural education, the Morrill Land Grant College Act of 1862 provided funding to set up agricultural and mechanical colleges in each state.

As agricultural research activities expanded, educators realized that research results could benefit not only students but also American farmers, and they recommended that agricultural extension departments be created at colleges to aid in getting this information to

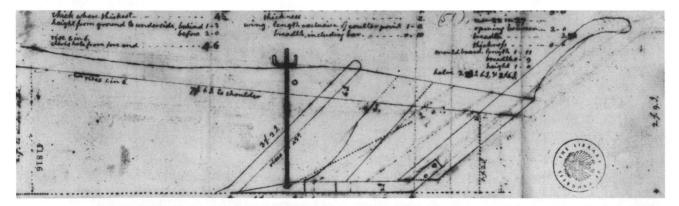

Thomas Jefferson, one of the most innovative planters in the American colonies, regarded agriculture as "a science of the very first order." His many inventions included a plow that could delve deeper than standard plows of the time and help limit soil erosion. *(Library of Congress, LC-MSS-27748–64)*

farmers. The first agricultural experiment stations were established in Connecticut and California in 1875, and, by 1887, there were experiment stations connected with eight land grant colleges, operating exclusively with state funding. Thirteen other states conducted less formalized work.

On March 2, 1887, led by Missouri Representative William Hatch, the U.S. Congress approved the Hatch Experiment Station Act to provide federal funding to each land grant university to create a network of agricultural experiment stations. Under the legislation, the stations would continue to be owned and controlled by the states and operate on state and federal monies. The purpose of the act was to elevate agriculture to a science by promoting scientific research and experimentation using the principles of agricultural science, to create uniformity in methods and results among the stations, and to distribute these research findings to farmers. Under the act, stations would conduct research on plant and animal physiology and diseases, crop rotation, new plant and tree acclimation, soil and water analysis, composition of manures, adaptation of grasses and forage plants, comparability of food for domestic animals, and methods of production.

In 1888, the Office of Experiment Stations was established to coordinate state and federal agricultural research and to publish important station research. By 1913, most of this publishing was done by the state stations themselves through periodic bulletins and annual reports. The Office of Experiment Stations became part of the USDA Agricultural Research Administration in 1942.

Since the passage of the Hatch Act in 1887, work done at the agricultural experiment stations has significantly contributed to solving many problems in agriculture by revising farming methods and making American farms more productive. Experiment stations also have influenced other areas, including the biological and medical sciences, which have used the research to cure diseases such as brucellosis, rabies, and tuberculosis. The use of antibiotics grew out of the research on soil bacteria at experiment stations.

Judith B. Gerber

Source

Marcus, Alan I. *Agricultural Science and the Quest for Legitimacy: Farmers, Agricultural Colleges, and Experimental Stations, 1870–1890.* Ames, IA: Blackwell, 1985.

AGRICULTURE

Agriculture is the management of land to grow crops or rear livestock for food, fiber, and other products such as skins. It has been a central factor in societal development and economic growth, requiring both heavy labor and the advancement of scientific research and technological innovation. The European discovery and colonization of the New World resulted in an exchange of crops and animals that transformed indigenous agricultural systems. Wheat, barley, various legumes and fruits, cattle, goats, and sheep were introduced to the New World, while squash, corn (maize), tomatoes, potatoes, and tobacco were introduced to Europe. The forests of America's eastern coast were opened up for prized wood and agriculture.

As the frontier moved west, further modification of the natural vegetation cover ensued. Wheat, corn, cotton, tobacco, rice, and sugarcane production dominated arable agriculture. Cattle and sheep herds came to occupy the central rangelands, where large-scale ranching was established. Market gardening for vegetables and fruit and dairy herds were established in urban hinterlands.

Ideas and innovations were exchanged between America and Europe, which was becoming increasingly reliant on America's food and fiber products. Examples of such innovations are the invention of the cotton gin by Eli Whitney in 1793, the development of food canning in the early 1820s, the invention of the mechanical harvester by Cyrus McCormack in 1831, the construction of the grain elevator in 1842, the pioneering of refrigerated meat transport by ship beginning in 1847, and the first steam tractors in 1868.

Science and scientists in the United States played a major role in the industrialization of agriculture and the emergence of large agribusinesses. This involved the increasing use of fossil fuels in mechanization, artificial fertilizer production, crop protection, and the development of chemicals to protect animal health. Related industries have developed, including those involved with food processing and agrochemical production for herbicides, insecticides, fungicides, and fertilizers. Along with animal and plant breeding programs, agricultural science has contributed to increasing yields, reduced labor costs, and wealth generation. At the same time, agricultural science has contributed to many environmental problems, such as soil erosion, water pollution, loss of habitats, and the invasion of alien species.

The United States today is a major world producer, consumer, and exporter of agricultural products, notably wheat, corn, soy, fruit, cotton, and meat. Its agricultural science continues to lead the world, notably in biotechnology, which involves the genetic modification of crops to create, for example, insect-resistant cotton and herbicide-resistant crops such as corn, soy, and wheat. The United States is also the world's major grower and consumer of genetically modified crops.

A.M. Mannion

Sources

Hurt, R. Douglas. *American Agriculture: A Brief History.* Ames, IA: Blackwell, 1994.

U.S. Department of Agriculture. http://www.usda.gov.

AGRONOMY

Agronomy, also known as agricultural science, is the practical use of knowledge about plants and soil to increase crop yields and crop production. It incorporates biology, botany, chemistry, and physics into an applied science that forms the foundation of agricultural practice. The state agricultural experiment stations established by the Hatch Act in 1887 provided American farmers with innovative new production techniques and soil management methods. This was illustrated by the research that the stations did on soybean production, which resulted in widespread distribution of soybeans for U.S. farmers to use as animal feed by the turn of the twentieth century.

One of the most influential practitioners of agricultural science was George Washington Carver, who, in 1896, became the director of agricultural research at Tuskegee Institute in Alabama. Carver conducted pioneering work in the field, finding more than 300 new uses for peanuts, sweet potatoes, and soybeans, helping to diversify the South's agriculture. His work contributed to the growing recognition of the field of agronomy.

As a result of the new trend in scientific agricultural research, agronomy was recognized as a branch of the agricultural sciences in the United States in the early 1900s. The American Society of Agronomy was established in 1907 and began publishing its *Agronomy Journal* later that year. The society, still active today, supports scientific, educational, and professional activities to enhance communication and technology transfer among agronomists and those in related disciplines on topics of local, regional, national, and international significance.

Agronomists help crop management by studying propagation, care, and management of cereal, field, and forage crops. Agronomy includes research on production techniques (such as irrigation) and on improving production through methods that increase quality and quantity of crops (such as selecting drought-resistant crops). Agronomists also study the prevention and correction of adverse environmental effects such as soil degradation.

Agricultural scientists classify crops into a number of agronomic categories. These include cereals (e.g., wheat, oats, barley); legumes (peanuts, beans); forage crops (grasses grown for animal feed); root crops (beets, carrots); tuber crops (potatoes); oil crops (flax, soybeans, sunflowers, mustard); sugar crops (sugarcane, corn); vegetable crops (potatoes, carrots, and so on); pharmaceuticals (tobacco, mint, hemp); fiber crops (flax, cotton); rubber crops (guayule); and special-purpose crops (cover crops, companion crops).

The study and practice of agronomy have changed significantly since the first half of the twentieth century. In the 1960s, intensified agriculture led to advances in selecting and improving crops and animals for high productivity and in developing new products like artificial fertilizers. Intensive agriculture has led to widespread environmental damage, however, and many agronomists started new fields—such as integrated pest management and waste treatment—to address the problems. Other new research fields include genetic engineering and precision farming.

Some of agronomy's contributions to crop production over the last fifty years include plant breeding, chemical fertilizers, and irrigation techniques. Agronomy also has led to better management of cropping systems such as double-cropping and soil conservation measures to minimize soil erosion.

Judith B. Gerber

Sources

American Society of Agronomy. http://www.agronomy.org.

Hurt, R. Douglas. *American Agriculture: A Brief History.* Ames, IA: Blackwell, 1994.

AMERICAN INDIAN CANOES

Native Americans over the centuries experimented with a variety of canoe types for use on rivers, streams, ponds, lakes, and coastal waters. The three principal types invented by the native peoples were skin canoes, dugout canoes, and bark canoes. The word "canoe" is of native origin, specifically from the Indians of the Caribbean, from whom it spread to North America by means of Spanish and French soldiers, traders, and hunters.

Dugout canoes were the most widespread type in America because of their rugged simplicity and adaptability to a variety of forest environments. Softwood trees such as the cottonwood (*Populus deltoides*), particularly those with a wide girth, found along the upper Missouri River, were best. Dugouts were heavier and less maneuverable than canoes made with wood frames and animal skin or bark, but they could be made quickly with a few simple tools, such as an adze, an axe, and flint to start a fire. The canoe maker would use the axe to fell a tree. The adze would be used to strip its bark and begin the process of alternately scraping and burning the log to form a hollow core. Hot coals softened the wood for easy removal. Dugouts were the canoes of choice for the southern Plains Indians and hunters and trappers of the southern Louisiana Territory.

American Indians and settlers of the northern Plains and northeastern forests had more choices in type of canoe to build because of the presence of the white birch (paper or canoe birch, *Betula papyrifera*). The white birch is found in northern climates and, along with pine, is a dominant tree of the northeastern forest. Birch bark peels off in long strips; it is flexible yet strong, and it looks like thick paper. Indians would fit the sheets of bark around a frame of spruce or cedar, then use black cedar roots, which are long, stringy, and strong, to sew the bark onto the frame. Such canoes were light yet sturdy, could hold much cargo and passengers, and could be easily repaired.

Similar were the skin canoes some native tribes made. These used a frame of pliable wood made soft in hot water; over this was stretched a buffalo or similar animal skin. The Inuits of Baffin Island used sealskins to cover their framed canoes and kayaks.

Russell Lawson

Source

Fichter, George. *How to Build an Indian Canoe.* New York: McKay, 1977.

APOLLO, PROJECT

On July 21, 1969, *Apollo 11* astronaut Neil Armstrong became the first human to set foot on the moon, declaring, "That's one small step for a man, one giant leap for mankind." This lunar landing, one of the biggest scientific achievements of the twentieth century, was televised to the world.

Project Apollo was begun by the National Aeronautical and Space Administration (NASA) in the early 1960s; it enjoyed huge success, cost $25.4 billion, and employed 400,000 people. The scientific goals of the project were to determine the history of the moon, to precisely map its surface, and to locate magnetic fields, sources of heat, and volcanic and seismic activity. Apollo was inaugurated by President John F. Kennedy's call to land a man on the moon before the decade of the 1960s was over. During the Cold War, the United States sought to keep ahead of the Soviet Union in science, technology, and discovery.

There were seventeen Apollo missions in all, ranging from success to disaster. *Apollo 1* was destroyed by fire on January 27, 1967, killing the crew of three men. *Apollo 7* (October 11–22, 1968)

The goal of the seventeen-mission, $25 billion Apollo program—and the entire U.S. space effort of the 1960s—was achieved with the moon landing by *Apollo 11* astronauts Neil Armstrong and Buzz Aldrin (pictured here) in July 1969. *(Neil A. Armstrong/NASA/Time & Life Pictures/Getty Images)*

was the first manned spacecraft that orbited the Earth, launched from the Kennedy Space Center in Florida, and the first to be televised from aboard the spacecraft. *Apollo 8* (December 21–27, 1968) marked the first time humans orbited the moon. *Apollo 10* observed the moon from ten miles out and orbited repeatedly as a rehearsal for a lunar landing.

On July 16, 1969, *Apollo 11* was launched with the crew of Neil Armstrong, Michael Collins, and Edwin Aldrin. The craft for landing on the moon consisted of two parts: the Command Service Module (CSM), which was the vehicle for the round trip to the moon; and the Lunar Module (LM), which was detached from the CSM in lunar orbit, carrying Armstrong and Aldrin to the surface of the moon. The LM *Eagle* landed on the Sea of Tranquility. Armstrong and Aldrin were on the lunar surface for about two hours, engaged in extra-vehicular activity (EVA). They collected rock samples and planted a seismometer. Upon their return to Earth, the USS *Hornet* collected the astronauts and the command module from the Pacific Ocean on July 24.

U.S. astronauts performed more experiments on the moon's surface during the *Apollo 12* mission (November 14–24, 1969). During the *Apollo 13* mission (April 1970), an oxygen tank exploded; the mission was aborted and the astronauts returned safely to Earth. The running of a battery-operated Jeep-like lunar vehicle was the highlight of the *Apollo 15* mission (July 26–August 7, 1971).

The *Apollo 17* mission (December 7–19, 1972) was the last in the program. The other planned missions were canceled for financial reasons. Thereafter, NASA turned its attention to the Space Shuttle program.

Patit Paban Mishra

Sources

Beattie, Donald A. *Taking Science to the Moon: Lunar Experiments and the Apollo Program.* Baltimore: Johns Hopkins University Press, 2001.

Chaikin, Andrew. *A Man on the Moon: The Voyages of the Apollo Astronauts.* New York: Viking, 1994.

Light, Michael. *Full Moon.* New York: Alfred A. Knopf, 2002.

Murray, Charles, and Catherine B. Cox. *Apollo: The Race to the Moon.* New York: Simon and Schuster, 1989.

ARMY CORPS OF ENGINEERS, U.S.

The U.S. Army Corps of Engineers has for more than 200 years been involved in engineering tasks for the United States, such as building military fortifications and monuments, bridging rivers, damming rivers and streams, surveying lands, building roads, and clearing forests. The Corps was created on June 16, 1775, by the Second Continental Congress for the purpose of building military structures for war with England.

Following the American Revolution, the Corps was temporarily disbanded, then reorganized in 1794 as the Corps of Artillerists and Engineers. In 1802, President Thomas Jefferson and Congress made the Army Corps of Engineers a separate entity from the Artillerists.

The task of the Army Corps of Engineers initially was to provide military fortifications. During the years leading up to the War of 1812, fortifications were established along the Atlantic coast to improve coastal defenses. Following that war, civilian responsibilities (such as surveying and road building) were given to the Corps by an act of Congress. During the 1820s, Congress commissioned the engineers and topographers of the

Corps to survey lands west of the Appalachian Mountains and along the Mississippi River to determine the best routes for roads, where canals were needed, what rivers needed dredging, and where rivers and streams should be bridged.

The Cumberland Road, which extended from Cumberland, Maryland, to Vandalia, Illinois, was completed by the Corps in 1841. The Corps was also responsible for building the Lincoln Memorial, the Washington Monument, and the Library of Congress. Other domestic activities of the Corps during the nineteenth century included surveys of coastlines, river deltas, and the Great Lakes; constructing lighthouses; building river levees; and flood control.

Historically, the U.S. Army Corps of Engineers has also been actively involved in the logistical requirements of war. In 1847, during the Mexican War, the Corps was considered crucial to the siege of the Mexican Army at the Battle of Vera Cruz. During the Civil War, the Corps acted on behalf of Union forces by building bridges and roads, ensuring the movement of supplies, constructing defensive fortifications, and erecting siege works.

During World War I, engineers were sent to France to help build roads and bridges and to supply lumber, especially for railroad tracks. Corps engineers and scientists were also involved in chemical weapons production and deploying tanks in battle. During World War II, the Corps again played a critical role in battle logistics, building airfields, military hospitals, munitions plants, supply depots, and bridges, and helping to destroy comparable enemy constructions. The efforts of the Corps continued in the Korean War and in Vietnam, with bridge building and forest cutting foremost among their contributions.

During the twentieth century, the Corps has been especially engaged in flood control and construction of hydroelectric dams, as in the Tennessee Valley and Oklahoma. The Army Corps of Engineers continues to operate many hydroelectric facilities. In recent years, the Corps has been particularly involved in relief from such natural disasters as floods and hurricanes.

Research and development has been an Army Corps of Engineers priority since its inception. Corps scientists and engineers have made particular contributions to the technology of hydraulics, such as developing meters to gauge river currents, devising techniques to increase water current to remove sandbars, and inventing hydraulic dredges for rivers.

Nicholas Katers

Sources

Reynolds, Terry S. *The Engineer in America: A Historical Anthology from Technology and Culture.* Chicago: University of Chicago Press, 1991.

Schubert, Frank N. *The Nation Builders: A Sesquicentennial History of the Corps of Topographical Engineers 1838–1863.* Washington, DC: U.S. Government Printing Office, 1980.

Shallat, Todd. *Structures in the Stream: Water, Science, and the Rise of the U.S. Army Corps of Engineers.* Austin: University of Texas Press, 1994.

ATOMIC BOMB

Atomic bombs are weapons that derive their explosive energy from nuclear fission, a process whereby certain radioactive elements split into lighter elements when bombarded by neutrons. The process initiates a chain reaction that releases tremendous amounts of energy. The term is actually somewhat of a misnomer, given the fact that nuclear fusion, the force behind the much more powerful hydrogen bombs, is no less atomic than nuclear fission. Still, the term is used primarily to identify the early pure fission bombs and occasionally as a catchword for all nuclear weapons.

The atomic bomb was a product of the Manhattan Project, a three-year, top-secret project involving the nations of Great Britain, Canada, and the United States. Atomic bombs were used on the Japanese cities of Hiroshima and Nagasaki in August 1945 in an effort to bring an end to World War II. The creation and use of atomic weapons forever changed the nature and course of America's place in international politics.

History

On August 2, 1939, physicist Albert Einstein wrote a letter to President Franklin D. Roosevelt concerning new theories by Enrico Fermi and Leo Szilard on using uranium to produce a nuclear chain reaction that might be harnessed into a bomb with unprecedented explosive power. The letter also hinted at possible efforts by Nazi Germany to use uranium from Czechoslovakian mines for experiments in nuclear fission. The letter urged Roosevelt to quickly support American university and perhaps even industrial research in this area to gain an advantage over the German effort.

Roughly a year later, in late 1940, a group of British scientists made a presentation to a select group of American physicists in Washington, D.C., on behalf of the MAUD Committee, a British group organized earlier in 1940 to study the possibility of developing a new weapon based on the fissioning of uranium-235 (U-235). The presentation led to the discovery that the two nations' research tracks were converging, and American and British scientists subsequently agreed to work together on the development of the atomic bomb.

Prior to the start of actual work on the design and construction of the atomic bomb, theoretical experiments had been carried out at a number of different research laboratories, including Columbia University and the University of Chicago. A number of work sites were later established in the United States, Britain, and Canada to support the massive scientific and technical effort needed to build the bomb. The U.S. and British governments worked together to purchase uranium ore from mines around the world, and uranium and plutonium production plants were built at Oak Ridge, Tennessee, and along the Columbia River at Hanford, Washington, respectively.

In 1943, physicist J. Robert Oppenheimer gathered the best scientific minds available and established Project Y at Los Alamos, a remote mesa in the Jemez Mountains north of Santa Fe, New Mexico. Because initial research had been conducted in New York at Columbia University, the Army Corps of Engineers Manhattan District was initially assigned to carry out any construction work relating to the project. Thus, "Manhattan Project" was the code name used for the project. Work at the principal Manhattan Project sites would eventually be supported by research and technical work at scores of universities and industrial corporations across North America.

Two Designs

The Manhattan Project research produced two distinct atomic bomb designs based on the principle of nuclear fission. Nuclear fission is the process whereby an atom splits, or fissions, into several smaller fragments after its nucleus is hit with a neutron. These fragments, or fission products, are less than the mass of the original atom because roughly 0.1 percent of the atom's original mass has been converted into energy. This conversion is the energy released during an atomic bomb explosion. The fission of a radioactive atom process produces several neutrons, which, in turn, bombard other nearby atomic nuclei to cause more fission and initiate a self-sustaining nuclear chain reaction. This all takes place during an atomic bomb explosion in less than a second.

One atomic bomb design, which would later be the basis for the "Little Boy" bomb used at Hiroshima, was a gun-type weapon that employed specially formulated high explosives to propel a mass of uranium along a gunlike barrel into the center of another mass of uranium. The impact of the two masses initiates a rapid nuclear fission and the violent release of energy in the form of an atomic explosion. The second bomb design, which would become the basis for the "Fat Man" bomb, used high explosives surrounding a plutonium sphere to compress it from all directions under millions of pounds of pressure to a point at which a nuclear chain reaction became self-sustaining and resulted in an atomic explosion. The scientists were so certain that the gun-type bomb would work that it was not tested before it was used on Hiroshima; the design of the implosion device was so speculative at the time that it had to be tested.

Probably the greatest challenge faced by the makers of the first atomic bombs was a lack of significant amounts of the U-235 and plutonium-239 (Pu-239) needed for the two bomb designs. Not only were uranium isotopes very difficult to extract, but the amount of ore needed for just a few pounds of enriched uranium was substantial.

Based on research conducted by Nobel laureate Harold Urey and his colleagues at Columbia

The explosion of atomic bombs on Hiroshima and Nagasaki (shown here) in August 1945 brought the end of World War II and marked the beginning of the nuclear age. The mushroom cloud became a symbol of the destructive power of modern technology. *(Keystone/MPI/Hulton Archive/Getty Images)*

University, a massive gaseous diffusion plant was constructed at Oak Ridge as the first step in the enrichment process. Enrichment then required the use of magnetic separation and gas centrifuge processes to further separate the useable lighter U-235 isotopes from the heavier, nonfissionable U-238 isotopes. Plutonium-239, an isotope of the man-made element plutonium, is found naturally in only minute amounts. The Pu-239 used in the implosion bomb was produced from uranium processed in the reactors at Hanford, Washington.

The testing of the "Gadget," the nickname used for the implosion device in Los Alamos during its development, took place at just a few seconds before 5:30 A.M. on July 16, 1945, at a desolate desert site called Trinity near Alamogordo, New Mexico. The detonation produced an explosion equivalent to nearly 20,000 tons of TNT and temperatures of millions of de-

grees Fahrenheit. The intense heat of the blast melted the sandy soil at the site, creating an olive-green radioactive glass called trinitite.

Use and Consequences

On August 6, 1945, the Japanese city of Hiroshima became the target of the world's first atomic bomb attack when the Little Boy bomb was exploded roughly 2,000 feet above the city with a force equivalent to nearly 12,500 tons of TNT. Three days later, the United States detonated a Fat Man atomic bomb over the port city of Nagasaki, destroying much of the urban center. The detonation produced an explosion equivalent to nearly 22,000 tons of TNT, but because of the steep slopes surrounding the city, the destructive effect of the blast was less than that of Hiroshima. Roughly one-quarter of a million people were killed in Japan by the world's first two atomic bombs. Many died well after the initial blast from burns and radiation sickness.

The atomic bombs used on Japan would undergo further refinements in the years that followed, based on data from Hiroshima and Nagasaki and on atomic tests conducted at Enewetak atoll in the South Pacific and at the Nevada Test Site. On November 1, 1952, an atomic bomb would be exploded as part of a larger, more powerful experimental hydrogen bomb device, code-named Ivy Mike, to usher in the current era of fusion-based nuclear weapons.

The development of the atomic bomb is considered one of the most controversial events in American science. Arguments over the bomb's invention and use have gone on for more than half a century and will no doubt continue. Proponents argue that the bombings of Hiroshima and Nagasaki halted a terrible and bloody war, perhaps saving many thousands of American and Japanese lives. Opponents contend that the use of two bombs on unprepared civilian populations was, among other things, reckless and cruel.

Perhaps the greatest consequence of the invention of the atomic bomb, and the generations of hydrogen bombs that followed, is that its development is now widely considered to have been a principal factor in the emergence of the

United States as a world superpower during the twentieth century.

Todd A. Hanson

Sources

Groves, Leslie R. *Now It Can Be Told: The Story of the Manhattan Project.* New York: Da Capo, 1983.

Rhodes, Richard. *The Making of the Atomic Bomb.* New York: Simon and Schuster, 1995.

Serber, Robert. *The Los Alamos Primer: The First Lectures on How to Build an Atomic Bomb.* Berkeley: University of California Press, 1992.

ATOMIC ENERGY COMMISSION

The Atomic Energy Commission (AEC) was formed in 1946 after the Atomic Energy Act, passed by the U.S. Congress that year, was signed into law by President Harry S. Truman. The mission of the AEC was to assume the scientific and technical activities of the Manhattan Engineer District, the organization responsible for supervising the design, development, and testing of the atomic bomb during World War II. Specifically, the AEC was directed to sustain the advancement of theoretical and practical knowledge relating to American nuclear science, including the potential peaceful uses of the atom; in the 1950s, President Dwight D. Eisenhower referred to the initiative as "Atoms for Peace."

Throughout much of its twenty-eight-year history, the AEC maintained a uniquely privileged status with Congress and the American public. Its monopoly on nuclear materials and nuclear research, coupled with the secrecy required of nuclear weapons work, made it one of the most well-funded and influential agencies of the period.

Truman appointed lawyer and former head of the Tennessee Valley Authority David E. Lilienthal to head the commission, along with New England businessman Sumner T. Pike, Iowa newspaper editor William T. Waymack, reserve admiral Lewis L. Strauss, and, as the only scientist on the commission, Los Alamos physicist Robert F. Bacher. Initially, the AEC oversaw scientific and technical work at the three Manhattan Project facilities: Hanford, Washington; Los Alamos, New Mexico; and Oak Ridge, Tennessee.

Over the next few decades, the Commission's scientific and industrial assets grew to include nuclear facilities and laboratories in twenty-one states.

Much of the AEC's scientific work in nuclear physics and energy was unprecedented and exploratory. The work included, among other activities, the development and testing of nuclear weapons, nuclear power plant design and development, radioisotope production, and the development of radioisotopic power sources for space exploration and medical applications. It also funded research on fusion reactors, high temperature superconducting power transmission systems, energy storage, solar energy, coal gasification, and geothermal resource development.

The Atomic Energy Commission was abolished on December 31, 1974, by the Energy Reorganization Act. The end of the AEC was part of a broader series of energy-related national policy initiatives that separated nuclear energy development from regulatory matters by assigning to the new Nuclear Regulatory Commission (NRC) responsibility for the regulation of civilian use of nuclear materials, principally materials used in medicine and the nuclear energy and manufacturing industries. The reorganization was primarily a response to growing criticism that the AEC could not properly regulate the same nuclear energy sources that it helped research, develop, produce, and operate.

The Energy Reorganization Act placed much of the basic research programs in atomic, nuclear, and radiation physics, and related disciplines of chemistry and applied mathematics, under the auspices of the newly created Energy Research and Development Administration (ERDA). As part of the reorganization, some 6,000 AEC employees went to ERDA, and almost 2,000 became part of the NRC. In October 1977, ERDA became part of the U.S. Department of Energy.

Todd A. Hanson

Sources

Duncan, Francis. *Atomic Shield: A History of the United States Atomic Energy Commission.* Vol. 2, *1947–1952.* University Park: Pennsylvania State University Press, 1969.

Hacker, Barton C. *Elements of Controversy: The Atomic Energy Commission and Radiation Safety in Nuclear Weapons Testing,*

1947–1974. Berkeley: University of California Press, 1994.

Hewlett, Richard G., and Oscar E. Anderson, Jr. *The New World: A History of the United States Atomic Energy Commission.* Vol. 1, *1939–1946.* University Park: Pennsylvania State University Press, 1962.

BELL, ALEXANDER GRAHAM (1847–1922)

Best known for his speaking telegraph, or telephone, Alexander Graham Bell also invented several other notable medical devices, helped develop the phonograph and the airplane, and was an avid proponent for the deaf.

Bell was born in Edinburgh, Scotland, on March 3, 1847, and came to the United States as a young man in his early twenties. He was initially interested in designing devices to help the deaf understand and speak. In 1876, his experiments resulted in the invention of a device that revolutionized communications and transformed business transactions. "Mr. Watson, come here, I want to see you," became famous as the first words heard over the newly devised speaking telegraph. By 1900, there were more than 1.5 million telephones in America, and by 1915, a transcontinental telephone line connected Washington, D.C., and California.

Although the telephone brought fame and fortune to Bell, he spent many years in litigation trying to protect his patent rights from challenge by corporate competitors. Bell was also a leading figure in educating the deaf; he married one of his deaf students, Mabel Hubbard. The couple produced two daughters, as well as one son who died as a newborn from respiratory failure in 1881. Spurred by the death of his son, Bell invented the vacuum jacket, an early version of the iron lung that aided victims of poliomyelitis in the twentieth century.

In 1881, Bell also attempted to help President James A. Garfield, who had been shot in an assassination attempt. Doctors, unable to find the bullet, turned to Bell and his associates in the hope that they could invent a mechanism to do so in time to save the president's life. In response, Bell invented the telephonic needle probe. Although the invention was unable to save Garfield, it was used extensively to save other lives during the Sino-Japanese War, the Boer War, and World War I. Bell received an honorary medical degree from the University of Heidelberg in 1886 for his contribution to surgery.

Bell's passion in working with the deaf continued throughout his life. In 1890, the American Association for the Promotion of the Teaching of Speech to the Deaf was incorporated in New York. Bell made monetary contributions to the fledgling organization, served as its president for eight years, and invented the audiometer, a device used to identify the partially deaf. Bell also advocated a day school for the deaf, which allowed them to interact with the hearing during various parts of the day. In 1885, he took his case to Madison, Wisconsin, and succeeded in passing legislation implementing his plan. By 1900, Wisconsin had fifteen day schools for the deaf, and other states were using Wisconsin's program as a model.

Bell also aided in shaping the U.S. census policy. In 1890, he made suggestions for making the census of the deaf more accurate and useful, and he persuaded Congress to retain census documents for future studies.

A cofounder of the magazine *Science* in 1880, Bell served as president of the National Geographic Society from 1898 to 1904. He died on August 2, 1922, in Nova Scotia, Canada.

Stacy L. Smith

Sources

Bruce, Robert V. *Alexander Graham Bell and the Conquest of Solitude.* Boston: Little, Brown, 1973.

Grosvenor, Edwin S., and Morgan Wesson. *Alexander Graham Bell: The Life and Times of the Man Who Invented the Telephone.* New York: Harry N. Abrams, 1997.

BOEING

The world's largest commercial aircraft manufacturer as of 2007, the Boeing Company has been at the forefront of aviation and aerospace technology through much of the twentieth and early twenty-first centuries. Among the innovative aircraft designed and built by Boeing are the 1938 Model 307 Stratoliner, the first transport aircraft with a pressurized cabin; the workhorse

B-17 bomber of World War II; the 707, America's first commercial jetliner, introduced in 1958; and the 747, the world's first jumbo jet, introduced into commercial service in 1970.

Boeing also has been deeply involved in American space exploration since the 1960s, producing the *Lunar Orbiter I,* the first U.S. craft to circle the moon in 1966. With McDonnell Douglas and North American Aviation, Boeing jointly designed and manufactured the Saturn V, the rocket that would eventually carry the *Apollo* spacecraft to the moon.

Early Years

The Boeing Company was founded in Seattle, Washington, on July 15, 1916, by William Boeing, a logging executive with deep pockets, a passion for aviation, and an understanding of wooden structures, critical to the manufacturing of airplane fuselages in the early years of aviation. Joining forces with former U.S. Navy engineer George Westervelt, Boeing founded the Pacific Aero Products Company to build the B&W, a seaplane they hoped to sell to the U.S. Navy. The company changed its name the following year to the Boeing Airplane Company.

Although the navy turned down the B&W seaplanes, the aircraft were bought by the New Zealand Flying School, representing Boeing's first international order. In New Zealand, the plane would set an altitude record of 6,500 feet (1,981.2 meters) in 1919.

Boeing was more successful with its Model C seaplane. The U.S. Navy purchased the Model C to use as a flight trainer during World War I in Boeing's first contract with the U.S. military. In December 1919, Boeing launched its B-1 mail plane, the company's first commercial aircraft.

With commercial aviation replacing military contracting as the most lucrative segment of the industry in the 1920s, William Boeing launched his own airline, Pacific Air Transport, in 1927. In 1929, the two companies merged to form the United Aircraft and Transport Corporation (UATC). The corporation also acquired several other aircraft engineering companies that year; among them was Pratt and Whitney, a manufacturer of engines. With the passage of the Air Mail

Act of 1934, which prohibited corporations from being in both the airplane manufacturing business and the airline industry, the UATC was split into three companies: Boeing Airplane, United Airlines, and United Aircraft, the precursor to United Technologies, which owns Pratt and Whitney.

Boeing's first deal with a commercial airline was manufacturing the Boeing 314 Clipper seaplane for Pan American World Airways (Pan Am); it was the largest civil aircraft at that time, with a ninety-seat (or forty-bed) capacity. In 1939, Pan Am used the plane to launch the first regular passenger service across the Atlantic Ocean.

In the mid-1940s, Boeing also launched the 307 Stratoliner. With its pressurized cabin, the plane became the first commercial aircraft designed to fly at 20,000 feet, above weather conditions in the lower atmosphere—thus avoiding excessive air tubulence, a major impediment to air travel at the time. Although only 10 of these aircraft were built, the Stratoliner represented a significant breakthrough in commercial aviation.

America's entry into World War II postponed further innovation in commercial design and construction, as Boeing and its competitors turned to the manufacture of military planes. Under government aegis, Boeing joined forces with Lockheed Aircraft and Douglas Aircraft to build the B-17 Flying Fortress, and with Glenn Martin and Bell Aircraft to manufacture the B-29 Superfortress. These two long-range bombers were responsible for much of the aerial destruction of targets in the European and Pacific theaters of the war.

With the end of World War II, Boeing, like other U.S. aircraft manufacturers, was hit hard by the loss of military contracts. Moreover, its first commercial aircraft in the postwar era, a modified version of the B-29 called the Stratocruiser, failed to catch on with airlines.

At the onset of the Cold War in the late 1940s, Boeing returned to military contracting, building the C-97 troop and freight transport, the B-50 bomber, and the KC-97 aerial tanker. The company also launched a ground-to-air pilotless aircraft, a precursor to anti-aircraft missiles, in 1949.

The Jet Age

While its commercial and military aircraft in the first half of the twentieth century established Boeing as a major innovator in the industry, its reputation was cemented with its advances in commercial jet aviation during the 1950s and 1960s. Boeing's 707 was not the first commercial jet aircraft to go into operation—that distinction lies with the ill-fated de Havilland Comet of Britain, which crashed several times in the early 1950s due to metal fatigue. But the 707, which first went into operation with Pan Am in 1958, soon came to dominate the world of commercial aviation.

The 707 carried passengers to destinations around the world in a matter of hours rather than the days that propeller-driven aircraft had taken. With its cruising speed of 550 miles (880 kilometer) per hour and a range of 4,500 miles (7,200 kilometers), the sleek, four-engine, 156-seat aircraft became the flying symbol of jet-age modernity and jet-set elegance. While production of the 707 ceased in 1984, its record of reliability ensured that more than 1,000 of the planes would continue to be used in commercial flight through the early 2000s.

Boeing introduced other jet aircraft in the 1960s, including the three-engine, medium-range 727 in 1963 and the two-engine, short-range 737 in 1967. But it was the 747 jumbo jet, built at the company's mammoth new factory in Everett, Washington, that captured the public's imagination. Introduced in 1968 and first flown commercially in 1970, the 747 dwarfed all other jetliners in operation. The double-decker, 416-passenger plane was so large that many airports had to modify their terminals and lengthen their runways to accommodate it. No less impressive was the 747's range. Able to fly more than 7,000 miles (11,200 kilometers) without refueling, it made the first nonstop trans-Pacific flights possible. In its various configurations, the 747 continues to be built today; more than 1,400 are in operation around the world.

By the 1970s and early 1980s, Boeing dominated the commercial aircraft manufacturing industry, overshadowing its main competitors, Lockheed and McDonnell Douglas. But developments in Europe and the Soviet Union would dramatically alter the company's strategy and commercial prospects. In the early 1970s, several European aircraft companies joined forces to establish the Airbus consortium. Its first product, the medium-range, wide-body A300, went into commercial operation in 1974. By the 1980s, the A300 and larger A320 were capturing a growing share of airline orders, with Airbus ultimately surpassing Boeing in overall commercial jet orders in the early 2000s.

By mid-decade, however, Airbus stumbled, plagued by design-related delays in the introduction of its superjumbo A380. In 2006, Boeing regained its title as the world's largest manufacturer of commercial aircraft, partly through its introduction of the 787 Dreamliner, whose increased fuel efficiency made it popular with airlines hard-hit by rising fuel prices.

Meanwhile, the collapse of the Soviet Union and the end of the Cold War in the early 1990s was affecting the U.S. defense business, as the nation's military budget shrank significantly. With military contracts dried up, the defense manufacturing industries underwent massive consolidation.

Flush with revenues from its commercial aviation division, Boeing—a somewhat minor player in military aviation during much of the Cold War—became a major player in the field in the late 1990s and early 2000s, acquiring the aerospace and defense units of North American Rockwell in 1996 and merging with McDonnell Douglas in 1997. Boeing also acquired another major aerospace innovator with its purchase of Hughes Space and Communications in 2000. These acquisitions made Boeing the second-largest defense contractor in the world, after Lockheed Martin, itself a product of 1990s mergers and restructuring.

By the mid-2000s, Boeing enjoyed revenues in excess of $60 billion annually and employed more than 150,000 people in seventy countries. In 2001, Boeing moved its corporate headquarters from its longtime home in Seattle to Chicago.

James Ciment

Sources

Bowers, Peter M. *Boeing Aircraft Since 1916.* Annapolis, MD: Naval Institute Press, 1989.

Lawrence, Philip K., and David W. Thornton. *Deep Stall: The Turbulent Story of Boeing Commercial Airplanes.* Burlington, VT: Ashgate, 2005.

BROOKLYN BRIDGE

Designed by civil engineer John Augustus Roebling and completed in 1883, the Brooklyn Bridge spans the East River between the boroughs of Manhattan and Brooklyn in New York City. The world's longest suspension bridge (1,595 feet) at the time of its completion, it was an engineering and artistic masterpiece regarded as a symbol of technological progress and the modern age.

Shortly after receiving the commission, Roebling declared that it would "not only be the greatest bridge in existence, but it will be the greatest engineering work of the Continent and of the age." Its towers, he said, "will be entitled to be ranked as national monuments." Unfortunately, while scouting the location for the towers, Roebling injured his foot and eventually died of tetanus, leaving his son, Washington Augustus Roebling, to oversee the project.

Spanning the East River presented a number of problems. To begin with, the sheer distance called for two interior supports. Two caissons, large brick structures providing support for the towers, were therefore sunk into the river as anchors. The construction of the caissons pre-

sented dangers for the workers. By the time they were complete, three workers had been killed by "the bends," or caissons disease, and many others had been paralyzed, including Washington Roebling himself.

Work began on the Brooklyn caisson on January 2, 1870. Weighing 3,000 tons when launched, it was lowered at the rate of 6 inches per week, reaching a final depth of 44 feet, 6 inches below the surface. On December 2, 1870, however, fire struck and work was halted for several weeks, as the caisson was flooded to douse the flames. Work on the Brooklyn caisson finally ended in March 1871.

The New York caisson was launched in May 1871, weighing 3,250 tons. In the summer of 1872, Washington Roebling had to be carried off the caisson, stricken with the bends. Paralyzed, he remained bedridden for the remainder of construction, directing the project from his bedroom window. After this, fear of caissons disease became so great that Roebling was forced to halt construction of the second caisson prematurely, estimating that over eighty more deaths would have occurred if work continued.

With the underwater work completed, construction focused on the elegant towers, on which four steel cables 15.75 inches in diameter

The Brooklyn Bridge, connecting the New York City boroughs of Manhattan and Brooklyn, was an engineering marvel at the time of its opening in 1883. The first steel-wire suspension bridge, it continues to carry a heavy flow of traffic across the East River. *(Hulton Archive/Getty Images)*

were hung to support the bridge. Going against accepted practices, Roebling chose steel over iron for its greater strength—which did not prevent a terrible accident in June 1878. The wires snapped, killing or injuring several workers, because the contractor in charge of producing the steel substituted cheaper materials. Roebling was furious, but he decided against reworking the cables, since they were already five times thicker than necessary.

Costing more than $15 million and twenty to thirty deaths, the Brooklyn Bridge was finally opened on May 24, 1883, fourteen years after construction began. Eighty-five feet wide and weighing 14,680 tons, it remains an engineering and artistic marvel—as well as a busy thoroughfare.

Benjamin Lawson

Source

McCullough, David. *The Great Bridge: The Epic Story of the Building of the Brooklyn Bridge.* New York: Touchstone, 1972.

CLOCKS AND TIMEPIECES

Early mechanical clocks, lacking faces and dials, signaled the time by means of sound. The term "clock" comes from the German word for "bell," *Glocke*. The machines in use in England in the late thirteenth century lost up to fifteen minutes each day, but clocks were improved through a spring developed by the Italian architect Filippo Brunelleschi in the early 1400s and a pendulum devised by the Dutch astronomer and physicist Christiaan Huygens in 1656. The uncoiling of the spring and the swinging of the pendulum provided a periodicity that could drive a clock's gears more reliably than previous methods.

In the 1760s, the chronometer, invented by the English carpenter John Harrison, supplanted the pendulum clock as the most accurate timekeeper. Its precision was proven on several transatlantic voyages, during which sailors used it to calculate the difference in longitude between Portsmouth, England, and Barbados in the West Indies. Their estimates were well within the acceptable margin of error of thirty

nautical miles; the chronometer was shown to lose less than three seconds a day.

Throughout the late eighteenth century and into the nineteenth, establishments cropped up in American port towns from Boston to New Orleans, making, selling, and testing chronometers. In 1807, when the U.S. Congress approved a request from President Thomas Jefferson to conduct a study of the nation's coastal lands, the Treasury and Navy departments, both of which led the study, relied heavily on chronometers.

The production of other types of timekeepers flourished in the United States. In 1807, two Connecticut investors commissioned Eli Terry, a Plymouth-based clockmaker who had received the first American patent for a clock mechanism, to build 4,000 long-case clocks in three years. During the first year, Terry used the down payment to construct machinery that helped him complete the order on time. In 1850, having been given workspace in Edward Howard's Massachusetts factory, Aaron L. Dennison conceived techniques for mass-producing watches. Both Terry's and Dennison's work helped bring these once costly machines within the reach of more people.

In the nineteenth century, inventors brought electricity to the craft of clock building, both to drive machine gears and to relay a time signal between timepieces. In 1852, Moses Farmer was granted the first U.S. patent for an electric clock for his work in synchronizing Boston's new fire-alarm system.

In 1884, Chester Pond unveiled his most famous invention, the self-winding clock, a battery-powered instrument that could go for a year untouched. A clock driven by the oscillations of a quartz crystal first appeared in 1929 after Warren A. Marrison, an engineer at New York's Bell Laboratories, discovered that the crystal vibrated at a regular frequency. By the end of World War II, the quartz clock had been refined to drift only one second every thirty years.

The greatest revolution in timekeeping was the atomic clock, a device driven by the oscillations of individual atoms. The idea for such a machine was born in 1945, when Columbia University professor I.I. Rabi theorized that a clock could be made in accordance with his theory of atomic beam magnetic resonance. Several years later, the National Bureau of Standards—now the National Institute of Standards and Technology—developed the world's first atomic clock, which used the ammonia molecule at first, then later used the cesium atom.

With the advent of atomic clocks, precision timekeeping improved exponentially. In 1967, the second (as a unit of measurement) was redefined; this fundamental unit was now derived not from the rotation of Earth, which could fluctuate in its speed, but from the oscillation of the cesium atom. Whereas a cesium clock in the 1950s would lose a second over a few hundred years, the cesium fountain atomic clock, put into use in 1999, is accurate to one part in 10^{15}—or about one second in 20 million years.

Danny Kind

Sources

Andrewes, William J.H. "A Chronicle of Timekeeping." *Scientific American* (September 2002): 76–85.

Bartky, Ian R. *Selling the True Time: Nineteenth-Century Timekeeping in America.* Stanford, CA: Stanford University Press, 2000.

Boorstin, Daniel. *The Discoverers: A History of Man's Search to Know His World and Himself.* New York: Vintage Books, 1985.

De Carle, Donald. *Watch and Clock Encyclopedia.* 2nd ed. London: N.A.G. Press, 1976.

Gibbs, W. Wayt. "Ultimate Clocks." *Scientific American* (September 2002): 86–93.

DEERE, JOHN
(1804–1886)

The inventor of the self-scouring plow, which revolutionized farming in the nineteenth century, John Deere was the founder of one of the world's largest manufacturers of agricultural implements and machinery,

John Deere was born in Rutland, Vermont, on February 7, 1804, to a tailor father and seamstress mother. He received little formal schooling and apprenticed in a blacksmith shop when he was seventeen. During his twenties, Deere opened several shops in Vermont, but all of them failed. Seeking richer agricultural lands, where farmers had more money and more need for blacksmithing services, he moved to Grand Detour, Illinois, in 1836.

Deere soon became familiar with a problem unknown to farmers in the stony New England countryside. So rich was the thick earth of the prairies that Midwest farmers were often forced to stop their plowing to pull the mud off the moldboard—the large, curved parts of the plow that turned the soil. Within a year of his move to Illinois, Deere had designed an all-steel plow that did not have this problem with mud buildup. By 1839, he had started production and sold his first ten plows. Word soon spread of the invention, and orders poured into his shop, allowing Deere to end his blacksmithing work and devote himself to the manufacture of plows.

In 1848, Deere moved operations to Moline, Illinois, to take advantage of the Mississippi River for water power and transportation. The John Deere company has been headquartered there ever since. As the nation's economy boomed in the 1850s, and as ever larger sections of the Midwestern prairie came under the plow, the company prospered. By the middle of that decade, it was producing some 13,000 plows a year. Deere also devised a number of other innovative tools during this period, including the double plow and shovel plow, as well as various kinds of cultivators and harrows.

Deere was also an innovator in production and marketing techniques. He was among the first manufacturers to recognize the potential savings of using interchangeable parts in the production process. And he employed a large team of sales representatives, known as "travelers," who established wholesale and retail dealerships in every state and in Canada. In 1858, he turned over day-to-day operations of the company to his son Charles so that he could devote himself to research and development of new equipment. An aggressive businessman in his own right, Charles Deere had the business incorporated as Deere and Company in 1868 and oversaw the further expansion of the firm, including its expansion into overseas markets.

Deere had five children with his first wife, Demarius Lamb, who died in 1865. The following year, he married her sister Lucenia Lamb. In his later years, John Deere devoted increasing amounts of his time and fortune to philanthropic causes, as well as local politics, becoming mayor of Moline in 1873. He died in Moline on May 17, 1886.

Aside from Cyrus McCormick's mechanical reaper, the Deere plow did more than any other invention to turn America into the world's most productive agricultural nation by the end of the nineteenth century. It helped transform the vast expanses of the Midwestern prairie and the Great Plains into the breadbasket of the world.

James Ciment

Sources

Broehl, Wayne G. *John Deere's Company.* Hanover, NH: University Press of New England, 1992.
Deere and Company. http://www.deere.com.

DURYEA, CHARLES (1861–1938), AND FRANK DURYEA (1869–1967)

The brothers Charles and J. Frank Duryea produced the first working automobile sold in America. The vehicle was driven by a one-cylinder, gasoline-powered engine. Two years later, the Duryea brothers established the first U.S. automobile manufacturing company, Duryea Motor Wagon of Springfield, Massachusetts.

Raised on a family farm in Illinois, the Duryea brothers had a passion for mechanical devices, and they began to build bicycles near the end of the cycling craze of the late 1880s. By 1890, Charles was determined to build an engine-driven rather than horse-drawn carriage, and he enlisted Frank (by far the better mechanic) to help.

The motor-carriage project was centered in Springfield. There, on September 23, 1893, Frank completed and drove a primitive vehicle a few hundred feet. Recognizing the inherent flaws in his prototype, Frank abandoned the horse-buggy form and designed an automobile from the ground up.

The new vehicle was completed by the summer of 1895 and was entered by Charles in a Chicago motor race held in November of that year. Driven by Frank on Thanksgiving Day, one day after a major blizzard, the Duryea car came in first, ahead of a German-built Mueller-Benz and four other entrants. The publicity gained in this and other races allowed the brothers to bring their automobile to full production in 1896.

The Duryea Motor Wagon Company produced and sold thirteen automobiles that year, making it the first commercial automobile factory in the United States.

Within just a few years, however, the Duryea cars were no more than antiquated museum pieces. By 1899, the brothers had gone their separate ways. Charles moved on to other pursuits as a consulting engineer, while Frank continued in the infant automobile industry. From 1904 to 1914, he served as vice president and chief engineer of the Stevens-Duryea company, a competitive firm that produced a number of high-quality, four- and six-cylinder cars. With Frank's retirement from the business in 1914, the Duryeas' influence on automotive history came to an end.

Like fellow bicycle mechanics Wilbur and Orville Wright, whose airplane designs were also quickly bypassed by others, the Duryea brothers are celebrated today mainly for contributing to the birth of a new industry and mode of transportation. One of the thirteen original Duryea cars made in Springfield is on view at the Henry Ford Museum in Dearborn, Michigan.

William M. Shields

Sources

Berkebile, Don. "The 1893 Duryea Automobile." *United States National Museum Bulletin* 240 (1964): 1–28.
May, George. *Charles E. Duryea: Automaker.* Ann Arbor, MI: Edwards Brothers, 1973.
Sharchburg, Richard P. *Carriages Without Horses: J. Frank Duryea and the Birth of the American Automobile Industry.* Warrendale, PA: Society of Automotive Engineers, 1993.

EASTMAN, GEORGE (1854–1932)

The inventor and industrialist George Eastman, the great popularizer of photography and founder of the Eastman Kodak Company, was born in Waterville, New York, on July 12, 1854. His father owned and operated a business school but traveled to nearby Rochester, where he sold roses and fruit trees to supplement his income. When his father died in 1862, George left school to work and help his mother. Eastman worked as an office boy and a clerk at Rochester

Savings Bank before he began to experiment with photography.

At the time, the common process was wet-plate photography, which involved coating a plate of glass with a substance called collodion, a proteinaceous substance made from egg white. Then, a layer of guncotton and alcohol mixed with bromide salts was added. Before the emulsion dried, but after it had set, the plate was dipped into silver nitrate and loaded into the camera in the dark. While taking a picture, the photographer had to spend a long time exposing the image to light. Any movement would cause blurring.

Eastman's original dry-plate process was advertised as the gelatino-bromide dry plate. But as he continued to experiment with dry plates—ever researching what others had done—he came up with a lighter-weight camera and a faster way to coat the plate. His methodical personality and ability to think ahead gave him an advantage, as he envisioned a coating machine, a step that greatly improved production efficiency. Eastman applied for a patent on a coating machine in 1879. The photographic dry plate was patented as both an apparatus and a process the following year.

In 1881, Eastman went into partnership with Henry Strong, a roomer in his mother's boarding house who recognized the opportunity to become prosperous, and they formed the Eastman Dry Plate Company with $2,000. The firm incorporated in 1884 to manufacture what it called American Film. Now that film was less bulky, a smaller and less expensive camera could be used.

Four years later, Eastman invented the first box camera. With William Hall Walker, he designed a paper film that could be rolled through the camera and advanced for each picture. The film and box camera could take up to 100 pictures, changing the paradigm of photography from a professional's realm to anyone's hobby. Word spread, and the company began receiving orders from all over Europe.

Eastman hired a chemist named Henry Reichenbach to improve the paper film, and, in 1889, Reichenbach succeeded in developing a transparent, flexible film, similar to the kind now used in standard cameras. In 1892, the same year that Eastman developed the daylight-loading film,

he changed the name of the company to Eastman Kodak. The Brownie, a low-priced, easy-to-use camera, made photography even more popular and pulled the company out of a temporary depression. The camera cost $1.

Eastman gave generously of his time and money to institutions in the city of Rochester, including the Eastman School of Music, the Eastman Dental Dispensary, the University of Rochester, Strong Memorial Hospital, and Kilbourn Hall, a performance auditorium named after his mother. An avid hunter, he traveled to Africa and brought back many exotic species to display in his home, which later became the George Eastman House, one of the world's leading museums of photography. His enduring love for Africa and its people spurred him to donate millions of dollars to the Tuskegee and Hampton Institutes.

As the infirmities of aging compromised his lifestyle, Eastman became more depressed and alienated from social encounters. His modest, quiet personality on the outside belied the aggressive energy he channeled into his inventions and business enterprise. No longer willing to cope with the ravages of arthritis and other degenerative health problems, he committed suicide on March 14, 1932, in Rochester.

Lana Thompson

Sources

Ackerman, Carl W. *George Eastman: Founder of Kodak and the Photography Business.* Delaware, OH: Beard, 1930.

Brayer, Elizabeth. *George Eastman: A Biography.* Baltimore: Johns Hopkins University Press, 1996.

Life Library of Photography: Light and Film. New York: Time-Life Books, 1975.

EDISON, THOMAS ALVA (1847–1931)

America's most prolific and celebrated inventor, Thomas Alva Edison turned the very act of invention from an enterprise of inspired individuals to an industry in itself, founding the Edison laboratory in Menlo Park, New Jersey, in 1876.

Edison was born on February 11, 1847, in Milan, Ohio, and lived most of his life in New Jersey. Receiving little formal education, he was tutored at home by his schoolteacher mother. Coming from a family of limited means, Edison went to work at the age of twelve, selling newspapers and food at local railroad stations. He became a telegraph operator at sixteen, working for Western Union and other companies through the late 1860s. Fired from several jobs for constantly tinkering with the equipment, Edison gradually shifted from telegraph operator to telegraph innovator, developing an improved printing telegraph for the financial industry in 1869.

In 1870, Edison moved to Newark, New Jersey. With a machinist named William Ungar, he founded the Newark Telegraph Company to manufacturer stock tickers, but also to experiment in other aspects of telegraphy. Given his growing reputation as an innovator, Edison was able to attract a coterie of bright young assistants from around the United States and across Europe. With industry expanding rapidly after the American Civil War, Edison's company won contracts to develop and manufacture new kinds of telegraph equipment, making it possible for newly emerging national firms to communicate between their headquarters and far-flung factories.

As his business grew, Edison, a tireless worker who put in extraordinarily long hours throughout his professional career, began to distance himself from the company's day-to-day operations. He increasingly devoted his time to inventing, establishing a separate laboratory for himself in 1875. He also began keeping copious notes of his work, to make sure that his patent claims were judged to be legitimate and predate those of any competitor.

Among the first inventions Edison developed in his new laboratory was the acoustic telegraph, which allowed multiple messages to be transmitted simultaneously on a single wire. In adapted form, the acoustic telegraph would emerge, through the work of Scottish-born inventor Alexander Graham Bell, as the telephone. In subsequent years, Edison would develop a number of innovations in telephony, including

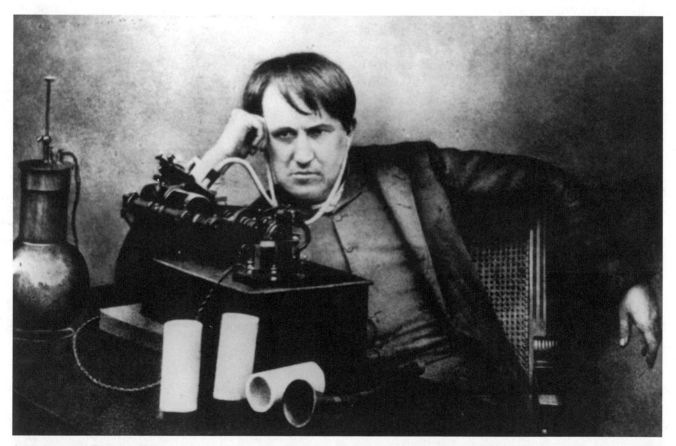

An exhausted Thomas Edison listens to a prototype of the phonograph after spending five continuous days and nights perfecting the device—which finally came to fruition in 1877. Edison transformed invention itself from an individual enterprise into an industry. *(Hulton Archive/Getty Images)*

the carbon-button transmitter, which Bell Telephone licensed from Edison for use in its products.

In 1876, Edison took a further step away from the operations of his company by moving to Menlo Park, where he set up another laboratory. In the next five years, Edison and his team of researchers developed one new invention after another, most notably the phonograph in 1877 and the incandescent light bulb in 1879. While the former won him worldwide acclaim as the "wizard of Menlo Park," including an invitation to demonstrate the device for President Rutherford B. Hayes at the White House, it was the latter invention that would make his fortune.

Electrical incandescence had been demonstrated as early as the first decade of the nineteenth century, and arc lights were already being used in street and theater illumination by the time Edison set out to develop a safe, long-lasting, inexpensive, and small-scale light bulb that could be used in homes and offices. Winning financial backing from executives at Western Union and other businesses, Edison expanded his Menlo Park facility and hired new assistants to help him with the project. Based on his work with the acoustic telegraph, which involved the use of high-resistance wires, Edison approached the problem differently from others working in the field of electric incandescence, most of whom subscribed to the idea that using a low-resistance filament was the key to inventing small-scale lighting. After much experimentation, Edison and his team developed the high-resistance carbon filament, the essential breakthrough that made the modern light bulb possible.

Edison was also working on a system for generating and transmitting the electricity necessary to make the light bulb work, and to create the business organization necessary to exploit his new invention. During the 1880s, Edison and his team, most notably physicist Francis Robbins Upton, developed a direct-current (DC) electrical transmission system. Edison, along with a number of investors, including many who had financed his light bulb experiments, organized the Edison Electric Illuminating Company of New York, which built the first Edison electric generating station in Lower Manhattan; the company would undergo several name changes before becoming General Electric (GE) in 1892.

While GE would go on to become one of the world's most successful manufacturing companies in the twentieth century, Edison's insistence on direct-current electrical transmission would prove a failure. Although safer than alternating current (AC), being developed by rival George Westinghouse in the late nineteenth century, DC could not be transmitted effectively over large distances, making it impractical for large-scale electrical utilities. Ultimately, Westinghouse's AC would become the backbone of the world's electrical systems.

In 1888, Edison embarked on the development of motion pictures. Over the next twenty years, his laboratory would produce a number of inventions in the field, including the first motion-picture camera and the first kinetoscope, a primitive kind of projector, in 1889 and 1894 respectively. But in the race to create a screen projector for mass viewing, Edison lost to French inventor Charles Lumière, and he also failed in his efforts to develop an integrated talking picture system.

Although Edison's greatest achievements were behind him by the turn of the twentieth century, his wide-ranging knowledge and curiosity led him to experiment with X-rays, the large-scale manufacture of iron and cement, and other ventures. During World War I, he advised the U.S. Navy on military technology.

By this time, Edison was extraordinarily wealthy, and he was often seen in the company of presidents and industry leaders. He enjoyed a particularly close friendship with auto manufacturer Henry Ford, and the two would take highly publicized automobile tours together. While Edison's frantic pace slowed as he entered his seventies, he continued to work at his laboratory nearly until his death.

Edison was married twice; his first wife died in 1884, and he remarried in 1886. He had six children, three with each wife. He died in Orange, New Jersey, on October 18, 1931.

James Ciment

Sources

Josephson, Matthew. *Edison: A Biography.* New York: McGraw-Hill, 1959.

Pretzer, William S., ed. *Working at Inventing: Thomas Edison and the Menlo Park Experience.* Dearborn, MI: Henry Ford Museum and Greenfield Village, 1989.

Wachhorst, Wyn. *Thomas Alva Edison, an American Myth.* Cambridge, MA: Harvard University Press, 1981.

ELECTRICITY

The development of electricity during the nineteenth century defined how Americans consume physical energy, which transformed American society. Nineteenth-century American scientists and inventors built on the discoveries of Europeans to produce, use, and distribute electricity. This led to revolutions in communications, transportation, industry, lighting, and the electronics industry, as well as a proliferation of electrical devices taken for granted in the twenty-first century.

At the beginning of the nineteenth century, Alessandro Volta of Italy invented the electric battery, which he called the wet cell. The volt, the unit of electrical pressure, is named after him in tribute to his work to find "flowing electricity" to complement existing static electricity. The voltage of each cell in the electric battery is low, usually between one and two volts, and the electron flow produces direct current as it flows along a single path. The reason for this unidirectional current is that the battery maintains the same polarity and output voltage across its two terminals.

The application of direct current in America enabled development of both the telegraph by Samuel F.B. Morse and the telephone by Alexander Graham Bell. Electric on and off pulses were used to send telegraph messages across great distances, while voice-modulated electricity on copper wires enabled speech to be sent via telephone.

During the latter half of the nineteenth century, American inventors applied the European understanding of alternating current. Alternating-current sources change their direction of flow for a defined number of times per second. This property of frequency is often expressed in terms of Hertz, named after the German physicist and engineer Heinrich Hertz, who verified the wave aspects of electromagnetic radiation in 1888. Alternating current sources also produce direct current for those devices that operate better without voltage variations.

To overcome the mass of electric batteries needed for power, late nineteenth-century American industry increasingly relied on the alternating current electric motor, invented by Nikola Tesla. Since direct-current motors are eas-ier to run at a variety of speeds, they were commonly found on subway and light-rail systems; however, electric power can be sent more economically and efficiently over long distances as high-voltage alternating current. Consequently, power companies use alternating-current motors. The ability to transmit electricity over great distances ended the need to locate power plants close to where products were produced. Industries were liberated from traditional power sources—wood, coal, and water.

Thomas Edison and Nikola Tesla pioneered the development of modern electrical power systems. In 1879, Edison introduced direct-current electric light by improving generators, patenting electric distribution, and developing the first electric meter for recording customer consumption. With this comprehensive approach, Edison knew how much to charge each customer for electric service.

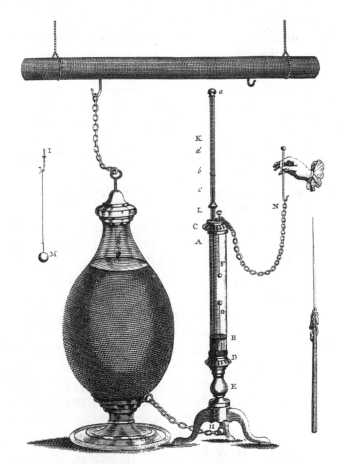

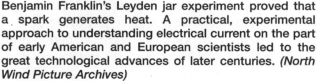

Benjamin Franklin's Leyden jar experiment proved that a spark generates heat. A practical, experimental approach to understanding electrical current on the part of early American and European scientists led to the great technological advances of later centuries. *(North Wind Picture Archives)*

Electrical distribution was becoming big business. Tesla, building on the contributions of Thomas Edison, Michael Faraday, and others, was able to visualize working inventions; his implementation of them led to the fusion of electrical theory with electrical engineering.

The Edison approach to electricity favored direct-current generation, with distribution limited to about a mile from a noisy generating plant, and suitable only for lighting. Tesla understood these pitfalls. His signature invention was the alternating-current motor, which allowed for the widespread generation and distribution of electrical power. He fought to introduce the alternating-current approach, whereby electrical power could be transmitted over hundreds of miles. In addition to lighting, alternating current could be used for powering residential appliances and industrial machinery. Tesla's vision for alternating-current power was ultimately realized during the twentieth century.

Samuel Insull made possible the widespread integration of America's electrical infrastructure. One of Edison's lieutenants, he became an advocate of alternating current in electrical production and distribution. Insull used economies of scale to overcome market barriers through the inexpensive production of electricity with large steam turbines. This technology made it easier to put electricity into more American homes. Insull drove demand for electricity by charging for the user's share of peak demand at one rate and for kilowatts used during off-peak hours at another rate. He encouraged and promoted through incentives the notion of load management to ensure a high level of continuous consumption to match the turbines' steady generating power. Insull's legacy was the democratization of electricity in America during the early twentieth century.

Cost-effective generation of electricity and the electric light bulb changed the quality of life for millions of Americans. Inexpensive and safe light, brought to the home as alternating current, replaced unsafe flame from candles and noxious fumes from gas. And because houses now could have lower ceilings, since no extra vertical headroom was needed to collect and vent gas, home construction costs plummeted.

Meanwhile, corporations were formed to both drive and serve the demand for increasing consumption of electricity. Westinghouse, founded in 1886, and General Electric, founded in 1892, focused primarily on kitchen and other household appliances. Both promoted convenience through ceiling fans (1882), desk fans (1886), coffeepots (1891), electric stoves (1893), and toasters (1909), to name a few. Convenience bolstered consumption, and these corporations thrived by serving the customer. Electricity also provided illumination for nighttime social events and leisure activities.

Robert Karl Koslowsky

Sources

Grob, Bernard. *Basic Electronics.* 8th ed. Westerville, OH: McGraw-Hill, 1997.

Jonnes, Jill. *Empires of Light: Edison, Tesla, Westinghouse, and the Race to Electrify the World.* New York: Random House, 2003.

Koslowsky, Robert. *A World Perspective Through 21st Century Eyes.* Victoria, Canada: Trafford, 2004.

ELECTRON MICROSCOPE

Unlike traditional microscopes, in which light passes through a series of magnifying lenses, electron microscopes use magnetically focused beams of electrons to create the magnified image.

Electron microscopes have two major advantages over light-based microscopes. First, they can be used to scan solid objects—opaque objects are not suitable for viewing with traditional light-based microscopes. Second, electron microscopes provide magnification hundreds of times greater than that of light-based microscopes. While light-based microscopes can magnify an image by a factor of several thousand, electron microscopes can magnify an image by a factor of 1 million or more, allowing the study of objects at the atomic level.

Much of the early work in electron microscopy took place in Europe and Canada during the years between World Wars I and II. The theoretical physics that would make the electron microscope possible was formulated by University of Paris scientist Louis de Broglie in the 1920s. The first prototype was developed in

1933 by German engineers Max Knoll and his student Ernst Ruska at the Technical University of Berlin. Physicists at the University of Toronto built the first practical electron microscope in 1938.

The first major U.S. work in electron microscopy was conducted at the research laboratories of the Radio Corporation of America (RCA) in the late 1930s and early 1940s. Led by Russian émigré physicist Vladimir Zworykin, the RCA scientists, many of whom were also conducting pioneering research on cathode ray tubes and television transmission and reception, developed early prototypes of a scanning electron microscope (SEM).

In the earlier transmission electron microscope (TEM), a beam of electrons was focused on an object, causing an enlarged version to appear on a fluorescent screen or photographic film. When using the TEM, the object was moved, a technique that was soon supplanted by instruments in which the electron gun moved and the object under scrutiny remained static. For example, the SEM used a moving electron beam to scan an object.

SEMs, however, did not have the same powers of resolution as TEMs. The earliest TEMs could resolve objects of less than 50 angstroms (one angstrom equals one ten-billionth of a meter), while the first SEMs had a resolution of about 2,000 angstroms. Because of their lesser resolution, SEMs did not attract as much attention in the scientific community as TEMs. But they had other advantages. By scanning objects, they allowed for more three-dimensional imaging, making them especially useful for examining the surface structures of specimens. As the power of SEMs improved in the post–World War II era, they increasingly supplanted TEMs in research laboratories.

Over time, scientists and engineers at RCA and other research facilities improved both types of electron microscopes, making them smaller, more powerful, and easier to use. And large-scale production made them more affordable. In 1969, RCA got out of the business of manufacturing electron microscopes, but a number of other U.S. firms continue to produce them.

James Ciment

Sources

Hawkes, Peter W., ed. *The Beginnings of Electron Microscopy.* Orlando, FL: Academic Press, 1985.

Newberry, Sterling P. *EMSA and Its People: The First Fifty Years.* Milwaukee, WI: Electron Microscopy Society of America, 1992.

ELIOT, JARED (1685–1763)

The Reverend Jared Eliot, an agricultural scientist, inventor, and graduate of Yale College, served as the pastor of the Congregational church of Killingworth, Connecticut, from 1707 to his death. He was a mostly self-taught physician, one of the last New England clergymen-physicians. In 1739, he presided over a meeting of medical practitioners at New Haven to regulate practice in Connecticut.

Eliot's travels to visit his parishioners and patients helped turn his mind to the agricultural challenges facing Connecticut, and he published the first of six short but influential treatises, *Essays on Field-Husbandry in New England,* in 1748. The last pamphlet in the series, which adapted the new agricultural ideas of Jethro Tull and other English "improvers" to the different circumstances of New England, was published in 1759.

Eliot tried to write in a more plain, accessible style than Tull, one appropriate to American farmers. The fifth treatise in the series actually takes the form of a traditional New England sermon, with Tull's writings replacing the Bible as the text being expounded. Eliot endorsed natural theology, pointing out that agriculture should turn human minds to their creator.

Eliot's endorsement of Tull's ideas was selective. He approved of Tull's method of planting seeds in even rows and, along with the mathematician and president of Yale College Thomas Clap and the wheelwright Benoni Hylliard Eliot, devised a set of improvements on Tull's "drill-plough" to mechanize the process (though the machine did not come into common use in Eliot's time). Eliot disagreed with Tull's argument against manuring, which Tull thought ineffective. Eliot believed that manuring was necessary to enrich the poor soils of Connecticut. He also endorsed some of the traditional, astrology-based beliefs about the phases of the moon influencing the proper time to carry out agricultural operations.

Eliot hoped that his work would initiate better communications and dissemination of innovations and best practices among farmers. He hoped that American farmers would form agricultural societies similar to those already existing in Scotland and Ireland, but his hopes were frustrated. Along with Clap, he served as an agent for the London Society of Arts in a project to promote silk culture in Connecticut.

In 1762, he published *An Essay on the Invention, or Art of Making Very Good, if not the Best Iron, from Black Sea Sand*. The following year, the London Society of Arts awarded Eliot a gold medal for his success in extracting iron from sand. Word of the prize came only days before his death, which occurred on April 22, 1763.

William E. Burns

Sources

Grasso, Christopher. "The Experimental Philosophy of Farming: Jared Eliot and the Cultivation of Connecticut." *William and Mary Quarterly* 3rd ser., 50 (1993): 502–28.

Wilson, Philip K. "Jared Eliot." In *American National Biography*, vol. 7. New York: Oxford University Press, 1999.

ERIE CANAL

Completed in 1825, the Erie Canal connected Lake Erie with the Hudson River, giving New York merchants access to upstate markets and the Great Lakes region. Connecting cities such as Albany, Syracuse, Rochester, and Buffalo, the Erie Canal brought considerable prosperity to inland New York and established New York City as the most important seaport in America. Spanning over 360 miles, the Erie Canal was the largest towpath canal of the time period, requiring eighty-three locks and eighteen aqueducts to bypass the Adirondack Mountains.

Begun in 1817 by order of the New York legislature, the Erie Canal marks a milestone of American engineering. With the backing of Governor De Witt Clinton and wealthy New York merchants, work on the canal proceeded rapidly, helped by the relatively gentle terrain of the proposed pathway. Clinton persuaded the state legislature to finance the undertaking with tolls, tax revenues, and bond sales to foreign investors interested in the project. The investment paid off, and the canal soon generated revenue beyond its initial cost.

New York's upstate cities also benefited from increased commerce due to the success of the Erie Canal. In 1818, a year before the canal opened, Rochester processed only 26,000 barrels of flour from local wheat; by 1828, with the canal, Rochester was processing 200,000 barrels and, by 1840, almost 500,000. The canal resulted in Rochester's population increasing from a few hundred residents to 20,000.

The success of the Erie Canal sparked a national canal boom, though few reached the heights of New York's system. The importance of these new canals led to widespread use of steamboats, a much faster alternative to the horse-drawn freight barges immortalized in popular song.

Three men oversaw the building of the Erie Canal: Benjamin Wright, David S. Bates, and Canvass White, the youngest and least experienced of the three. Wright, as chief engineer, oversaw much of the project, while Bates served as assistant engineer and worked on the middle section, including the aqueduct near Rochester.

Building the canal was an amazing feat of engineering and resolve, especially as much of the work was done by hand. By the end of the project, millions of cubic yards of dirt had been dug, thousands of tons of rock quarried to construct the massive locks, and giant reservoirs of water filled to provide a steady supply of water. Tens of thousands of workers, many of whom were immigrants, provided the manual labor, working under dismal conditions. In a marsh near Syracuse, for example, nearly a thousand workers died due to an outbreak of fever.

Today, the Erie Canal is no longer used commercially. Despite renovations from 1836 to 1862, the role of the canal was eclipsed in the twentieth century by new developments such as railroads, the interstate highway system, the St. Lawrence Seaway, and the advent of commercial flight, all of which provided more effective means of transporting goods. Today, the

Spanning New York State from Buffalo on Lake Erie to New York City on the Hudson River, the 360 mile (580 kilometer) Erie Canal—shown here in the upstate town of Lockport—was an audacious undertaking and an economic boon to the young republic. It opened in 1825. *(Kean Collection/Hulton Archive/Getty Images)*

canal is part of the New York State canal system, and it has limited use, primarily for pleasure craft.

Benjamin Lawson

Sources

Bernstein, Peter L. *Wedding of the Waters: The Erie Canal and the Making of a Great Nation.* New York: W.W. Norton, 2005.

Henretta, James A., David Brody, and Lynn Dumenil. *America: A Concise History.* New York: Bedford/St. Martin's, 1999.

Johnson, Paul E. *A Shopkeeper's Millennium: Society and Revivals in Rochester, New York, 1815–1837.* New York: Hill and Wang, 1978.

FACTORIES

A factory is a place of production where powered machinery is used by workers to transform or assemble materials. It may consist of one or more separate buildings on a single site. The term factory is generic; other names are steel mills, automotive plants, machine shops, and locomotive works.

Factories had their antecedents in preindustrial manufactories where artisans and laborers using hand tools were gathered together. In the case of some shipyards and arms production facilities, these could be very large. It was not until the Industrial Revolution, however, that the true factory was born. Here, workers sold their labor, working in buildings and using tools owned by others. The factory thus became the place of production best able to exploit the technology of the Industrial Revolution and the social relations that accompanied the technology. The subsequent history and current status of the factory cannot be disentangled from the general history of industry and manufacturing technology.

It is impossible to say what was America's first factory. Important early manufacturing sites included Oliver Evans's automated gristmill in Delaware in the 1790s; New England textile mills based on English technology and brought to America by Samuel Slater, culminating in his cotton mill at Pawtucket, Rhode Island, in 1793; and the federal armories at Springfield, Connecticut, and Harper's Ferry, Virginia, where the concept of mass production using interchangeable parts began to be worked out.

New technology during the late 1800s, especially electrification, radically changed the nature of factories. Rather than just a shell to enclose production, the factory became part of the production process, designed by experts to facilitate standardized mass production. Electrically operated machinery (with alternating current motors) allowed plants to be laid out more rationally. Electrical lighting replaced daylight or more dangerous forms of artificial illumination.

Whenever possible, engineers built control of production into the design and layout of the machines. Skilled workers became machine tenders, while innovations in material-handling equipment eliminated the need for large numbers of unskilled laborers. These developments culminated shortly before World War I in the moving assembly line at Henry Ford's Highland Park, Michigan, factory designed by architect Albert Kahn.

In recent decades, factory work has been further transformed by the introduction of industrial robots, computer-assisted manufacturing tools and techniques, and new management ideas aimed at making the factory workplace less alienating. However, such developments as the outsourcing of production, competing offshore manufacturing, and the rise of the information and service economy have resulted in a continuing decline in factory-based jobs in the United States. The National Association of Manufacturers estimates that only 11 percent of the labor force is currently employed in industrial plants.

This is seen most vividly in stark industrial "brownfields" of abandoned manufacturing sites and the stagnating economies of the former industrial heart of the nation, now termed the Rust Belt. The American factory, though far from a thing of the past, is no longer the characteristic place of work and production it once was.

James Hull

Sources

Biggs, Lindy. *The Rational Factory: Architecture, Technology, and Work in America's Age of Mass Production.* Baltimore: Johns Hopkins University Press, 2003.

Fingleton, Eamonn. *In Praise of Hard Industries.* Boston: Houghton Mifflin, 1999.

Licht, Walter. *Industrializing America.* Baltimore: Johns Hopkins University Press, 1995.

FORD, HENRY (1863–1947)

With his assembly-line manufacture of automobiles and the 1908 introduction of the Model T, Henry Ford put America on wheels. His innovations and successes helped to jump-start the modern car industry. Ford was born on a farm in Springwells (now Greenfield), Michigan, on July 30, 1863. Although a less than average student, he showed an aptitude for mathematics and was fascinated with machinery. Determined to avoid farm life, Ford moved to the nearby town of Detroit when he was sixteen, and he landed an apprenticeship position in the engine shop of a shipbuilding company.

In 1882, Ford went to work for the Westinghouse Engine Company, selling and servicing steam engines. After several other jobs, he was hired in 1891 as an engineer with the Edison Illuminating Company, where his mechanical abilities got him promoted to chief engineer within two years.

While working at Edison, Ford became fascinated with the internal combustion engine, a power source that had been developed in Europe in the middle years of the nineteenth century. Smaller and more powerful for their size than steam engines, internal combustion engines were the leading candidates as power sources for the newly emerging technology of automobiles.

In 1896, Ford completed his first vehicle, a car he called the Quadricycle. Improved prototypes followed, and, in 1899, with the backing of fifteen investors, Ford opened the Detroit Automobile Company. Unable to turn a profit, however, the company went out of business. Ford turned to building racing cars, setting his sights on speed records in the first years of the twentieth century.

With his reputation as an innovator more firmly established, Ford opened his second company, the Ford Motor Company, on June 16, 1903, with $28,000 in capital, a dozen workers, and a remarkably small 12,000-square-foot (1,100-square-meter) assembly plant. Like most early automobile assemblers, Ford bought many of his components from outside contractors. The engines and chassis came from John and Horace

Dodge, who became majority shareholders in Ford's company.

A major stumbling block Ford faced was patent infringement. A patent for the gasoline-powered automobile was held by the Edison Electric Car Company, as a backup in case its electric-powered cars proved to be failures. In 1899, several makers of gasoline automobiles formed the Association of Licensed Automobile Manufacturers (ALAM) to license all automobile manufacturers. Ford refused to comply, continuing to build automobiles in defiance of the ALAM. Ultimately, the dispute would end up in federal court in 1911. Ford won.

Meanwhile, in 1906, Ford had introduced his Model N, an 800-pound (360-kilogram), 18-horsepower runabout. But Ford was not happy with the car's high price or its mechanical problems. One of his main backers, Detroit coal merchant Alexander Malcolmson, believed that bigger profits could be made with heavier and more expensive automobiles. With the money generated by the highly successful Model N, Ford bought out Malcolmson, clearing the way to focus on the car that he wanted to build—a cheap and reliable vehicle for use by farmers and workers.

Beginning in 1906, Ford assembled a team of engineers at a separate plant to build the prototype of the Model T, a crank-started, 20-horsepower vehicle with a four-cylinder engine made of lightweight vanadium steel. Originally intending to build several models, he instead created a standard chassis upon which the various body types could be fitted. In 1908, the car was finished, and Ford notified his dealers that it was ready for sale. Ford used interchangeable parts and mass production to keep the prices down—$825 for the convertible model and $1,000 for the hardtop. Those prices put the cars out of the reach of the small farmers Ford had intended as his market, but the Model T's reliability and relatively low cost—most cars at the time sold for $2,000 and up—made it a hit. In 1910, Ford opened a massive 62-acre (25-hectare) plant at Highland Park.

At Highland Park, Ford put into practice two elements that would bring the Model T's price down substantially, making it truly a car of the average American. The first innovation was the time-motion study, perfected by industrial engi-

Henry Ford's first vehicle powered by a combustion engine, the two-cylinder Quadricycle introduced in 1896, helped him arrange the funding to start his first automobile manufacturing company—which failed. *(Hulton Archive/Getty Images)*

neer Frederick Taylor, by which everything workers did was studied closely to find ways to cut out extraneous movement and wasted time. The second innovation was the moving assembly line, an idea borrowed from the great meatpacking plants of Chicago, where animal carcasses were moved from worker to worker, each of whom had a single and simple task to perform. The process saved enormous amounts of time and effort by keeping workers from having to move around the factory to assemble automobiles. So successful were these ideas that, by the time the last of some 15 million Model Ts rolled off the assembly lines in 1927, the cost of the basic model had dropped to $290.

The Model T made Ford the largest automobile company in the world, and the efficiency of his production process allowed him to introduce, in 1914, the then unheard-of wage of $5 a day. In many people's minds, Ford had invented mass production, and the word "Fordism" entered the vocabulary as a synonym for it.

Despite this success, the Ford Motor Company was outstripped by rival General Motors as the world's largest automobile manufacturer in the late 1920s, partly because Ford failed to follow GM's lead in marketing a variety of models at different price ranges. In 1927, Ford replaced the Model T with the more powerful and better-equipped Model A. By that time, this new model was relatively primitive compared to cars being built by GM's Chevrolet division, and it proved far less successful than the Model T.

Meanwhile, Ford had moved on to other pursuits, attempting to act as peacemaker during World War I and going into the airplane business in the 1920s. He also bought a newspaper, the *Dearborn Independent,* in the 1920s, which gained notoriety for publishing anti-Semitic literature.

In 1938, Ford suffered a stroke and turned over the day-to-day management of the company to his son Edsel. When Edsel died in 1943, Ford returned to the helm at the age of eighty. Two years later, he turned over the reins to his grandson Henry Ford II in 1945. Henry Ford died in Dearborn, Michigan, on April 7, 1947.

James Ciment

Sources

Brinkley, Douglas. *Wheels for the World: Henry Ford, His Company, and a Century of Progress, 1903–2003.* New York: Viking, 2003.

Nevins, Allan, and Frank Ernest Hill. *Ford.* New York: Arno, 1976.

Watts, Steven. *The People's Tycoon: Henry Ford and the American Century.* New York: Alfred A. Knopf, 2005.

FULLER, R. BUCKMINSTER (1895–1983)

Ever original and innovative, R. Buckminster Fuller was described in a 1964 *Time* magazine cover story as the "first poet of technology." An autodidact whose works refused the disciplinary boundaries of academia, he made important contributions as a designer, architect, inventor, artist, mathematician, poet, and futurist.

He was born Richard Buckminster Fuller on July 12, 1895, in Milton, Massachusetts; his father died when he was twelve. After attending the local Milton Academy, he briefly went to Harvard; later, he received a degree from Bates College.

Fuller married in 1917 and served in World War I. For a time in the 1920s, he owned a company that designed versatile buildings. After experiencing a personal crisis during the 1930s, Fuller took a faculty position at Black Mountain College in Asheville, North Carolina. There, he devised the geodesic dome.

As a solid yet lightweight structure that allows for a wide coverage of space without internal supports, the geodesic dome represented a breakthrough in shelter construction. With the unveiling of the golf-ball-shaped structure that housed the U.S. Pavilion at Montreal's Expo '67, Fuller achieved the status of a pop culture icon. While the cost-effective and easily constructed domes never gained the widespread use as low-cost housing that he envisioned, they have indeed been used to provide shelter for low-income families in poorer countries. An estimated 300,000 geodesic domes are believed to exist throughout the world today, with uses ranging from homes, radar installations, and weather stations to industrial and sporting facilities.

Fuller was an early advocate of renewable energy sources, including solar, wind, and wave power. Through the use of such alternative sources, he argued, human societies could meet all of their energy needs while discontinuing the use of fossil fuels and atomic energy. Perceiving that the global energy crisis was brought about by ignorance, Fuller envisioned the practicality of seemingly fantastic alternatives. He theorized, for example, that fixing a wind generator to every energy transmission tower in the United States would generate almost four times the country's power output at the time.

Fuller's other works include the Dymaxion Map, which displays the world's continents on the flat surface of a polyhedron with minimal distortion; in 1946, he received the first new patent for a cartographic system in 150 years. Fuller is also remembered for designing the Dymaxion House, a low-cost, energy efficient, prefabricated shelter. His life's achievements include twenty-five U.S. patents, twenty-eight books, and forty-seven honorary doctorates in various disciplines. Despite having no formal architectural training, he received the Gold Medal of the American Institute of Architects in 1970, its highest honor, as well as the Gold Medal of the Royal Institute of British Architects in 1968.

Fuller viewed his work as an integrated effort to anticipate and solve the major problems facing the planet and its inhabitants. He sought greater "life support" for all people through using fewer and fewer natural resources. His holistic approach was reflected in Fuller's attempt to develop what he called a comprehensive anticipatory design science to recognize and address future ecological and social problems. He stressed what he saw as the humanitarian purpose of technology. By providing insights into the essential unity of the natural world, technology could offer a guide to solving social problems.

Fuller coined the term "Spaceship Earth" to convey his view of the unity of planetary life and the need for all people—its passengers—to act cooperatively. People everywhere, he believed, are "local problem solvers" who are connected to each other, the rest of the world, and the entire universe by working for the betterment of each other. Fuller summed up his approach in the now familiar exhortation "Think globally; act locally."

His progressive vision and concern for issues of social and environmental justice have extended Fuller's influence well beyond the realms of science, design, and architecture. His commitment to improving life on Spaceship Earth resonated with the counterculture movements of the 1960s and 1970s, and his ideas have continued to inspire contemporary social movements of diverse interests and philosophies. He died in Los Angeles on July 1, 1983.

Jeff Shantz

Sources

Sieden, Lloyd S. *Buckminster Fuller's Universe: His Life and Work.* New York: Perseus, 2000.

Zung, Thomas T.K. *Buckminster Fuller: Anthology for a New Millennium.* New York: St. Martin's, 2002.

GEMINI, PROJECT

The Gemini space program, officially designated Project Gemini, was designed by America's National Aeronautics and Space Administration (NASA) to bridge the technology gap between the Mercury program and the Apollo lunar missions. Authorized by Congress in 1961, the primary objectives of the program were to subject humans, support systems, and equipment to long-duration flights orbiting Earth; successfully execute rendezvous and docking techniques; prove that astronauts could perform tasks outside the spacecraft; and perfect methods of entering the atmosphere and landing at a preselected landing point.

NASA established the Gemini Project Office at the Manned Spacecraft Center in Houston, Texas, to manage the program and chose the McDonnell Aircraft Corporation to construct the spacecraft. The technology and experience of Project Mercury were drawn on to create the two-crew-member Gemini spacecraft, which was designed for increased aerodynamic lift and improved maneuverability. The new spacecraft shared the basic design configuration of the Mercury capsule, but it included a number of enhancements: onboard computers that provided for precision navigation; fuel cells that generated enough electricity for long-duration missions; ejection seats that served as the primary escape system; rendezvous radar; and storable propellant fuels. Weighing nearly 8,000 pounds, each of the twelve Gemini spacecraft relied on its Titan II launch rocket to supply the necessary thrust to carry it into orbit.

The first initial Gemini missions were unmanned tests of the vehicle. Manned missions occurred from March 1965 to November 1966; Gemini astronauts orbited Earth in flights ranging from five hours to fourteen days. On March 23, 1965, Virgil I. "Gus" Grissom and John W. Young launched into space aboard *Gemini III*, in which they orbited Earth three times. Three months later, *Gemini IV* astronaut Edward H. White II became the first American to perform an extravehicular activity. Using a tether and hand-held maneuvering unit, he moved about outside the capsule during a twenty-two-minute space walk. In December 1965, a major objective

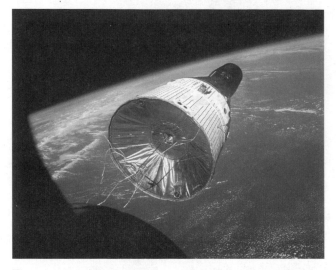

The successful rendezvous of two spacecraft, the *Gemini 7* capsule (pictured here) and the *Gemini 6* craft (from which this photo was taken), in December 1965 satisfied one of the major objectives of the entire Gemini program. *(NASA/Time & Life Pictures/Getty Images)*

of the Gemini program was achieved when crewmembers of *Gemini VI* were able to rendezvous with the *Gemini VII* spacecraft. On November 11, 1966, astronauts James A. Lovell, Jr., and Edwin E. "Buzz" Aldrin, Jr., launched into orbit aboard *Gemini XII*. This final Gemini flight included a five-hour space walk by Aldrin and a successful docking operation with an unmanned Agena upper-stage booster.

Although the Gemini program was essentially a technological learning experience for NASA, it also included a program of experiments in space science. Gemini astronauts conducted studies in astronomy, atmospheric sciences, biology, medicine, space environment, and radiation effects. They also carried out special communication tests, took Earth terrain color photographs, observed weather patterns, and researched the effects of prolonged weightlessness and immobilization on humans.

The Gemini program did not represent the pathbreaking endeavor of Project Mercury, nor was it as ambitious as the later Apollo flights to the moon. The twelve-flight program, consisting of ten manned missions, developed the necessary techniques for advanced lunar missions, including orbital rendezvous and docking operations and extravehicular activities. These missions played an important role in achieving President John F. Kennedy's goal of landing humans on the moon and returning them safely to Earth.

Kevin Brady

Sources

Grimwood, James M., Barton C. Hacker, and Peter J. Vorzimmer. *Project Gemini Technology and Operations: A Chronology.* Washington, DC: NASA, 1969.

Hacker, Barton C., and James M. Grimwood. *On the Shoulders of Titans: A History of Project Gemini.* Washington, DC: U.S. Government Printing Office, 1977.

GIRDLING

To "girdle" a tree means to cut a strip around the circumference of the trunk, slicing through both bark and cambium layers in order to deny the crown sufficient nutrients to sustain leafy growth. A fully girdled tree will eventually die as a result of the cut, but partial or temporary girdling has long been employed by horticulturists to increase the size of fruit on orchard trees. Silviculturists (forest managers) use girdling to rid woodlots of undesirable invasive species.

Girdling is best known, however, as a method of land clearance in the "long fallow" or "slash and burn" farming typical of subsistence agricultural societies throughout history, though many contemporary groups also employ the practice. In the absence of either large pools of available labor, animal power, or metal tools such as axes, agricultural societies in America found it more economical to clear temporary farming plots in forest areas by using the slash and burn technique (of which girdling was a primary component). This method was used rather than the full land clearance typical of Anglo-European farmers who settled in such heavily forested areas as eastern North America.

In the traditional agricultural system, small farming parties entered a mature forest area and pulled, cut, piled, and then burned the shorter underbrush. After this, they used their available cutting tools (for example, axes or knives of stone or animal bone) to girdle the remaining larger trees. Girdling removed the leaf canopy and allowed sunlight to stream down to the forest floor where crops were planted; farmers used ash from the burned brush as fertilizer.

Since many mature forest areas have relatively thin layers of topsoil (especially in tropical rain forests, where much of the organic material is locked up in the plants themselves), the accumulated fertility of humus on the forest floor would be depleted within just a few years, resulting in declining crop yields. At that point, traditional agriculturists would move on to a different area of the forest and repeat the process. The depleted plot would be left to lie fallow long enough (sometimes for decades) for the natural vegetation to rejuvenate itself, after which the area could be farmed again in the same manner.

Several American Indian societies practiced girdling and long-fallow agriculture in the temperate forests of eastern North America. Many Anglo-European settlers learned of the technique from Native Americans in the colonial era, later adopting it for their own farms. The system proved particularly attractive to Scots-Irish settlers in the heavily wooded "backcountry" valleys of the Appalachian and Allegheny highlands. As members of this large ethnic group moved south and west after the American Revolution, they transferred long-fallow methods such as girdling to the Cumberland plateau of eastern Kentucky and Tennessee, the Ozark hill country of Arkansas, and other areas, including the dense woods of central Indiana and Illinois. In these latter areas, however, slash and burn farming was typically a prelude to the development of a permanently cleared landscape suitable for a more intensive agricultural system, where farmers intended to work the same land for many years, maintaining fertility through the application of manure and artificial fertilizers rather than following the long-fallow method.

Nonetheless, long after the Civil War, both girdling and long-fallow agriculture persisted in many isolated areas of southern Appalachia and the Ozarks well into the twentieth century. Travelers reported the appearance of "deadenings"—open areas of girdled and dying trees in the middle of healthy forests.

As early as the first decades of the nineteenth century, however, both girdling and long-fallow agriculture were passing from the American scene. Girdled trees had always presented a hazard to farmers—they were known as "widow-makers" on the frontier—as dead limbs or entire trees might fall without warning. The long-fallow agriculture in which girdling played such a central role also required that farmers have access to sufficiently large property holdings to accommodate the prolonged periods of idleness necessary for "tired" land to renew itself. But the descendants of the Scots-Irish subsistence farmers had a long tradition of big families and partible inheritance (equal division of property among all heirs), making it increasingly difficult to keep the large farms intact, especially when southern Appalachia began to be overrun by timber and coal companies around the turn of the twentieth century.

Thus, like many other traditional skills developed in rural America, the knowledge of girdling is now preserved mainly as a cultural artifact. The exceptions are those instances noted above where the practice is still in use for forest maintenance and horticulture.

Jacob Jones

Sources

Doolittle, William E. "Agriculture in North America on the Eve of Contact: A Reassessment." *Annals of the Association of American Geographers* 82:3 (1992): 386–401.

Farragher, John Mack. *Sugar Creek: Life on the Illinois Prairie.* New Haven, CT: Yale University Press, 1986.

Otto, John S. "Forest Fallowing in the Southern Appalachian Mountains: A Problem in Comparative Agricultural History." *Proceedings of the American Philosophical Society* 133:1 (1989): 51–63.

Otto, John S., and N.E. Anderson. "Slash-and-Burn Cultivation in the Highlands South: A Problem in Comparative Agricultural History." *Comparative Studies in Society and History* 24:1 (1982): 131–47.

GODDARD, ROBERT HUTCHINGS (1882–1945)

Robert H. Goddard took the theory of propulsion and turned it into reality. Goddard's thirty-plus years in rocketry produced an impressive list of achievements and inventions, including application of a nozzle to more efficiently propel rockets, proof that a rocket works in a vacuum, production of mechanical designs for both multistage and liquid-fuel rockets, the first inertial guidance system, the first implementation of blast vanes for directional-thrust control, and the first powered projectile to exceed the speed of sound. His accomplishments are reflected in a total of 214 patents.

Goddard was born on October 5, 1882, in Worcester, Massachusetts; he attended Worcester Polytechnic University and Clark University, from which he earned a Ph.D. in 1911. Goddard's life was shaped by his reading of H.G. Wells's *The War of the Worlds* (1898), a science fiction fantasy about the invasion of Earth by Martians. He imagined building a device that could ascend to Mars. In 1914, Goddard transformed from fantasist, dreaming about flights into space, to physicist, determining the size of a rocket needed to raise 1 pound of payload into space.

The most significant aspect of Goddard's first patent, in 1914, was the adaptation of a cone-shaped nozzle to the rocket. This innovation allowed combusting gases to apply pressure against the cone as they streamed from the rocket chamber, thereby increasing the rocket's power and its lift potential. The ability of Goddard's solid-fuel rocket motors to convert energy into thrust provided a tenfold increase in efficiency from 2 to 20 percent.

Also important were the many basic operational experiments of his rocket designs. In 1915, Goddard proved that a rocket could provide thrust in a vacuum; in fact, working in a vacuum improved rocket thrust by roughly 20 percent over thrust in air. The impact of this achievement revealed that rocket operation was independent of its surroundings. This discovery was significant in the success of the future space age.

In 1919, Goddard wrote *A Method of Reaching Extreme Altitudes,* in which he described liquid-fuel and solid-fuel rocket motors and multistage rockets. His rockets of the mid-1920s included a combustion chamber and nozzle for propulsion, an igniter to ensure constant combustion, and a pumping mechanism to effectively mix the fuel and liquid oxygen in the chamber. Initially, the motor was on top of the rocket so that it would drag the rest of the assembly behind it. Goddard believed this configuration would ensure straight flight. Later, he would discover that as long as the motor was located on the rocket's axis, stability would be maintained.

On March 16, 1926, in Auburn, Massachusetts, Goddard launched the world's first liquid-fuel rocket, using a combination of liquid oxygen and gasoline propellants. The rocket weighed 10 pounds and did not lift off until the excess fuel burned off, so the motor's 9 pounds of thrust could propel it skyward. The rocket rose in an arc, peaking at a height of 41 feet, and completed a semicircle before hitting the ground 184 feet away. The flight lasted just 2 seconds, with an average speed determined to be 60 miles per hour.

During 1930, Goddard and his team experimented with a number of design improvements and added to their growing list of successes. A 10 foot rocket, featuring high-pressure gas to force the propellants into feed lines, was successful in reaching an altitude of more than 1,800 feet and achieving a maximum speed of just over 400 miles per hour. At this time, the state of aerodynamics was so young that the world's fastest airplane had not yet reached such speeds. Even more significant was the fact that this design was the precursor of primary space vehicle propulsion employed thirty years later.

Goddard built the "A" series rockets from September 1934 to October 1935, experimenting with gyroscopes to perfect a flight stabilization system. These rockets were roughly 14 feet tall and 9 inches in diameter. The flight of A-14 achieved a height of 2,000 feet before nose-diving into the ground. After this experience, Goddard built the "K" series motors and performed nine static tests from November 1935 to February 1936. From here, the "L" series rockets were built. They were shorter, at less than 13 feet tall, but thicker, with 18 inch diameters. In 1938, improvements allowed for the flight of the L-30, which achieved a record height of 3,294 feet.

During the period between the two world wars, Goddard was not only the most famous American scientist but also the most publicized rocketeer. The only repetitive mistake of his career was his emphasis on using gyroscopes for flight stability. The laws of mechanics dictate that stability could be achieved with sufficient acceleration, in addition to reaching higher altitude; however, Goddard never came to discover this engineering fact.

The U.S. government invested millions of dollars in rocket development at the conclusion of World War II and used a number of Goddard's inventions in violation of his patent rights. The misappropriation of his inventions was cloaked in secret government classifications to avoid paying royalties. As Goddard's inventions were found in virtually every jet plane or rocket that entered the atmosphere and beyond, it finally became clear that the U.S. government should compensate Goddard for his patents.

Goddard died on August 10, 1945. It was not until 1960, well after his death, that the government admitted to pilfering Goddard's ideas and paid his heirs $1 million in compensation.

Goddard believed rockets could carry humans into space, yet most of his work was funded by the military to build missiles as weapons. This was because the only way to generate government interest in rocketry at that time was to design weapons of war. In memory of this brilliant scientist who wanted to see human space travel realized, NASA's Goddard Space Flight Center in Greenbelt, Maryland, was established on May 1, 1959. Goddard's vision of reaching other worlds came to pass in 1969 with the *Apollo 11* moon landing.

Robert Karl Koslowsky

Sources

Baker, David. *The Rocket: The History and Development of Rocket and Missile Technology.* New York: Crown, 1978.

Clary, David A. *Rocket Man: Robert H. Goddard and the Birth of the Space Age.* New York: Hyperion, 2003.

Koslowsky, Robert. *A World Perspective Through 21st Century Eyes.* Victoria, Canada: Trafford, 2004.

GOODYEAR, CHARLES (1800–1860)

Charles Goodyear invented the thermoset (vulcanization) process that made rubber commercially applicable. Vulcanized rubber greatly improved the efficiency, durability, and performance of nineteenth-century engines, both internal combustion and steam, and made possible the development of higher performance engines and machines capable of sustained operations over longer periods with far greater temperature and pressure tolerances.

Goodyear was born in New Haven, Connecticut, on December 29, 1800. He began his working career by partnering with his father in the family hardware business. When the business went bankrupt in 1830, Goodyear began experimenting with altering the properties of India rubber so that it would be useful across a wider range of temperatures than it was in its natural state.

In 1835, he patented a process that created a new form of rubber by boiling a compound of the natural gum (latex) base from which rubber is derived and magnesium oxide in a mixture of quicklime and water. Goodyear soon learned that even a small amount of a weak acidic substance, apple juice for example, so reduced the durability of this form of rubber as to render it useless for practical application. He began using nitric acid as a curative agent and in 1837 produced mailbags created through this process for the U.S. government. It was discovered, however, that the rubberized fabric degraded quickly in the hot climate common to the southern United States.

Seeking another solution, he used a sulfur-based curative process developed by Nathaniel M. Hayward, who briefly worked for Goodyear. In 1839, Goodyear accidentally discovered vulcanization when he dropped an experimental mixture of rubber and sulfur on a hot stove. The sulfur acted as a curing agent (connecting repeating molecular chains with intermolecular chemical bonds) that, when heated, irreversibly cross-linked the rubber molecules. The result of this process was a strong, flexible, durable rubber that was more resistant to solvents and less affected by temperature variations. Vulcanized rubber can be formulated to maintain an elasticity range within specific temperature tolerances and molded and conformed to seal gaps in engines and to serve many other purposes.

Vulcanization results in a thermoset rubber that does not melt on reheating and does not become brittle at cold temperatures. The rubber molecule has a number of sites, called cure sites, where a sulfur atom attaches and forms a chain of two to ten sulfur atoms that bridge to other rubber molecules, forming a polymer. A higher ratio of sulfur to rubber in the mixture produces a higher number of these bridges. The greater the number of these bridges, the harder the rubber—and vice versa. Such variations are particularly important in applications such as tires.

Goodyear spent the ensuing years consumed with determining the correct mixture and conditions of vulcanization, even selling his children's textbooks to provide cash for purchasing needed materials. He finally deemed his process sufficiently perfected to apply for a patent in 1844. By then, however, others—most notably Horace H. Day—also claimed the discovery. In 1852, the Third U.S. Circuit Court (in Trenton, New Jersey) adjudicated Goodyear's patent infringement case against Day and declared Goodyear the sole inventor of vulcanization.

In addition to promoting his discovery in the United States, Goodyear exhibited articles made from his invention throughout Europe. He was awarded the Great Council Medal at England's Crystal Palace Exhibition (the Great Exhibition of 1851) and the Grand Medal at the Paris Exposition of 1855.

Goodyear chronicled the discovery and perfection of vulcanization in his autobiography, *Gum-Elastic and Its Varieties,* originally published in two volumes (1853–1855). Even though he was granted sixty patents and created a process that today is used to create products essential to industries ranging from health care to space exploration, when Goodyear died in New York City on July 1, 1860, his only financial legacy to his family was a debt of $200,000.

Richard M. Edwards

Sources

Alliger, G., and I.J. Sjothun. *Vulcanization of Elastomers: Principles and Practice of Vulcanization of Commercial Rubbers.* Melbourne, FL: Krieger, 1978.

Korman, Richard. *The Goodyear Story: An Inventor's Obsession and the Struggle for a Rubber Monopoly.* San Francisco: Encounter, 2002.

Peirce, Bradford. *Trials of an Inventor: Life and Discoveries of Charles Goodyear.* Seattle, WA: University Press of the Pacific, 2003.

Slack, Charles. *Noble Obsession: Charles Goodyear, Thomas Hancock, and the Race to Unlock the Greatest Industrial Secret of the Nineteenth Century.* New York: Hyperion, 2003.

GUN MANUFACTURING

Gun manufacturing in America altered with the changes in technology and production brought about by the Industrial Revolution. Primitive guns in the late 1700s had smoothbore barrels and used the flintlock firing mechanism. Eighteenth-century gunmakers learned their craft serving as apprentices to master gunsmiths, employing a variety of specialized tools to build guns by hand with nonuniform parts. During the nineteenth century, gunsmiths followed Eli Whitney's model of building standardized firearms using interchangeable parts.

One of the first American contributions to gun manufacturing was the development of the Kentucky rifle toward the end of the eighteenth century. The Kentucky rifle used the European invention of "rifling" or cutting spiral grooves in the bore of the barrels to stabilize the passage of the bullet, creating a highly accurate, long-range weapon. After the invention of the weatherproof percussion cap by Englishman Joshua Shaw in 1814 (patented by him in 1822, after he had immigrated to America), the rifle was more reliable and not limited to use in dry weather (a requirement for igniting the black powder in muzzle-loaders). Moreover, the new rifles, known as "breech loaders," were loaded at the rear part of the firearm, rendering obsolete the cumbersome action of loading a ball and powder down the muzzle.

In 1835, Samuel Colt of Connecticut invented the revolver, or repeating pistol. This design was improved by subsequent manufacturers such as Horace Smith and Daniel Wesson. There also were advances in ammunition. Primed brass cartridge cases enabled gunmakers to build a powerful handgun for the famed Colt 45 and 38 special cartridges. Later, bottleneck brass cases were used in rifles, and plastic casings were used in shotguns.

In 1862, American inventor Richard Gatling patented his namesake repeat-firing weapon and founded the Gatling Gun Company in Indianapolis, Indiana. This hand-cranked machine gun had six barrels and a drum magazine, and used percussion caps, firing 200-plus rounds per minute (when it did not jam). In 1865, Gatling demonstrated an improved model of the weapon, incorporating copper-cased cartridges and a vertical magazine. The redesigned gun was adopted by the U.S. Army in 1866 and soon thereafter came into use by governments around the world.

In 1870, Gatling relocated to Connecticut, where he continued to improve the design

Fig. 251.—The original Colt Revolver.

BREECH-LOADING REVOLVERS.

The original Colt revolver, the first practical repeating firearm, was patented by Samuel Colt in 1836 and went into mass production the same year. His factory pioneered the moving assembly of interchangeable, machine-manufactured parts. *(Hulton Archive/Getty Images)*

throughout his life; developing a ten-barrel model, .45, .50, and 1 inch caliber models (for naval use), and an engine-driven model, the first "mini-gun." In 1907, the Gatling Gun Company merged with Colt Firearms. Four years later, the U.S. Army declared the hand-cranked guns obsolete, replacing them with fully automatic machine guns.

Gun technology continued to advance, along with innovations and improvements in barrels, firing mechanisms, priming chemistry, primer design, smokeless powder development, ballistics science, bullet development, and cartridge case manufacture. John Browning of Utah was a leading innovator during the late nineteenth and early twentieth centuries. His innovations included the use of recoil and gases from the gun to open the action and load subsequent rounds for automatic operation. Browning invented a repeating rifle in 1884, a machine gun in 1890, and an automatic pistol in 1911.

American gunsmiths such as Browning worked on upgrading old designs and developing new actions in accordance with the demands of modern warfare. Browning's developments in automatic weapons were adopted by the U.S. Army; his .30- and .50-caliber machine gun designs were used in both World War I and World War II. The army also adopted his semi-automatic and automatic pistols and rifles.

As the sophistication of military weapons changed, so did the tactics employed by military forces. Gun manufacturing became more specialized during the world wars, and the military used company-based and military-based gunsmiths. At the same time, civilian gunsmiths worked on sporting rifles, shotguns, and pistols. Sporting applications reflected the growing interest in shooting by the general public.

Today, gun manufacturers work to upgrade a weapon's action (including the firing pin), the barrel or choke components on shotguns, and the carving and assembly of wooden or synthetic stocks. Specialized military and private gun manufacturers generally focus on maintenance and accuracy of various small-arms weapons. Sport and military cartridge developers work in concert with gun manufacturers to refine shooting performance based on powder burn rate, bullet weight and caliber, velocity, wind drift, and energy specifications.

James Steinberg and Phoenix Roberts

Sources

Berk, Joseph. *The Gatling Gun: 19th Century Machine Gun to 21st Century Vulcan.* Boulder, CO: Paladin, 1991.

Van Zwoll, Wayne. *America's Great Gunmakers.* South Hackensack, NJ: Stoeger, 1992.

Wahl, Paul. *The Gatling Gun.* New York: Arco, 1965.

Whisker, James B. *The Gunsmith's Trade.* Lewiston, NY: Edwin Mellen, 1992.

HOOVER DAM

One of the most ambitious engineering projects in American history, the Hoover Dam, also known as the Boulder Dam, was constructed between 1931 and 1936. A concrete, gravity-arch structure, 1,244 feet (379 meters) long, the dam is located on the lower Colorado River between Nevada and Arizona.

Standing 726 feet (221 meters) high, the Hoover Dam is the second-highest dam in the United States, after California's Oroville Dam. Its hydroelectric capacity is just over 2,000 megawatts, providing electricity for communities in Arizona, Nevada, and southern California. Behind the dam sits Lake Mead, an artificial reservoir with an expanse of roughly 158,000 acres (64,000 hectares).

As conceived in the early 1920s, the dam had three purposes: 1) to prevent flooding when melting snows from the Rocky Mountain headwaters engorged the river each spring and threatened farmland; 2) to provide water for irrigation on farms and for use in cities in rapidly growing southern California; and 3) to generate electricity for those same communities. The fears in other Southwestern states that California would take most of the water and electricity were alleviated with the Colorado River Compact of 1922. Still, it took more than six years for the federal government, which financed construction of the $49 million dam, to sign off on the project. The contract to build the dam was given in 1931 to a consortium of six Western contracting firms, known as the Six Companies.

The site of the dam in a sun-blasted, hard rock canyon presented a number of engineering problems. To prevent flooding at the site, two temporary coffer dams had to be built. Then, the Colorado River had to be diverted. This was achieved by digging four diversion tunnels, each more than 50 feet (15 meters) in diameter and collectively more than 3 miles (4.8 kilometers) in length, through solid rock. With that achieved by early 1933, construction of the dam itself began in June.

The Hoover Dam would be the largest concrete structure ever built up to that time, and that scale presented the engineers with some new challenges. Perhaps most formidable was the cooling and contracting process. If workers poured all of the concrete in one solid form, so much heat would be generated that the material would take more than a century to cool. And, as the concrete cooled, it would contract, undermining the stability of the dam.

To avoid this problem, the engineers came up with a novel solution. Rather than create a single massive block of concrete, they constructed numerous interlocking, trapezoidal columns, which would allow heat to dissipate at a much faster rate. To further speed up the cooling, metal coils were placed in each column and river water was run through them. These solutions allowed the bulk of the dam to be built in just over two years.

Even as the dam was being built, excavations were being dug for the powerhouse, where the hydroelectric generators would be installed. In addition, transmission lines were laid over a dis-

The Hoover Dam, located on the Arizona–Nevada border, harnessed water from the Colorado River for power generation and flood control. At the time of its completion in 1936, it was the world's tallest dam and largest concrete structure. *(Keystone/Hulton Archive/Getty Images)*

tance of nearly 300 miles (480 kilometers) to the Los Angeles metropolitan area.

Despite the problems and scale of the project, the dam was completed on time and on budget. The first power transmitted on October 26, 1936, less than five years after construction had begun. But the dam had other costs. Nearly 100 workers died in its construction.

James Ciment

Sources

Dunar, Andrew J., and Dennis McBride. *Building Hoover Dam: An Oral History of the Great Depression.* New York: Twayne, 1993.

Stevens, Joseph E. *Hoover Dam: An American Adventure.* Norman: University of Oklahoma Press, 1988.

HYDROELECTRICITY

Hydroelectricity is electrical power produced by turbines using water falling or flowing from a natural or artificial source. It is also called hydroelectric power, hydropower, or simply hydro.

Most hydroelectric facilities either divert water from a natural source, such as at Niagara Falls, or use water from a natural source impounded by a dam, such as Lake Mead behind the Hoover Dam on the Colorado River. In limited cases, a power plant can use excess capacity to pump water to a reservoir that can then be used to generate hydroelectric power during peak usage periods. Also, relatively small amounts of electricity can be generated using the force of flowing water.

In a typical hydroelectric generating site, water enters a conduit known as a penstock, which delivers it to a turbine. Water striking the blades of the turbine causes the blades to turn. This causes a shaft connected to a generator to turn. There, magnets rotating past copper coils induce electrical current. A transformer takes this current and delivers it to transmission wires at the high voltages necessary for the economical long-distance transmission of electricity. Power lines then carry the current to stations, where it is progressively stepped down to usable voltages and delivered to homes, businesses, and other end users.

Individual electricity-producing sites are linked through power grids to produce and distribute power throughout large regions. While economically and technically efficient, such grids are vulnerable to disruptions, causing massive power outages. This was seen notably in 1965 and 2003 when small technical failures caused cascading outages, leaving millions in the northeastern United States and adjacent areas of Canada without power for hours or in some cases days.

During the 1880s, the Westinghouse Electric Company—founded by inventor and industrialist George Westinghouse—acquired European-developed alternating current (AC) technology, direct current having already been implemented by Thomas Edison in New York City. With the development of the induction coil, a practical AC transformer by William Stanley, and an effective AC motor by Nikola Tesla, the technical superiority of AC was established by the end of the decade. The first large-scale transmission of AC power by hydroelectricity began at Niagara Falls in 1895.

Subsequently, the availability of large amounts of relatively inexpensive hydroelectric power greatly assisted the growth of North American industry. The enormous costs associated with hydroelectric generation and frequent location of generation sites on public lands have made government agencies, notably the U.S. Bureau of Reclamation in the Western states, the Tennessee Valley Authority in Appalachia, and Ontario Hydro at Niagara Falls, the major players in this sector.

The Department of Energy reports the total U.S. hydroelectric capacity at more than 100 gigawatts. This represents about 10 percent of the country's electrical power, down from 40 percent in the early twentieth century and about half the world average. Proponents of hydroelectricity point to its renewable and nonpolluting nature, but the environmental and social impacts of developing hydroelectric sites have been considerable. With few important untapped sites for adding hydroelectric capacity, this source is unlikely to grow in importance, but it will remain a part of the nation's energy mix for the foreseeable future.

James Hull

Sources

Hughes, Thomas Parke. *Networks of Power: Electrification in Western Society.* Baltimore: Johns Hopkins University Press, 1983.

Nye, David E. *Electrifying America: Social Meanings of a New Technology, 1880–1940.* Cambridge, MA: MIT Press, 1990.

HYDROGEN BOMB

The hydrogen bomb is a thermonuclear device that derives its explosive energy from nuclear fusion. An atomic (fission) bomb inside the device triggers the fusion reaction, causing the nuclei of light atoms—typically an unstable isotope of hydrogen—to fuse under extremely high temperatures.

Hydrogen bombs are called "thermonuclear" because of the intense heat needed to overcome the electrical repulsion between positively charged hydrogen nuclei and cause them to fuse into helium atoms. The fusion reaction converts hydrogen into helium; in the process, it explosively releases tremendous amounts of energy.

The hydrogen bomb was first used on November 1, 1952, when an explosion equal to more than 10 million tons of TNT vaporized the island of Elugelab in Enewetak Atoll in the Pacific Ocean's Marshall Islands. Within 90 seconds after detonation, the bomb's fireball reached 57,000 feet in height; the subsequent mushroom cloud was nearly 100 miles across by the time it reached its farthest extent. The explosion (codenamed Ivy Mike) was created by a thermonuclear explosive device built according to the theory of staged radiation implosion developed by Edward Teller and Stanislaw Ulam at Los Alamos. (Andrei Sakharov in the Soviet Union and other nuclear scientists in Britain, China, and France were also working with the theory of staged radiation implosion in independent efforts.)

The hydrogen bomb, or Super, as it was originally called, was developed in spite of serious differences of opinion within the nuclear weapons science community and American political leadership. The Super had been considered possible during the Manhattan Project, but the work at the time had focused specifically on developing the fission bomb. Once the Manhattan Project had ended, some of the scientists began to reconsider the Super.

Other Manhattan Project scientists, however, given the existence of powerful fission weapons, saw no need for the vastly more powerful fusion-based hydrogen bombs. One of these was Los Alamos laboratory director J. Robert Oppenheimer, who lost his top-secret security clearance in December 1953 over his opposition to the hydrogen bomb and his 1930s Communist Party connections.

A typical hydrogen bomb works by using the energy of an atomic fission device, called the hydrogen bomb "primary," to compress and then ignite a physically separate mass of fusion fuel, called the "secondary." The detonation of the primary device creates high levels of X-ray energy that is used to compress and heat an unstable isotope of hydrogen, typically deuterium or tritium, and pack the hydrogen atoms closer together. As the hydrogen isotope is compressed to extremely high densities and temperatures, a second fission device, called the "spark plug," located in the secondary's center, ignites the mass, creating a fusion reaction like that found in the sun.

Todd A. Hanson

Sources

Herken, Gregg. *Brotherhood of the Bomb: The Tangled Lives and Loyalties of Robert Oppenheimer, Ernest Lawrence, and Edward Teller.* New York: Henry Holt, 2002.
Rhodes, Richard. *Dark Sun: The Making of the Hydrogen Bomb.* New York: Simon and Schuster, 1995.

IRONWORKS, COLONIAL

The American Northeast during the colonial period possessed a combination of readily available iron ore and wood. Thus, Anglo-American colonial promoters envisioned an iron industry from the beginnings of permanent colonization.

Iron in North America could be refined out of surface iron ore rather than deep mined. It was smelted out of the ore by heating with charcoal made of hardwoods. English ironworks switched from charcoal to coal in the mid-eighteenth century, but Americans, with a relative abundance of wood, continued using charcoal throughout the colonial period.

The cheapest and crudest way to refine iron was "blooming," heating ore in a charcoal fire and pounding it to eliminate the "slag" or waste. Bloomeries produced only small amounts of iron and demanded heavy labor, but they met the needs of small communities seeking iron for domestic and farming uses.

High-volume iron operations used the blast furnace. Blast furnaces stood between 20 and 30

The first American ironworks, fired by abundant supplies of wood, date to the mid-1600s in New England. The simple "blooming" process was labor-intensive and produced only small amounts of iron for tool making and other domestic uses. *(New York Public Library, New York)*

feet high, used charcoal as fuel, and a rock with a low melting temperature, such as limestone, as a flux, which combined with impurities in the iron to form the slag calcium silicate. The carbon monoxide gas produced by the charcoal reduced the iron oxide to iron, and the extreme heat of the blast furnace, the oxygen for which was provided by water-powered bellows, liquefied the metal. Then, the metal could be poured out in long bars called "pigs." Pig iron was further refined to remove the carbon it had absorbed in the furnace by being melted repeatedly in a refinery hearth and then hammered into shape.

Beginning in the 1640s, John Winthrop, Jr., was an active promoter of ironworks in Massachusetts, where the General Court issued him a monopoly in 1644. His ironworks at Lynn and Braintree used some of the most advanced technology of the time but proved economically unsuccessful. The Saugus Ironworks in Massachusetts, established in 1648, also was unsuccessful, despite its producing a ton of pig iron per day for four years before it closed.

Iron production increased in America during the eighteenth century. Thomas Rutt led the establishment of the eastern Pennsylvania iron industry, while Colonel Alexander Spottswood set up a blast furnace near Fredericksburg, beginning the Virginia iron industry. Two large mills were set up in Maryland to serve the British market—the Principio Company, founded around 1720, and the Baltimore Company, founded in 1731.

By mid-century, the colonies surpassed England in the production of pig iron. The expansion of American cast iron and steel production, however, disturbed English authorities, who passed the Iron Act in 1750. By removing duties, the measure encouraged Americans to export bar and pig iron to Britain, but it forbade the manufacture of cast iron or steel. The Iron Act was largely ignored in the colonies, however; cast iron and steel continued to be produced. By 1775, American manufacturers produced about 15 percent of the world's iron.

William E. Burns

Sources

Bridenbaugh, Carl. *Cities in Revolt: Urban Life in America, 1743–1776.* New York: Oxford University Press, 1955.

Mulholland, James A. *A History of Metals in Colonial America.* University: University of Alabama Press, 1981.

Robbins, Michael W. *The Principio Company: Iron-Making in Colonial Maryland 1720–1781.* New York: Garland, 1986.

KETTERING, CHARLES F.
(1876–1958)

One of the driving forces in advancing automobile technology in the twentieth century was Charles Franklin Kettering, known as "Boss Ket."

Born in Loudonville, Ohio, on August 29, 1876, Kettering graduated from Ohio State University in 1904 with a degree in engineering. He went to work for the National Cash Register Company (NCR) in Dayton. There, he helped develop the first electric cash register and became the chief of NCR's inventions department.

In 1909, he resigned to design automotive electrical equipment at the Dayton Engineering Companies (Delco), which he founded with Edward A. Deeds. While at Delco, Kettering developed the all-electric automobile lighting and ignition systems; in 1911, he invented the first electric self-starter, introduced by Cadillac in 1912. He also built a gasoline engine-driven generator, the "Delco," which allowed farms in isolated areas to provide their own electricity.

Building on the purchase of the Wright brothers' Wright Company, Kettering founded the Dayton-Wright Airplane Company (DWA) in 1914. During World War I, he developed a propeller-driven "aerial torpedo" (cruise missile) with a 200-pound bomb load.

In 1916, Kettering sold Delco to the United Motors Corporation, later General Motors Corporation (GM). From 1920 to 1947, he served as GM's vice president and director of research for the General Motors Research Corporation. In this position, Kettering was instrumental in the development of quick-drying automobile lacquers (Duco paint), the high-speed, two-cycle diesel engine, and the modern, high-compression automobile engine, introduced in 1951.

Kettering's home, Ridgeleigh Terrace, located in the town named for him, Kettering, Ohio, was air-conditioned in 1914, making it the first home so equipped in the United States. Kettering's air-conditioning system used toxic gases (ammonia, methyl chloride, and sulfur dioxide) as refrigerants, and in 1928, Kettering collaborated with chemist Thomas Midgley, Jr., to create freon, the nontoxic chlorofluorocarbon that became the primary refrigerant in air conditioning and refrigeration units. Under Kettering's leadership, General Motors and Dupont formed the Kinetic Chemical Company in 1930 to produce the freon. Freon was used by Frigidaire in the first widespread consumer refrigerators and by the Carrier Engineering Corporation in the first self-contained home air-conditioning unit introduced in 1932. Midgley and Kettering also collaborated on the creation of high-octane anti-knock fuels.

Kettering held more than 200 patents, including those for a portable lighting system, a treatment of venereal disease, and the prototype of the modern incubator for maintaining the temperature of premature infants. Though he received no patents for his work on magnetism, his research on its application to potential diagnostic medical technologies forms the basis of Magnetic Resonance Imaging (MRI), which revolutionized medical diagnostics.

In 1927, Kettering created the C.F. Kettering Foundation for the Study of Chlorophyll and Photosynthesis "to sponsor and carry out scientific research for the benefit of humanity." The foundation sponsored research into cancer and photosynthesis and promoted cooperative and scientific education in the United States. Now renamed the Kettering Foundation, it is committed to answering the question "What does it take to make democracy work as it should?"

In 1945, Kettering and Alfred P. Sloan, Jr., the longtime head of GM, co-founded the Sloan-Kettering Institute for Cancer Research at the Memorial Cancer Center in New York City. Two years later, Kettering retired from GM. He died on November 25, 1958, in Dayton. The General Motors Institute was renamed Kettering University in 1982.

Richard M. Edwards

Sources

Boyd, Thomas Alvin. *Charles F. Kettering: A Biography.* Washington, DC: Beard, 2002.

Zehnpfennig, Gladys. *Charles F. Kettering: Inventor and Idealist; a Biographical Sketch of a Man Who Refused to Recognize the Impossible.* Men of Achievement Series. Minneapolis, MN: T.S. Denison, 1962.

LAND, EDWIN
(1909–1991)

Edwin Herbert Land, developed a one-step, sixty-second process for developing and printing photographs known as Polaroid photography

Land was born on May 7, 1909, in Bridgeport, Connecticut. He became interested in polarized light, light that vibrates or aligns in a single plane not visible to the human eye, while a freshman at Harvard in 1926. Land invented a method of producing inexpensive plastic or polymeric sheets or film called Polaroid J sheets that polarized light. He did this with iodoquinine sulfate (herapathite) crystals that had been reduced to the submicroscopic level and placed in a homogeneous suspension designed to prevent the finely divided hexagonal particles from settling rapidly. These crystals were first discovered in 1852 by William Herpath and were known to show different colors when viewed from different axes (dichroism).

Land found that when these small crystals were forced through narrow slits, they polarized, that is, aligned in two conflicting or contrasting patterns. When light passed through Land's sheets or through film, certain wavelengths of light were absorbed, creating an image. The absorption of wavelengths of light also allowed unwanted light to be removed by filters and lenses treated with Land's process. The sheets, film, and filters could be varied, absorbing light at specific wavelengths.

Land and George Wheelwright III, a Harvard physics instructor, in 1932 formed the Land-Wheelwright Laboratories in Boston. By 1936, Land was applying his polarizing (absorption) principle to camera filters, sunglasses, automobile headlights, and other optical devices. In 1937, Land created the Polaroid Corporation, based in Cambridge, Massachusetts. Just prior to World War II, he introduced a polarized light, three-dimensional, motion-picture process. During the war, Land used his polarizing principle to invent infrared filters, dark-adaptation goggles, and target finders.

Land also improved his polarizing sheets. Quinine was essential in making the iodoquinine sulfate crystals and was in short supply due to its use in treating soldiers with malaria. Land stretched large sheets of polyvinyl alcohol (clear plastic) and thereby aligned the molecules that he dyed with iodine. This improved polarizing material continues to be used in sunglasses and camera filters, and as a component in liquid crystal displays, found in such devices as digital watches and portable computers.

Land's Polaroid Land Camera (demonstrated in 1947, first sold in 1948) was capable of producing finished black and white, and later color photographic prints in sixty seconds. This instant photography worked by squeezing chemicals stored on the border of the film through the camera's two rollers. The chemicals first coated the film with an opaque layer, creating a mini-darkroom; then dyes were released. As the film developed, the opaque layer cleared, revealing the photograph.

Land also demonstrated that color pictures are composed of white and pink light rather than blue, green, and red light. He explained the effect, the mechanism of which is still not completely understood, in his "retinex" theory of color perception, postulating that a minimum of three independent image-forming mechanisms (retinexes) make the effect possible. Among his many other accomplishments are a microscope for viewing living cells in natural color and the optics for the U-2 spy plane made during the 1950s by Lockheed (today Lockheed Martin).

After his retirement from the Polaroid Corporation in 1980, Land founded the Rowland Institute for Science with funding from his Cambridge-based Rowland Foundation, a charitable organization Land had established in 1960. While working at the Rowland Institute, he was involved in the discovery that light and color perception are regulated by the brain and not the retina. In 1957, Land was awarded an honorary doctorate from Harvard, and he received the Medal of Freedom from President John F. Kennedy in 1963. Land's more than 500 patents in light and plastics are second in number only to those of Thomas Edison. Land died on March 1, 1991, in Cambridge.

Richard M. Edwards

Sources

Earls, Alan R., Nasrin Rohani, and Marie Cosindas. *Polaroid.* Mount Pleasant, SC: Arcadia, 2005.

McElheny, Victor K. *Insisting on the Impossible: The Life of Edwin Land.* New York: Perseus, 1998.

Olshaker, Mark. *The Instant Image: Edwin Land and the Polaroid Experience.* Briarcliff Manor, NY: Stein and Day, 1978.

LAND GRANT UNIVERSITIES

The first American colleges and universities emphasized the study of theology and languages. These disciplines were too arcane for reformers in the mid-nineteenth century who wanted the states to establish colleges that would teach agriculture, engineering, and other disciplines with a practical orientation. Between 1855 and 1858, Michigan, Pennsylvania, Maryland, and Iowa all founded agricultural colleges.

In 1862, Congress passed the Morrill Act, granting each state 30,000 acres of land per congressional representative; with the proceeds from the sale of this land, each state was to endow an agricultural and mechanical college. Historians refer to these institutions as land grant colleges or universities, to mark their origin in the granting of land rather than of money.

From the outset, farmers wanted these universities to serve a regulatory function, charging them in the 1860s with analyzing fertilizers and in the 1870s with analyzing insecticides to verify their chemical composition. Scientists, however, sought a more ambitious research agenda. Tensions peaked in 1878, when the governing board of the Ohio Agricultural and Mechanical College changed its name to the Ohio State University. Farmers saw in this decision a betrayal of the university's commitment to applied science in general and to agriculture in particular.

Congress sought to satisfy the proponents of both agriculture and a broad research agenda. In 1887, it codified the agriculture-first strategy of farmers in the Hatch Act, which gave each state $15,000 annually to establish and maintain an agricultural experiment station as the research arm of its land grant university. In 1890, Congress placated those who advocated breadth of instruction and research by passing the Second Morrill Act, which gave each land grant university $25,000 a year for teaching and research in agriculture but also in English, mathematics, and other disciplines with no obvious connection to agriculture.

The wrangling over the focus of the land grant institutions was part of a larger debate regarding science. The founders of the land grant universities shared the Jeffersonian belief that science should yield practical results. By the early twentieth century, however, a core of scientists and university administrators had come to regard the focus on applied science as enervating. Not all science needs be utilitarian to be legitimate, they asserted; the pursuit of science for its own sake was both an end in itself and a foundation for applied research.

In this view, basic science and applied science were seen as the two poles of a single continuum. In that spirit, Congress in 1906 gave the agricultural experiment stations, the citadels of applied science at the land grant universities, an additional $15,000 for basic research. Yet the thrust of science at the land grant universities, from hybrid corn to biotechnology, had been utilitarian. If agriculture were to benefit from this research, the land grant universities would need to communicate the results to farmers. To this end, Congress in 1914 passed the Smith-Lever Act, giving each land grant university $10,000 a year to establish a Cooperative Extension Service.

From the early days, the federal government and the states have cooperated in research at the land grant universities. Federal agencies have stationed scientists at each institution to conduct research of regional and national scope. In this system, duplication is inevitable; several land grant universities, for example, breed new varieties of wheat. The federal government has tried to minimize duplication by concentrating research at centers affiliated with several of the land grant universities.

The research also tends to be interdisciplinary; research on insect-borne corn viruses, for example, requires cooperation among agronomy, plant pathology, entomology, microbiology, and genetics. The trend toward increased interdisciplinary research at America's land grant universities is expected to continue in the twenty-first century.

Christopher Cumo

Sources

Brunner, Henry S. *Land-Grant Colleges and Universities, 1862–1962.* Washington, DC: Department of Health, Education, and Welfare, 1962.

Eddy, Edward D. *Colleges for Our Land and Time: The Land-Grant Idea in American Education.* New York: Harper, 1957.

LASER

Laser is an acronym for "light amplification by stimulated emission of radiation." Laser light differs from sunlight; the former is a single color at a specific wavelength, while the latter is a mixture of colored light. In addition to being monochromatic, laser light is also coherent, since all of its light waves travel parallel to one another in a phase relationship where wave crests and troughs reinforce each other.

American physicist Gordon Gould invented and coined the term laser in 1957, extending the work of another American physicist, Charles Townes, who invented the maser ("microwave amplification by stimulated emission of radiation") in 1953. The laser replaced microwaves with visible light. Gould determined that a laser, which emits concentrated photons, could heat matter to the temperature of the sun's surface in only one-millionth of a second. He prophetically noted its potential uses in communications, radar, and heating. In 1960, yet another American physicist, Theodore Maiman, became the first to demonstrate a functioning laser, generating 10,000 watt, high-energy pulses of red light, a ruby laser. From that point on, an entire family of lasers was developed—gas lasers (1961), semiconductor lasers (1962), liquid lasers (1966), quantum-well lasers (1975), and X-ray lasers (1985) —which are linked by the common action of concentrated electromagnetic emissions.

Because the laser travels in a narrow, parallel beam, it is ideal for use in measurement systems, such as surveying. Because of lasers' high degree of precision, their use in manufacturing environments ensures quality.

In surgery, the concentration of energy in a narrow laser beam makes it possible to excise a very small area and leave the surrounding region untouched. Lasers are used in surgical pro-

Highly concentrated beams of light, lasers have found hundreds of scientific and commercial applications in just half a century. Here, an engineer works on what he hopes will be a superfast computer that uses lasers to transmit and process data. *(John Chiasson/Getty Images)*

cedures that annihilate cancer cells, remove kidney stones, unclog blood vessels, and improve eyesight. Because a laser beam is less intense at its edges than at its center, the laser can be used as a tweezer-like tool to move microscopic objects—a capability of great value in biological studies where single cells must be manipulated without damage.

High-powered laser can be channeled with precise control for such other functions as welding operations, drilling diamonds, and cutting metal. Low-power applications find many uses in communication systems, bar-code scanners, and consumer electronics.

Fiber optic systems, employed in today's telecommunication systems, commonly employ a laser light source. Digital information transmitted as electrical current modulates the light beam, turning the laser on when a digital pulse is present and turning the laser off when the digital pulse is absent. This modulated light beam is coupled into a glass fiber at a specific wavelength and sent over the design distance. At the receiving end, a photodiode detector converts the laser pulses into electrical signals representing the original digital information. Because of the widespread use of fiber optic systems, semiconductor diode lasers are the most commercially exploited class of lasers.

Robert Karl Koslowsky

Sources

Edwards, Terry. *Fiber Optic Systems: Network Applications.* New York: John Wiley and Sons, 1989.

Hecht, Jeff, and Dick Teresi. *Laser: Light of a Million Uses.* Mineola, NY: Dover, 1998.

Silfvast, William T. *Laser Fundamentals.* New York: Cambridge University Press, 1996.

LATIMER, LEWIS HOWARD (1848–1928)

The African American inventor Lewis Howard Latimer was among an elite group of scientific pioneers who significantly advanced the Industrial Revolution.

He was born on September 4, 1848, in Chelsea, Massachusetts, to George and Rebecca Latimer, who had escaped slavery in Virginia only six years before. A contemporary of such other distinguished African American inventors as Norbert Rillieux, Elijah McCoy, and Granville Woods, Latimer served two years in the Union navy during the American Civil War.

After his discharge from the navy in 1865, Latimer worked as a drafter at a patent firm in Boston. His creative intellect and ability for drafting were readily apparent early on, and it was not long before he was offered a job by Alexander Graham Bell. Working in the same laboratory, Latimer and Bell became friends, and Bell asked Latimer to draw the patent design for the communications machine he had invented, which he called the telephone. Latimer's drawings were submitted to the Patent Office, and Bell was granted the patent in 1876.

By 1879, Herman Maxim, the inventor of the machine gun, recognized Latimer's talent and hired him to work for the U.S. Electric Lighting Company in Bridgeport, Connecticut. Latimer's work there would center around improving Thomas Edison's lightbulb design; specifically, he aimed to lengthen the life span of the bulb. Between 1880 and 1881, Latimer devised a way of encasing the carbon filament within a cardboard envelope, which prevented the filament from breaking and extended the life of the light while reducing its cost and significantly expanding its use among the general public. Unfortunately for Latimer, the invention was legally patented to the company rather than to him personally, and he never benefited financially from it.

In 1882, Latimer joined his friend Charles Weston at the Westinghouse Electric Company and supervised several major public lighting projects, including some of the first street lights in New York City. The following year, he joined forces with Edison at the Excelsior Electric Company (later part of General Electric), where he was appointed chief drafter. He also functioned as an expert legal witness for the Board of Patent Control, a patent protection organization formed by the two largest electric companies, Westinghouse and General Electric.

Latimer continued to design and invent throughout his life. Among his inventions are a safety elevator, a locking rack for hats, coats, and umbrellas, and a book supporter. In addition, he wrote the first electrical lighting textbook, *Incandescent Electric Lighting: A Practical Description of the Edison System* (1890), taught at the Henry Street Settlement for recent immigrants, and was asked to be a founding member of the Edison Pioneers, a group of men who had created the electrical industry. He died on December 11, 1928, in Flushing, New York.

Paul T. Miller

Sources

Clarke, John Henrik. "Lewis Latimer—Bringer of the Light." In *Blacks in Science: Ancient and Modern,* ed. Ivan Van Sertima. Somerset, NJ: Transaction, 1987.

Fouche, Rayvon, and Shelby Davidson. *Black Inventors in the Age of Segregation: Granville T. Woods, Lewis H. Latimer, and Shelby J. Davidson.* Baltimore: Johns Hopkins University Press, 2003.

LAWRENCE LIVERMORE NATIONAL LABORATORY

For over half a century, the Lawrence Livermore National Laboratory has been involved in the creation, maintenance, testing, and refurbishing of the U.S. nuclear weapons stockpile. It also conducts research in nuclear fusion, the biological effects of radiation, and non-nuclear weapons of mass destruction (WMD), and addresses homeland security issues.

Located in Livermore, California, the facility is named for physicist Ernest O. Lawrence, who founded the Lawrence Berkeley National Laboratory in 1931. The Lawrence Livermore National Laboratory is one of the three national security laboratories of the National Nuclear Security Administration of the U.S. Department of Energy. Its staff includes 2,700 scientists and engineers.

This national laboratory's innovations include submarine-launched, megaton-class warheads (explosive power equal to 1 million tons of TNT) fitted to underwater-launched Intercontinental Ballistic Missiles (ICBMs); a single, high-yield, multiple nuclear warhead package (MIRV); technologies for the removal of radiation and other contaminants from groundwater; linking computer processors performing the same task (multiple parallel computer processing) in the Advanced Simulation and Computing Program; and modeling of regional and global climate conditions and change.

The laboratory works to ensure the safety, reliability, and utility of the U.S. nuclear arsenal through the Stockpile Stewardship Program, which also refurbishes weapons and their components as necessary. The lab also seeks to detect, prevent, and reverse nuclear, chemical, and biological WMD proliferation through the Strengthening Homeland Security and Countering WMD Proliferation and Use program. In 1991, the lab established the Nonproliferation, International Security, and Arms Control directorate, which develops large-scale, long-term, reliable, sustainable, and affordable clean energy production. In addition, the lab seeks ways to dispose of nuclear waste, and analyzes atmospheric plume analysis of any release of radioactive or other hazardous materials.

Scientific research areas include bioscience, chromosome mapping, high-speed cell sorters for research, and biodetectors to detect potentially hazardous molecules in the air and water. The Center for Accelerator Mass Spectrometry researches atomic and subatomic accelerator technologies used to study subatomic particles. The National Ignition Facility uses laser technology to study the physics, ignition, and fusion burn of weapons systems. The Forensic Science Center is involved in chemical and forensic analysis. The Superblock facility researches materials science.

Richard M. Edwards

Sources

Gusterson, Hugh. *Nuclear Rites: A Weapons Laboratory at the End of the Cold War.* Berkeley: University of California Press, 1996.

Lawrence Livermore National Laboratory. http://www.llnl.gov.

LINDBERGH, CHARLES A. (1902–1974)

The most celebrated aviator in American history, Charles Lindbergh achieved fame in 1927 by becoming the first person to complete a solo, nonstop flight across the Atlantic Ocean. The flight was closely followed in the American and European press, and the public on both continents hailed Lindbergh as a hero.

The son of a congressman, Charles Augustus Lindbergh was born in Detroit on February 4, 1902, and was raised on a farm in Minnesota. From an early age, he took an interest in the new field of aviation, but he was turned down as a pilot in World War I because of his young age. Enrolling as an engineering major at the University of Wisconsin in 1920, he soon dropped out to pursue a career in aviation.

In 1923, Lindbergh bought his first airplane and flew it around the South and West on barnstorming tours, a popular entertainment of the day, in which pilots flew stunts above admiring crowds. A year later, he entered the U.S. Army Air Service, precursor to the U.S. Air Force. He graduated at the top of his class in training as a pursuit pilot in 1925 before being appointed a captain in the Missouri National Guard. While serving in the National Guard, Lindbergh also flew the U.S. mail between St. Louis and Chicago.

Advances in aircraft technology and the exploits of military aces in World War I had generated enormous interest in the field of aviation after the war, both among the general public and in the business community, which was beginning to recognize the commercial potential of flight. When Raymond Orteig, a New York City hotelier, offered a $25,000 prize in 1919 to anyone who could fly solo from New York to Paris

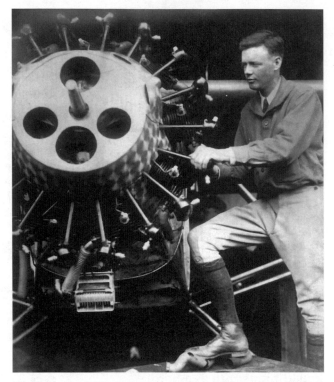

The Spirit of St. Louis monoplane, which carried Charles Lindbergh on his historic nonstop solo flight across the Atlantic in 1927, was powered by an air-cooled, 220-horsepower, nine-cylinder engine. The engine had a special device to keep it greased for the entire flight. *(Hulton Archive/Getty Images)*

without stopping, Lindbergh convinced some St. Louis businesspeople to finance the construction of an airplane to enter the contest.

Designed by aviation engineer Donald Hall, the single-seat, single-engine monoplane—equipped with a 223-horsepower, air-cooled, nine-cylinder Wright radial engine—was built in just sixty days in the spring of 1927, at a cost of $10,000. When Hall complained that he needed more time to iron out stability problems, Lindbergh responded that he preferred a somewhat unstable craft, as it would keep him awake for the transatlantic flight, which was estimated to take more than forty hours. To accommodate the 450-gallon fuel tank in the front, the plane was stripped of all excessive weight and had no windshield, which required Lindbergh to use a periscope to navigate.

At 7:52 A.M. on May 30, 1919, Lindbergh took off from Roosevelt Airfield on Long Island, New York, in a plane dubbed *The Spirit of St. Louis.* Over the next thirty-three and one-half hours, he battled winds, icy conditions, and fatigue, fi-

nally landing in front of a crowd of more than 100,000 people at Le Bourget Field outside Paris. The flight was celebrated around the world. Lindbergh sailed home aboard a U.S. Navy cruiser sent specifically to pick him up, receiving a huge ticker-tape parade upon his return to New York City.

Lindbergh's great triumph would be followed by equally great tragedy. On the night of March 31, 1932, the infant son of Lindbergh and his wife, writer Anne Morrow Lindbergh, was kidnapped from their home in Hopewell, New Jersey. The child was found dead on May 12.

In the late 1930s, Lindbergh was embroiled in controversy when he became the spokesperson for the isolationist movement, which was against U.S. involvement in World War II, and made anti-Semitic remarks. Nevertheless, with U.S. entry into the war following Pearl Harbor, Lindbergh volunteered to serve, flying some fifty combat missions in the Pacific. After the war, he became an adviser to the U.S. military.

Lindbergh won a Pulitzer Prize for his autobiography, *The Spirit of St. Louis,* published in 1953. He died in Maui, Hawaii, on August 26, 1974.

James Ciment

Sources

Hixson, Walter L. *Lindbergh, Lone Eagle.* New York: Harper-Collins, 1996.

Lindbergh, Charles A. *The Spirit of St. Louis.* New York: Scribner's, 1953.

Ross, Walter S. *The Last Hero: Charles A. Lindbergh.* New York: Harper and Row, 1967.

Manhattan Project

The Manhattan Project was the largest integrated scientific research and development project in American history. Started in 1942, the once top-secret project combined the intellectual and financial resources of several nations—principally the United States, Great Britain, and Canada—in a race to develop the world's first atomic bomb. At the time, Nazi Germany was believed to be working on a bomb of its own. The historic project not only would usher in the Atomic Age but also would change the very nature of American science.

Supervised by the U.S. Army Corps of Engineers' Manhattan Engineer District, the project had three principal work sites: the Clinton Engineer Works at Oak Ridge, Tennessee; the Hanford Engineer Works in Washington State; and Project Y, a site at Los Alamos, New Mexico. Teams at these three sites worked in relative secrecy to provide the effort and materials necessary for the development, design, and construction of the atomic bomb. In addition to the government work, a number of prominent American companies, such as Chrysler Corporation, DuPont, Eastman Kodak, and Monsanto, also made valuable contributions to several peripheral but nonetheless critical aspects of the project.

Two distinct atomic bomb designs were produced. The first design, "Little Boy," used explosives to propel a mass of uranium into another piece of uranium. The collision initiated a state of nuclear fission and the subsequent violent release of energy known as an atomic explosion. The second design, nicknamed "Fat Man," used explosives to compress a plutonium sphere from all directions. This explosion initiated a self-sustaining nuclear chain reaction that resulted in an atomic explosion.

The research, design, and assembly were done at Los Alamos, while the supporting sites at Oak Ridge and Hanford produced the enriched uranium and plutonium, respectively. The first atomic device was tested on July 16, 1945, at a desert site near Alamogordo, New Mexico. In the weeks that followed, the Little Boy and Fat Man atomic bombs would be used, respectively, on the Japanese cities of Hiroshima and Nagasaki.

The work of the Manhattan Project advanced American science in a number of fields and provided much of the underpinning for future scientific work in such areas as criticality research, high explosives research, radionuclide chemistry, and theoretical physics. From the project would come technical advances in computers, metallurgy, metrology, health physics, and radiation safety.

The scope and scale of the Manhattan Project remains unparalleled in American scientific history. No other scientific undertaking has ever marshaled so many resources in such a limited time for such a specific task. By the time the project ended, it had employed thousands of individuals and cost the governments involved nearly $2 billion. "Manhattan Project" has since become a catchword for any large science effort of short duration involving vast resources and bright scientific minds.

Todd A. Hanson

Sources

Groves, Leslie R. *Now It Can Be Told: The Story of the Manhattan Project.* New York: Da Capo, 1983.

Rhodes, Richard. *The Making of the Atomic Bomb.* New York: Simon and Schuster, 1995.

MASSACHUSETTS INSTITUTE OF TECHNOLOGY

The Massachusetts Institute of Technology (MIT) is among the preeminent institutions of higher learning in America in the fields of pure and applied science. Incorporated by an act of the state legislature in 1861, its principal founder was geologist William Barton Rogers, supported by other leading Bostonians who recognized the need for a college devoted to the sciences and practical arts. MIT opened its doors in Boston in 1865 with fifteen students. Six years later, it admitted its first woman student, Ellen Swallow, who graduated in 1873 with a degree in chemistry. "Boston Tech" expanded steadily in size and enrollment at its location on Boylston Street. By 1910, it was apparent that new structures and more space were needed; in June 1916, the university moved across the Charles River to Cambridge, into a new campus designed by MIT graduate W. Welles Bosworth. A major portion of the funds for the project were provided by inventor George Eastman, founder of the Kodak Corporation.

In 1917, MIT lost its annual appropriation from the Commonwealth of Massachusetts and was forced to look for new ways of funding its education and research programs. Under the leadership of university president Richard MacLaurin, electrical engineer Dugald Jackson, and chemical engineer William Walker, MIT developed a "Technology Plan" that aimed at developing closer connections between the university's research program and the needs of industry. The plan succeeded in securing MIT's financial future.

Major contributions to the U.S. effort in World War II put MIT in a position to benefit from the government's massive postwar defense programs. The university's close ties to the defense industry came under fire during the turbulent Vietnam War years, however, forcing it to divest control of several military-funded laboratories.

Today, the MIT campus in Cambridge is densely packed with dormitories, education and research buildings, a sports complex, and student service facilities. Total enrollment exceeds 10,000 students. MIT has an annual budget of nearly $2 billion and invested assets valued at nearly $8 billion. More than sixty Nobel Prizes have been awarded to former or current faculty members and students. In 2004, MIT named its first woman president, neurobiologist Susan Hockfield.

William M. Shields

Sources

Jarzombeck, Mark. *Designing MIT: Bosworth's New Tech.* Boston: Northeastern University Press, 2004.

Prescott, Samuel C. *When MIT Was "Boston Tech."* Cambridge, MA: Technology Press, 1954.

Stratton, Julius A., and Loretta H. Mannix. *Mind and Hand: The Birth of MIT.* Cambridge, MA: MIT Press, 2005.

Wylie, Francis. *M.I.T. in Perspective.* Boston: Little, Brown, 1975.

McCormick, Cyrus Hall (1809–1884)

A farmer turned inventor, Cyrus Hall McCormick devised several agricultural implements, the most important of which was a reaper for cutting grain. The device represented a major step forward in the mechanization of American agriculture.

Born to a farm family in Rockbridge County, Virginia, on February 15, 1809, McCormick took an interest in agricultural innovation at a young age. In 1831, he patented a plow and designed a reaper. Drawn by a team of two horses, the reaper vibrated a blade across a toothed platform that enmeshed the grain to be cut, much as a comb enmeshes hair. Over the next two years, McCormick tested his reaper on farms in Lexington, Virginia, garnering interest from local farmers.

McCormick patented his reaper in 1834, but he did not immediately begin widespread production, insisting that improvements were needed before it should be manufactured in quantity. He was, moreover, in debt during the late 1830s, having invested heavily in an iron production company that failed during the Panic of 1837.

By 1841, McCormick had turned around his finances and raised the capital to begin manufacturing reapers in Lexington. In 1844, he licensed production in Brockport, New York, Cincinnati, Ohio, and New York City. Dissatisfied with the workmanship at these factories, he restructured and concentrated production in 1847 in Chicago.

Although he patented improvements to his reaper in 1845 and 1847, the original patent expired in 1848, opening the manufacture and sale to competition. By 1850, no fewer than thirty manufacturers vied for the market, a number that increased to more than 100 by 1860.

Meanwhile, McCormick exhibited his reaper at the London World's Fair in 1851, winning the Council Medal and accolades from the London *Times*. Favorable reviews generated sales. McCormick sold a total of seven reapers in 1842, 500 in 1847, and 4,561 in 1858. In the latter year, he grossed $466,659.

McCormick owed his success as much to business savvy as to inventiveness; he was among the first to insist on field trials as the measure of performance. As Henry Ford would later do with the automobile, McCormick mass-produced reapers of a single, uniform type. Never content with the current model, he also introduced a series of innovations: a mower for cutting grass, a raker for gathering grain or grass into bundles, and two types of binders for tying together bundles of grain or grass. McCormick advertised ahead of production. He guaranteed his reaper against defects and allowed farmers to buy on credit.

Profits from the sale of his reaper allowed McCormick to become a philanthropist, journalist, and politician. In 1859, he made a significant donation to the Presbyterian Theological Seminary of the Northwest and, in 1866, he donated to the Union Theological Seminary at Hampden-Sidney, Virginia. In 1860, he bought the *Presbyterian Expositor* and the Chicago *Times*.

McCormick used these papers to campaign against secession and for the peace wing of the Democratic Party. An advocate of free trade and the westward expansion of the railroad, Mc-Cormick supported Stephen A. Douglas in the 1860 presidential election. Abraham Lincoln's election led McCormick to discontinue the *Expositor*, though in 1872 he bought a second Presbyterian newspaper, the *Interior*.

In 1864, he ran for Congress as the Democratic candidate but lost to Republican challenger John Wentworth. Between 1872 and 1877, McCormick was the chair of the Democratic central committee in Illinois. His international reputation reached its apex in France, where he was inducted into the Legion of Honor in 1869, and into the French Academy of Science in 1879.

McCormick and his wife, Nancy Maria of Jefferson County, New York, whom he married in 1858, had seven children. He died in Chicago on May 13, 1884.

Christopher Cumo

Sources

Aldrich, Lisa J. *Cyrus McCormick and the Mechanical Reaper.* Greensboro, NC: Morgan-Reynolds, 2002.
Judson, Clara. *Reaper Man: The Story of Cyrus Hall McCormick.* Boston: Houghton Mifflin, 1948.

MERCURY, PROJECT

America's first manned spaceflight program, Project Mercury was initiated by the National Aeronautics and Space Administration (NASA) on October 7, 1958. The objective of the program was to place humans into orbit around Earth and return them safely to the ground. In two phases, suborbital and orbital flights, Project Mercury provided the space agency with new knowledge and operational experience in human spaceflight.

In November 1958, NASA established a Space Task Group to manage the Mercury project. While the group defined mission goals and technical requirements, the space agency awarded the McDonnell Aircraft Corporation a contract to construct the vehicle. Devised from existing technology and equipment, the one-person Mercury spacecraft was a wingless capsule designed

The original NASA astronauts, known as the Mercury Seven, were: (back row, left to right) Alan B. Shepard, Virgil I. "Gus" Grissom, and L. Gordon Cooper, Jr.; (front row, left to right) Walter M. Schirra, Jr., Donald K. "Deke" Slayton, John H. Glenn, Jr., and M. Scott Carpenter. *(NASA/Getty Images)*

to protect its human passenger from the vacuum and radiation of space. With dozens of controls, electrical switches, fuses, and levers, the capsule interior had just enough space to include the astronaut. The vehicle also had an ablative heat shield, which safeguarded the astronaut from extreme temperatures as the capsule reentered Earth's atmosphere. Redstone and Atlas launch vehicles were used to place the spacecraft in suborbital and orbital flights, respectively. Prior to the manned flights, the space agency first conducted unmanned tests to evaluate the integrity of the launch vehicles and capsule.

The first Americans to pilot a Mercury capsule were drawn from a group of 110 military test pilots. The astronaut selection process was based on the pilots' flight experience, a long array of physical requirements, and psychological testing. On April 9, 1959, NASA announced the seven original Mercury astronauts: L. Gordon Cooper, Jr., Virgil I. "Gus" Grissom, M. Scott Carpenter, Alan B. Shepard, Jr., Walter M. Schirra, Jr., John H. Glenn, Jr., and Donald K. "Deke" Slayton. On May 5, 1961, astronaut Alan Shepard became the first American in space when a Redstone rocket carried him in a Mercury capsule on a fifteen-minute suborbital flight. Nine months later, John Glenn became the first American to orbit Earth in the *Friendship 7* as it successfully circled the planet three times. On May 15, 1963, Gordon Cooper was launched into space aboard *Faith 7*, where he remained in orbit for a day. The success of the mission led NASA to cancel a seventh manned Mercury flight.

The Mercury spacecraft was not designed for performing experiments due to operational limitations and volume and weight constraints, but the space agency recognized the scientific value of conducting research in an orbiting vehicle. During the Mercury flights, therefore, the astronauts carried out a series of experiments in biomedicine, physical science, and engineering. They also performed perception studies, observed weather conditions, took terrain photographs, collected radiation measurements, examined liquids in zero gravity, and deployed a tethered balloon experiment.

Lasting nearly five years, the twenty-five-flight program, including six manned space missions in the Redstone and Atlas rockets. The program achieved its primary objectives of placing a manned spacecraft in orbit and demonstrating that humans could rocket into space, operate a vehicle in space without any negative consequences, and return safely to Earth.

Kevin Brady

Sources

Grimwood, James M. *Project Mercury: A Chronology.* Washington, DC: U.S. Government Printing Office, 1963.

Swenson, Loyd S., Jr., James M. Grimwood, and Charles C. Alexander. *This New Ocean: A History of Project Mercury.* Washington, DC: NASA, 1966.

MILLS

The Industrial Revolution in America is said to have begun with the founding of Slater's Mill in Pawtucket, Rhode Island, in 1793. Established by the English textile manufacturer Samuel Slater, it was the first facility in America to spin cotton mechanically. Setting out to replicate England's thriving industry, Slater reconstructed an English plant and machines from memory and smuggled plans. Located near the great falls at the junction of the Blackstone and Seekonk rivers, Slater's mill was a great success. Slatersville and other mill villages sprang up nearby. Separate communities reliant on the mill, these villages were built, maintained, and sustained directly by the mill's owners. Many mill workers were children: All of Slater's original nine employees were between the ages of seven and twelve. Over the course of succeeding decades, textile mills and mill towns spread throughout New England.

Francis Cabot Lowell, head of the Boston Manufacturing Company, developed mechanical looms for his Waltham, Massachusetts, factory in 1814–1815. The first vertically integrated factory in America, the Waltham mill conducted all operations, from spinning to finished goods, in the same building. Further improvements were implemented in a second Waltham mill in 1816–1818, finalizing the systems later used in his namesake town, Lowell, Massachusetts, founded near the Pawtucket Falls of the Merrimack River in 1821. Touted as "the first large, planned, industrial city in America," Lowell grew to accommodate 33,000 people by mid-century.

Like the mills at Waltham, Lowell's Boott mills were four stories high, rectangular, 150–160 feet long by 40–50 feet wide, with centrally located exterior stair towers reaching from the ground to the top floor. Originally built in the 1830s, the buildings were later improved with fire protection systems, such as ceiling sprinklers.

Requiring cheap labor, the Boott mills hired young farm women to cut costs, but the mills also provided them with housing and education. Such innovations established Lowell as America's premier model of an industrial city, as its system fostered both economic and social benefits to members.

As time passed, working conditions changed, as did the workforce. New Irish, Portuguese, French Canadian, and Greek immigrants were all desperate for work. Also, Massachusetts passed legislation overseeing child labor, working conditions, and education, reducing total profit.

By the 1920s, as New England's aging mills battled high taxes, unionized labor, and costly transportation expenses, many investors began focusing their efforts on new textile plants in the South, causing many Northern factories to reduce operations or close. Today, abandoned mills and factories are a common sight in New England industrial cities and towns.

Benjamin Lawson

Sources

Lincoln, Jonathan Thayer. "The Beginnings of the Machine Age in New England: David Wilkinson of Pawtucket." *New England Quarterly* 6 (1933): 716–82.

Prude, Jonathan. "Capitalism, Industrialization, and the Factory in Post-Revolutionary America." *Journal of the Early Republic* 16 (1996): 237–55.

MORSE, SAMUEL F.B.
(1791–1872)

Samuel Finley Breese Morse was an artist and inventor of the telegraph, which used electrical impulses to transmit messages in code.

Morse was born in Charlestown, Massachusetts, on April 27, 1791. His father, Jedidiah Morse, was a congregational minister and geographer. After an education at Yale and art studies in England, Morse settled down to the life of a professor and artist. A gifted and prolific painter, he produced more than 300 portraits and historical paintings during his first forty-one years. In 1832, Morse became the first fine arts professor of an American college—New York University—and he was one of the thirty cofounders of the National Academy of Design, serving effectively as president from 1826 to 1845.

His contributions as an artist were overshadowed by his fame as an inventor. Morse's Yale education exposed him to lectures on electricity and opportunities to assist with electrical experiments using wet-cell batteries. On a return ocean voyage from England in 1832, he conceived the idea of transmitting information by electricity over iron wire. The genesis of this idea came from shipboard talk of recent discoveries in electromagnetism and the development of the electromagnet. Due to his commitment to the arts, however, Morse was unable to complete his first telegraph apparatus until late 1835.

Morse learned how to employ more power using denser coils. He also developed the dot-dash Morse code, telegraph key, and sounder for telegraph operators to use. Morse was influenced by the work of British scientists William Cooke and Charles Wheatstone and was assisted by friends and associates Leonard Gale and Alfred Vail.

In 1842, Morse lobbied Congress to provide funds to set up a telegraph line. He demonstrated the telegraph between two committee rooms in the U.S. Capitol. Congress responded with a $30,000 appropriations bill in 1843 to build the first telegraph line, from Baltimore to Washington. Morse was named superintendent of telegraphs and was responsible for construction. The line was completed in 1844, and Morse invited a number of notable figures to the May 24 formal opening. The first message sent was a biblical quotation from Numbers 23:23: "What hath God wrought!"

After the initial demonstration, acceptance was rapid; thousands of miles of telegraph cable were strung along railway lines across North America. The word "telegram" entered the lexicon in 1852, and more than 200 million telegraph messages were sent annually at the height of its popularity. Today, telegraphy has largely been replaced by fax machines, e-mail, and text messaging.

Morse died in New York on April 2, 1872.

Robert Karl Koslowsky

Sources

Oslin, George P. *The Story of Telecommunications.* Macon, GA: Mercer University Press, 1992.

Silverman, Kenneth. *Lightning Man: The Accursed Life of Samuel F.B. Morse.* New York: Alfred A. Knopf, 2003.

NASA

The National Aeronautics and Space Administration (NASA) was created when President Dwight D. Eisenhower signed the National Aeronautics and Space Act into law on July 29, 1958. During the half-century of its existence, NASA has led the world in developing aircraft that can fly at supersonic speeds, carry humans into outer space, and take them to the moon; in conducting zero gravity research; and in developing reusable spacecraft such as the Space Shuttle.

NASA was created amid Cold War tensions between the United States and the Soviet Union. During the 1950s, research into missile technology by the two countries resulted in the first artificial satellite, *Sputnik,* launched by the Soviet Union into Earth orbit in October 1957. The United States, afraid that it was being left behind in the arms race, inaugurated NASA to narrow the technological gap between the two countries.

One of the first tasks of NASA was to respond to *Sputnik* with a satellite of its own. This goal was achieved on the last day of 1958 when *Explorer 1* rode a Jupiter C rocket leaving Earth's orbit. The satellite discovered the Van Allen radiation belt, a region of highly charged particles trapped in Earth's magnetic field.

NASA's early facilities included the Jet Propulsion Laboratory at the California Institute of Technology. The lab has been a prime mover in NASA's unmanned space programs. These programs included the Pioneer, Ranger, Surveyor, Viking, Magellan, Galileo, and Pathfinder missions that explored the moon, Mars, Venus, Mercury, Jupiter, and the limits of the solar system.

Pioneer missions encompassed twenty years of space exploration, from 1958 to 1978, including the most famous missions, *Pioneer 10* and *Pioneer 11,* which explored Jupiter and Saturn before traveling to the outskirts of the solar system. *Ranger* and *Surveyor,* from 1961 to 1965 and 1966 to 1968, were lunar probes gathering information in preparation for the manned Apollo program.

Voyager 2 and *Voyager 1* were launched in August and September 1977. *Voyager* 2 flew by, studied, and photographed Jupiter, Saturn, Uranus, and Neptune and is now heading for the edge of the solar system. *Voyager 1* studied and photographed Jupiter and Saturn, crossed the "termination shock" (a region of compressed, heated particles at the edge of the solar system), and is traveling toward interstellar space.

Pathfinder, launched in 1996, landed on Mars, and a remote rover, *Sojourner,* explored the Martian surface, collecting samples. At the same time, *Global Surveyor* orbited Mars and sent substantial data back to Earth about the red planet. *Exploration* rovers sent to Mars in 2004 traversed the planet's surface.

A highly publicized unmanned mission was *Deep Impact,* which flew by the comet Tempel 1 and released an "impactor" that collided with the comet on July 4, 2005, making a deep crater.

Deep Impact discovered new facts about the nature of comets, such as their porous nature, low gravity, and core material that dates back to the formation of the solar system.

NASA's initial manned space programs were Project Mercury, from 1959 to 1963; Project Gemini, from 1963 to 1966; and Project Apollo, from 1967 to 1972. These successful programs demonstrated that humans could function in space without any significant negative reactions, that rendezvous and docking operations between spacecraft were possible, that spacecraft could successfully reenter the atmosphere, and that spacecraft could successfully take humans to the moon.

Seeking to establish a permanent human presence in space, NASA followed the Apollo program with the *Skylab* space station, which served as an orbital laboratory where crewmembers conducted research and experiments. In 1975, NASA cooperated with the Soviet Union in the Apollo-Soyuz Test Project. This international endeavor included a crew exchange and the testing of rendezvous and docking maneuvers. In 1981, NASA launched the first reusable spacecraft, the Space Shuttle *Columbia,* which was used for more than a hundred missions.

In 1984, Congress authorized NASA to construct a new space station; however, budgetary and developmental constraints hindered the project. By 1993, Russia agreed to cooperate with the United States to build such a facility. Known as the International Space Station, it became operational on November 2, 2000, when the Expedition One crew docked with it.

Over the years, NASA has launched Intelsat, Echo, Telstar, and Syncom communications satellites to provide long-range communications on Earth. Additionally, NASA sent the *Ranger, Pioneer, Mariner, Viking,* and *Voyager* spacecrafts to survey the moon, investigate Venus and Mars, and explore the outer planets in the solar system. In April 1990, the deployment of the Hubble Space Telescope enabled researchers to make numerous discoveries about the origins of the universe. NASA also launched satellites such as TIROS and Landsat to research Earth's weather patterns and resources.

On February 1, 2003, the Space Shuttle *Columbia* disintegrated over Texas as it was returning to Earth from a mission, killing all seven crew members on board. The tragedy was the result of a piece of foam breaking off during launch from the orbiter's external tank and striking the Space Shuttle's left wing. The space fleet was grounded for nearly three years as NASA modified the Space Shuttles and implemented measures to improve safety. On July 26, 2005, the launch of Space Shuttle *Discovery* marked the resumption of U.S. space flights. In January 2004, President George W. Bush authorized NASA to complete the International Space Station, continue robotic exploration of the solar system, retire the Space Shuttle fleet, and develop a vehicle that would enable humans to return to the moon.

Kevin Brady and Russell Lawson

Sources

Anderson, Fred W. *Orders of Magnitude: A History of NACA and NASA, 1915–1980.* Washington, DC: NASA, 1981.

Heppenheimer, T.A. *Countdown: A History of Space Flight.* New York: John Wiley and Sons, 1999.

National Aeronautics and Space Administration. http://www.nasa.gov.

Nautilus

The USS *Nautilus* was the world's first nuclear submarine. In 1948, driven by the growing Cold War, former U.S. submariner Hyman Rickover gathered a team of scientists and engineers, established the Naval Reactors Branch of the Atomic Energy Commission, and developed a workable nuclear reactor, called S1W, which used fission instead of internal combustion to produce steam to power turbines. Nuclear power had a number of advantages over previous technology: It all but eliminated a submarine's need to surface for air, and it replaced over 630 metric tons of diesel with about one-half kilogram of uranium. Moreover, while previous submarines required two engines, diesel and electric, *Nautilus* had only the reactor.

The *Nautilus* keel was laid at General Dynamic's Electric Boat Division at Groton, Connecticut, in 1952, and the ship was christened by First Lady Mamie Eisenhower on January 21, 1954. After its final fitting and testing, Commander Eugene P. Wilkinson, *Nautilus*'s first captain,

The USS *Nautilus*, the world's first nuclear-powered submarine, makes an early sea trial in 1955. Its passage beneath the polar ice cap to the North Pole in 1958 demonstrated the potential—scientific as well as military—of nuclear submarines. *(Library of Congress, LC-USZ62–103120)*

ordered all lines cast off on January 17, 1955, and signaled the historic message "Underway on nuclear power."

Nautilus was not only the world's first nuclear-powered vessel but the first ship designed to operate almost entirely underwater. Previous submarines were actually submersibles, or surface ships that could submerge for short periods, usually from one to two days. On the *Nautilus*, submersion would be limited only by the food and oxygen needs of the crew. The vessel's shakedown cruise in May 1955 took it from Connecticut to Puerto Rico, a total of 1,376 miles (2,220 kilometers) in 89.9 hours. Underwater all the way, it was the longest and fastest submerged cruise ever, averaging 15 knots—World War II's like-named *Nautilus* (SS-168) had averaged 8 knots submerged.

In 1958, the sub made a transit of the Arctic Ocean, becoming the first ship to reach 90 degrees north latitude—the North Pole—and gathering more scientific data than had been obtained in all previous Arctic explorations combined. On returning to the United States, the crew received a ticker-tape parade in New York City and the first peacetime Presidential Unit Citation ever awarded.

Though much of its career was spent in development of new antisubmarine warfare techniques, *Nautilus* saw action in the Mediterranean Sea during the 1956 Suez Crisis, and it was part of the 1962 quarantine of shipping during the Cuban Missile Crisis. In twenty-five years on active duty, *Nautilus* sailed more than 500,000 miles (more than 800,000 kilometers), refueling only twice, before arriving at Mare Island Naval Shipyard in California, where it was decommissioned on March 3, 1980. Designated a National Historic Landmark in 1982, *Nautilus* was returned to Groton, where it remains permanently berthed as part of the Submarine Force Museum.

Phoenix Roberts

Sources

Anderson, William R., and Clay Blair, Jr. *Nautilus 90 North.* Blue Ridge Summit, PA: Tab, 1989.

Friedman, Norman, and James L. Christley. *U.S. Submarines Since 1945: An Illustrated Design History.* Washington, DC: Naval Institute Press, 1994.

Gillcrist, Dan. *Power Shift: The Transition to Nuclear Power in the U.S. Submarine Force As Told by Those Who Did It.* Lincoln, NE: Universe, 2006.

U.S. House of Representatives. *Advanced Submarine Technology and Antisubmarine Warfare.* Park Forest, IL: University Press of the Pacific, 2005.

U.S. Navy Submarine Force Museum. http://www.ussnautilus.org.

NUCLEAR ENERGY

Nuclear energy derives from European scientific developments and American engineering skill during the first half of the twentieth century. Building on Albert Einstein's theories of matter and energy, Ernst Rutherford's work with uranium, and Niels Bohr's work with the atomic structure, American physicist Ernest Lawrence invented the cyclotron in 1931. In a cyclotron, subatomic particles are accelerated to great speeds, colliding with the atomic nuclei of unstable elements.

After the discovery of fission in 1938, Italian American Enrico Fermi worked on the emission (from the uranium nucleus) of secondary neutrons and their associated self-sustaining chain reactions. He developed a series of experiments at the University of Chicago that culminated in the December 1942 creation of an atomic pile and the first controlled nuclear chain reaction. As one of the scientific leaders on the Manhattan Project for the development of nuclear energy and the atomic bomb, Fermi solved many of the physics problems associated with nuclear development. For nuclear fission to produce a continuous supply of energy, the reaction must be controlled such that it achieves a steady state of operation. The first large-scale nuclear power plant to take advantage of this principle was in Shippingport, Pennsylvania, which began operation in December 1957.

In a nuclear reaction, complete fission of uranium 235 produces 2.5 million times more heat than an equivalent weight of carbon in coal, oil, or natural gas. The controlled fission process generates heat that is used to produce high-pressure steam to rotate a mechanical turbine, which in turn generates electricity. Nuclear reactor efficiency is about 30 percent, with 70 percent of the heat from fission released locally into the atmosphere or a nearby river or ocean. The 30 percent of heat energy converted to electricity is similar to that of a coal plant, but nuclear reactors require much less fuel to produce comparably large amounts of energy. A single kilogram of uranium is equivalent to the energy found in 3,000 tons of coal, without producing the global warming effects of carbon dioxide or the pollution of sulfur dioxide and nitrogen oxides.

Recognizing the possibilities of this alternative energy source, Americans during the 1970s and 1980s built a series of nuclear reactors; by 1990, there were 112 reactors in the United States. By 1980, nuclear energy produced more electricity than oil, and, by 1983, it produced more electricity than natural gas. Hydroelectric power also fell behind nuclear energy, yielding its second place status behind coal in 1984.

Today, the United States still derives about 16 percent of its energy needs from nuclear power. Sixty-five locations around the country have 104 nuclear reactors in operation. The U.S. safety record is considered one of the best in the world and is continually improving. Opponents of nuclear power cite the risks of nuclear contamination of the environment and humans by means of accidental release of radioactive contaminants released into the air and water. Also, as the technology becomes more common, some fear the possibility of proliferation of nuclear technology to countries that sponsor terrorism.

Nevertheless, the federal government encourages the development of nuclear power because of the nation's ever increasing need for energy. The U.S. Department of Energy has plans to develop a dual-purpose next-generation nuclear power plant that will produce both electricity and hydrogen. In the future, nuclear energy may offer the least polluting and most available source to satisfy the growing American demand for energy.

Robert Karl Koslowsky

Sources

Garwin, Richard L., and Georges Charpak. *Megawatts and Megatons: A Turning Point in the Nuclear Age?* New York: Alfred A. Knopf, 2001.

Seaborg, Glenn T. *Adventures in the Atomic Age: From Watts to Washington.* New York: Farrar, Straus and Giroux, 2001.

OPPENHEIMER, J. ROBERT (1904–1967)

J. Robert Oppenheimer's leadership in the Manhattan Project during World War II resulted in the successful development of the world's first atomic bomb. The project lasted twenty-seven months and, was spread geographically throughout the United States; it culminated in the dropping of two atomic bombs on Japan in August 1945, which brought a rapid end to World War II.

Julius Robert Oppenheimer was born in New York City, the eldest son of German immigrants, on April 22, 1904. After undergraduate studies at Harvard University and postgraduate work in England and Germany, where he studied with such leading European physicists as J.J. Thompson and Max Born, Oppenheimer returned to the United States in 1929 to become a physics professor at the University of California at Berkeley.

At Berkeley, Oppenheimer investigated the energy processes of subatomic particles. He identified the process of quantum tunneling, whereby a particle moves from one point to another without passing through intermediate points. He also predicted the existence of astronomical black holes, which occur when a massive star collapses under its own gravitational force.

Upon the formation of the Manhattan Project in 1942, Oppenheimer was appointed director of the new weapons laboratory at Los Alamos, New Mexico. The work there lasted from the spring of 1943 to the fall of 1945, leading to the first and successful atomic bomb test on July 26, 1945, at Alamogordo, New Mexico, some 300 miles south of Los Alamos. Oppenheimer codenamed the test "Trinity," for the three planned atomic bomb detonations using the fission process. The other two bombs were detonated over Hiroshima and Nagasaki, Japan, on August 6 and August 9, respectively.

In 1945, at the peak of activity, 125,000 people were employed on the Manhattan Project across all sites. In New Mexico, the average age of the employees was just over twenty-nine. Costs totaling about $2.2 billion made the Manhattan Project the largest research and development

J. Robert Oppenheimer (left), the director of the Manhattan Project, and U.S. Army General Leslie Groves, the military officer in charge of the program, examine the remains of a tower in Los Alamos, New Mexico, from which an atomic test bomb was set off in 1944. *(Keystone/Hulton Archive/Getty Images)*

undertaking until the International Space Station in the 1990s. As fast as the project team was formed, however, it was disbanded; most of the key players moved on to other projects or joined the Atomic Energy Commission (AEC).

After the war, Oppenheimer became chair of the General Advisory Committee of the AEC, which was authorized by federal legislation in April 1946 to assume control of research and development of nuclear weapons. One of the first tasks of the commission was to improve the fission process and build an atomic arsenal. Oppenheimer aggressively pursued these objectives, but he opposed developing more powerful nuclear weapons. After the Russian success in detonating an atomic bomb in 1949, however, President Harry S. Truman mandated the AEC to build an even more powerful weapon. The result was the hydrogen bomb.

Oppenheimer resigned his post shortly thereafter and became the director of Princeton's Institute for Advanced Study, a position he held from 1947 to 1966. During the McCarthy era in the early 1950s, Oppenheimer became the object of investigation and persecution by the government for his contacts with members of the Communist Party; his security clearance was revoked in 1953.

A decade later, however, President Lyndon B. Johnson awarded Oppenheimer the Enrico Fermi Award and restored his security clearance. Oppenheimer died of cancer on February 18, 1967.

Robert Karl Koslowsky

Sources

Bernstein, Jeremy. *Oppenheimer: Portrait of an Enigma.* Chicago: Ivan R. Dee, 2004.

Bird, Kai, and Martin J. Sherwin. *American Prometheus: The Triumph and Tragedy of J. Robert Oppenheimer.* New York: Alfred A. Knopf, 2005.

Cassidy, David C. *J. Robert Oppenheimer and the American Century.* New York: Pi Press, 2005.

Seaborg, Glenn T. *Adventures in the Atomic Age: From Watts to Washington.* New York: Farrar, Straus and Giroux, 2001.

PATHFINDER, MARS

The *Mars Pathfinder* spacecraft was launched on December 4, 1996, from Cape Canaveral, Florida. Its seven-month, 309 million mile journey to Mars was NASA's greatest success in decades. *Pathfinder* landed on the surface of Mars on July 4, 1997, with millions of television viewers watching.

En route to the Red Planet, the spacecraft made four course correction maneuvers as it spun at a rate of two rotations per minute along its spin axis oriented toward Earth. NASA used an innovative approach for *Pathfinder*'s atmospheric entry and landing on Mars. A four-and-a-half minute automated sequence controlled the descent through the thin Martian atmosphere and touchdown on the rocky surface. *Pathfinder*'s entry velocity of 16,600 miles per hour was 80 percent faster than earlier (Viking mission) landers. The spacecraft entered the atmosphere of Mars directly from interplanetary space and not from orbit.

Pathfinder underwent 20 g's of force during peak aerodynamic deceleration, roughly one minute after atmospheric entry. To reduce the speed of descent, a large parachute was deployed. Giant airbags, fifteen feet in diameter, inflated seconds before touchdown, absorbing the shock of impact and protecting the lander from the rugged terrain as it made initial contact. The lander bounced for several minutes along the planet's surface. When the landing was over, the airbags were deflated and retracted, and the planetary rover—named *Sojourner,* after American abolitionist Sojourner Truth—was deployed to explore the Martian landscape.

Ares Vallis, a former flood basin, was chosen as the landing site because of its relatively safe terrain and proximity to a diversity of rocks deposited from a catastrophic flood. During *Sojourner*'s excursions across the surface, it relied on the lander for Earth communications and imaging support. *Sojourner*'s close-up photos and instrument data obtained from studying rock composition were fed to *Pathfinder* for transmission to Earth scientists. Besides the rover's photographs, the lander sent such data as the planet's temperature, winds, and atmospheric pressure. A camera on the lander took panoramic images of the landing site and monitored the lowering of the lander's ramp and initial deployment of *Sojourner.* The mission lasted for three months, much longer than anticipated. Mission control lost radio contact with *Pathfinder* on September 27, 1997, and the mission officially ended on March 10, 1998.

One of the public relations objectives for the *Mars Pathfinder* was to highlight NASA's "faster, better, cheaper" directive and its ability to deliver results. The scientific results of the mission, and the public response to the robotic rover traveling the barren Martian surface, transcended all scientific objectives and exceeded expectations. To the public, the neighboring planets of the solar system never seemed so close.

Robert Karl Koslowsky

Sources

Benjamin, Marina. *Rocket Dreams: How the Space Age Shaped Our Vision of a World Beyond.* New York: Free Press, 2003.
Godwin, Robert, ed. *Mars: The NASA Mission Reports.* Toronto: Apogee, 2000.

PHOTOGRAPHY

The science of photography began with the work of nineteenth-century European chemists who experimented with recording various shades of light on metal, glass, and paper. Early twentieth-century American inventors turned European photographic science into a widespread art form and easy-to-use technology for the masses.

The camera is a device that records images that can be reproduced as photographs or electronically on display screens. The origins of the camera date to the seventeenth century, when the camera obscura was developed by artists to depict natural scenes. Its components include a box, a viewfinder and a lens to project the vertical scene, and a mirror behind the lens that deflects the image ninety degrees and projects it on a horizontal surface.

This early invention was eventually connected to advances in chemistry by German, French, English, and American chemists who developed various means of capturing and fixing an image on a flat medium such as a metal or glass plate. Englishman Thomas Wedgwood in 1802 invented the negative by using silver compounds applied to paper; when exposed to light, the dark areas of the image turned light and the light areas turned dark. In the 1820s, French physicist Joseph Nicéphore Niépce used a similar chemistry to produce the world's first photographs, called heliographs.

In the 1830s, French artist Louis-Jacques-Mandé Daguerre collaborated with Niépce to produce the earliest production photographs, called daguerreotypes. Plates were inserted in the back of the box-shaped camera, and the shutter opened while the subject sat motionless; chemical treatments were used to develop the images. In America, Mathew Brady used the daguerreotype with inexpensive tin plates to document the images of the Civil War, which established the field of photojournalism

In 1841, Englishman William Talbot in 1841 patented the technique (calotype) of recording a negative in the camera; in the developing process, a positive was made on a separate chemically treated paper. His fellow Englishman Frederick Scott Archer in 1848 began using glass plates in a process called wet-plate or collodion photography to produce finely detailed negatives, which had to be developed after exposure.

In the early 1900s, the Englishman Richard Maddox and American George Eastman perfected the dry-plate process that placed an emulsion on a paper or plastic film base that could be processed later. These innovations in film processing led to the mass production of roll film cameras. Use of negative film that was later converted to a positive picture became the industry standard. Eastman, who in the 1880s founded the company that would become Kodak, established factories in both Great Britain and Germany, overtaking a more established but, by then, outdated European photo industry.

In 1895, the Lumière brothers in France developed the movie camera and projector. American

Kodak's Brownie camera—named for inventor Frank Brownell—revolutionized photography in the first decade of the twentieth century. Handheld, easy to operate, and inexpensive, the Brownie introduced the concept of the amateur "snapshot." *(Library of Congress, LC-USZ62–70574)*

movie makers soon produced early narrative motion pictures such as *The Great Train Robbery* (1903), and the advent of talking pictures in the late 1920s gave rise to the modern motion picture industry in the United States. In 1925, the German Leitz Company produced a still camera using 35 mm film, the same used in movie cameras. Over time, camera technology was refined for specialized uses, such as high speed, high altitude, space, microscopic, and medical diagnoses photography.

In the 1940s, American inventor Edwin Land developed a film processing method that allowed a developed (positive) picture to emerge directly from the camera. This technique was further sophisticated by his Polaroid Company and sold widely on the international market from the 1950s to the 1970s.

By the early 2000s, the digital chip began replacing chemical film processing for still and motion picture cameras. Using computer technology, pictures in the form of image files are stored in camera memory cards, from which the photographer can transfer image files to a computer and use graphics software to modify the image. The greater the size of the image sensor chip, and the greater the number of megapixels on it, the finer the image. Since digital photographs are printed using ink-jet and dry ink paper technology, the quality is similar to print chemistry. The ability of photographers to alter their photos, however, no longer assures viewers that the photograph is a genuine depiction.

James Steinberg and Oliver Benjamin Hemmerle

Sources

Brayer, Elizabeth. *George Eastman: A Biography.* Baltimore: Johns Hopkins University Press, 1996.

Newhall, Beaumont. *History of Photography: From 1839 to the Present.* Lebanon, IN: Bulfinch, 1982.

North, Michael. *Camera Works: Photography and the Twentieth-Century Word.* New York: Oxford University Press, 2005.

Sandler, Martin W. *Photography: An Illustrated History.* Oxford, UK: Oxford University Press, 2002.

Taft, Robert. *Photography and the American Scene: A Social History, 1839–1889.* New York: Dover, 1964.

Welling, William B. *Photography in America: The Formative Years, 1839–1900.* Albuquerque: University of New Mexico Press, 1987.

Wensberg, Peter C. *Land's Polaroid: A Company and the Man Who Invented It.* Boston: Houghton Mifflin, 1987.

PINCKNEY, ELIZA LUCAS (1722–1793)

Eliza Lucas was a South Carolina plantation manager and agricultural innovator best known for the introduction of indigo, the source of a popular blue dye.

Born in the West Indies in 1722, she immigrated to South Carolina with her family when she was 16. Her father, a British army officer from Antigua, took the unusual step of sending her to be educated in London. She played the flute, read Milton and Plutarch, spoke French fluently, and was an amateur astronomer, but her passion was botany.

While still in her teens, she managed her father's 600 acre plantation at Wappoo Creek near Charleston, corresponding with him about her planting decisions. She began growing indigo in 1739. By 1744, she had produced a commercially acceptable product suitable for export to Britain. That same year, she married Charles Pinckney, a leading statesman and planter.

Indigo was a particular challenge because of the complicated processing necessary to produce the dye, and Eliza Lucas Pinckney had much difficulty finding a reliable expert to oversee the process. With success, she distributed indigo seed among her fellow planters to encourage them to try the crop.

In the 1740s, as more dyes of all kinds were needed to give color to the ever growing quantity of textiles produced in England, indigo cultivation expanded dramatically, and the British Parliament agreed to subsidize its production in 1749. Indigo's growing season complemented that of rice, South Carolina's staple crop, and indigo became a crop second only to rice in the South Carolina economy.

Indigo was only one of the crops with which Pinckney experimented. Many derived from the Caribbean. She grew cotton, lucerne, ginger, and cassava. She also grew figs, hoping to dry them for export, and she tried packing eggs in salt to export them to the British Caribbean. On another family plantation, Belmont on the Cooper River, she tried to cultivate silk but, like the many other colonial Americans who tried to produce it commercially, she was unsuccessful. She also educated some of her young slaves to read, hoping to start a school for slave children, a highly unusual step for a South Carolina slaveowner.

Though increasingly busy with raising her four children, she nevertheless found time to continue experimenting with silk production. Accompanying her husband to England, she took along some Belmont silk, which was woven into a gold brocade dress. In England, she came into contact with a network of gardeners and agriculturalists interested in new plants from America.

After Charles Pinckney's death in 1758, Eliza Pinckney managed the family's plantations for the benefit of her young sons and daughter. During this time, she participated in networks of botanical exchange with fellow South Carolina planters and others across the Atlantic.

Although, like many South Carolinians, Pinckney had closer connections with the Caribbean and England than with the other continental colonies, she supported the American Revolution. Indeed, her sons Charles Cotesworth Pinckney and Thomas Pinckney played important roles in the war and its aftermath. Eliza Lucas Pinckney died of cancer on March 26, 1793.

William E. Burns

Source

Pinckney, Elise, ed. *The Letterbook of Eliza Lucas Pinckney.* Columbia: University of South Carolina Press, 1997.

PLUTONIUM

Plutonium is a radioactive, silver-colored, metallic element used principally as fuel in nuclear weapons and some nuclear reactors. Plutonium has been a source of both scientific challenge and political controversy since its discovery by Glenn Seaborg, Edwin McMillan, Arthur Wahl, and Joseph Kennedy in 1941 at the University of California, Berkeley, while bombarding uranium targets with deuterium neutrons from a 60 inch cyclotron. Although trace amounts are found naturally in uranium ores, plutonium must be manufactured.

During the Manhattan Project, plutonium was used to create the first implosion-type nuclear device tested in July 1945 and the atomic bomb dropped on Nagasaki, Japan, on August 9, 1945. Plutonium for this weapon was made from an isotope of uranium, uranium 238. (The bomb used against Hiroshima, Japan, on August 6 was constructed with uranium 235.) This was processed in nuclear reactors at the Hanford Engineer Works in Hanford, Washington; Hanford would remain America's largest producer of weapons-grade plutonium throughout much of the Cold War.

Scientific research on plutonium since the end of World War II has focused primarily on enhancing reactions in nuclear warheads and on the use of the metal in nuclear energy production. Although there are fifteen known isotopes of plutonium, the two most common ones are plutonium 239, which is used in nuclear weapons, and plutonium 238, used primarily in nuclear power sources.

Aside from nuclear weapons and nuclear reactors, plutonium has few uses. With the end of the Cold War, scientists have continued to explore potential scientific, commercial, or industrial uses for the metal. Currently it is used in radioisotope thermoelectric generators (RTGs) for space probes; one kilogram of plutonium 238 can provide 22 million kilowatt-hours of heat energy, making it possible for interstellar probes to operate far into the coldest reaches of outer space, where solar panels are ineffective.

Handling plutonium is a complicated affair. At the end of World War II, health physicists and physicians working under the auspices of the Atomic Energy Commission conducted a range of medical studies on the effects of plutonium on laboratory animals and human subjects, frequently without the informed consent of the human subjects. Plutonium is highly radioactive and requires only relatively small amounts to achieve a self-sustaining nuclear reaction. It also reacts chemically with oxygen and water, creating a pyrophoric compound that burns easily in room-temperature air. Because of the radiological and chemical hazards, as well as the risk of nuclear proliferation if it falls into the wrong hands, plutonium must be kept in specially designed containers in high-security storage facilities.

Because of the large stockpiles built up during the Cold War, the United States currently has more plutonium than it needs, creating a variety of environmental problems. With the cessation of plutonium production in the late 1980s, Hanford began an extensive environmental cleanup project that is rife with overlapping and sometimes conflicting regulatory, political, and technical challenges.

Plutonium 239 has a half-life—the time required for one-half of a quantity of the plutonium to become naturally nonradioactive through the process of radioactive decay—of roughly 24,000 years. Scientists continue to work on long-term technical solutions for the storage of used nuclear fuel and nuclear waste.

Todd A. Hanson

Sources

Greenwood, Norman Neill, and A. Earnshaw. *Chemistry of the Elements.* 2nd ed. Oxford, UK: Butterworth-Heinemann, 1997.

Seaborg, Glenn T., et al. *The Plutonium Story: The Journals of Professor Glenn T. Seaborg, 1939–1946.* Columbus, OH: Battelle, 1994.

POPULAR SCIENCE MAGAZINE

Popular Science magazine is one of the oldest American periodicals seeking to bring the wonders of science to the general public. Founded in 1872 as a philosophical journal covering theoretical science, it later became a leading resource for the lay public in the coverage of applied science.

First published as the *Popular Science Monthly,* the magazine was envisioned by founding editor Edward L. Youmans as a vehicle for presenting and explaining the scientific ideas and breakthroughs then occurring in Europe to the thoughtful but scientifically untrained reader in America. Youmans carried out this plan with a missionary zeal, and the magazine introduced the American public—including scientists—to the works of Herbert Spencer, Charles Darwin, Thomas Huxley, and John Tyndall, as well as, through translations, the works of other European scientists. The inclusion of information on Darwin and his supporters led some religious critics to refer to the magazine as the "evolutionist monthly." The natural sciences were featured prominently, but emphasis was also given to the emerging social sciences, with important articles on sociology, psychology, economics, and political science.

Later years found American authors appearing more frequently in the pages of *Popular Science Monthly,* including ichthyologist David Starr Jordan and inventor Alexander Graham Bell. The magazine also offered coverage of important scientific addresses and meetings. By the mid-1880s, circulation was nearly 18,000, an extraordinary number for a specialized publication.

Upon Youmans's death in 1887, his brother William took over the editorship; he carried on the magazine's mission until 1900, when Columbia University professor and psychologist James McKeen Cattell, already the publisher of *Science,* was named editor. A year later, the Cattell-owned Science Press purchased the magazine. Cattell, an outspoken proponent of science popularization, continued the journal along the same lines as the Youmans brothers. A 1904 merger with *Sanitarian,* a journal of health and public health, brought an increase in the number of health and medical articles.

By 1915, faced with declining circulation, Cattell concluded that the magazine could no longer attempt to reach both the lay public and readers with some scientific training (scientists, teachers, and the college educated). He sold the name to the publishers of the *World's Advance,* a popular journal devoted to mechanical devices and their development. Cattell retained the subscriber list and launched a new journal, *Scientific Monthly,* which continued the philosophy of the Youmans brothers.

The new editor of *Popular Science Monthly* was Waldemar Kaempffert, who had been associated with *Scientific American* and who would later become the first science editor of the *New York Times.* The magazine now featured copious illustrations and the decidedly practical bent that would become its hallmark. This included regular coverage of aviation, the automobile, radio, and photography. Advertising was now accepted. Well-known scientists continued to contribute articles; however, many came from industrial laboratories rather than universities, reflecting a growing trend in American (and European) science. George Eastman, Thomas Edison, Henry Ford, and Robert Goddard all wrote for the magazine. The emphasis was now on the products that science made possible (technology) rather than on new scientific knowledge.

World War II brought a shift in editorial emphasis to the new technologies of the battlefield, such as jet aircraft, radar, and the atomic bomb. And while coverage of automobiles and more familiar household technologies never disappeared, newer technologies continued to add to the magazine's appeal: satellites and transistor radios in the 1950s (when the name was shortened to *Popular Science*); space flight in the 1960s; solar and nuclear power in the 1970s; personal computers in the 1980s; and the Internet and digital photography since then.

Today, *Popular Science* continues to inform and educate lay readers about the basics of emerging science and invention. It also sparks the imagination of the next generation of scientists and engineers.

George R. Ehrhardt

Sources

Burnham, John C. *How Superstition Won and Science Lost: Popularizing Science and Health in the United States.* New Brunswick, NJ: Rutgers University Press, 1987.

Mott, Frank Luther. *A History of American Magazines.* Vol. 3, *1865–1885.* Cambridge, MA: Harvard University Press, 1938.

RADIO

Radio is the wireless transmission of electromagnetic signals through amplitude or frequency modulation. A radio is also an electronic device that can receive or send radio signals within a specific frequency (between 3 hertz and 300 megahertz), traveling in an oscillating wavelength that is below that of visible light. Additional types of electromagnetic radiation with frequencies near that of radio include microwave, infrared, ultraviolet, X-rays, and gamma rays.

A transmitted radio wave, which moves at the speed of light, can travel several thousand miles and be picked up by a radio or other device, such as a television or telephone. The discovery of radio signals and development of electrical equipment to transmit sound and images via radio waves brought profound changes to human communication.

Early Development

In 1844, Samuel F.B. Morse, an inventor and painter from New York, demonstrated a new electromagnetic communication system called telegraphy. In Morse's system, an electrical device made a series of dots and dashes that spelled out a coded message to a receiver on the other end of a telegraph wire. Although rudimentary and time-consuming, Morse's telegraph was the first reliable long-distance communication system using electrical signals. It was limited in that it relied on the physical connection of wires from one point to another, often over long distances. The wires were frequently damaged or broken and required constant maintenance to ensure a connection.

James Clerk Maxwell, a Scottish physicist whose groundbreaking publication *Treatise on Electricity and Magnetism* (1873) established the concept of radio waves, theorized that electromagnetic energy could be transmitted through space without connecting wires or cables. In 1888, the German physicist Heinrich R. Hertz, working from the University of Karlsruhe, devised a transmitting oscillator that emitted radio waves and detected them using a metal loop with a gap on one side. When the loop was placed within the transmitter's electromagnetic field, sparks were produced across the gap, thus proving that electromagnetic waves could be sent through space and physically detected.

In the final decade of the nineteenth century, inventors and scientists sought to advance the technology of wireless radio transmissions. The Serbian American engineer and inventor Nikola Tesla, working in New York, developed a wireless transmission technique in 1891—albeit beyond the range of human hearing. Granted a U.S. patent, Tesla lectured the following year at the Institution of Electrical Engineers in London on the transmission of intelligence without wires.

Meanwhile, Italian inventor Guglielmo Marconi had made key advances in wireless radio communication. He traveled to Great Britain in 1897 to perform a demonstration, sending and receiving a signal 4 miles (6.4 kilometers) and then nearly 10 miles (16 kilometers) away. Whereas Tesla used electric currents rather than electromagnetic radio waves for his wireless transmissions, Marconi believed if radio waves could be transmitted and detected over long distances, a practical form of wireless telegraphy would be possible. Using a telegraph key to modulate the electric signal, Marconi was able to transmit Morse code over a distance as great as 3.7 miles (6 kilometers). With U.S. Navy support, Marconi continued to increase radio transmission distances. By 1901, he detected a telegraphic signal in Newfoundland that had been transmitted from Cornwall, England, some 1,860 miles (2,990 kilometers) away.

First Broadcasts

American inventors led the effort to further extend Marconi's radio research. Working at Chicago's Armour Institute of Technology, Lee De Forest developed the audion, a three-element vacuum tube, in 1906. The audion amplified sound and enabled voice transmission, liberating wireless communication from dependence on Morse code to transmit messages. By 1913, the American Telephone and Telegraph Company (AT&T) had established coast-to-coast phone service, with a government-sanctioned monopoly.

On Christmas Eve 1906, Canadian inventor Reginald Fessenden, who had worked for both Thomas Edison and George Westinghouse and held over 200 patents, transmitted the first transatlantic radio signal from Brant Rock, Massachusetts, to Macrihamish, Scotland. For the test, Fessenden had built identical 420-foot (128-meter) towers in each location; for the transmission, he used a high-frequency alternator built for him by the Swedish American Ernst Alexanderson, who worked at General Electric in New York. Alexanderson went on to improve the high-frequency alternator, which led to the birth of commercial radio broadcasts.

The earliest practical application of radio technology was to provide communication between ships and shore locations, especially for weather broadcasts or military transmissions. In April 1912, a Marconi wireless set was aboard the *Titanic* when it struck an iceberg in the frigid North Atlantic, and late-night Morse code transmissions enabled the rescue of the few dozen survivors left adrift in lifeboats after the ship sank. Interference from amateur operators was said to have impeded communication with rescue ships, however, and the public pressured Congress in the aftermath to regulate the airwaves.

Before passage of the Radio Communications Act of 1912, there was no government regulation of radio transmitters in the United States. Amateur radio stations competed for airwaves, especially in and around big cities in the industrial Northeast, creating chaos on the airwaves. The Radio Communications Act defined the airwaves as a "collective national resource of the United States" and confined amateurs radio operators—known as "hams"—to the band above 1,500 kilocycles. The act also named the Department of Commerce and Labor as the licensing authority for radio operations, limited the number of licenses to be granted to commercial interests, and allowed the federal government to seize private stations in the event of war or a national disaster.

The Wilson administration exercised that right during World War I, ordering all private radio operations shut down on April 7, 1917—citizens could not own or operate radio transmitters or receivers. The U.S. military took over the radio industry, and the development of related technology was accelerated. Major electrical firms working for the armed forces made significant advances in vacuum-tube engineering and manufacturing, and research into oscillating crystal circuits was initiated. (The vacuum tube would not be used even on a limited basis in commercial receivers until the mid-1920s, and it was not until the 1940s that it became the dominant technology.)

During World War I, the belligerents used radio to enhance their information-gathering and operational capabilities. Radios were placed in aircraft and combat boats, and battlefield commanders were able to exchange messages in real time. As hubs of communication, radio stations became key battlefield targets.

Postwar Boom

In the decades following World War I, radio exerted perhaps the most radical influence on human communication since Gutenberg's invention of the moveable-type printing press in the fifteenth century. On August 31, 1920, the first American radio station—8MK in Detroit—began regular broadcasting, followed quickly by KDKA in Pittsburgh, the first licensed "commercial station" owned by Westinghouse.

By 1929, more than 600 radio stations, broadcasting from commercial studios, colleges, newspapers, and cities, were transmitting to an increasing number of receiver owners. Between 1923 and 1930, 60 percent of American families purchased a radio set. From the start of the industry, radio broadcasting was closely connected to the sale of radios; one could not exist without the other. As more radios were sold, more stations were set up to serve them and more programming was produced and broadcast.

Thus, radio was several industries in one. It was the producer of a household appliance that went from $10 million in sales in 1921 to an $843 million industry in 1929. Sets ranged from the inexpensive (around $50) to the very expensive (hundreds of dollars), which retailers sold to consumers on the installment plans. By the end of the decade, even though many Americans still did not own a radio, most had access to a set through friends, family, or community organizations.

News and entertainment were the staples of early radio. The fledgling medium was seen primarily as a means of giving people information faster than they could get it in newspapers, or to put them in the audience of a faraway or costly live performance. As the demands and cost of radio programming increased, small stations became affiliated with larger ones to form networks, which shared programming.

The Radio Corporation of America (RCA), one of the largest producers of radio sets, formed the first national network, the National Broadcasting Company (NBC) in 1926. The Columbia Broadcasting System (CBS) was formed in 1928, followed by the Mutual Broadcasting Company (MBC) in 1934. In addition to the major national networks, there were dozens of smaller regional networks, many owned by radio technology companies such as General Electric and Westinghouse. In the meantime, the federal government furthered regulation of the industry in 1927 by establishing the Federal Radio Commission, which later became the Federal Communications Commission (FCC).

In the 1930s, radio provided free information and entertainment during the Great Depression with established formats of news, music, drama, and comedy. American culture began a long process of homogenization, as urban and rural areas across the country heard the same voices. On the technological front, the year 1938 brought a significant breakthrough. Working in his basement laboratory at Columbia University in New York City, inventor Edwin Armstrong devised a receiver system that used frequency modulation (FM) to access the radio band, greatly reducing static and improving sound quality (at a shorter listening range) than AM radio. The emergence of FM radio challenged the existing structure of AM systems, and the number of stations continued to increase. By 1948, about 460 FM stations were broadcasting in the United States.

In the early 1940s, U.S. armed forces in World War II had benefited from another technological innovation. The Motorola company, based in Chicago, had developed the "handie-talkie," or "walkie-talkie," a portable two-way radio.

The development of technology to transmit pictures on radio bandwidths—television—was another sea change for mass media in the United States. The first designs were unveiled by Charles Francis Jenkins in 1925, when the images were modulated on AM and the sound on FM. The introduction of black-and-white commercial televisions in the late 1940s challenged the dominance of radio as a mass medium at the national level.

Radio programming emphasized recorded music, information, and talk-show programming rather than live events. The demand for improved sound quality in the 1970s led to an expansion of FM stations that could broadcast in stereo format; the number of FM stations soon exceeded that of AM stations.

The desire for commercial-free, educational, dramatic, and special programming led the FCC to authorize FM stations to form an organization called National Public Radio (NPR), which eventually became an independent nonprofit corporation supported by government funding and private donations. By the 1980s, NPR had more than 1,000 affiliate stations; as of 2007, its audience exceeded 16 million listeners per week.

The FCC implements and enforces rules designed to address market share, bandwidth allocation, and other communications issues. In 1941, the agency issued its first monopoly rule, preventing a single company from owning more than 25 percent of radio broadcasting systems in any market. At the time, only seven radio stations met or exceeded that limit.

In 1964, the FCC released its duopoly rule, which prevented any company from owning both radio and television stations in the same market, or two radio stations in a larger market and one in a smaller market. Further consolidation in the communications industry motivated the FCC to act again in 1970, limiting companies to ownership of only one radio station, television station, and newspaper all in the same market. In 2003, however, the FCC reversed its stand and voted to relax most of its ownership rules. Congress intervened and had the ownership rules restored, but several other deregulation measures were implemented.

Other Radio Technologies

Radio technology is used in a number of advanced electronic systems. Radio Detection and Ranging (RADAR) was developed by the U.S. military in the early 1940s to provide navigational positions and an image of objects and landforms.

Cooking with microwaves was discovered accidentally by an engineer at Raytheon labs in Waltham, Massachusetts. The company introduced the first microwave oven model in 1947.

Engineers at Bell Labs in New Jersey built the first portable wireless telephones in the 1960s. These were followed within three decades by the first consumer mobile (or cellular) phones. By the early twenty-first century, microwave radio signals were able to provide clear mobile phone communications through a network of towers across the United States. As of mid-2007, there were approximately 238 million cellular phones in use.

Radio as an entertainment and information medium maintains a strong cultural presence, with more than 12,000 AM and FM stations broadcasting in the United States. Significant developments of the twenty-first century include the implementation of digital and satellite technologies, which have combined to form a new medium of space-based radio at 2.3 GHz. Satellite radio, with a U.S. subscriber base numbering in the millions and growing steadily, is unconstrained by the physical limitations of Earth-based systems, thereby providing strong, clear signals where AM and FM do not reach. Although the FCC issues licenses to satellite radio providers, program content is unregulated—allowing for a wider choice of programming but looser standards.

James Fargo Balliett and Steven J. Rauch

Sources

Arnheim, Rudolf. *Radio.* New York: Arno, 1971.

Garratt, G.R.M. *The Early History of Radio.* London: Institution of Electrical Engineers, 1994.

Harlow, Alvin F. *Old Wires and New Waves: The History of the Telegraph, Telephone, and Wireless.* New York: D. Appleton-Century, 1936.

Shiers, George, ed. *The Development of Wireless to 1920.* New York: Arno, 1977.

Sterling, Christopher H., and John M. Kitross. *Stay Tuned: A History of American Broadcasting.* Mahwah, NJ: Lawrence Erlbaum, 2001.

SATELLITES

Artificial space satellites are a product of both the technological change that has taken place since 1945 and the rivalry between the United States and the Soviet Union during the Cold War. Since the late 1950s, satellites have come to play a vital part in American life and society, from defense and communications to weather forecasting and at-home entertainment.

The roots of the rocket technology that made satellites possible go back to Germany in World War II. After the end of the war, the Americans and Soviets raced to acquire such technology and enlisted the expertise of the German scientists who developed it. While the Americans succeeded in recruiting Wernher von Braun, who had headed the German rocketry program, both sides were able to acquire the technology to develop their own ballistic missiles.

The Americans and the Soviets also realized that the technology could be used to launch satellites into space. In 1955, both nations pledged to launch their own satellite as part of the International Geophysical Year (IGY). The IGY was an international effort to study physical phenomena on Earth, and it was scheduled to run from July 1957 to December 1958.

It was widely expected that the United States would be the first country to put a satellite in orbit. Therefore, it came as a surprise when, on October 4, 1957, the Soviet Union launched *Sputnik I* on an R7 rocket that had been developed under the leadership of the great Soviet scientist Sergei Korolev. The Soviets followed up this success a month later with the launch of *Sputnik II*, which carried the first living creature into space, a dog called Laika.

Under pressure to respond, the Americans rushed their preparations, resulting in the launchpad explosion of the Vanguard rocket on December 6, 1957. This failure was dubbed "Kaputnik" by the international press and was a humiliating blow to American prestige. The Department of Defense, which still ran the space program in early 1958, quickly regrouped. On January 31, 1958, it launched *Explorer I*, using a Jupiter C rocket that von Braun had developed for the U.S. Army. In light of the problems to date, the federal government created a civilian agency, the National Aeronautics and Space Administration (NASA), to more effectively coordinate U.S. space efforts.

After these first successes, each of the superpowers began to develop satellites that were increasingly more advanced and useful. In 1961, for instance, the United States launched the first spy satellite, called *CORONA*. Surveillance photos were taken in space, and then the film was ejected in a capsule that was retrieved in midair by the U.S. Air Force. Despite the limitations of this system, such satellites were able to provide extensive intelligence on the Soviet Union and prove that the "missile gap" of the late 1950s and early 1960s did not exist.

The United States also developed the first navigation satellite, *Transit,* in 1959; the first weather satellite, *TIROS,* in 1960; and the first experimental communication satellites, *ECHO 1, Telstar 1* and *2,* and *Syncom 1, 2,* and *3,* in the early 1960s. In 1965, the United States launched *Early Bird,* the world's first commercial communications satellite.

At the same time, however, the Soviets continued to develop new and more capable satellites of their own. They launched their first reconnaissance satellite, *Zenit,* in 1961; the world's first maneuverable satellite, *Polyot 1,* in 1963; and a system of communication satellites, known as Molniya, beginning in 1965.

Other countries joined the superpowers in space: Canada and Great Britain in 1962, France in 1965, and Japan and China in 1970. All launched satellites for telecommunications, navigation, surveillance, or weather-monitoring purposes.

The 1970s saw more improvements in satellite technology, with the United States launching the Landsat satellites to monitor changes in the Earth's landscape and the first Geostationary Satellite (GOES) to monitor weather patterns. (A satellite in geostationary orbit is one that remains fixed over a specific point on the Earth, usually at a distance of more than 20,000 miles from the surface.) The United States also developed the KH-11 series of reconnaissance satellites; these used an electro-optical system to transmit images to ground stations.

Another major technological advance was the development of satellite television. In 1965, the American Broadcasting Corporation (ABC) proposed to create a satellite system to serve the domestic television audience; however, it was Telstar Canada that launched the first domestic communications satellite in 1972. This satellite, *ANIK*, was quickly joined by Western Union's *WESTAR 1* in 1974 and RCA's *SATCOM F1* in 1975. One result of these satellites was the phenomena of so-called superstations, such as TBS out of Atlanta, whose programming could be transmitted throughout the United States. In addition, the rapid growth of cable television in the United States was a product of these satellites, since content could be immediately transmitted from one place (where a major sports event was taking place, for example) to local cable providers throughout the country.

With the Soviet invasion of Afghanistan in 1979 and the renewed arms buildup launched by the Reagan Administration in late 1980, Cold War tensions—dormant during the détente of the 1960s and 1970s—were revived. One effect was that, in 1983, the Reagan Administration proposed the Strategic Defense Initiative (SDI), more commonly known as Star Wars. This would have involved the development of a system of satellites to protect the United States from incoming Soviet intercontinental ballistic missiles (ICBMs). But technological difficulties and the end of the Cold War in the late 1980s prompted then President George H.W. Bush to scale back the program dramatically.

Throughout the 1990s and the early twenty-first century, more advanced communication, reconnaissance, and weather satellites have continued to be developed not only by the United States, but by Russia, Japan, and the European Space Agency. One example was the Hubble Space Telescope, which, after early technical problems, was launched by NASA in 1990, providing spectacular images of faraway galaxies. Other examples have included satellite television services such as Direct TV, which has provided increased competition to cable TV. The launch of *SpaceShip One*, the first commercial rocket to achieve suborbital flight, in 2004, promised an expansion of satellite launches in the twenty-first century.

Matthew Trudgen

Sources

Dickson, Paul. *Sputnik: The Shock of the Century.* New York: Walker, 2001.

Gavaghan, Helen. *Something New Under the Sun: Satellites and the Beginning of the Space Age.* New York: Springer-Verlag, 1998.

Heppenheimer, T.A. *Countdown: A History of Space Flight.* New York: John Wiley and Sons, 1997.

Peebles, Curtis. *The Corona Project: America's First Spy Satellites.* Annapolis, MD: Naval Institute Press, 1997.

Richelson, Jeffrey T. *America's Secret Eyes in Space: The U.S. Keyhole Spy Satellite Program.* New York: Harper and Row, 1990.

Walter, William J. *Space Age.* New York: Random House, 1992.

SHIPBUILDING

An integral part of American industry since the colonial period, shipbuilding has changed substantially with technological advances over the centuries. Changing trade patterns, urbanization, industrial developments, commercial growth, and political conflicts also have affected shipping practices and shipbuilding. Prior to the second industrial revolution of the late nineteenth century, ships were the most effective method of transporting people and goods, and all thriving seaports required efficient vessels to carry on their trade. The development of alternative forms of transportation—such as railroads, trucks utilizing interstate highways, and commercial aviation—in the mid-nineteenth and twentieth centuries forced the shipbuilding industry to specialize and consolidate in specific seaports.

From the start, shipbuilding was an important industry in America. During the colonial period, wind-powered wooden vessels were the most common, and America's shipbuilding industry flourished due to abundant forests, especially in the Northeast. The availability of wood allowed American shipwrights to produce vessels at a fraction of the cost of English-built ships. Skilled craftsmen, such as woodworkers, blacksmiths, riggers, caulkers, and sail-makers, all contributed to the building process. Early building techniques, arduous by today's standards, required workers to fit hand-cut wood into the framework manually.

Colonial vessels tended to be small by European standards. Most American merchants traded within the English colonial system, especially along the Atlantic seaboard and the Caribbean, and they relied heavily on trade in the English-controlled West Indies. After the Revolutionary War, however, American merchants sought new trade venues. The change in trade routes led to new shipbuilding techniques to accommodate more ambitious forms of enterprise.

In the nation's early decades, U.S. shipbuilding was not centered in a few major shipyards as it is now, and many small seaports built their own vessels. Even in the first decades of the nineteenth century, small local craft such as dories and gundalows were the most common vessels. Larger vessels such as brigs and full-rigged ships required more workers and deeper water, and most were built near the nation's largest urban centers: Philadelphia, New York, Boston, Baltimore, and Charleston. Major shipbuilding centers were located near the mouths of rivers or bays, in areas with easy access to deep water to aid the launching process. Smaller cities, such as Salem, Massachusetts, that lacked sufficient space for normal launching adopted alternative techniques like the "side launch," in which the ship was built and launched parallel to the shoreline to reduce the amount of space necessary for the launch.

International trade required larger and more efficient vessels. By the mid-nineteenth century, clipper ships, popular for their speed and sleek design, dominated. In the second half of the nineteenth century, demand for more spacious cargo holds led to the development of colossal schooners with four and five masts. Only the seaports with the best facilities, such as Philadelphia, Boston, and New York, and rapidly growing Western ports such as San Francisco, had shipyards large enough to accommodate these vessels. In general, the tonnage produced at shipyards in the mid-nineteenth century tended to grow in proportion to the city's population surge, as big cities like New York out-competed smaller ports, and American shipyards consolidated near the major seaports.

Technology aided the consolidation process. Industrialization led to the development of steam-powered vessels in the mid-nineteenth century, revolutionizing shipbuilding methods. The Bessemer process made iron and steel feasible materials for building ships; as wind ceased to be the primary means of propulsion, the shipbuilding process became increasingly mechanized. Instead of relying on skilled local craftsmen, American shipbuilding began to rely on semi-skilled workers to operate manufacturing machines. The rise of mass production both benefited and hurt the American shipbuilding industry: Workers took less pride in their work, and the working conditions were often dangerous, but the building process was more efficient, and new technology enabled larger, safer ships to be built.

In the early twentieth century, military shipbuilding supplanted the production of trading vessels as the most lucrative sector of the industry. Large, privately run shipyards like the Iron Works at Bath, Maine, and the New York Shipbuilding Company produced many of America's most technologically advanced warships of the mid-twentieth century, such as the aircraft carrier USS *Kitty Hawk*, completed in 1961 at the now-defunct New York shipyard. Similar shipyards established around the nation's coasts prior to World War I included the Chicago Shipbuilding Company near Lake Michigan and the Union Iron Works in San Francisco, both of which supplemented U.S. Navy yards during the wartime production boom.

After World War II, the American shipbuilding industry went into a prolonged recession, and many shipyards have closed over the years. Most contemporary shipbuilding enterprises rely on military contracts, producing high-tech vessels at a high prices, and there is steep competition for a limited number of contracts. As a result, many existing navy yards, such as the one at Portsmouth, New Hampshire, face possible closure, while others, such as the Charleston Navy Yard in Boston, have already shut down. Technological innovations have forced the modern shipbuilding industry to specialize and isolate production to specific locations, creating a very different scenario from the informal local production methods common in America's early decades.

Benjamin Lawson

Sources

Crowell, John Franklin. "Present Status and Future Progress of American Shipbuilding." *Annals of the American Academy of Political and Social Science* 19 (January 1902): 46–60.

Korndorff, L.H. "A Challenge to the American Shipbuilding Industry." *Proceedings of the American Academy of Political Science* 19 (May 1941): 28–35.

Morison, Samuel Eliot. *Maritime History of Massachusetts, 1783–1860.* Boston: Houghton Mifflin, 1961.

Petters, Mike. "American Shipbuilding: An Industry in Crisis." *U.S. Naval Institute Proceedings* 132 (February 2006): 15–19.

SIKORSKY, IGOR IVANOVICH (1889–1972)

Igor Ivanovich Sikorsky was one of the great pioneers in the history of aviation and famous as the inventor of the first practical helicopter.

He was born in Kiev, Russia, on May 25, 1889. His family was financially comfortable, and he was able to spend much of his youth reading and studying. Chemistry was a favorite subject. In 1903, he became a student at the naval academy, and, three years later, he went to Paris to learn more about airplanes and to acquire an engine.

Passionately interested in aviation, Sikorsky decided to build his own helicopter. By 1909, he managed to get a primitive model off the ground. His parents and sister supported his early work, giving him money whenever he asked for it.

By 1910, Sikorsky had shifted to the construction of airplanes and slowly perfected his own models. In 1911, he built an early version of a multiengine plane. He was so successful that the next year, at the age of just twenty-three, he became the head of the airplane division of a large Russian corporation. The director encouraged his work, and this allowed Sikorsky's inventiveness to flourish.

He began almost immediately to construct a plane propelled by four engines. The Grand, as it was called, was the world's first plane with four engines; it had a 92 foot wingspan and an enclosed cabin, among other unusual features. Next came various versions of another model called the Ilia Mourometz; soon this enormous plane, also powered by four engines, was put into production. Russia needed large bombers, because World War I had broken out. Working diligently, Sikorsky succeeded in building planes capable of flying at 10,000 feet.

When conditions in Russia began to degenerate because of the 1917 revolution, Sikorsky left behind his homeland, his fortune, and his young child Tania. After a brief stay in France, he immigrated to the United States.

On March 30, 1919, Sikorsky arrived in New York, an event that marked the beginning of the second stage of his life. He had little money and no command of English, but he was ever optimistic about the future of aviation and his own abilities. U.S. aviation was in its infancy, however, and finding a niche in the industry proved difficult despite Sikorsky's reputation. During 1921 and 1922, he was forced to earn an income as a teacher and lecturer.

After landing a short-term development contract with the government, he founded a company in 1923. The fledgling firm was on tenuous financial ground, relying on the support of small investors. A significant boost came from the Russian composer Sergei Rachmaninoff, who invested $5,000.

On January 27, 1924, Sikorsky married Elizabeth Semion, a Russian immigrant; the couple would have four sons. Sikorsky's sisters and his daughter Tania came from Russia to join his family on Long Island.

Sikorsky's company produced an all-metal transport plane, the S-35, that was used to haul cargo. One managed to cross the Andes at 19,000 feet. The S-35 was modified so that it could be used to cross the Atlantic, and Sikorsky hoped it would be the first model to make the trip; a crash set back the development schedule, however, and Charles Lindbergh, flying the custom-built *Spirit of St. Louis,* was the first to accomplish the feat in 1927.

In 1928, Sikorsky built a large plant in Stratford, Connecticut, where his company produced the S-38, an amphibian plane that proved to be a great success. These flying boats or clippers were, after the four-engine triumph, Sikorsky's second great accomplishment.

He then collaborated with Lindbergh on the S-40; at 17 tons, it was the largest American transport and could be equipped with either traditional landing gear or amphibian floats. It was used to carry passengers in South America. In 1933, the S-42, with a controllable-pitch propeller, crossed both the Atlantic and Pacific oceans. In just one flight, it managed to break eight world records. More developments followed, including nonstop flights to Europe.

Eventually the flying boat era came to an end. In 1939, Sikorsky returned to his earlier dream of creating a practical helicopter. This marked the beginning of his third and final period of accomplishment. His prescience and inventiveness resulted in the first successful machine that could be used for transport and rescue in emergencies where a plane could neither hover nor land. The VS-300 and subsequent models were the forerunners of today's helicopters.

Sikorsky retired in 1957 and died fifteen years later, on October 26, 1972. He was widely admired and frequently honored, receiving the National Defense Transportation Award, the Wright Brothers Memorial Trophy, the Copernican Citation, the Collier Trophy, and the National Medal of Science, among other awards.

Robert Hauptman

Sources

Cochrane, Dorothy, Von Hardesty, and Russell Lee. *The Aviation Careers of Igor Sikorsky.* Seattle: University of Washington Press, 1989.

Delear, Frank J. *Igor Sikorsky: His Three Careers in Aviation.* New York: Dodd, Mead, 1969.

Sikorsky, Igor I. *The Story of the Winged-S: An Autobiography.* New York: Dodd, Mead, 1939.

SINGER, ISAAC (1811–1875)

An innovator and shrewd businessman, Isaac Merritt Singer founded the Singer Sewing Machine, arguably the most recognizable American brand in the world by the early 1900s. He was not first to build a sewing machine; that distinction goes to inventor Elias Howe, who patented it in 1846. But Singer incorporated the basic features and functionalities that made sewing machines practical household appliances.

He was born in Pittstown, New York, on October 27, 1811. His parents were German immigrants; they divorced when he was ten years old, and Singer stayed with his father. Like many of the great nineteenth-century American inventors, Singer had little formal education. At the age of twelve, he left his father's house to move in with an older brother in Rochester, New York. There, he worked at odd jobs, including a brief apprenticeship in a mechanic's shop.

Although he was a natural mechanic with a gift for understanding the inner workings of machines, Singer became intrigued by stage acting and pursued a theatrical career through the mid-1840s. Unable to support himself as an actor, he took a job at his brother's contracting business, working on the Lockport and Illinois Canal. While employed there, Singer invented a rock-drilling machine and sold the patent for $2,000, which he invested in a theater company. When that failed, he turned once again to mechanical invention, developing a machine for carving metal and wood type. Failing to attract adequate investment to manufacture his carving machine, Singer turned to the sewing machine.

By the end of the 1840s, a number of early sewing machines were already on the market. In 1851, Singer acquired one of the machines being manufactured by Orson Phelps in Boston. Within a few days, Singer had determined what made the machine—and others like it—so inefficient and unreliable. The shuttle moved in a circular motion, taking a twist out of the thread with each movement. Singer's prototype replaced the circular motion of the shuttle with a to and fro motion that moved the thread in a straight line. Instead of having a bar push the needle horizontally through the fabric, Singer had his needle go up and down. These changes made the machines work much faster and reduced the propensity to snap the thread. Singer's innovations were so effective that they are still central to sewing machine design today.

Singer patented his invention and went into business with Phelps and several other partners in 1851. Ruthless in business, Singer used trickery and threats to get his partners to sell their shares in the company for token amounts. At the same time, he brought his lawyer, Edward Clark, into the business. It proved a wise choice, as Singer became caught up in a number of patent suits with other sewing machine inventors and manufacturers in the early 1850s. In 1856, Clark created a patent pool, which ended the sewing machine wars of the late antebellum era. Clark also would be the architect of I.M. Singer Company's rapid expansion throughout the United States and around the world in the late nineteenth century, establishing franchises for the manufacturing, distribution, and sale of Singer sewing machines on every continent except Antarctica.

The Singer sewing machine revolutionized the garment industry. Not only was it efficient and reliable, but it was easy to operate, allowing for the rapid growth of the ready-to-wear clothing industry. Prior to the spread of the sewing machine, most clothes were hand-sewn to order, a business that catered primarily to the wealthy. The inexpensive Singer sewing machines also allowed for the spread of textile sweatshops in New York City and other American metropolises, a system that brutally exploited immigrant workers in the late nineteenth and early twentieth centuries.

By the early 1860s, Singer was a multimillionaire who no longer took an active role in managing the company, instead devoting his time to various mistresses and life as a country gentleman. Singer's personal life was complicated. He married twice and lived with several other women, fathering twenty-four children by five different women. He died at his estate in Torquay, England, on July 23, 1875.

James Ciment

Source

Bissell, Don. *The First Conglomerate: 145 Years of the Singer Sewing Machine Company.* Brunswick, ME: Audenreed, 1999.

Brandon, Ruth. *A Capitalist Romance: Singer and the Sewing Machine.* London: Barrie and Jenkins, 1977.

SLATER, SAMUEL (1768–1835)

Samuel Slater established the first American textile mill to mechanically spin cotton, and Slater's innovations sparked the Industrial Revolution in America. Pawtucket, Rhode Island, the location of Slater Mill, initially became the leading city in industrial America, and the model for later mill cities in the Northeast.

Born on June 9, 1768, in Derbyshire, England, Slater became an apprentice in a cotton mill in 1782. By the time he left England for America, he had intricate knowledge of English mill designs. He arrived in New York in 1789 and moved to Pawtucket, Rhode Island, where he entered the employ of Moses Brown. A successful merchant, Brown had recently established a new textile mill in the city, and he hired Slater to help operate and improve the machinery.

Based on the designs of the English inventor Richard Arkwright, Slater made changes to the mill that enabled it to be water-powered. Brown was sufficiently impressed to make Slater a partner. In 1793, Slater built a new mill near the falls at the junction of the Blackstone and Seekonk rivers in Rhode Island (this mill is still in operation as a museum).

Before Slater's innovations, the American textile industry had been forced to rely on England for its goods, as it was able to produce only small quantities at a slow rate. The innovations introduced in Slater Mill allowed America to produce its own goods with less reliance on cross-Atlantic trade.

In 1797, because of disagreements over the operation of the facility, Slater built a new mill, the White Mill, across the river in Pawtucket. There, he opted to specialize in one particular process—producing yarn, rather than the standard practice of manufacturing finished products. This specialized approach allowed greater efficiency in production and yielded greater

Slater Mill in Pawtucket, Rhode Island, built in 1793, was America's first successful mill for spinning raw cotton into textiles. Founder Samuel Slater brought the technology from England by memory. *(Library of Congress, HAER RI, 4-PAWT, 3–8)*

profits, as less machinery was needed to perform the operations.

Slater's most ingenious innovation, however, lay not in mechanics but in the development of mill communities in which his employees could live. Built within walking distance of the mill, these company towns extended the power of the mill owner over employees. Slater and his associates controlled everything, from grocery stores to schools and churches.

One such community, founded in 1803, was Slatersville, located north of Pawtucket near the Massachusetts border. The compound included two tenement buildings in which employees and their families lived. Slater's control extended to the most minute details of his employees' lives, including not only their wages but their expenditures as well—making him a wealthy man. Samuel Slater died on April 20, 1835.

Benjamin Lawson

Sources

Conrad, James L., Jr. "'Drive That Branch': Samuel Slater, the Power Loom, and the Writing of America's Textile History." *Technology and Culture* 36 (1995): 1–28.

Penn, Theodore Z. "The Slater Mill Historic Site and the Wilkinson Mill Machine Shop Exhibit." *Technology and Culture* 21 (1980): 56–66.

Slater Mill. http://www.slatermill.org.

SPACE PROBES

Like many scientific and technical achievements of the Cold War era, human exploration of the solar system was initially propelled by the rivalry between the United States and the Soviet Union. From the early 1960s to the present day, U.S. and Soviet/Russian probes have visited seven of Earth's eight planetary neighbors, sending back valuable information and helping to promote public interest in outer space.

On October 4, 1957, the launch of the Soviet satellite *Sputnik* startled the American people and their leaders. The message was loud and clear: Soviet space science was more advanced than that of the United States. But even as several early milestones in the space race were cleared by the Russians, American preparations for the Apollo program were closing the technology gap and setting the stage for a rivalry in voyages to Earth's closest planetary siblings: Venus and Mars.

The Soviet Union opened the race with a determined effort. Moscow placed a great deal of pressure on Sergei Korolev, the head of the Soviet space program. With the United States moving closer to a successful moon shot during the 1960s, Korolev began a titanic effort in unmanned exploration to maintain Soviet prestige.

Korolev targeted both of Earth's immediate neighbors, but the alignment of the planets in the early 1960s meant that a journey to Venus would take considerably less time than a trip to Mars. At the time, little was known about Venus. Astronomers had long known that it was about the same size as Earth and that it had an atmosphere, but they knew nothing about its composition. Some Soviet scientists speculated that it could be a jungle planet with a thick cover of water vapor clouds, or possibly a semi-hospitable desert world. In the late 1950s, the American physicist Carl Sagan correctly theorized that Venus was in fact a scorched, inhospitable world, but his ideas were not widely accepted at the time.

The Soviet Union made the first of its attempted launches in 1960. In stunning contrast to the success of his Earth orbiters, however, Korolev's interplanetary efforts began with a series of dismal failures. After dealing with various flaws in rocket design, the USSR finally sent the *Venera 1* probe hurtling toward Venus on February 12, 1961. The Soviets lost contact with the probe seven days after the launch.

The U.S. National Aeronautics and Space Administration (NASA), acting through the Jet Propulsion Laboratory (JPL) in California, responded to Soviet efforts by sending the spacecraft *Mariner 2* to Venus. Launched on August 27, 1962, this probe had the advantages of a light and simple design. Its mission was also uncomplicated: to reach Venus without losing radio contact. On December 14, the probe successfully rendezvoused with Venus, confirming that the planet possessed a hostile environment and was unsuitable for visits by human beings.

The Soviet Union continued its efforts to reach Venus, eventually sending fifteen successful missions between 1967 and 1984, including numerous surface landings. The United States, however, shifted most of its attention to other planets. In July 1965, *Mariner 4* arrived at Mars, again beating the Soviets and providing the first useful images of the red planet.

In the 1970s, American successes in planetary exploration accelerated, giving NASA a permanent lead over the relatively idle Soviet program. In 1973, the United States took a leap forward by launching *Mariner 10*; it flew past Venus in February 1974 en route to several meetings with Mercury, the solar system's innermost planet. In 1976, the United States landed two Viking probes on the surface of Mars. These probes collected a vast amount of data, but their findings proved disappointing to some members of the scientific community, who had hoped to find signs of life. The year 1978 saw an American return to Venus in the form of two Pioneer probes, which used radar to scan and map the surface of the planet for the first time.

Buoyed by their triumphs in the inner solar system, NASA scientists and engineers also had begun work on ambitious plans to visit the giant gaseous planets of the outer solar system—Jupiter, Saturn, Uranus, and Neptune. In 1969, NASA announced that a "grand tour" of the gas giants would take place in the later 1970s. For this task, NASA began to plan new types of probes, equipped with a wide array of cameras and other sensory devices. They also planned the inclusion of plutonium power systems, replacing the solar panels that had been effective in the inner solar system but which would not provide enough power in the dim light conditions farther out.

In 1972, the "grand tour" was nearly canceled due to budgetary concerns. The Vietnam War continued to stretch the federal budget, and NASA was already funding other expensive projects. NASA responded by scaling back its plans. *Pioneer 10* provided a first look at Jupiter in March 1972, and *Pioneer 11* reached Saturn in April 1973.

The "grand tour" was finally begun in earnest in 1977, with the launch of NASA's twin Voyager probes. They were to travel first to Jupiter and Saturn and then continue the mission to Uranus, Neptune, and beyond. Both probes distinguished themselves as they visited the first two gas giants. In spite of an ongoing struggle to maintain the program's budget, *Voyager 2* was sent on to meet Uranus in 1986 and Neptune in 1989. As of 2007, NASA continued to maintain contact with the two *Voyagers*, which were then investigating the extreme edges of the solar system.

In the late 1980s and 1990s, NASA began to focus more on launching probes that would remain in orbit of their target planets instead of simply flying by. The probe *Magellan* arrived at Venus in August 1990 and began an extensive mapping project. The probe *Galileo* reached Jupiter in December 1995 and settled in for a multiyear mission. In 1997, NASA cooperated with European space agencies in launching the probe *Cassini* on a course for Saturn.

NASA also launched an ambitious series of probes to Mars in the 1990s. The *Mars Pathfinder* and its *Sojourner* rover, which landed successfully on Mars on July 4, 1997, captured public attention with its detailed photos of the planet's surface. But the *Mars Observer* was lost in 1992, and two additional Mars probes were lost in 1999.

These failures drew significant criticism. NASA redeemed itself with two successful landings of Martian rovers in January 2004. Named *Spirit* and *Opportunity*, the two robotic crafts explored different hemispheres of the planet. *Opportunity*, in particular, found convincing mineral evidence that Mars once had free-flowing water on its surface.

Recent NASA missions include the Cassini-Huygens Mission, operated in conjunction with the European Space Agency (ESA). Since 2004, *Cassini* has orbited Saturn, providing photos and data of the planet, its rings, and its moons; the *Huygens* probe landed on Saturn's moon, Titan, in 2005. The probe *Deep Impact* collided with the comet Tempel 1 in July 2005, providing photos and data on the nature of comets. The spacecraft *Dawn* lifted off on September 27, 2007, on an almost four-year mission to study the asteroids Vesta and Ceres.

David Stiles

Sources

Burrows, William E. *The New Ocean: The Story of the First Space Age.* New York: Random House, 1998.

"JPL Missions." NASA-Jet Propulsion Laboratory, California Institute of Technology. http://www.jpl.nasa.gov/missions.

Reeves, Robert. *The Superpower Space Race: An Explosive Rivalry Through the Solar System.* New York and London: Plenum, 1994.

SPACE SHUTTLE

The U.S. Space Shuttle, a reusable rocket-propelled space vehicle that lands like an aircraft, was developed by the National Aeronautics and Space Administration (NASA) during the 1970s and took its first flight in 1981. Six Space Shuttle vehicles have been built: *Enterprise* (used only for tests and not spaceworthy), *Columbia*, *Challenger*, *Discovery*, *Atlantis*, and *Endeavor*. *Challenger* broke up during lift-off in January 1986, and *Columbia* disintegrated during re-entry in February 2003. As of 2007, there have been 118 shuttle, or Space Transportation System (STS), missions.

NASA administrators had long desired a reusable vehicle for a variety of research and deployment missions in space, and the Space Shuttle program received congressional approval and funding in January 1972. The first mission, STS-01, piloted by John Young and Robert Crippen aboard the shuttle orbiter *Columbia*, was launched from Cape Canaveral, Florida, on April 12, 1981, and it landed in the California desert on April 14. Its payload was a package of sensors and measuring instruments, but its primary mission was to go up and come down safely. The next three missions also tested performance and systems.

Since mission STS-05 in November 1982, Space Shuttles have been used to conduct experiments in zero gravity, deploy and retrieve communications and observation satellites, carry parts for the International Space Station, map the Earth's surface, and repair and upgrade the Hubble Space Telescope. From 1995 to 1998, shuttle vehicles also were used to bring astronauts and supplies to the Russian space station *Mir.* While the orbiter can carry up to ten astronauts, the typical Space Shuttle mission has carried a crew of seven.

Among other equipment and devices, the Space Shuttle is equipped with a pressurized Spacelab, designed by the European Space Agency for research in zero gravity; the Remote Manipulator System, a robot arm that moves payloads in and out of the cargo bay; and the Manned Maneuvering Unit, a backpack that allows astronauts to fly short distances from the vehicle. Although the Space Shuttle itself is

The launch of the orbiter *Columbia* on April 8, 1981, from Kennedy Space Center in Florida, marked the beginning of the Space Shuttle program and the era of reusable manned spacecraft. *(Keystone/CNP/Hulton Archive/Getty Images)*

reusable, the external fuel tank, carrying the liquid hydrogen and oxygen propellant required for launch, is not. In addition, the Space Shuttle's external solid rocket boosters must be recovered from the ocean after each launch.

On January 28, 1986, after twenty-four flights, the Space Shuttle program came to a halt after the sudden explosion of *Challenger* some seventy-three seconds after liftoff from the Kennedy Space Center. The explosion destroyed the spacecraft and killed all seven crew members aboard, including the first civilian chosen to ride the Space Shuttle, teacher Christa McAuliffe. After extensive investigation, a failed O-ring seal on one of the solid rocket boosters was found to have been the cause of the explosion. The shuttle program was subsequently grounded until September 1988.

A second disaster occurred on February 1, 2003, when the shuttle *Columbia*, nearing completion of STS-107, inexplicably broke apart over Texas, just a few minutes before its scheduled landing. Following extensive investigations, it was ultimately determined that a piece of foam insulation had been dislodged during liftoff and had damaged one of the vehicle's wings. The Space Shuttle returned to space on July 26, 2005.

Space Shuttle missions are expensive. By 2006, they were costing more than $500 million each, far in excess of the original projections in the 1970s. In addition to the expense of individual missions, the tragic events that brought all planned flights (and their satellite launches) to a halt cast a pall over the entire shuttle program. The X-33 program, conceived as the next generation of reusable launch vehicle, was canceled as early as 2001.

No matter when U.S. space flight resumes, NASA has announced that the Space Shuttle program would end by 2010, complying with President George W. Bush's long-term plan. Other programs designed for other purposes, such as the proposed Orbital Space Plane, are expected to replace it.

Vickey Kalambakal

Sources

Columbia Accident Investigation Board. *Report.* Washington, DC: National Aeronautics and Space Administration, 2003–2004.

Harland, David M. *The Space Shuttle: Roles, Missions, and Accomplishments.* New York: Wiley, 1998.

SPACE STATION

The Russian *Salyut* and *Mir* space stations and the United States's *Skylab* were forerunners to the establishment of more permanent research facilities and human habitations in space. Manned artificial satellites such as *Skylab*, which was launched by the National Aeronautics and Space Administration (NASA) in 1973, are a class of satellites sent into space for the performance of scientific tasks.

NASA's objective for *Skylab* was to establish an orbiting space laboratory, increasing the time astronauts could spend in space, and allowing for the study of the effects of space on the human body. Superior astronomical observations of the moon, planets, sun, and stars were also achieved. Three different crews manned the $2.5 billion space station before its usefulness was outlived. After 34,981 orbits, the abandoned *Skylab* burned up in a fiery blaze as it fell into Earth's atmosphere in 1979.

The next generation of space stations began when President Ronald Reagan approved the Freedom project, which was intended, in part, to gain political advantage over the Soviets during the Cold War. With the subsequent collapse of the Soviet Union, however, President Bill Clinton pushed for international collaboration and cost sharing among many countries, but especially with Russia, due to its decades of experience with spaceflight and space station design. The Russian–U.S. space alliance was an essential step in making a next-generation space station a reality. A 1993 multinational agreement ended the antagonism between East and West and opened up a new era of cooperation in space.

Since 1993, fifteen countries and five space agencies have been involved in the construction of the International Space Station (ISS). In addition to the United States and Russia, they include Japan, Canada, and member countries of the European Space Agency such as Belgium, Denmark, France, Germany, Italy, the Netherlands, Norway, Spain, Sweden, Switzerland, and the United Kingdom. The participants in this program are contributing components and human resources to build a station that is visible to the naked eye from Earth.

The concept for the ISS is akin to Tinkertoy construction, requiring about fifty American and Russian launches to ferry all the modules for assembly in space. NASA's Space Shuttle plays a key role, ferrying large parts and personnel to and from the station. On November 2, 2005, the International Space Station reached its fifth anniversary of continuous human presence in space. Since the first crew arrived in 2000, fifteen Americans and fourteen Russians have lived and worked there.

The ISS continues to evolve with each crew change and every supply visit from Earth. The space station has become a state-of-the-art laboratory for scientific research. It features a microgravity environment that exists nowhere on Earth, which helps scientists understand how the body functions for extended periods of time in space. Experience gained from studies, such as bone loss measurements and radiation shielding, will be critical in long-duration missions to Mars.

Upon completion, the football-field-size ISS will comprise two laboratory modules from Russia and the United States and one laboratory module each from Europe and Japan. An American habitation module and two Russian-built life-support system modules are the foundational components. All these modules connect to a 290-foot beam, which also supports solar panels to generate electricity and a robot manipulator arm that traverses its length to perform heavy lifting. The station has a width of 240 feet across its solar arrays and a height of 90 feet. Currently, there are 15,000 cubic feet of living space. Forty computers connected by an extensive fiber optic network run the ISS to sustain the crew and safely navigate the station's orbit of Earth.

The ISS is part of a larger vision inaugurated by President George W. Bush in 2004, currently being planned and implemented by NASA. The "Vision for Space Exploration" seeks to use the ISS as a base for experiments, planning, and manned and unmanned flights to the moon, Mars, and farther into the solar system.

Robert Karl Koslowsky

Sources

Caprara, Giovanni. *Living in Space: From Science Fiction to the ISS.* Buffalo, NY: Firefly, 2000.
Launius, Roger D. *Space Stations.* Washington, DC: Smithsonian, 2003.

STEAM ENGINE

A steam engine uses combustion to heat water until it pressurizes into steam and generates force by moving a piston up and down or turning a turbine. Steam engines convert this force into mechanical action to pump liquids or spin a shaft to turn wheels, conveyors, looms, gristmills, paddles, propellers, and electric motors.

The first steam engine was developed by 1730 in England by Thomas Newcomen for pumping water out of mines. James Watt, a Scottish inventor, improved early designs by developing the double-acting steam engine in 1782. His company, Boulton and Watt, produced steam engines for nearly 120 years in the Soho Foundry in England.

By 1804, the American Oliver Evans had upgraded Watt's designs with a high-pressure steam engine for the American market. The use of stationary steam engines became widespread in American factories, breweries, mills, and farms, contributing significantly to the Industrial Revolution. Designs were continually improved, as power output increased and uses expanded to firefighting, ships, and other transportation.

Maritime Applications

In 1787, the Connecticut inventor John Fitch had developed a reliable boat equipped with a steam engine and propelled by paddles. Robert Fulton improved on the early designs and in 1807 built the *Claremont*, the first steam-driven sternwheeler. Steamboat manufacture expanded and provided extensive transportation for passengers along the Mississippi and other rivers from the 1830s on. By 1812, the Washington Navy Yard was building steam engines for war boats, and private firms were building marine engines for commercial use.

Moses Rogers, Fulton's associate, applied the steam-driven paddlewheel to oceangoing sailing ships. This dramatically shortened the time it took to cross the Atlantic Ocean; in 1819, Roger's ship the *Savannah* made the trip to England in a month's time. Given the disadvantages of paddlewheel propulsion, the Scottish John Ericsson and English Francis Smith in 1836 jointly developed the screw propeller, which further improved the speed and reliability of ship propulsion. In 1843, the U.S. Navy launched its first screw-propeller steam warship, the USS *Princeton,* designed by John Ericsson. Commercial steam-powered passenger ships using Ericsson's screw propellers were able to offer reliable transportation across the Atlantic and beyond.

By the late 1850s, U.S. commercial and military vessels were expanding the nation's shipping routes into Asia. In 1854, Commodore Matthew Perry sailed into Tokyo Bay, and the Japanese government signed the Treaty of Kanagawa, which opened Japanese ports to American merchants. The famed American black steamships, trailing steam and smoke, were technological mysteries in the Asian world.

During the American Civil War, Ericsson was responsible for the construction of the North's screw-driven ironclad, the USS *Monitor.* Following the war, the U.S. Navy converted to steel-hull ships and initially used powerful steam reciprocating engines.

By the early 1900s, the new steam-turbine engine was in production. The steam turbine uses a series of nozzles to releases superheated, high-pressure steam on a turbine with blades, causing them to spin and produce considerable energy.

The steam turbine was initially developed in 1892 by Charles Parsons of England. The design was improved upon by American Charles Curtis with a velocity-compounded impulse stage turbine; Curtis's engine was patented in 1896. The Parsons steam turbines were used in many American ships, beginning in 1910 with the USS *North Dakota.* The last traditional steam-powered warship was the USS *Texas* in 1912. Curtis steam turbines were used primarily as motors for electrical generation in America. The Hendy Iron Works in California produced hundreds of triple reciprocating steam engines and later steam turbine engines. In 1947, Westinghouse bought the company to produce steam and gas turbines for electrical generation.

Land Transportation and Other Uses

With western expansion in America during the early 1800s, the need for overland transportation increased. In England, Richard Trevithick developed a steam-powered vehicle that traveled on wheels positioned on a steel track. Trevithick's ideas soon generated rail transport companies in Britain and the United States.

In the mid-1800s, the Baltimore and Ohio Railroad Company and others began constructing tracks. By 1860, the United States had more than 30,000 miles (48,000 kilometers) of track installed, and by 1869, the transcontinental railroad was completed. Early locomotives were imported from England, but American companies soon developed their own steam locomotives, improving on the original design. Baldwin Locomotive Works, the American Locomotive Company, and the Lima (Ohio) Locomotive Works produced thousands of locomotives from the 1830s to the 1900s. The locomotive required an engineer to drive and a fireman to stoke the boiler and provide maintenance. African American inventor Elijah McCoy conceived the lubricator cup, which allowed steam-engine lubrication during operation. Skeptical of imitations, American engineers came to want only "the real McCoy" product.

Ohioan Moses Latta is credited with developing the first American-built steam firefighting engine in Cincinnati in 1852. The steam boiler enabled pressure for suction from a water supply and allowed crews to spray water more than 100 feet (30 meters). Many fire departments across the country adopted the machine, which was much more effective in extinguishing fires than the previous hand pumpers.

By 1897, the Stanley Steamer and Fulton steam cars grew in popularity for personal transportation. A Stanley prototype in 1906 set an automobile land speed record of 127 miles (204 kilometers) per hour. The development of the gasoline-powered internal combustion engine cut short uses of such steam-powered cars, although Japan continued production until the 1950s. In the United States, other steam engine innovations included the steam-powered bicycle, introduced in 1867. The steam tractor, produced by a variety of companies, was widely used in American agriculture during the late 1800s. And the experimental Besler steam biplane reported successful flights in 1934, though it was never put into production.

By the 1920s, American electrical utility companies were rapidly constructing electrical generating stations. These were made possible by massive steam turbine engines, generating thousands of horsepower; these were coupled to large electrical motors that generated electricity for sale to the public. Power plants enabled consumers and industry to use newly developed electrical devices and machines at home and at work. While most steam-engine technology has been replaced, its legacy lives on in the network of electrical power generating stations around the world.

James Steinberg

Sources

Hindle, Brooke, and Steven Lubar. *Engines of Change: The American Industrial Revolution, 1790–1860.* Washington, DC: Smithsonian, 1986.

Kras, Sara Louise. *The Steam Engine.* Philadelphia: Chelsea House, 2004.

Sutcliffe, Andrea. *Steam: The Untold Story of America's First Great Invention.* New York: Palgrave Macmillan, 2004.

SYNTHETIC RUBBER

Unlike natural rubber, derived from the tree *Hevea brasiliensis* found abundantly in Southeast Asia, synthetic rubber is an artificially made polymer material acting as an elastomer, that is, a substance that undergoes stress then returns to its original size. Synthetic rubbers such as butyl, neoprene, nitrile, polysulphide, and styrene-butadiene are resistant to heat, sunlight, and oil. They are used in electrical cable insulation, telephone wiring, sheet products, roofing, fuel hoses, and roadways. The raw materials of synthetic rubber include petroleum, coal, and natural gas.

The impetus for the development of synthetic rubber came from the limited supply of natural rubber during the 1920s, in part because of the Stevenson Act, which was imposed on the rubber trade by England to protect its rubber industry. In the United States, DuPont scientist Elmer Bolton headed a team that began researching polymers. Bolton brought Harvard chemist Wallace Carothers to DuPont and hired Notre Dame chemist Julius Nieuwland, who actively researched polymerization, as a consultant. In 1929, DuPont researchers used polymerization to create choroprene. This product was introduced by the company as Duprene; it was successfully mass-produced in the 1930s and is now called neoprene.

Demand for synthetic rubber increased dramatically during the late 1930s and 1940s because of World War II, as Japan controlled the regions where most of the world's supply of natural rubber was produced. The U.S. government began production of the monomer styrene, which was used to create a synthetic substance called Buna rubber. Ultimately, fifty U.S. factories producing synthetic rubber were established during the war, producing double the amount of worldwide natural rubber made before the attack at Pearl Harbor.

Today, synthetic rubber is used in a variety of products; for example, automotive tires manufactured in the United States are made of approximately two-thirds synthetic rubber. Advances in technology have led to the development of numerous synthetic rubbers, such as nitrile rubber, that can withstand high temperatures and are used in such applications as gasoline hoses, automotive seals, adhesives, sealants, and examination gloves. The elastomer industry of the United States is the largest in the world, producing about a quarter of the world's rubber output.

Patit Paban Mishra

Sources

Herbert, Vernon, and Attilio Bisio. *Synthetic Rubber: A Project That Had to Succeed.* Westport, CT: Greenwood, 1985.

Hofmann, Werner. *Rubber Technology Handbook.* New York: Hanser, 1989.

Korman, Richard. *The Goodyear Story: An Inventor's Obsession and the Struggle for a Rubber Monopoly.* San Francisco: Encounter, 2002.

TELEGRAPH

As the Industrial Age dawned in the early nineteenth century, a new communications medium emerged based on the transmission of electromagnetic energy over wire. The ability to send messages by electromagnetic telegraph overcame the tyranny of distance that had plagued previous incarnations of telegraph technology.

The first successful application of the new technique is credited to Samuel F.B. Morse, a

Yale-educated artist and painter who first heard of electromagnetism during a conversation about new scientific experiments during a return voyage from Europe to the United States in 1832. Morse pursued the development of an electromagnetic telegraph by integrating myriad scientific concepts of early telegraph pioneers into a usable and practical mode. Prior to 1832, electrical telegraphy was impractical because of limitations in the electric current conveyed from a remote transmitter. Morse explored ways to enhance and improve the method by collaborating with several American and European telegraphy experts, particularly Alfred Vail and Leonard Gale, who became his close colleagues.

By 1835, Morse had developed a working telegraph model, and, in 1838, he invented a code based on a system of dashes and dots to communicate messages. This new language of communication, called Morse code, was a major breakthrough, because it was a simple and easily learned system that made for efficient transmission of messages. The electromagnetic telegraph operated by breaking the circuit transmission between the machine's battery and receiver, with the breaks being measured in dashes and dots.

On January 6, 1838, in Morristown, New Jersey, Morse successfully operated his device for the first time. On February 21, 1838, he demonstrated his telegraph to President Martin Van Buren and cabinet members in Washington, D.C. Morse sought financial support from the U.S. government for his invention and, in 1843, received $30,000, with which he built a small telegraph system between Washington and Baltimore. On May 24, 1844, the first telegraph message, "What hath God wrought!" was transmitted along the line.

Morse's invention was quickly copied and became a widely used instrument of business and personal communication, as well as a competitive commercial enterprise. By 1851, more than fifty companies were operating telegraph lines in the United States. Continuous improvements in technology and application lead to type-printing telegraphs, multiplex machines, and facsimile devices, all designed to carry a higher message density and faster transmission. The electric telegraph had a far-reaching impact,

transforming social interaction, culture, politics, economics, and military affairs. In 1868, the first telegraph message was successfully transmitted across the Atlantic Ocean, opening a new era in global communications.

Steven J. Rauch

Sources

Harlow, Alvin F. *Old Wires and New Waves: The History of the Telegraph, Telephone, and Wireless.* New York: Appleton-Century, 1936.

Shiers, George, ed. *The Electric Telegraph: An Historical Anthology.* New York: Arno, 1977.

TELEPHONE

Credit for successfully applying the principles of transmission of speech over wire is generally given to Alexander Graham Bell, a professor of vocal physiology at Boston University. Bell combined an understanding of the nature of sound with knowledge of the established electric telegraph, leading him to conceive of sending sound along a wire based on a variance in pitch.

Along with his assistant Thomas Watson, Bell explored the idea of a "harmonic telegraph" device and, by 1875, had proved that different tones could be sent over wire. After the development of an effective transmitter and receiver, Bell and Watson achieved their goal on March 10, 1876, when Bell spoke to his assistant in another room through the instrument: "Mr. Watson, come here, I want to see you."

The basic principle for operation of the telephone is based on the air-pressure changes caused by human speech, which in turn cause a thin iron diaphragm to vibrate in front of an iron core surrounded by a coil of wire. The vibrations on the electromagnetic field are then conducted by electrical current through wires connected to a receiver. At the receiver, another electromagnet translates the fluctuations in current into vibrations on a diaphragm that the listener hears as human speech.

Bell's achievement enticed many scientists to copy and improve upon the device. In 1877, Thomas Edison patented the carbon button transmitter, a small device that was highly sensitive to the pressure of modulating sound waves. Edison's invention greatly increased the distance of speech transmission, and the telephone soon challenged the telegraph as the primary means of long-distance communication. When Bell's patents expired in the early 1890s, more than 6,000 independent telephone companies sprang up across the United States.

A major challenge was to establish a network that would allow communication between cities, and that could carry multiple conversations on the same set of wires. The telephone network was achieved in 1913, when the American Telephone and Telegraph Company (AT&T) was granted a regulated monopoly status by the U.S. government to run a national telephone network to deliver "universal" telephone service. By 1923, there were over 23 million telephones connected to the network for government and private use.

During World War II, technology improvements included mobile phones using radio signals, as well as advances in component and system packaging. By mid-century, the telephone had become an integral part of the American economy and society, enabling people to more effectively and efficiently communicate. By the early 1990s, almost 94 percent of U.S. households had at least one telephone.

As technology advanced, the telephone was no longer limited to wires. In 1946, AT&T began development of mobile car-phone systems. In 1973, Motorola's head researcher, Martin Cooper, successfully demonstrated a wire-free, handheld phone that allowed him to call his counterpart at rival Bell Labs while walking down a New York City street. This first cell phone was based on a network of multiple base stations located in an overlapping pattern of "cells," or coverage areas; in this system, calls are automatically handed over from one station to another when the phone moves from one coverage area to the next.

The proliferation of this cell phone technology was restricted at first due to the limited allocation of the frequency spectrum bandwidth by the FCC. By 1983, the FCC allocated the bandwidth to establish the first cellular system in Chicago, but even that was not enough when the number of cell phone subscribers passed 1 million in 1987.

In the mid-1990s, cell phone technology became commercially viable, as more spectrum was allocated and microchip technology improved the ability to transfer data within a cellular system. The use of communications satellites orbiting Earth allowed line-free communications almost anywhere on the planet. As of 2006, there were more than 60 million cell phone customers within the United States, and the technology was being enhanced by the use of Internet protocols and fiber optics to allow the personal computer and Internet to provide wireless telephone communications.

Steven J. Rauch

Sources

Harlow, Alvin F. *Old Wires and New Waves: The History of the Telegraph, Telephone, and Wireless.* New York: Appleton-Century, 1936.

Shiers, George, ed. *The Telephone: An Historical Anthology.* New York: Arno, 1977.

TELEVISION

Television is an electronic system for transmitting still or moving images and sound to receivers that recreate them. To create television broadcasts, video signals are broken up, transmitted through the air, and then reassembled by the technology inside the television set.

The first television sets were made possible by the development of the cathode ray tube (CRT), the basis of the picture tube found in all early televisions. The CRT is a specialized vacuum tube that uses electron beams to produce an image. The tube shoots beams of electrons at a phosphorescent surface, causing it to glow. By controlling where each beam strikes, a glowing picture can be created.

The cathode ray tube was invented by the German scientist Karl Ferdinand Braun in 1897. Then, in 1927, a young American engineer named Philo T. Farnsworth developed the dissector tube, a type of cathode ray tube that uses a magnetic field to sweep the electrons vertically and horizontally. Using this technique, the tube can put a dissected picture back together again.

Later that year, on September 7, Farnsworth demonstrated the first electronic television in San Francisco using his dissector tube. He also applied for and received a patent for his invention. At about the same time, however, the Radio Corporation of America (RCA) was trying to patent a similar device. This led to a patent battle that lasted over ten years, resulting in RCA paying Farnsworth for patent licenses.

Many became interested in television technology when it was featured in an RCA exhibition at the 1939 World's Fair in New York. By 1941, there was enough interest that the National Broadcasting Company (NBC) and Columbia Broadcasting System (CBS) were both able to launch commercial television stations in New York City. Television did not really catch on, however, until after World War II.

The first U.S. coast-to-coast broadcast took place in 1951, as President Harry S. Truman addressed the opening of the Japanese Peace Treaty Conference in San Francisco. At the time, there were an estimated 13 million television sets in the United States. The year 1951 also saw the debut of the *I Love Lucy* show, which broke new ground by being the first program to be produced on film instead of broadcast live. It also established Lucille Ball as TV's first major female star.

There have been many advances in television technology over the years. Today, signals are delivered via cable or satellite dish receiver as well as through the airwaves. The cathode ray tube system also has been improved upon. Some modern televisions use a liquid crystal display (LCD) instead of a CRT. The LCDs create the image on a layer of liquid crystal material sandwiched between two sheets of glass. The newest technologies include plasma television, a type of flat-panel display made up of a layer of gas between two glass plates, and HDTV, which allows for a wide-screen experience on a home television.

Beth A. Kattelman

Sources

Hilmes, Michelle, and Jason Jacobs. *The Television History Book*. London: British Film Institute, 2004.

Schatzkin, Paul. *The Boy Who Invented Television: A Story of Inspiration, Persistence, and Quiet Passion*. Burtonsville, MD: Teamcom, 2002.

THREE MILE ISLAND

Early on the morning of March 28, 1979, Unit 2 of the Three Mile Island (TMI) nuclear power plant, located on an island in the Susquehanna River near Harrisburg, Pennsylvania, experienced a minor malfunction that caused an automatic shutdown of the nuclear reactor.

In the first few minutes of the shutdown, a malfunctioning valve led plant operators into a series of catastrophic errors that eventually defeated all of the plant's automatic protection systems. The result was severe damage to the core of the reactor, a minor release of radiation into the environment, and, for a time, concern that hydrogen buildup inside the reactor could cause an explosion.

Poor planning for such unexpected events also led to a communications breakdown with state and local officials. To gain control of communications in the confused, near-panic situation, President Jimmy Carter ordered the Nuclear Regulatory Commission (NRC) to send a senior regulatory official to TMI to ensure that accurate and timely information about the accident was provided to government officials, the press, and the public.

After several uneasy days, in which control of the reactor was in doubt and Pennsylvania's governor ordered a precautionary evacuation of pregnant women from the area, the reactor's condition was stabilized and the threat to the public reduced. The name Three Mile Island, however, became associated with public concerns about the safety of nuclear power plants.

The accident at TMI revealed a number of flaws in reactor design, operator training, and emergency planning. In the course of the next decade, all U.S. nuclear power plants were required by the NRC to retrofit improved safety and monitoring equipment, retrain and requalify plant operators, rewrite control-room procedures, redesign control-room layouts and instrumentation, and cooperate with state and local officials in preparing emergency plans for each site.

As of 2006, U.S. reactors have not experienced another accident as serious as the one that occurred at Three Mile Island. New reactor designs employ a passive safety approach that would prevent a repeat of accidents of this type.

William M. Shields

Sources

American Chemical Society. *The Three Mile Island Accident: Diagnosis and Prognosis.* Washington, DC: American Chemical Society, 1986.

The Report of the President's Commission on the Accident at Three Mile Island. New York: Pergamon, 1979.

Three Mile Island: A Report to the Commissioners and to the Public. Washington, DC: U.S. Government Printing Office, 1980.

TRANSCONTINENTAL RAILROAD

Technically, the first transcontinental railroad was only 48 miles in length, crossing the Isthmus of Panama. Antedating the completion of the Panama Canal (1914) by almost six decades, this five-year project, finished in 1855, posed many of the same formidable challenges faced by the canal's engineers, including malarial swamps and rugged mountains.

While the Panama Railroad was remarkable in its own right, spanning the hundreds of miles of desert, plains, and mountains through the western expanse of the North American continent was far more extraordinary. On May 10, 1869, the "Golden Spike" ceremony at Promontory, Utah, marked the completion of this 1,776 mile transcontinental railroad—and the triumph over numerous obstacles, both natural and human made. The event showcased the intersection of big dreams, politics, war, business, government, ethnic strife, human pathos, and engineering skill.

As early as 1845, Americans such as Asa Whitney, a New York merchant, urged that a transcontinental railroad be built. The Gadsden Purchase in 1853 was designed to facilitate a southerly route for a railroad, but Illinois Senator Stephen A. Douglas, a senator representing Illinois in the U.S. Congress, attempted to win Southern support for a Chicago terminus in exchange for his support of the Kansas-Nebraska

Completion of the transcontinental railroad—linking America's eastern and western rail networks in a nationwide mechanized transportation system—was marked by the driving of a golden spike at Promontory, Utah, on May 10, 1869. *(MPI/Hulton Archive/Getty Images)*

Act (1854)—legislation that promoted the concept of "popular sovereignty" on the question of extending slavery to U.S. territories. But neither the Midwest nor the South could garner the necessary congressional votes for where the line would be built.

Abraham Lincoln's election in 1860, Republican ascendancy, and the withdrawal of Southern members of Congress during the Civil War greased the rails for a Midwestern route. The Pacific Railroad Act of 1862 authorized the Union Pacific and the Central Pacific to build rail lines from east and west and to meet in the middle.

Finding a viable path through the Sierra Nevada Mountains—one that would not employ too steep a gradient—was perhaps the greatest engineering challenge. Railroad engineer Theodore Judah was able to solve the problem, plotting a path through Donner Pass. His 1857 treatise *A Practical Plan for Building the Pacific Railroad* sketched for the public what would be needed for the enterprise: a transit party (mapping and marking the route), a leveling party (determining the vertical profile), and a survey engineer, to calculate the gradient as well as all the logistical details (amount of masonry and timbers for bridges, the manner of crossing rivers, etc.). Judah's plan was intended to give investors an accurate estimate of construction cost.

Besides the extremes of craggy mountains and dusty deserts, the land presented other challenges. A lack of trees on the plains necessitated the importation of wood for railroad ties and led to improvisations, such as coating more readily available cottonwood with a zinc solution to help prevent the soft wood from rotting quickly. Snow sheds were devised to protect against snowdrifts, and gigantic mechanical shovels were introduced for scraping snow away. A steam engine, hauled to the site by men and oxen, drilled through granite during the completion of a tunnel. Nitroglycerin, newly discovered and highly volatile, was used to blast through stubborn rock far more effectively than black powder.

The Union Pacific's Irish workers fought the elements and Native Americans as the line snaked its way through Nebraska and Colorado. Chinese workers risked life and limb as they were lifted in buckets across rock faces or as they dug tunnels, carving the Central Pacific through California, Nevada, and into Utah. As the railroads approached their rendezvous, a still-standing record was set by the Central Pacific workers, as they laid more than 10 miles of track in one day.

As the hammer struck the ceremonial "Golden Spike," an electric circuit was completed across the continent—a signal via telegraph that set off celebrations in major American cities. The nation was bound together, not just with the copper of the telegraph wires, but now with rails of iron—a reality that manifested a stronger union not only politically but also financially.

Frank J. Smith

Sources

Ambrose, Stephen E. *Nothing Like It in the World: The Men Who Built the Transcontinental Railroad 1863–1869.* New York: Simon and Schuster, 2001.

Bain, David Haward. *Empire Express: Building the First Transcontinental Railroad.* New York: Penguin, 2000.

U.S. MINT

Although the United States declared its independence in 1776, Americans continued to use various foreign and state-issued coins in the absence of a national mint. After years of effort by Thomas Jefferson, Alexander Hamilton, Robert Morris, and others, Congress passed the Mint Act of 1792, which created the U.S. Mint. With the stroke of a pen, the new nation discarded the British system of coinage and adopted the decimal system we know today.

The U.S. Mint was established in Philadelphia, Pennsylvania, in 1792 under the leadership of one of America's leading scientists, David Rittenhouse, and it issued its first coins in 1793. The early process of striking coins was labor intensive and time consuming. Horses, oxen, and humans supplied the energy needed to drive the various machines. Originally, the Mint relied on human-powered mill and screw presses that required several minutes to produce each coin. Steam-powered presses, capable of producing 120 coins per minute, were introduced in 1836. Over the years, the Mint grew with the nation, adopting new technology and opening new branches to meet demand.

Minting coins is a precise science. One hundred years ago, the Mint required the services of many technicians to produce a quality product. The value of a coin was based on the amount of gold or silver it was made of. These precious metals had to be certified for purity by an assayer before they could be refined, cast into ingots, alloyed, and rolled several times to produce strips of precise thickness. Disks called "planchets" were cut from these strips, which were then stamped into coins. Coins exceeding one dollar in value were struck from gold; those valued at one dollar and below were struck from an alloy of silver and copper. Pennies were struck from copper alloyed with small amounts of other metals.

Gold and silver are no longer used in circulating coins. Although today's coins appear silver, they are actually made from a layer of copper sandwiched between two layers of nickel. A blanking press punches out disks, which are annealed to soften them for striking. After a ridge is created around the edge of the blank, the disk is struck in the coining press, which imprints the design of the coin. Finally, the coins are inspected for quality, counted, bagged, and distributed to banks for circulation. Pennies are produced in a similar manner from precut copper-coated zinc blanks.

Since 1792, the U.S. Mint has produced circulating and commemorative coins to feed the nation's growing economy. The Mint has always relied on technology and mechanization to keep pace with this important charge. Today, the U.S. Mint's facilities in Philadelphia and Denver are sufficient to supply all the circulating coins for the United States—up to 50 million coins every twenty-four hours.

Charles Delgadillo

Sources

Evans, George G., ed. *Illustrated History of the United States Mint.* Philadelphia: George G. Evans, 1892.

Schwarz, Ted. *A History of United States Coinage.* New York: A.S. Barnes, 1980.

Stewart, Frank H. *History of the First United States Mint: Its People and Its Operations.* Philadelphia: Frank H. Stewart Electric Co., 1924.

U.S. Mint. http://www.usmint.gov.

URANIUM

Uranium is a silver-colored radioactive metal used in nuclear power reactors and nuclear weapons. It is the heaviest naturally occurring metal element in Earth's crust, and it is found in fourteen known atomic forms, or isotopes. The most prevalent of these are the naturally occurring isotopes uranium 234, uranium 235, and uranium 238. Of these, uranium 238 is by far the most abundant. The world's principal uranium sources are found in the Congo River basin in Africa, northern Saskatchewan in Canada, and Colorado and Utah in the United States. Historically, uranium ores have also been mined in Australia, Europe, Brazil, Kazakhstan, Ukraine, and Russia.

The discovery of uranium involved European and American scientists. Martin Klaproth, a Berlin apothecary, first characterized "uranit" as an element in 1789. In 1841, French chemist Eugene-Melchior Peligot isolated the element uranium, and, in 1896, physicist Antoine Becquerel discovered its radioactivity.

The uranium ore pitchblende was first discovered in the United States in 1871 in waste from abandoned gold mines near Denver, Colorado. Yellow carnotite uranium ores were discovered in 1881 on the Colorado Plateau in southwestern Colorado and eastern Utah in the late 1890s and used chiefly in the production of yellow-colored glass and ceramics. Fueled by growth in the medical uses of radium, ore processors began extracting radium from the region's carnotite ore shortly after the turn of the twentieth century. By the late 1920s, much of the uranium ore mined was being processed for its vanadium content. Vanadium was used as an additive in steelmaking to stabilize carbide for producing rust-resistant and high-speed tool steels.

With the discovery of atomic fission in 1938, uranium became a potential source of fissionable material for use in atomic bombs and to generate nuclear power. In 1942, the top-secret Manhattan Project used uranium to produce one of the world's first atomic bombs, an atomic fission bomb that exploded over Hiroshima, Japan, in August 1945. A plutonium-based implosion bomb had also been developed and was used on Nagasaki three days later. Because it is easier to convert uranium 238 to plutonium in breeder reactors than it is to extract uranium 235 from ores through a process called enrichment, plutonium became the principal material for nuclear weapons. Today, both uranium and plutonium are used in American thermonuclear weapons.

Uranium is also used as a nuclear reactor fuel in nuclear power generating facilities around the world. In nuclear reactors, a controlled fission chain reaction caused by the splitting of uranium 235 atoms creates intense heat that is used to convert water into steam. The steam is fed through a turbine, which in turn drives an electricity-producing generator. One ton of natural uranium can produce 40 million kilowatt-hours of electricity, which is equivalent to burning more than 78,000 barrels of oil or 15,000 tons of coal.

The first nuclear reactors went online in 1954, when the world's first nuclear-powered submarine, the USS *Nautilus*, was launched. In 1955, Arco, Idaho, became the first town in the United States to be powered by electricity produced by nuclear energy.

Some of the greatest challenges to more widespread use of uranium for power generation in the United States lie in problems related to its production, processing, and disposal. Uranium mining, milling, conversion, enrichment, and fuel fabrication have historically created production and processing wastes. Likewise, the storage, reprocessing, and disposal of spent uranium fuel from reactors have grave environmental, political, and nuclear nonproliferation implications given that the half-life—the time required for one-half of a quantity of the uranium to naturally become nonradioactive through the process of radioactive decay—of uranium 235 is more than 700 million years.

Todd A. Hanson

Sources

Bickel, Lennard. *Deadly Element: The Story of Uranium.* New York: Stein and Day, 1981.

Gittus, John H. *Uranium.* London: Butterworths, 1963.

Hofman, Sigurd. *On Beyond Uranium: Journey to the End of the Periodic Table.* New York: Taylor and Francis, 2002.

VON BRAUN, WERNHER (1912–1977)

German-born rocket engineer Wernher von Braun led the creation of powerful booster rockets (Saturn series) that supported the U.S. space program to send satellites and then humans into space. With hundreds of other top German scientists, von Braun emigrated to the United States after World War II, became a U.S. citizen, and went to work for the U.S. government.

Wernher Magnus Maximillian von Braun was born on March 23, 1912, in Wirsitz, Prussia, to politician Magnus Freiherr von Braun and Emmy von Quistorp. When he was young, his mother gave him a telescope, and he used it to spend numerous hours looking into the stars. In 1920, Wirsitz became part of Poland and his family moved to Berlin, Germany.

At first, von Braun did not enjoy math, but upon obtaining a copy of *The Rocket into Interplanetary Space* (1923) by Hermann Oberth, he focused intensely on physics. As a teenager, he experimented with a rocket-powered wagon, which did not sit well with the local police. Von Braun and a few friends began using the local dump for launching rockets. The German Army noticed and visited him in 1930 to arrange a citizen advisory role to the military. That same year, he attended the Berlin Institute of Technology and assisted his mentor Oberth in the first liquid-fueled rocket motor tests. By 1934, von Braun had earned a doctorate in physics in aerospace engineering from the Technical University of Berlin.

Von Braun joined the Nazi party in 1937 and used schematics from American rocket engineer Robert Goodard to begin building a rocket for the German government. As the head of German rocket technology during World War II, von Braun led the development of the V-2 (*Vergeltunswaffe* 2), the first ballistic missile, able to travel up to 200 miles (320 kilometers) and carry a 2,200-pound (1000-kilogram) warhead. By 1942, military rocket testing was under way. The project was set back by targeting problems and devastating bombing of the manufacturing facilities by Allied forces in 1944, although scores of V-2s were fired at London that same year.

After the war, von Braun was brought to the United States in a secret program called Operation Paperclip, in which nearly 1,600 German scientists, technicians, and other personnel were extricated from Germany. Relocated to Fort Bliss, Texas, and provided with 118 staff members and unused parts from the V-2 program, von Braun began building a U.S. rocket program in 1946. Four years later, the endeavor was moved to Huntsville, Alabama, where von Braun oversaw construction of the Redstone rocket. This surface-to-surface ballistic missile, used for the first live nuclear weapons tests, was able to fly 200 miles and carry a 3.75-megaton nuclear warhead.

Von Braun became a naturalized U.S. citizen in 1955. With his childhood ambitions still strong, he began writing about the possibilities of a space station and helped Walt Disney produce three television programs about space exploration. In 1957, after the successful launch of the Soviet space satellite *Sputnik,* von Braun's team modified a Redstone rocket to create the Juno-C, which effectively launched the U.S. space program on January 1, 1958, by lifting the first U.S. artificial satellite, *Explorer 1,* into space.

The public had mixed reactions to seeing von Braun's name in the headlines, but his place in modern American science and technology was secure. Von Braun joined the National Aeronautics and Space Agency (NASA) in 1960, and he was appointed a deputy administrator in 1970. He served as director of the new Marshall Space Flight Center in Alabama, and ultimately headed development of the Saturn V rocket that carried the *Apollo 11* astronauts to the moon. The reduction of funding for the Apollo program in 1972 led to his retirement.

Von Braun continued his work in the space industry, helping create the National Space Institute in 1976, consulting with several aerospace companies, and touring the world to speak to university audiences about his experiences and passion for space exploration. He died at the age of sixty-five on June 16, 1977, from a combination of intestinal cancer and injuries sustained in a car crash. President Jimmy Carter called him "a man of bold vision."

James Fargo Balliett

Sources

Piszkiewicz, Dennis. *Wernher von Braun: The Man Who Sold the Moon.* Westport, CT: Praeger, 1998.

von Braun, Wernher. *Conquest of the Moon.* New York: Viking, 1953.

———. *The Mars Project.* Chicago: University of Illinois, 1991.

Ward, Bob. *Dr. Space: The Life of Wernher von Braun.* Annapolis, MD: Naval Institute Press, 2005.

WEST, JOSEPH (?–CA. 1692)

As an agricultural scientist and one of the first governors of colonial South Carolina, Joseph West helped preserve the fledgling colony through its difficult early years. At the same time, he attempted, with less success, to organize the colony's agricultural systems in conformity with the plans of the proprietors in Great Britain, including their erstwhile secretary, the philosopher John Locke.

While West was dispatched to the Carolinas (via Barbados) in 1669 with three ships of supplies and colonists, Locke and like-minded pro-colonization contemporaries were engaged in a strenuous debate at home about whether such colonization schemes were feasible or even advisable during a time of acute economic and social distress. The Great Plague of 1665 and the London fire of 1666 had wreaked havoc in England, while an ongoing war with the Dutch severely strained the nation's finances, military, and workforce. It was certainly not the best time, opponents of colonization argued, to commit able-bodied men and scarce resources to questionable colonial enterprises thousands of miles away.

Locke and his supporters argued, to the contrary, that the colonies would not only provide new trade opportunities for the home country, including markets for manufactures, but they could also supply commodities such as timber and naval stores currently purchased from competing countries. The colonies might also serve as a repository for the more disaffected citizens of England, particularly criminals, paupers, and the unemployed, thus reducing social tensions at home while simultaneously giving the castoffs a chance to improve and redeem themselves through honest labor in the soil.

For Locke and other like-minded agrarians ("physiocrats" in the parlance of the day), agriculture, in particular crop-growing, would not only convert the colonial wilderness into productive land, but it would convert the colonists into productive members of a stable society. The leaders of the new South Carolina colony—including West, who served as governor from 1671 to 1672, and again from 1674 to 1682—were therefore instructed to insist that colonists till the soil to the exclusion of less socially valuable activities such as slave trading or even raising livestock. This directive ignored the fact that these were both common pursuits in the Barbados colonies from which Carolina drew many of its early settlers.

The instructions provided to the colony's leaders were often suprisingly specific. As historian John Otto notes, West received detailed orders about how to tend the colony's cattle, which were at all times to be looked after by "one or more" herders who would "bring [the cattle] home at night, & putt them in yor enclosed Groud, otherwise they will grow wild & be lost."

But the realities on the ground in Carolina did not often conform to the theories propounded back in England. Transplanted Barbadians continued to engage in the businesses that had proved profitable for them in the islands, particularly slave trading and the provisioning of pirate ships. Almost all of the Carolina settlers, meanwhile, soon learned that in the semitropical environment of South Carolina, it made more sense to allow branded cattle to roam freely in the woods for most of the year, rather than go to the trouble of rounding them up each day. Moreover, even with West's stern demand for crop planting to the exclusion of other activities, the colony at first proved incapable of feeding itself, requiring the proprietors to pay for expensive resupply operations.

Growing increasingly exasperated with the enterprise, the proprietors forced West out of the governorship in 1682. Despite a proprietor-sponsored wave of non-Barbadian settlement in the colony, South Carolina would continue to evolve into a center of slave trading and the open-range cattle business, rather than the society of crop-growing yeomen envisioned by the proprietors. In fact, the year of West's departure also witnessed the colony's first export of cattle to Barbados, the beginning of a lucrative trading network with the sugar colonies of the Atlantic and Caribbean.

Jacob Jones

Sources

Arneil, Barbara. "Trade, Plantations, and Property: John Locke and the Economic Defense of Colonialism." *Journal of the History of Ideas* 55:4 (October 1994): 591–609.

Otto, John. "Open-Range Cattle-Ranching in the Florida Pinewoods: A Problem in Comparative Agricultural History." *Proceedings of the American Philosophical Society* 130:3 (September 1986): 312–24.

Sirmans, M. Eugene. "Politics in Colonial South Carolina: The Politics of Proprietary Reform, 1682–1694." *William and Mary Quarterly* 3rd ser., 23:1 (January 1966): 33–55.

WHITNEY, ELI
(1765–1825)

Eli Whitney, renowned for inventing the cotton gin, was born in Westboro, Massachusetts, on December 8, 1765. He attended Yale, graduating in 1792. In addition to his famous invention, Whitney introduced the service contract and the American system of mass-production, assembly-line manufacturing, using prefabricated interchangeable parts.

In the late eighteenth century, English textile mills were the heart of the Industrial Revolution in Britain. The British textile mills used cotton to produce the thread for the weaving process, and their demand for cotton far exceeded the available supply. The American South, a major producer of cotton, exported only a small amount of its total available volume.

The problem was that removing the seeds from raw cotton made the conversion of cotton into thread a highly labor intensive process. Because the plentiful green-seed, short-staple cotton grown on inland plantations in the United States required this labor-intensive process, it was too expensive for use in the English mills. The black-seed, long-staple cotton grown in the southern coastal regions was more easily cleaned of its seed, but it was less plentiful.

Whitney's cotton gin was designed to mechanically remove the seeds with substantially less labor. The machine featured a simple, four-part design. Workers fed raw cotton into a hopper, from which a revolving cylinder with wire hooks pulled the cotton through an iron barrier with narrow slots smaller than the seeds. The seeds were blocked by the barrier, allowing only the soft cotton fibers to pass through. A bristled cylinder, revolving in the opposite direction, then cleared (brushed) the seedless cotton from the teeth of the first cylinder. The centrifugal force of the second cylinder flung the cleaned cotton out of the gin.

Whitney secured a patent for the cotton gin in 1794 and formed a partnership with Phineas Miller to manufacture the machine. From the beginning, Whitney marketed the machine in tandem with a separate and renewable service contract. The business failed three years after the patent was issued, however, partly because the domestic demand for the cotton gin was limited by the relatively small number of Southern planters who dominated the plantation system. Moreover, the simplicity of the cotton gin's design made it easily reproduced by mechanics and planters who ignored Whitney's patent and built their own gins. Those who did buy Whitney's device were unwilling to pay the service fees, because they could easily make repairs themselves. After Congress decided not to renew his cotton gin patent when it expired in 1807, Whitney refused to patent his later inventions.

Whitney's creation of the American system of manufacturing is perhaps his greatest legacy. The conventional method of manufacturing at the time, the English system, created individualized products as one or more craftspeople worked on production from beginning to end. If a replacement part was needed, it had to be crafted to fit the specific product. Though each craftsperson might create finished products with similar characteristics, the parts and system of assembly were so particular to each worker that it was difficult and time consuming for anyone other than the original maker to repair the product. Whitney reasoned that standardized parts assembled in a repeated process would reduce the skill, labor, and time necessary to manufacture or repair any product.

Federal armories in the 1790s were capable of producing only 1,000 muskets a year but, anticipating war with France, sought thousands more. Whitney won the bid to manufacture 10,000 muskets over two years. His plan was to use interchangeable parts mass-produced on an assembly line. Whitney, however, underestimated the time that he needed to invent and build the new machines necessary to produce standardized parts.

Whitney invented a milling machine to shape metal and other materials in exactly the same dimensions. He also invented a router that replaced the hand chisel for carving and shaping wooden components in the same dimensions. He designed job-specific machine tools and jigs (templates) that allowed efficient assembly of the muskets by a team (line) of unskilled or semiskilled workers, as opposed to craftspeople who had learned their trade over a number of years.

Because all of the parts were standardized and assembled in the same repetitious manner, the time and cost of manufacturing the muskets decreased. The interchangeability of the parts also meant that needed repairs could be made without returning the product to the manufacturer. Whitney's plea for additional time on the project was granted when he demonstrated the effectiveness of his new system of manufacturing to president-elect Thomas Jefferson and other government officials in 1801, assembling operational muskets from stacks of interchangeable parts.

Whitney succeeded in creating a system that radically improved the production time and quality of manufactured goods. It was Whitney's American system of manufacturing that Henry Ford used to create the first assembly-line automobile, and it is Whitney's system that dominates global manufacturing to this day. Eli Whitney died in New Haven, Connecticut, on January 8, 1825.

Richard M. Edwards

Sources

Bagley, K., and Ray Douglas Hurt. *Eli Whitney: American Inventor.* Mankato, MN: Capstone, 2003.

Green, Constance. *Eli Whitney and the Birth of American Technology.* Upper Saddle River, NJ: Pearson Education, 1997.

WOODS, GRANVILLE
(1856–1910)

The African American inventor Granville T. Woods, who obtained patents for more than fifty electrical and mechanical devices, was born on April 23, 1856, in Columbus, Ohio. He attended school until the age of ten, when he went to work at a local machine shop that repaired railroad equipment. Fascinated by electrical power and the railroads, he paid close attention to how different pieces of equipment were used and even paid some workers to explain electrical concepts to him.

In 1872, at age sixteen, Woods took employment with the Danville and Southern Railroads in Missouri, but he quit in the face of racist policies that denied him promotions. For the next several years, he traveled around the Midwest and East, working in machine shops and furthering his studies. In 1878, he secured a job with the British steamship *Ironsides*; however, he soon became disenchanted with his limited role. Understanding that the only company that would not discriminate against African Americans would be his own, Granville Woods and his brother, Lyates, began the Woods Electrical Company in Cincinnati, Ohio, in 1880.

The Woods's company made a number of significant advances in telephone, telegraph, and electrical equipment, including an improved steam boiler, an automatic air-brake system for trains, an improved telephone transmitter, and a device that combined the telephone and telegraph. The latter technology, called "telegraphony," allowed a telegraph station to send voice and telegraph messages over a single line. The invention became so successful that the American Bell Telephone Company purchased it from Woods. General Electric and the Westinghouse Air Brake Company also purchased some of Woods's innovations.

Woods also invented and patented the electric third rail system, which enabled the development of the overhead railroad system found in many metropolitan cities, such as Chicago, St. Louis, and New York. However, Woods is best known for his 1887 creation, the synchronous multiplex railway telegraph. This device, a variation of the induction telegraph, allowed messages to be sent between railway stations and moving trains, thereby dramatically increasing railway safety by facilitating easier and more accurate communication between train conductors and railway controllers.

Unfortunately for Woods, other inventors attempted to lay claim to his creations and innovations. Most notable among these disputes was one brought by Thomas Edison, who stated that he had created a telegraph device before Woods and that he was therefore entitled to its patent. Woods defended himself against Edison's claim on two occasions, proving that he alone made the device.

In the end, Edison admitted defeat and offered Woods a position with the Edison Company. It was ironic, therefore, that Woods became known as the "Black Edison." By the time of his death on January 30, 1910, he had become one of America's most respected inventors in his own right.

Paul T. Miller

Sources

Fouche, Rayvon, and Shelby Davidson. *Black Inventors in the Age of Segregation: Granville T. Woods, Lewis H. Latimer, and Shelby J. Davidson.* Baltimore: Johns Hopkins University Press, 2003.

Hayden, Robert. "Black Americans in the Field of Science and Invention." In *Blacks in Science: Ancient and Modern*, ed. Ivan Van Sertima. Somerset, NJ: Transaction, 1987.

WRIGHT, ORVILLE
(1871–1948),
AND WILBUR WRIGHT
(1867–1912)

Orville and Wilbur Wright were the inventors and builders of the first successful self-propelled, heavier-than-air craft—flown at Kitty Hawk, North Carolina, on December 17, 1903. Their invention revolutionized not only transportation but warfare as well in the twentieth century.

Wilbur was born in Millville, Indiana, on April 16, 1867, and Orville was born in Dayton, Ohio, on August 19, 1871. From childhood, the brothers were always close. They claimed that their interest in aviation dated from the moment, in 1878, when their father presented them with a toy helicopter. But it was the death of German glider pioneer Otto Lilienthal in 1896 that inspired them to pursue the idea of building a motor-driven aircraft. By that time, the brothers, each having dropped out of high school, had started printing and bicycle businesses together.

Around the turn of the twentieth century, the Wright brothers were not the only ones interested in flight technology. Inventors and scientists around the world were experimenting with aircraft, and the Wrights set themselves the task of reading everything they could on the subject. Their research led them to conclude that the major problems of aerodynamics and propulsion already had been solved by others. As a result, they focused on the chief remaining problem in building a successful airplane: control of the aircraft.

By the late 1890s, the two were focusing on how to allow the operator to control the three motions of an aircraft: pitch (up and down movement), roll (rotation of the aircraft), and yaw (side to side movement). Their first success came with the idea of wing-warping, or twisting the wing across its span, to control roll. To try

out their idea, they used their skills as bicycle mechanics to build a test glider.

The brothers studied U.S. Weather Bureau reports to find the best place to test their glider. They ultimately decided on the windy sand dunes on North Carolina's Outer Banks. Between 1900 and 1902, they flew their gliders more than 250 times at Kitty Hawk, achieving a flight of more than 600 feet (183 meters). The extensive testing helped them overcome other problems in control and aerodynamics. They then returned to Dayton to deal with the final challenge, propulsion. At their bicycle shop, they built a small, lightweight, four-cylinder engine, which they mounted on their test plane.

Returning to Kitty Hawk in the fall of 1903, they quickly achieved success, with Orville making a 120 foot (36.5 meter), 12 second flight on December 17. Several more attempts produced an 856 foot (261 meter), 29 second flight, with Wilbur at the controls. At first, they tried to keep their achievement a secret, fearful that rivals would steal their ideas. But by the late 1900s, a number of aviators in Europe were able to copy their design and fly aircraft of their own. In 1908, the brothers won their first contract to build an

The Aviation Age began with the first controlled, powered flight in a heavier-than-air craft by Wilbur and Orville Wright on December 17, 1903. The Wright Company, which they founded six years later, eventually produced nineteen airplane models. *(Fox Photos/Hulton Archive/Getty Images)*

airplane for the U.S. Army, and they formed the Wright Company the following year.

Wilbur spent most of his remaining years fighting patent infringement suits, while Orville sold his shares in the company in 1915, becoming an adviser to the military during World War I. Orville retired from aviation shortly thereafter and spent the rest of his life promoting aviation and defending against rivals' claims to have invented the first airplane. So upset was he by the Smithsonian Institution's decision to award the distinction of airplane inventor to one Samuel Langley that he donated the brothers' first 1903 craft to the London Science Museum.

Neither brother ever married, and both lived out their lives in Dayton. Wilbur died on May 30, 1912, and Orville died on January 30, 1948.

James Ciment

Sources

Crouch, Tom D. *The Bishop's Boys: A Life of Wilbur and Orville Wright.* New York: W.W. Norton, 1989.

Heppenheimer, T.A. *First Flight: The Wright Brothers and the Invention of the Airplane.* Hoboken, NJ: John Wiley and Sons, 2003.

Tobin, James. *To Conquer the Air: The Wright Brothers and the Great Race for Flight.* New York: Free Press, 2003.

DOCUMENTS

The Telegraph Explained

The invention of the telegraph by Samuel F.B. Morse (and others) revolutionized communication in America, setting the stage for further progress in communications, such as the telephone and radio. The device was explained to American readers in 1881.

This telegraph is based upon the principle that a magnet may be endowed and deprived at will with the peculiarity of attracting iron by connecting or disconnecting it with a galvanic battery; all magnetic telegraphs are based solely upon this principle. The telegraphs bearing the names of the several inventors, as Morse (who may be called the pioneer in this invention), House, Bain, etc., are simply modifications in the application of this great principle.

It is by breaking off the magnetic circuit, which is done near the battery, that certain marks are produced by means of a style or lever, which is depressed when the current is complete, and of the length of the interval of the breaking of this current, that signs of different appearances and lengths are produced and written out upon paper, making in themselves a hieroglyphic alphabet, readable to those who understand the key. This is the entire principle of electromagnetic telegraphing.

It was formerly considered necessary to use a second wire to complete the magnetic circuit, now but one wire is used, and the earth is made to perform the office of the other.

Where the distance is great between the places to be communicated with a relay battery is necessary to increase the electric current, and in this manner lines of great length may be formed.

The House apparatus differs from the Morse only that by means of an instrument resembling a piano-forte, having a key for every letter, the operator, by pressing upon these keys, can reproduce these letters at the station at the other end of the line, and have them printed in ordinary printing type upon strips of paper, instead of the characters employed on the Morse instrument to represent these letters.

The Bain telegraph differs from either of the two preceding methods, simply in employing the ends of the wires themselves, without the means of a magnet or style to press upon the paper, the paper being first chemically prepared; so that when the circuit of electricity is complete, the current passes through the paper from the point of the wires, and decomposes a chemical compound, with which the paper is prepared, and leaves the necessary marks upon it. There is not the same need for relay batteries upon this line as upon the others.

The greatest and most important telegraphic attempt is the successful laying of the cable across the Atlantic Ocean, which was finally completed and open for business July 28th, 1866. The cable lost in mid ocean in the unsuccessful attempt of the summer of 1866, has been recovered, and now forms the second cable laid, connecting the Eastern with the Western Continent.

The operation of telegraphing is very simple, and can easily be learned, being purely mechanical.

Source: Henry Hartshorne, *The Household Cyclopedia of General Information* (New York: Thomas Kelly, 1881).

The Science of Agriculture

During the eighteenth and nineteenth centuries, Americans focused on the practical aspects of science. Ensuring the fertility of the soil was of utmost importance, requiring specific techniques of fertilization and crop rotation, as revealed in the following excerpt from The Household Cyclopedia of General Information, *published in 1881.*

Vegetation, in its simplest form, consists in the abstraction of carbon from carbonic acid, and hydrogen from water; but the taking of nitrogen also, from ammonia especially, is important to them, and most of all, to those which are most nutritious, as the wheat, rye, barley, &c., whose seeds contain gluten and other nitrogenous principles of the greatest value for food. Plants will grow well in pure charcoal, if supplied with rain-water, for rain-water contains ammonia.

Animal substances, as they putrefy, always evolve ammonia, which plants need and absorb. Thus is explained one of the benefits of manuring, but not the only one as we shall see presently. Animal manure, however, acts chiefly by the formation of ammonia. The quantity of gluten in wheat, rye, and barley is very different; and they contain nitrogen in varying proportions. . . . During the putrefaction of urine, ammoniacal salts are formed in large quantity, it may be said, exclusively; for under the influence of warmth and moisture, the most prominent ingredient of urine is converted into carbonate of ammonia.

Rotation of Crops. The exhaustion of alkalies in a soil by successive crops is the true reason why practical farmers suppose themselves compelled to suffer land to lie fallow. It is the greatest possible mistake to think that the temporary diminution of fertility in a field is chiefly owing to the loss of the decaying vegetable matter it previously contained: it is principally the consequence of the exhaustion of potash and soda, which are restored by the slow process of the more complete disintegration of the materials of the soil. It is evident that the careful tilling of fallow land must accelerate and increase this further breaking up of its mineral ingredients. Nor is this repose of the soil always necessary. A field, which has become unfitted for a certain kind of produce, may not, on that account, be unsuitable for another; and upon this observation a system of agriculture has been gradually formed, the principal object of which is to obtain the greatest possible produce in a succession of years, with the least outlay for manure. Because plants require for their growth different constituents of soil, changing the crop from year to year will maintain the fertility of that soil (provided it be done with judgment) quite as well as leaving it at rest or fallow. In this we but imitate nature. The oak, after thriving for long generations on a particular spot, gradually sickens; its entire race dies out; other trees and shrubs succeed it, till, at length, the surface becomes so charged with an excess of dead vegetable matter, that the forest becomes a peat moss, or a surface upon which no large tree

will grow. Generally long before this can occur, the operation of natural causes has gradually removed from the soil substances, essential to the growth of oak leaving others favorable and necessary to the growth of beech or pine. So, in practical farming, one crop, in artificial rotation with others, extracts from the soil a certain quantity of necessary materials; a second carries off, in preference, those which the former has left.

We could keep our fields in a constant state of fertility by replacing, every year, as much as is removed from them by their produce. An increase of fertility may be expected, of course, only when more is added of the proper material to the soil than is taken away. Any soil will partially regain its strength by lying fallow. But any soil, under cultivation, must at length (without help) lose those constituents which are removed in the seeds, roots and leaves of the plants raised upon it. To remedy this loss, and also increase the productiveness of the land, is the object of the use of proper manures.

Land, when not employed in raising food for animals or man, should, at least, be applied to the purpose of raising manure for itself; and this, to a certain extent, may be effected by means of green crops, which, by their decomposition, not only add to the amount of vegetable mould contained in the soil, but supply the alkalies that would be found in their ashes. That the soil should become richer by this burial of a crop, than it was before the seed of that crop was sown, will be understood by recollecting that three-fourths of the whole organic matter we bury has been derived from the air: that by this process of ploughing in, the vegetable matter is more equally diffused through the whole soil, and therefore more easily and rapidly decomposed; and that by its gradual decomposition, ammonia and nitric acid are certainty generated, though not so largely as when animal matters are employed. He who neglects the green sods, and crops of weeds that flourish by his hedgerows and ditches, overlooks an important natural means of wealth. Left to themselves, they ripen their seeds, exhausting the soil, and sowing them annually in his fields: collected in compost heaps, they add materially to his yearly crops of corn.

Source: Henry Hartshorne, *The Household Cyclopedia of General Information* (New York: Thomas Kelly, 1881).

The Lowell Mills

The textile mills that opened in Lowell, Massachusetts, in the 1830s made a sanguine attempt to find a pastoral balance between technology and nature. Mill owners sought to avoid the negative consequences of the Industrial Revolution, which included an entrenched working class and dirty factory towns. The Lowell Mills were to avoid such pitfalls by hiring young farm girls who would work for a few years, earning enough money to build a proper dowry.

The mill was set next to the flowing stream, surrounded by the beauties of nature. What could be a more ideal scenario for the happy blending of nature and technology? A more realistic description is provided by the recollections of Lowell factory girl Harriet Robinson.

In what follows, I shall confine myself to a description of factory life in Lowell, Massachusetts, from 1832 to 1848, since, with that phase of Early Factory Labor in New England, I am the most familiar—because I was a part of it.

In 1832, Lowell was little more than a factory village. Five "corporations" were started, and the cotton mills belonging to them were building. Help was in great demand and stories were told all over the country of the new factory place, and the high wages that were offered to all classes of work-people; stories that reached the ears of mechanics' and farmers' sons and gave new life to lonely and dependent women in distant towns and farm-houses. . . . Troops of young girls came from different parts of New England, and from Canada, and men were employed to collect them at so much a head, and deliver them at the factories.

At the time the Lowell cotton mills were started the caste of the factory girl was the lowest among the employments of women. In England and in France, particularly, great injustice had been done to her real character. She was represented as subjected to influences that must destroy her purity and self-respect. In the eyes of her overseer she was but a brute, a slave, to be beaten, pinched and pushed about. It was to overcome this prejudice that such high wages had been offered to women that they might be induced to become mill-girls, in spite of the opprobrium that still clung to this degrading occupation. . . .

The early mill-girls were of different ages. Some were not over ten years old; a few were in middle life, but the majority were between the ages of sixteen and twenty-five. The very young girls were called "doffers." They "doffed," or took off, the full bobbins from the spinning-frames, and replaced them with empty ones. These mites worked about fifteen minutes every hour and the rest of the time was their own. When the overseer was kind they were allowed to read, knit, or go outside the mill-yard to play. They were paid two dollars a week. The working hours of all the girls extended from five o'clock in the morning until seven in the evening, with one half-hour each, for breakfast and dinner. Even the doffers were forced to be on duty nearly fourteen hours a day. This was the greatest hardship in the lives of these children. Several years later a ten-hour law was passed, but not until long after some of these little doffers were old enough to appear before the legislative committee on the subject, and plead, by their presence, for a reduction of the hours of labor.

Those of the mill-girls who had homes generally worked from eight to ten months in the year; the rest of the time was spent with parents or friends. A few taught school during the summer months. Their life in the factory was made pleasant to them. In those days there was no need of advocating the doctrine of the proper relation between employer and employed. *Help was too valuable to be ill-treated. . . .*

The most prevailing incentive to labor was to secure the means of education for some *male* member of the family. To make a *gentleman* of a brother or a son, to give him a college education, was the dominant thought in the minds of a great many of the better class of mill-girls. I have known more than one to give every cent of her wages, month after month, to her brother, that he might get the education necessary to enter some profession. I have known a mother to work years in this way for her boy. I have known women to educate young men by their earnings, who were not sons or relatives. There are many men now living who were helped to an education by the wages of the early mill-girls.

Source: Harriet H. Robinson, "Early Factory Labor in New England," in Massachusetts Bureau of Statistics of Labor, *Fourteenth Annual Report* (Boston: Wright & Potter, 1883).

Section 14

HISTORY AND PHILOSOPHY
OF SCIENCE

History as Science

Science—the means by which humans seek understanding without bias, or knowledge that is concrete and immutable—is a relative term. The American understanding has changed over the centuries: from a nonspecific view (science as the rational approach to intellectual inquiry) to an esoteric view (science as a highly specialized method of achieving empirical results). The study of history is a case in point.

To the eighteenth-century Enlightenment thinker, historical inquiry was as much a scientific endeavor as inquiry into nature in all of its forms. Indeed, human and natural history were kindred studies, subject to the same standards of research and analysis. The discovery and exploration of North America brought forth new opportunities for scientists to collect and analyze information and to create narrative profiles of the human and natural history of America. Hence much of the nonfiction literature of the colonial period was descriptive of America's unique and hitherto unknown lands and peoples.

Early American histories tended to be local in focus but broad in conception, adopting as the definition of history the ancient Greek version of the word, *historia*, which meant literally "inquiries." The works of historians such as John Smith, William Stith, Robert Beverley, William Strachey, William Douglass, Increase Mather, Cotton Mather, Jeremy Belknap, William Gordon, James Sullivan, Hannah Adams, and Mercy Otis Warren were general inquiries, broadly conceived, of peoples, places, and natural phenomena past and present.

Early American Historiography

One of the first great practitioners of the natural and human historical narrative was John Smith, who explored the James River, Chesapeake Bay, and New England coast in 1607, 1608, and 1614, respectively. Smith referred to his works of geography, ethnography, and cartography as *history*, by which he meant an inquiry into the immediate past of the people and places he observed. His works were not chronological accounts of past events, but contemporary accounts of what he observed in the *terra incognita* of America.

Similarly William Strachey's *Historie of Travell in Virginia Britannica,* written about 1613, was a contemporary human and natural history. In 1705, Robert Beverley likewise combined the observations of a natural historian with a narrative of human affairs in his *History and Present State of Virginia.*

Eighteenth-century Enlightenment thinkers continued to see the natural and human history of America as irrevocably linked. Examples of this approach are numerous. Thomas Jefferson's *Notes on the State of Virginia* (1784) sublimates human affairs to a broad portrait of the natural history of America. James Sullivan, in his *History of the District of Maine* (1795), likewise could not but see the history of Maine settlements within the context of the massive Northern forest.

Mastery of this approach to history culminated in 1792 with the publication of Jeremy Belknap's three-volume *History of New-Hampshire.* The first two volumes narrate New Hampshire political history, while the third volume provides a natural portrait of the colony and state: geography, landscape, rivers and mountains, agricultural product, flora and fauna, trade and shipping, demographics, social and political institutions, and the character of the people. Belknap recorded significant and remarkable geographic discoveries and natural events such as human confrontations with animals, meteorological occurrences, floods, earthquakes, storms, and the Dark Day of May 1780.

Belknap founded the Massachusetts Historical Society in 1791 as an organization to express and promote his definition of science, which encompassed all aspects of human experience, especially human and natural history. Belknap, who became the corresponding secretary of the new organization, wrote a circular letter to "every Gentleman of Science in the Continent and Islands of America" to enlist their aid in acquiring information on "the *natural, political,* and *ecclesiastical* history of this country." He suggested that corresponding members of the Massachusetts Historical Society focus their attention on town and church histories; political events, including wars; "biographical anecdotes of persons in your town," especially those "who have been remarkable for ingenuity, enterprise, literature, or any other valuable accomplishment"; a "topographical description of your town or county, and its vicinity; mountains, rivers, ponds, animals, vegetable productions"; agriculture; "monuments and relicks of the ancient Indians"; "singular instances of longevity and fecundity from the first settlement to the present time"; "observations on the weather, diseases, and the influence of the climate, or of particular situations, employments and ailments, especially the effect of spirituous liquors on the human constitution"; manufacturing; fisheries; education; and "remarkable events."

The accounts of nineteenth-century travelers adopted a similarly broad approach to the relationship between history and science. Scientists such as Thomas Nuttall and John Bradbury combined natural history and human history with a narrative of their own adventures descending rivers and crossing the Great Plains. The *Journals of Lewis and Clark* and Francis Parkman's *The Oregon Trail* (1849) are the best expressions of this genre.

"New History"

After the Civil War in America, the development of the professional disciplines in the social, physical, and life sciences threatened the long and fecund marriage of history and science. A revised, positivist definition of science based on the assumption that it has little to do with human subjectivity and individual personalities resulted in the judgment that the narrative histories of the past were impressionistic and "qualitative," rather than scientific and "quantitative." Hence,

at the same time that the historical profession was emerging in the late 1800s and early 1900s, historians such as Charles Harvey Robinson advocated a scientific approach to historical inquiry that he termed "new history."

Henry Adams's nine-volume *History of the United States During the Administrations of Adams and Jefferson* (1889–1891) provides an example of the approach. In this great work, Adams tried to construct a total history that examined political and institutional as well as economic and demographic change. To determine the wealth of the state of Massachusetts, for example, Adams examined shipping tonnage, banking, import and export duties, taxation, and even postal receipts. He calculated that the amount taxed increased 70 percent from 1800 to 1817. From similar statistics, he concluded that wealth in America doubled every twenty years, while population doubled every twenty-three years.

In the 1960s and 1970s, American historians embraced the assumptions and methods of the social sciences to create an interdisciplinary approach to history that relied heavily on quantitative analysis of surviving local records such as tax, probate, and census records. This "new history" focused on the masses rather than the elite, the community as a whole rather than just its leaders. Fields of research included the new social history, the new urban history, psychohistory, demographic history, family history, the new economic history, the study of *mentalitie,* and the new quantitative history, focused on research techniques such as sampling, statistical analysis, theoretical models based on available data, and computer programs such as the Statistical Package for the Social Sciences.

Academic journals emerged in the 1960s and the 1970s to accommodate the new history. Among them were *Historical Methods, Social Science History, Journal of Interdisciplinary History, Journal of Social History, Journal of Urban History,* and *Computers and the Humanities.* Leaders of this movement included Darrett Rutman, Philip Greven, Jackson Turner Main, and Robert Fogel.

Rutman was one of the first identifiable proponents of the "new social history." In studies such as *Winthrop's Boston: A Portrait of a Puritan Town 1630–1649* (1965), he called into question the qualitative methods of "historians of early New England, and particularly the intellectual historians who have dominated the field in the last generation"—those who "limit themselves to the study of the writings of the articulate few, on the assumption that the public professions of the ministers and magistrates constitute a true mirror of the New England mind." Historians, Rutman argued, gain a truer reflection of a past society by examining concrete sources, such as wills, deeds, tax records, and other local documents.

Greven borrowed the methodology of European demographers to study family life in colonial America. "The principal value of quantification of demographic data," he wrote, "is that it ought to enable historians to make precise distinctions between the life experiences of individuals and groups in different places and times." Others, such as Michael Zuckerman, John Demos, and Kenneth Lockridge, turned to anthropology, sociology, computer science, and the theories of Robert Redfield, Erik Erikson, Eric Wolf, Clifford Geertz, Claude Levi-Strauss, Max Weber, and Emile Durkheim to guide their studies of colonial history.

Main's *The Social Structure of Revolutionary America* (1965) is an examination of local tax and probate records in the thirteen colonies as evidence of the social and economic structure in America. Fogel's work, often called "cliometrics," led to significant reinterpretations of colonial slavery, as in his book, co-written with Stanley Engerman, *Time on the Cross* (1974).

The new history is social scientific. Jerome Clubb, for example, argues that systematic analysis of the past complements the analysis of contemporary societies by sociologists, anthropologists, and economists. "The task of a genuine social scientific historian would be to use the past to construct empirical social theory to describe and explain specific events of the past." This would result in "an improved social science," he maintains, that "would increase the utility of historical evidence for the pursuit of scientific knowledge of human affairs, and the study of the past can contribute to that improvement."

Russell Lawson

Sources

Clubb, Jerome M., and Erwin Scheuch, eds. *Historical Social Research: The Use of Historical and Process-Produced Data.* Stuttgart, Germany: Klett-Cotta, 1980.

Kammen, Michael, ed. *The Past Before Us: Contemporary Historical Writing in the United States.* Ithaca, NY: Cornell University Press, 1980.

Landes, David, and Charles Tilly. *History as a Social Science.* Englewood Cliffs, NJ: Prentice Hall, 1971.

Lawson, Russell M. *The American Plutarch: Jeremy Belknap and the Historian's Dialogue with the Past.* Westport, CT: Praeger, 1998.

Riley, Stephen T. *The Massachusetts Historical Society 1791–1959.* Boston: Massachusetts Historical Society, 1959.

The Philosophy of Science

The philosophy of science has undergone telling changes in America. Scientists of the colonial period accepted the Aristotelian approach of systematic logic applied to the induction and deduction of truth. The challenges to Aristotle during the sixteenth and seventeenth centuries in Europe, however, led to an increasing emphasis on the empirical method.

American scientists of the eighteenth century were more apt to quote Francis Bacon, the English empiricist, than Aristotle. In *Novum Organum* (1620), Bacon argued for a "new method": knowledge acquired systematically by means of experimentation. The Baconian method implied that meaning is achieved by verifiable theories based on the precise observation of natural phenomena.

Positivism

During the nineteenth and into the twentieth centuries, Western science was thought to be the means by which humans could achieve objective knowledge. The philosophy of positivism argued that science can overcome the limitations of human perspective by providing stable interpretive schemes and models built on empirical evidence. Facts—accumulated evidence gathered under controlled conditions shown to be valid many times over—form the basis for a breadth of perspective that can potentially apply to all times and all places rather than to isolated instances. Likewise, science does not have to rely on the experiences of one person and his or her personal interpretation. Rather, theory transcends individual subjectivity to acquire the status of objective reality.

Positivist scientists believe in the validity of reason and confirmable evidence; if they use imagination, it is within the confines of a verifiable methodology. Such scientists assume that they can accumulate sufficient, verifiable data pertaining to human and natural experience to explain the various aspects of this experience according to standards of logic, reason, and science.

Scientists therefore try to separate the subject from the object. The only way to know the characteristics of the object under study is for the scientist to analyze it as an observer rather than as a participant. Positivists see science as cumulative, in that theory is rarely overturned; instead, it is explicated or reduced to a more fundamental explanation of real phenomena.

Even social science, as David Landes's and Charles Tilly's *History as Social Science* (1971), "is problem-oriented. It assumes that there are uniformities of human behavior that transcend time and place and can be studied as such. . . . The aim is to produce general statements of sufficiently specific content to permit analogy and prediction." The social scientist "states his hypothesis in the form of an explanatory model, preferably in mathematical language and so framed that the criteria of proof are measurable."

Marvin Harris's, in *Cultural Materialism* (1979), adds that science seeks "to restrict fields of inquiry to events, entities, and relationships that are knowable by means of explicit, logico-empirical, inductive-deductive, quantifiable public procedures or 'operations' subject to replication by independent observers."

Subjectivity

In the wake of World War II and its examples of human irrationality and destructiveness, theorists began to rethink the philosophical bases of science. Irrationality, anomalies, imagination, and emotions now seemed even more a part of human life, including even the realm of scientific inquiry.

Before the war, historian of science George Sarton had argued that explanations often have an aesthetic quality for theoretical scientists, who have a subjective desire to find out what is true, "to understand more deeply and more fully the whole of nature, including ourselves and our relations to it. An intense curiosity to find the truth about things in general and himself in particular is as much a characteristic of man as his thirst for beauty and justice."

Albert Einstein in 1931 declared "the cosmic religious experience is the strongest and noblest driving force behind scientific research." He believed that "the only deeply religious people of our largely materialistic age are the earnest men of research."

One of the first postwar theoreticians to propose the subjectivity of science was Stephen Toulmin, who argued that the scientist's own *weltanschauung,* or conceptual worldview, forms the fundamental assumptions of scientific method and theory. Scientific laws, in Frederick Suppe's words, "are methods of representing regularities already recognized, being methods of representing phenomenal deviation from ideals of natural order." The truth or falseness of a law depends on how well it explains the phenomena. A scientific theory comprises "laws, hypotheses, and ideals of natural order," the truth or falseness of which depends on how successful the theory is at "representing phenomena." Toulmin believed that science is cumulative, because divergent theories, old and new, can be understood according to the explanatory order of the *weltanschauung.*

Thomas Kuhn, however, author of *The Structure of Scientific Revolutions* (1962), argued that science is not cumulative because of the revolutionary aspect of the paradigm. Kuhn believed that the paradigm (*weltanschauung* or worldview) is a conceptual framework of fundamental assumptions that directs the methods, theories, and organization of the scientific community. The paradigm becomes the means of explanation, a scientific dogma, for a generation of scientists who refuse to consider alternatives, as long as the paradigm satisfactorily explains phenomena. Trust and belief in the paradigm hinders the objective pursuit of truth and disinterested science, as defined by the Baconian and positivist theories of science.

In the wake of the challenge to positivism by theorists of science such as Toulmin and Kuhn, science has adjusted its lofty goals of objective truth to more realistic aims. Probability, rather than certainty, is the emphasis today.

Science, in Marvin Harris's words, "claims to be able to distinguish between different degrees of uncertainty. In judging scientific theories one does not inquire which theory leads to accurate predictions in all instances, but rather which theories lead to accurate predictions in more instances. Failure to achieve complete predictability does not invalidate a scientific theory; it merely constitutes an invitation to do better."

Russell Lawson

Sources

Harris, Marvin. *Cultural Materialism: The Struggle for a Science of Culture.* New York: Random House, 1979.

Kuhn, Thomas S. *The Structure of Scientific Revolutions.* 1962. Chicago: University of Chicago Press, 1996.

Landes, David S., and Charles Tilly. *History as Social Science.* Englewood Cliffs, NJ: Prentice Hall, 1971.

Rosenberg, Alex. *The Philosophy of Science: A Contemporary Introduction.* New York: Routledge, 2000.

Sarton, George. *The History of Science and the New Humanism.* Bloomington: Indiana University Press, 1937.

Suppe, Frederick, ed. *The Structure of Scientific Theories.* 2nd ed. Urbana: University of Illinois Press, 1977.

The Emergence of an American History of Science

The history of the philosophy and methodology of science has intrigued students of nature and humanity for centuries. In America, the study of the history of science was part of the natural history emphasis of the early explorers. John Smith and other colonial naturalists and historians wrote about what Native Americans had discovered about the use of the land, what they knew about geography, and how they understood the universe.

Eighteenth-century Enlightenment historians explained American history according to the impact of the natural environment on human affairs. Robert Beverley, for example, in *History and Present State of Virginia* (1705), examined Virginia political and social history in the context of natural history. Thomas Jefferson likewise, in *Notes on Virginia* (1782), could not but see the history of America as one of humans interacting with nature. Jefferson believed that nature was the product of a benevolent creator who provided a universe of order, harmony, and perfection based on natural laws that, upon discovery, would allow humans to create a society similarly ordered and harmonious.

The ultimate expression of this point of view was Jeremy Belknap's masterful *History of New-Hampshire* (1784, 1791, 1792); in three volumes, this work encompassed the political, social, institutional, cultural, economic, and natural history of the colony and state of New Hampshire. Belknap, like other eighteenth-century thinkers, believed that human progress in science depended on working with the environment, accommodating nature rather than forcing or conquering it.

The American Enlightenment thinker, because of the uniqueness of the colonial experience and the constant reminders of the Puritan "errand into the wilderness," believed that humanity and nature (hence the creator) worked in concert to bring about the best society. Even the most skeptical eighteenth-century thinkers, such as Thomas Paine (in *The Age of Reason,* 1794), could appreciate the divine reason that formed the universe. How then could humans presume to control what was wrought by God, the creator?

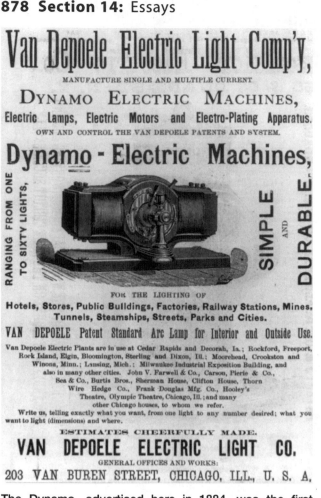

The Dynamo, advertised here in 1884, was the first electric generator capable of producing power for industry. To historian Henry Adams, it symbolized the dynamic force of modern technology and its effect on civilization. *(Library of Congress, LC-USZ62–67964)*

Henry Adams

One of America's historians of science was Henry Adams, who created in his autobiographical *Education of Henry Adams* (1918) a view of the historical process of the victory of the sciences in the war waged to gain dominance over nature. Adams's metaphors to describe this centuries-long conflict were the Virgin, representing traditional society, religion, and human dependence on nature, and the Dynamo, or electric motor, which represents science, technology, secularism, and materialism. "The Virgin," he wrote, "had acted as the greatest force the Western world ever felt, and had drawn man's activities to herself more strongly than any other power, natural or supernatural, had ever done." But the Dynamo was supreme: "very slowly the accretion of these new forces, chemical and mechanical, grew in volume until they acquired sufficient mass to take the place of the old religious science."

The victory of the Dynamo cast humanity into a different world, with new structural components forged by science—industry, transportation, communications, and more powerful weapons—that resulted in new ideological foundations. People in the early twentieth century, when Adams was composing the *Education,* were forming their ideas in response to an understanding of the scientific laws of the universe, rather than forming the scientific laws of the universe out of their own ideas. The latter approach, devoted to an understanding of the supernatural and subjective (the Virgin), had resulted in a sense of security, unity, and truth, whereas the victory of science and the Dynamo led people to see the apparent chaos and uncertainty inherent in the universe.

George Sarton

The formal, academic study of the history of science in America began with the creation of the History of Science Society in 1924. George Sarton, a native of Belgium, who had founded the journal *Isis* in 1912, established the society in part to support the academic journal dedicated to the study of the history of science. Sarton served as editor of *Isis* from 1913 to 1952.

Sarton conceived of the modern study of the history of science as a humanistic discipline. "Science," he wrote in *The History of Science and the New Humanism* (1937), "is nothing but the human mirror of nature." While noting, "we can see nature only through man's brain," he added, "we are always studying nature, for we cannot see man without it." The History of Science Society continues to be a dominant organization worldwide, orienting and focusing the attention of scholars on the interdisciplinary study of the history of science.

Russell Lawson

Sources

Adams, Henry. *The Education of Henry Adams.* 1918. Boston: Houghton Mifflin, 1961.

Becker, Carl L. *The Heavenly City of the Eighteenth-Century Philosophers.* New Haven, CT: Yale University Press, 2003.

Sarton, George. *The History of Science and the New Humanism.* Bloomington: Indiana University Press, 1937.

The Sociology of Science

Historians and theorists of science in the years after World War II began to recognize that science involves more than methodology and philosophy—that it relies as much on the collective pursuit of knowledge. Philosophers such as Frederick Suppe, Steven Toulmin, Thomas Kuhn, Karl Popper, and Max Feyerabend developed an approach to understanding science that relied on examining subjective elements, including human interaction. How scientists relate to one another and organize themselves has a profound impact on the acquisition of knowledge, the agreed-upon theories of physical, biological, and social scientists.

Sociologists of science study the modes of organization and means of communication among scientists. Early American scientists, typically amateurs, lawyers, clergy, and merchants who practiced science as an avocation, organized themselves by means of various institutions. Early colleges such as Harvard and Yale, like elite colleges and universities of today, had restrictive and competitive admission, a difficult curriculum, and a developing camaraderie among graduates. There were exceptions, of course, such as Benjamin Franklin, but by far most scientists during the eighteenth-century Enlightenment were college-educated and exclusionary. The leading intellectuals in any given community were generally the lawyers, merchants, clergy, and physicians who had graduated from European universities or American colleges. Scientific correspondence involved letters exchanged among the well-educated. Jeremy Belknap called such intellectuals the "sons of science," or the inheritors of the scientific revolution that had begun in Europe during the previous two centuries.

Belknap, who resided in a small town in northern New Hampshire, where he often felt cut off from the scientific communities of Boston and Philadelphia, was nevertheless a leader in the development of organizations to bring scientists together to pursue their inquiries collectively. During the American Revolution, Belknap wrote to his friend, Ebenezer Hazard: "Why may not a *Republic of Letters* be realized in America as well as a Republican Government?

Why may there not be a Congress of Philosophers as well as of Statesmen? And why may there not be subordinate philosophical bodies connected with a principal one, as well as separate legislatures, acting in concert by a common assembly? I am so far an enthusiast in the cause of America as to wish she may shine Mistress of the Sciences, as well as the Asylum of Liberty."

Belknap wrote his letter in 1780, the same year that the American Academy of Arts and Sciences was organized in Boston; already the American Philosophical Society in Philadelphia had been in operation on and off for more than thirty years. In 1791, Belknap himself organized the first historical society in America, the Massachusetts Historical Society. His plan was that "each Member on his admission shall engage to use his utmost endeavors to collect and communicate to the Society, Manuscripts, printed books and pamphlets, historical facts, biographical anecdotes, observations in natural history, specimens of natural and artificial Curiosities, and any other matters which may elucidate the natural, and political history of America from the earliest times to the present day." Membership in the Massachusetts Historical Society was by invitation only. Its journal, the *Collections*, sent to members, included correspondence and reports that were for a limited audience of American literati. The conduct of American science by 1800, in short, lacked only one crucial element of twenty-first-century science: professionalization.

In the mid-nineteenth century, American science became the endeavor of professionals who pursued it for a living and taught in universities. The model for scientific research in the American academy was the German university system. The decentralization of higher education in Germany fostered competition for the top scholars of academic disciplines that were becoming rapidly more specialized. Although the German university system emphasized the search for pure science, scholarship, and specialization of disciplines, its offspring—the American university system—emerged at a time when applied science was in great national demand. The Morrill Act of 1862 set aside land and funds for the

It was not until the twentieth century that the ranks of professional scientists expanded to include significant numbers of African Americans. Here, pharmaceutical students do lab work at all-black Howard University about 1900. *(Library of Congress, LC-USZ62–35750)*

creation of land-grant universities devoted to research, instruction, and extension of practical knowledge. The Johns Hopkins University in Baltimore was the first to take the lead in the professionalization of science by creating graduate schools where students worked directly with professors who trained them in the ways and means of science.

Thus, American universities at the end of the nineteenth century were encouraging, in the words of Joseph Ben-David, "the establishment of specialized research roles and facilities" for professors and "large-scale systematic training" of students. The emergence of universities as the centers of pure and applied scientific knowledge led them to assume "an important function in the growing professionalization of occupational life, and in making research an increasing permanent aspect of business, industry, and administration."

After World War II, America philosophers of science such as Thomas Kuhn began to think systematically about the sociology of science. In his groundbreaking book *The Structure of Scien-*

tific Revolutions (1962), Kuhn argued that the unifying force in the organization of science is the "paradigm"—the commonly accepted beliefs, theories, and rules that unite scientists and direct scientific research within a field. Once a paradigm is established, Kuhn argues, scientists can engage in "normal science"—specialized research and analysis, using accepted methodologies, to seek solutions to unsolved problems.

Another important component of paradigmatic science is its proliferation through the education of students in the methods, assumptions, and common goals of the scientific establishment. The rite of passage is the Ph.D., a symbol that the student has mastered the techniques of normal science and the assumptions of the paradigm. The new holder of a Ph.D. is ready to become an equal member in the professional enterprise of science. Research is presented at conferences and submitted to esoteric, peer-reviewed journals. Conferences and journals are the means by which communication among peers is accomplished. The academic world, moreover, observes a set, formal ranking that tends to be preserved by an "old guard." As long as a young scientist conforms to the paradigm and performs normal science in the proper way, according to professional standards, he or she is likely to rise in the ranks—from assistant professor to associate professor to professor.

Kuhn's concept of "normal science" embraces the accepted methods and common assumptions of a scientific discipline. Modern science, for example, is organized around the belief that science involves the systematic accumulation and verification of knowledge: *systematic* in that it is organized, unified, and directed; *accumulated* by experience using the empirical method; and *verified* according to the standards of the scientific method, the proving or disproving of hypotheses by controlled experiment, as well as the judgment of one's scientific peers.

The scientist tries to use the most objective methods at his or her disposal, yet subjectivity can never be avoided, and truth can be chimerical. Still, scientists search for the most probable reflection of reality. Yet a probable truth agreed on by similar-thinking puzzle solvers is a far cry from the positivist ideal of the seeker of truth acquiring objective knowledge of human and natural phenomena.

Russell Lawson

Sources

Harris, Marvin. *Cultural Materialism: The Struggle for a Science of Culture.* New York: Random House, 1979.

Kuhn, Thomas S. *The Structure of Scientific Revolutions.* 1962. Chicago: University of Chicago Press, 1996.

Suppe, Frederick, ed. *The Structure of Scientific Theories.* 2nd ed. Urbana: University of Illinois Press, 1977.

Tucker, Louis Leonard. *Clio's Consort: Jeremy Belknap and the Founding of the Massachusetts Historical Society.* Boston: The Massachusetts Historical Society, distributed by Northeastern University Press, 1989.

ADAMS, HENRY (1838–1918)

Henry Brooks Adams, the grandson of John Quincy Adams and great-grandson of John Adams, was a Harvard history professor, philosopher, writer, and critic. He is best known for his autobiography, *The Education of Henry Adams* (1918); histories such as the nine-volume *History of the United States During the Administrations of Adams and Jefferson* (1889–1891); his novel *Democracy* (1880); and collections of essays such as *Mont-Saint-Michel and Chartres* (1913).

A prominent theme in Adams's life and work was the overwhelming technological change wrought by science that occurred over the course of his life. He declared that he was born into a pastoral and faith-filled world like that of the seventeenth and eighteenth centuries and was ending his life in a secular world that was dominated by machines. "I firmly believe," he declared in an 1862 letter to his brother Charles, "that before many centuries more, science will be the master of man. The engines he will have invented will be beyond his strength to control. Some day science may have the existence of mankind in its power, and the human race commit suicide by blowing up the world."

At the time, Henry was serving as secretary to his father, Charles Francis Adams, the U.S. ambassador to England. It was the period of the Civil War, when the weapons of war could indeed seem daunting. But even after the war, and until the end of his life, Henry Adams repeated the same theme—destruction awaits a world that creates mechanistic forces it cannot control.

The Virgin and the Dynamo

Adams felt pulled in two different directions: the past of a pastoral world confident in God's control and human knowledge, and the future of a mechanistic, secular world where God is unknown and science reveals multiplicity rather than simplicity. After a stint teaching at Harvard, the suicide of his wife in the 1880s, and financial ruin in the 1890s, Adams sank into depression, unsure of the importance of science and history.

It was at this time that he came to recognize two themes of human history, which he termed the Dynamo and the Virgin. The Dynamo is human technology, the product of reason and science. It represents the human aim to control nature. The Virgin is the pastoral and the divine, representing religion and a reliance on nature.

These two forces, Adams believed, have been in constant conflict throughout human history, particularly in the several centuries since the beginning of the Scientific Revolution in the 1500s. Humans, once childlike in their dependence on nature, their humility before the forces of nature and the supernatural that they could not comprehend, were now, because of scientific and industrial revolutions, coming to think that nature is subject to human will and knowledge. But Adams concluded that such knowledge, such control, is illusory and elusive. Humans believe they control, but they do not.

The Science of History

Adams likewise was equivocal about history as a science. In some respects, he believed that through statistical analysis of economics, social change, and politics, a precise historical narrative—like his own *History of the United States During the Administrations of Adams and Jefferson*—could approach science. As he wrote in *The Degradation of the Democratic Dogma* (1919): "Any science assumes a necessary sequence of cause and effect, a force resulting in motion which cannot be other than what it is. Any science of history must be absolute, like other sciences, and must fix with mathematical certainty the path which human society has got to follow." Adams discovered, however, that such absolutism in the understanding of the human past is elusive.

As he wrote at the conclusion of *The Education of Henry Adams:* "The child born in 1900 would . . . be born into a new world which would not be a unity but a multiple, where no one had ever penetrated before; where order was an accidental relation obnoxious to nature; artificial compulsion imposed on motion; against which every free energy of the universe revolted." Science seemed to have degenerated into chaos, knowledge into ignorance. "All that a historian won was a vehement wish to escape. He saw his education complete and was sorry he ever began it. As a matter of taste, he greatly preferred his eighteenth-century education when God was a father and nature a mother, and all was for the best in a scientific universe." But the ongoing benevolent design of the Enlightenment worldview was not to be in the industrial world of the nineteenth and twentieth centuries.

Caught between two worlds, yet still attempting to impose a scientific formula to explain the dialectic of nature and humanity, Adams enlisted mathematical theory and empirical methodology to measure humanity's increasing control over nature. To achieve a science of history comparable to the natural sciences, he needed a way to measure "the law of reaction between force and force—between mind and nature—the law of progress."

According to Adams, the acceleration of human progress is an "invariable law" in human history. The manipulation of nature by humans is a standard proportional increase that is uniform and predictable. The force of nature, he believed, is the independent variable and the force of humanity the dependent variable: the human force, reacting with nature, "increases in the direct ratio of its squares" over time. Yet human progress is not natural, rather forced and artificial, and Adams feared for the future of accelerating control of human technology over the natural environment.

Russell Lawson

Sources

Adams, Henry. *The Degradation of the Democratic Dogma.* New York: Macmillan, 1919.
———. *The Education of Henry Adams.* 1918. Boston: Houghton Mifflin, 1961.

AFRICAN AMERICAN SCIENTISTS

Many early African Americans were pioneers in the fields of science and technology, developing methods for working with tools and crops with limited resources, instituting medical practices based on herbal remedies, and even performing minor surgical procedures under inhospitable conditions. Many of the tens of millions of Africans kidnapped and forced into labor in the Americas had previous knowledge of herbal medicine, crop cultivation, animal husbandry, and textile fabrication. Low-country rice planters in South Carolina preferred captured workers from the Congo-Angola region, because these Africans were experienced rice growers and would be able to put their knowledge to work directly upon arrival. From these beginnings, African Americans would go on to make significant contributions to the scientific advancement of the nation and to its industry, natural sciences, medicine, and high technology.

Inventors

One of the best-known African American inventors of the eighteenth century was Benjamin Banneker. Born in rural Maryland just outside Baltimore, he excelled in mathematics and quickly became known for his intellectual abilities. One of his admirers was George Ellicott, a mathematician who lent him books on astronomy. Banneker mastered the material and, by 1783, devoted himself to his astronomical studies. Among Banneker's achievements were assisting the planning of Washington, D.C., producing an annual almanac from 1792 to 1802, and carrying on a prolonged correspondence with Thomas Jefferson concerning the equality of races, a correspondence that would eventually sway Jefferson toward recognizing the intellectual abilities of all peoples.

In the nineteenth century, with the onset of the Industrial Revolution, African American inventors contributed mightily to technological innovation. Creative and brilliant men such as Elijah McCoy (for whom the phrase "the real McCoy" was coined), Jan Matzeliger, Granville

T. Woods, Lewis Latimer, Garrett Morgan, and Norbert Rillieux created timesaving and safety-promoting devices that changed the face of industry. Among the hundreds of inventions developed and patented by African Americans during the 1800s were the locomotive lubricating cup (McCoy), which made it possible for trains to be continually lubricated without stopping; the shoe lasting machine (Matzeliger), which made mass production of shoes possible; the railway induction telegraph (Woods), which made communication between moving trains and the train yard possible; the cotton-thread light-bulb filament (Latimer), which made electric lighting affordable for domestic use; the automatic traffic signal (Morgan), which prevented countless traffic-related accidents and fatalities; and the vacuum evaporator (Rillieux), which revolutionized the processing of sugar.

Natural Sciences

Many African Americans have had distinguished careers in the natural sciences. George Washington Carver, Charles Henry Turner, and Mathew Henson stand out for excellence in their fields.

Carver, forced to grow up without his parents, both of whom were enslaved in Missouri, was one of America's most extraordinary scientists. He showed a keen interest in nature early in life and experimented with plants even as a boy. After being rejected by Highland University in Kansas on the basis of his race, Carver was admitted to Simpson College in Indianola, Iowa, in 1887. By 1894, he had not only earned his undergraduate degree at Iowa State College but was offered a faculty position as well. In what would become one of the most important moments in the history of American agricultural science, Booker T. Washington, the founder of Tuskegee Institute in Alabama, implored Carver to join the school as director of agriculture. Beginning with only a crude laboratory and twenty acres of land, Carver would transform Tuskegee into a world-famous agricultural sciences institution. He became known for his work on improving soil conditions, instituting effective crop rotations, and finding multiple uses for peanuts, sweet potatoes, and pecans. He eventually produced more than 300 products from the peanut alone, including facial cream, ink, cooking oil, soap, and cheese. In addition, Carver developed over 115 products from the sweet potato and over 75 products from the pecan.

Between the late 1800s and early 1900s, Charles Henry Turner's research into animal behavior, especially that of insects, made him one of the premier scientists in his field. He published more than fifty scientific papers, taught classes at Sumner High School in St. Louis, and earned the respect of scientists throughout America and Europe. Although he was offered a job as professor at the prestigious but predominately white University of Chicago, he turned it down, saying he could do more good for African Americans by staying among them and teaching scientific skills to young people in his community.

During the late nineteenth century, Matthew Henson was driven to explore unknown parts of the world. Although he had little formal schooling, he was trained onboard a cargo ship during his teens and then as an associate to Robert Peary on their many survey expeditions. As a team, Peary and Henson traveled to the far reaches of the world for some twenty-three years. Their most noteworthy trip was to the North Pole in 1909, when both men became the first explorers ever to reach this historic point. He earned his place in scientific history as a kind of polar astronaut whose knowledge and skill in navigating the frozen tundra of the Arctic helped open a new world.

Medicine

A number of African American doctors have pioneered medical procedures, perfected new techniques, and excelled in the practice of medicine. Among them are former U.S. Surgeon General Joycelyn Elders; Percy Lavon Julian, who invented a synthesis of cortisone from soybeans; and Carlton B. Goodlet, a family doctor who also practiced law and edited the San Francisco *Sun-Reporter* newspaper.

In the early twentieth century, one medical scientist stands out for his life-saving accomplishments in the field of blood plasma. Charles Drew pioneered the process of blood storage that was instrumental in saving lives during

World War II and thereafter. Upon graduating from Amherst College, he decided to pursue his career in medical science and enrolled at McGill University in Montreal, Canada. By 1933, he had earned degrees in medicine and surgery; he was appointed a resident at Montreal General Hospital, where he became interested in hematology (blood science) research. In 1938, Drew worked with John Scudder, a leading scientist in the field, and devised a way to make a blood bank—a place where blood could be stored for long periods of time without breaking down. Drew's first experiments succeeded in storing blood for seven days, and he continued to work toward developing a method that would last even longer. Early in World War II, Drew was appointed medical director of the Blood Transfusion Association of the Red Cross; however, due to his refusal to segregate the blood of blacks and whites, a standard practice at the time, he was forced to step down and returned to Howard University in 1941. Drew spent the last years of his life training promising black doctors and speaking out against racial prejudice, especially in medicine.

Daniel Hale Williams was another African American doctor who advanced medical practice. Even as a youngster, he had a steady hand and a vibrant intellectual curiosity. In 1877, Dr. Henry Palmer recognized this while Williams was cutting his hair and agreed to have Williams apprentice in his office. In 1880, after only two years of apprenticeship, Williams entered Chicago Medical College. After graduation, he taught anatomy at his alma matter and opened his own office on Chicago's South Side in 1883. Williams also opposed the racism of his time; in 1891, he opened the Provident Hospital and Training School for Nurses in Chicago, where African Americans were able to receive high-quality medical training and care. Surrounded by top-tier medical professionals and armed with an in-depth knowledge of anatomy, Williams quickly became one of the leading surgeons in Illinois.

Williams is perhaps best known for an 1893 operation he performed on James Cornish, a stabbing victim. Without the aid of modern medical devices such as heart-rate or respiratory monitoring machines, he opened Cornish's chest, repaired his pericardium (the sac that en-velops the heart), and sutured him up for recovery. Cornish went on to live an active life for twenty more years. Williams was the first person to perform such a successful open-heart surgery, an operation that seemed beyond the reach of even the most confident and skilled surgeons of that time.

High Technology

In the latter part of the twentieth century, science and technology became primary fields of inquiry—and of educational investment—in the Western world. Nuclear energy, space exploration, and computer technology led the way, producing many well-known and respected African American scientists. Among them were Lloyd Quarterman, Mae Jemison, and William Northover.

In the 1940s, Quarterman was one of six African American scientists who worked on the Manhattan Project to develop the atomic bomb. After college at St. Augustine's in North Carolina, Quarterman was hired by the U.S. War Department to work on atomic research at Columbia University in New York and the University of Chicago's Argonne National Lab. Combining forces with Enrico Fermi, he worked with radioactive materials to develop nuclear reactors. In addition to his work in radioactive and fluoride chemistry, Quarterman was a spectroscopist (a scientist who studies the interaction of radiation and matter). He is credited with inventing a new type of cell, the diamond cell, which enabled scientists to see the vibrations of different molecules in a solution. In this way, he was able to manipulate molecules to form new compounds.

Mae Jemison was raised in Chicago and developed an early love of astronomy. She earned a chemical engineering degree from Stanford and a medical degree from Cornell before joining NASA in 1987. Jemison was the science mission specialist on the STS-47 *Spacelab J* flight, on which she conducted experiments in life sciences and material sciences, including bone cell research. After serving for six years as a NASA astronaut, she established the Jemison Group to focus on the beneficial integration of science and technology into daily life. In addition, after teaching environmental studies at Dartmouth College from 1995 to 2002, she began directing the Jemison Institute for Advancing Technology in Developing Countries.

The chemist William R. Northover conducted pioneering work in fiber optics at Bell Labs from the 1960s to the 1980s. His work gave birth to some of the most important developments in research on the science of glass fiber light guides, which can transmit digitally coded information. The research and application of glass properties technology that Northover and his colleagues worked on have had far-reaching benefits in the telecommunications industry. Northover has also worked directly on projects involving semiconductors and lasers.

African American scientists and inventors have made significant contributions and important breakthroughs in fields as diverse as agronomy, nuclear science, medicine, and information technology. Even though many of the first African Americans arrived in bondage, they came with a collective pool of skills and intellect that enabled them to forge a new life and create mechanisms that would make their situations more tolerable. Many early pioneers would improve methods of industrial work and safety, making work less grueling and dangerous over time. Industrial improvements were followed by innovations in science and medicine, improving the quality of life not just for themselves, but for all Americans.

Despite racial prejudice, African American scientists and inventors have stood firmly for equality of intellect and opportunity. They have shown the nation and the world that it is the quality of people, not their appearance, that makes greatness.

Paul T. Miller

Sources

Black Inventor Online Museum. http://www.blackinventor. com.

Diggs, Irene. *Black Inventors.* Chicago: Institute of Positive Education, 1975.

Fouche, Rayvon, and Shelby Davidson. *Black Inventors in the Age of Segregation: Granville T. Woods, Lewis H. Latimer, and Shelby J. Davidson.* Baltimore: Johns Hopkins University Press, 2003.

Hayden, Robert C. *Seven Black American Scientists.* Reading, MA: Addison-Wesley, 1970.

Van Sertima, Ivan, ed. *Blacks in Science: Ancient and Modern.* Somerset, NJ: Transaction, 1987.

AMERICAN ACADEMY OF ARTS AND SCIENCES

The American Academy of Arts and Sciences (AAAS), founded in 1780, is the second oldest science society in America (after the American Philosophical Society) and the first successful science society founded in New England. Initially housed on the campus of Harvard College, the academy still maintains its headquarters in Cambridge, Massachusetts.

The AAAS was begun during the American Revolution by leaders such as John Hancock and John Adams, who believed that a successful republic required the gathering of scientific information and the promotion of scientific knowledge. The AAAS encouraged members to contribute papers and other communications on natural and human history, which would be communicated to other thinkers and scientists throughout the country by means of its periodical, the *Memoirs.* During the nineteenth and twentieth centuries the means of communication and membership altered, but not the focus on all aspects of thought and culture in America, particularly the sciences.

Fellows of the AAAS have included scientists, thinkers, policymakers, and writers such as Daniel Webster, Asa Gray, Louis Agassiz, Ralph Waldo Emerson, Percival Lowell, Albert Einstein, and Talcott Parsons. The AAAS continues to be an exclusive organization, inviting only recognized experts and prizewinners into its ranks as fellows. Recent inductees include journalists, poets, academics, researchers, artists, architects, actors, corporate executives, politicians, and musicians; natural scientists in the fields of mathematics, computer science, physics, chemistry, astronomy, engineering, molecular biology, biochemistry, psychology, ecology, public health, medicine, and surgery; and social scientists in economics, anthropology, sociology, demography, political science, public policy, law, and history.

For the past half-century, the journal of the AAAS, *Daedalus,* has published scholarly articles dealing with a host of topics in the sciences, humanities, education, and public policy. Recent issues have focused on such subjects as inequality in America, public education, science and religion, aging, imperialism, justice, time, bioethics, diversity, and modernity.

The American Academy of Arts and Sciences supports scholarship through various prizes named in memory of famous American scientists and writers. The Rumford Prize, for example, named for the physicist Benjamin Thompson, Count Rumford, is awarded for contributions to the understanding of heat. The Talcott Parsons Prize, named in honor of the American sociologist, awards work in the social sciences. The Emerson-Thoreau Medal recognizes lifetime achievement in literature.

Russell Lawson

Source

American Academy of Arts and Sciences. http://www.amacad.org.

Daedalus (formerly *Journal of the American Academy of the Arts and Sciences*). 1955–2007.

Oleson, Alexandra, and Sanborn C. Brown, eds. *Pursuit of Knowledge in the Early American Republic: American Scientific and Learned Societies from Colonial Times to the Civil War*. Baltimore: Johns Hopkins University Press, 1976.

AMERICAN ANTIQUARIAN SOCIETY

Founded in 1812 and located in Worcester, Massachusetts, the American Antiquarian Society (AAS) is both a scholarly association and a research library for the study of American culture, history, arts, and sciences through the year 1876. The AAS is dedicated to collecting and preserving printed records, from the first European settlement of North America through the American Civil War and Reconstruction era, and making documents available to scholars, writers, artists, genealogists, and the general public.

The society's library contains more than 3 million items, including books, pamphlets, periodicals, visual records, and works of art. The AAS is also a publisher of bibliographies and books on antiquarianism, and it produces the scholarly journal *Proceedings of the American Antiquarian Society* twice a year.

The AAS was the brainchild of printer and publisher Isaiah Thomas, whose periodical the *Massachusetts Spy* was one of the leading patriot newspapers of the Revolutionary War era. In 1812, the Massachusetts legislature responded to a petition by Thomas and passed an act incorporating the AAS as America's first national historical society. According to the measure, the society was to "encourage the collection and preservation of the Antiquities of our country, and of curious and valuable productions in Art and Nature [that] have a tendency to enlarge the sphere of human knowledge."

The seed of the AAS's collection was Thomas's bequest of some 8,000 books, as well as funds to build the society's first library building in 1820. The collection grew as historians, many of them well-heeled amateurs, donated their own libraries and collections of artifacts. The first catalog of the society's collections, published in 1837, was produced by Christopher Columbus Baldwin, Thomas's successor as AAS president.

Over the course of the nineteenth and twentieth centuries, the society added new buildings and vastly expanded its collection. Under the leadership of Waldo Lincoln in the early twentieth century, the AAS began to systematically collect copies of newspapers and journals from the colonial, early republic, antebellum, and early post–Civil War era in American history. Beginning in the 1950s, this extensive collection of early American publications was put on microfilm and made available to libraries around the country. In 2002, the society began putting the first installments of a digital version of the collection online.

The AAS offers a host of scholarly and educational programs, including undergraduate seminars in American Studies and American history, fellowships at its Center for Historic American Visual Culture and in its Program in the History of the Book in American Culture, and a public lecture series and performances on topics in American history. The organization also provides primary source materials for students and continuing-education programs for history and social science teachers.

Many of the twenty-four members of the AAS's governing council are historians and antiquarians. The council appoints a president and advises on policy and program initiatives. Other society officers are elected by the nearly 800 AAS members from across the United States and around the world.

James Ciment

Sources

American Antiquarian Society. http://www.american antiquarian.org.

Burkett, Nancy H., and John B. Hench, eds. *Under Its Generous Dome: The Collections and Programs of the American Antiquarian Society.* Worcester, MA: American Antiquarian Society, 1992.

AMERICAN ASSOCIATION FOR THE ADVANCEMENT OF SCIENCE

The American Association for the Advancement of Science (AAAS) is a nonprofit organization of scientists and members of the public. Unlike various other scientific societies, such as the National Academy of Sciences, the AAAS does not restrict membership, preferring to let anyone join who has an interest in science. Vast in the scope of its aims, the AAAS facilitates the exchange of information among scientists, engineers, mathematicians, and members of the public; fosters international cooperation among scientists in the pursuit of research that crosses national boundaries; encourages high school students to pursue careers in science, engineering, and mathematics; advocates government funding of science; and shares the knowledge of and appreciation for science with the public.

A group of geologists met in 1840 at the Franklin Institute in Philadelphia, Pennsylvania, to form the American Society of Geologists. Intent on expanding its base to include scientists from all disciplines, eighty-seven members of the society gathered on September 20, 1848, at the Academy of Natural Sciences in Philadelphia to found the AAAS. In 1851, Alexander Dallas Bache, the great-grandson of Benjamin Franklin and AAAS president, called for the association to organize scientists with the aim of persuading the federal government to fund science. In doing so, the AAAS has helped forge the long-standing link between the government and scientific community in America. The federal government began funding agricultural research in the nineteenth century, expanded its funding of science during the two world wars, and today funds research in innumerable disciplines of science.

As important as the establishment of an agenda for the association was the recruitment of scientists. William Barton Rogers, a geologist and founder of the AAAS, wanted an association in step with the democracy of Jacksonian America, but Bache, a patrician by upbringing, hoped the AAAS would be an elite organization of the most distinguished scientists. In its early days, the AAAS was nearer the vision of Rogers than that of Bache. In 1848, it required no scientific credentials of its members. In the early 1870s, however, the AAAS moved toward Bache's position by creating the category of *fellow* for its most distinguished members.

At first, the AAAS welcomed generalists with open arms, but by the last quarter of the nineteenth century, the AAAS had become an association of specialists. In 1873, insect researchers formed the Entomological Club within the AAAS; by 1882, the association had a total of nine sections: mathematics and astronomy, physics, chemistry, mechanical science, geology and geography, biology, microscopy and histology, anthropology, and economic science and statistics.

The annual meeting each December was not only a forum for the presentation of research and the discussion of issues at the forefront of science but also a means to forge links across disciplines, and the association urged all of the sectional organizations to attend. The 1850 meeting included a session on the fashionable field of craniometry. In 1887, physicists Albert Michelson and Edward Morley presented the findings of their groundbreaking ether-drift experiments. In 1908, chemists Gilbert N. Lewis and Richard C. Tolman lectured on Albert Einstein's special theory of relativity. In 1916, physicist Robert Millikan lectured on the Bohr atom.

Another forum for the dissemination of science is the journal *Science*. In 1894, James McKeen Cattell, a professor of experimental psychology at Columbia University, began to edit *Science* with the aim of publicizing both the latest scientific research and the agenda of the AAAS. Cattell intertwined the content of the journal with the business of the association, which, in 1895, began to subsidize publication. In 1900, AAAS members began to receive *Science* as a benefit of membership, and, in 1944, the AAAS acquired the journal. During his fifty-year tenure at *Science*, Cattell emphasized the uniqueness of the AAAS, reminding readers that, unlike the National Academy of Sciences, it did not receive federal funding. Certain that bigger was better, Cattell made a virtue of the AAAS's size, its geographic diversity, and its disciplinary breadth.

As permanent chair of the Executive Committee of the AAAS between 1925 and 1941, Cattell saw in President Franklin D. Roosevelt's New Deal an opportunity to strengthen government support of science. Mindful that the Great Depression had hurt scientists as well as the working class, the AAAS in 1934 urged Congress to create the Project for Scientific Aid for Public Works and the Recovery Program for Scientific Progress within the framework of New Deal agencies. The proposal called for $2.6 million to create jobs for unemployed scientists for six months. In addition to this initiative, the AAAS urged the government and universities to grant fellowships to unemployed scientists. Beyond these efforts the AAAS joined physicist Karl Compton in seeking $16 million in congressional aid to scientific research on the grounds that scientific knowledge was essential to national welfare. In these ways, the New Deal invigorated the AAAS's call for government support of science.

The Great Depression gave way to the affluence of post–World War II America, but the AAAS was slow to benefit from prosperity. The 1947 meeting attracted 2,700 attendees, down from nearly 5,000 the previous year, as the association suspended the presentation of technical papers in favor of papers that drew broad connections among scientific disciplines. In 1948, a group of biologists voted to establish the American Institute of Biological Societies with no formal ties to the AAAS, and, in 1951, institute members voted to hold their annual meeting apart from that of the AAAS.

The decision prompted the AAAS to convene that year at the Arden House Conference at Columbia University. Conferees acknowledged that the old ideal of the AAAS as a forum for the exchange of the latest science could no longer be the chief aim. This function would need to be secondary, given that the disciplinary organizations were now the clearinghouses for the latest science. Conferees therefore committed the AAAS to increasing public awareness of the importance of science.

In keeping with this ideal, the AAAS began in the mid-1950s to hold seminars for the media and to publish books with popular appeal rather than just for a small community of scholars. In 1955, the association accepted $300,000 from the Carnegie Corporation to create a Science Teaching Improvement Program, which evolved into its education department. Especially popular was the AAAS's television show *Nova*.

As the AAAS has grown, it has undertaken new initiatives. In the 1990s, its presidents F. Sherwood Rowland and Jane Lubchenco directed the AAAS to publicize the dangers of ecological degradation and global warming. In a broad concern for human rights, the AAAS condemned the governments of Cambodia and Nicaragua. Eager to persuade a new generation of children and adolescents to pursue an education in science, the AAAS in the mid-1990s created the radio program *Kinetic City Super Crew* and a companion Web site.

At the same time, the AAAS broadened the diversity of its members. It had enrolled women as early as 1850 but was slow to promote them to leadership positions. In 1969, AAAS members elected their first female president, mathematician Mina Rees. Since then, women have totaled one-quarter of AAAS presidents and members of the board of directors. In 1972, the AAAS created the Office of Opportunities in Science to encourage women and minorities to pursue careers in science, engineering, and mathematics. From its nucleus of eighty-seven men in 1848, the AAAS in 2007 has grown to 120,000 members and 262 affiliate organizations, serving millions of scientists worldwide.

Christopher Cumo

Sources

American Association for the Advancement of Science. http://www.aaas.org.

Kohlstedt, Sally G. *The Formation of the American Scientific Community: The American Association for the Advancement of Science, 1848–1860.* Urbana: University of Illinois Press, 1976.

Kohlstedt, Sally G., Michael M. Sokal, and Bruce V. Lewenstein. *The Establishment of Science in America: 150 Years of the American Association for the Advancement of Science.* New Brunswick, NJ: Rutgers University Press, 1999.

AMERICAN HISTORICAL ASSOCIATION

The American Historical Association (AHA) is a private organization of historians dedicated to preserving America's historical resources and disseminating historical knowledge and research to a wide audience.

The AHA was founded in response to the increasing professionalization of the study of history in the late nineteenth century. In September 1884, a group of historians met in Saratoga Springs, New York, and decided to form the AHA as an independent organization. Prior to this time, the main national organization for historians was the American Social Science Association. The AHA grew in membership during the 1880s and became incorporated by the U.S. government in 1889.

Today, the organization is headquartered in Washington, D.C., and has 15,000 members. Throughout its history, it has played important roles in promoting historical research, archiving historical manuscripts and artifacts, and developing tools to assist history teachers at all levels.

Publishing historical research is an important goal of the AHA. Although this is accomplished through a variety of journals and monographs, the flagship of the AHA since 1895 has been the *American Historical Review,* a journal devoted to all aspects of world and American history. In the early years of the AHA, the compilation of bibliographies of historical works grew in importance because of the need to organize scattered historical resources and writings. The AHA produced a *Guide to Historical Literature* in 1931 (revised in 1961), which provided an extensive bibliography of historical writings about all historical periods.

The AHA collects and preserves important historical documents and artifacts by promoting the creation and expansion of historical archives at the state and local levels. AHA members such as John Franklin Jameson, concerned that the U.S. government neglected its own historical records, which were haphazardly organized and often stored in buildings that were not fireproof, pushed for a national archives beginning in 1890. Eventually, Congress appropriated funds for the task in 1926. Construction of the National Archives in Washington, D.C., began in 1931 and was completed in 1935.

The AHA has also promoted secondary education in the United States by advocating a four-year history curriculum as well as encouraging the training of high school history teachers. The AHA has been a trendsetter for social science professional organizations. It has shown the influence that a collection of individuals who share similar goals can have on the larger society.

Wade D. Pfau

Sources

American Historical Association. http://www.historians.org.

Jameson, J. Franklin. "The American Historical Association, 1884–1909." *American Historical Review* 15:1 (1909): 1–20.

Link, Arthur S. "The American Historical Association, 1884–1984." *American Historical Review* 90:1 (1985): 1–17.

AMERICAN MUSEUM OF NATURAL HISTORY

The American Museum of Natural History—located on the west side of Manhattan's Central Park at 79th Street—began in 1869 when Albert Smith Bickmore, a student of biologist Louis Agassiz, managed to convince some wealthy patrons and government officials to found a world-class museum. Those involved began by purchasing individual collections composed of thousands of mounted birds, mammals, fish, and reptiles. These objects were stored in a temporary location until the museum was built. The museum also sponsored collecting and research expeditions.

By the middle of the twentieth century, the museum held the largest natural history collection in the United States, supplied by some 1,000 major expeditions and maintained by more than 500 employees. The building, which also contained dozens of laboratories and a 200,000-volume library, had been visited by more than 100 million people.

Today, the museum's holdings are so rich and extensive that they are housed in a complex consisting of twenty-five buildings. These include the Hayden Planetarium, the Rose Center for Earth and Space, an IMAX theater, exhibit halls, a working library, and the Theodore Roosevelt Memorial Building. The impressive main entrance features four 54 foot columns, surmounted by statues of Daniel Boone, James Audubon, and Lewis and Clark; it opens into a large, classical interior.

The world-famous collection of the museum includes countless life-like dioramas of bear, elk, wolves, tigers, and gorillas; dinosaurs and their eggs (the tyrannosaurus rex skeleton is 18.5 feet high); elephants; a giant squid; a 76-foot-long blue whale replica, which hangs from the ceiling; amphibians and reptiles; fish; insects; invertebrates; fossils; shells; coral; Native American artifacts such as headdresses, totem poles, a Hopi pueblo, and a 64 foot Haida canoe; artifacts of the Maya and Aztec peoples; pottery; extraordinary rocks, minerals, and gems (including the Star of India sapphire and the De Long star ruby); carved jade, ivory, and amber; astronomical, meteorological, and geological exhibits such as models of volcanoes and earthquakes; thousands of meteorites; and some 32 million other items.

A number of prominent Americans have been associated with the museum in one way or another. Theodore Roosevelt and J. Pierpont Morgan were instrumental in its founding. Henry Fairfield Osborn's research in paleontology earned him its presidency. Franz Boas, Ruth Benedict, Margaret Mead, and George Gaylord Simpson all worked there, and Robert E. Peary, Vilhjalmur Stefansson, A.L. Kroeber, Colin Turnbull, and Napoleon Chagnon did fieldwork or led expeditions throughout the world (including Mongolia, Patagonia, New Guinea, the North Pole, Africa, the North Pacific, India, Burma, and the western part of the United States).

Like the National Geographic Society, the American Museum of Natural History sponsors both scientific research, sometimes with living creatures and carried out on the premises, and expeditions. It also supports *Natural History*, a periodical that keeps readers apprised of happenings in the natural world. Each of the year's ten issues includes sections on the museum and listings of special exhibits (e.g., live frogs or butterflies), lectures, and educational programs.

Robert Hauptman

Sources

American Museum of Natural History. http://www.amnh.org.

Hellman, Geoffrey. *Bankers, Bones, and Beetles: The First Century of the American Museum of Natural History.* Garden City, NY: Natural History Press, 1969.

Rexer, Lyle, and Rachel Klein. *American Museum of Natural History: 125 Years of Expedition and Discovery.* New York: Harry N. Abrams, 1995.

Saunders, John Richard. *The World of Natural History as Revealed in the American Museum of Natural History.* New York: Sheridan House, 1952.

AMERICAN PHILOSOPHICAL SOCIETY

The American Philosophical Society was the first scientific society in America. It was founded in Philadelphia in 1743, the same year that Benjamin Franklin published *A Proposal for Promoting Useful Knowledge Among the British Plantations in America*. Franklin was the first secretary of the society, and Thomas Hopkinson, a signer of the Declaration of Independence, was the first president. The seed from which the American Philosophical Society sprang was a group of young men called the Junto that Franklin had formed in 1727 to promote inquiry, experiment, and the exchange of ideas on a wide range of subjects.

The society merged in 1769 with the American Society for Promoting Useful Knowledge, also founded by Franklin, who became the first president of the combined American Philosophical Society and held the office until his death. Hopkins and Samuel Vaughan helped revive the society after the American Revolution, in part because of a charter granted by the state of Pennsylvania in 1780 that allowed the society to correspond with learned individuals and institutions "of any nation or country" on its legitimate business at all times, "whether in peace or war." In 1789, the first woman was elected to the society: Russia's Princess Dashkova, the president of the Imperial Academy of Sciences in St. Petersburg.

Franklin was succeeded as president in 1791 by the astronomer David Rittenhouse who, in the 1760s, had won international recognition for the society by plotting the transit of Venus from telescopes mounted on a platform behind what is now Independence Hall in Philadelphia. Thomas Jefferson served as the society's third

The American Philosophical Society Held at Philadelphia for Promoting Useful Knowledge, formally established as such in 1769, is the oldest learned society in the United States. Benjamin Franklin was elected as its first president. *(Library of Congress, HABS PA, 51-PHILA, 46–1)*

president, from 1797 to 1814. Along with others in the society, Jefferson helped prepare Lewis and Clark for the scientific, linguistic, and anthropological elements of their exploration of the newly acquired lands of the Louisiana Purchase. The society sponsored the publication of the papers of Lewis and Clark, as well as those of Benjamin Franklin, Joseph Henry, and William Penn, among many others.

The society also promoted America's economic independence by seeking to investigate, understand, and improve agriculture, manufacturing, transportation, and other areas of endeavor. An introductory page in the first *Transactions of the American Philosophical Society* (published since 1771 and currently published in five issues a year) states: "The Promoting of useful Knowledge in general, and such branches thereof in particular . . . being the express purpose for which the American Philosophical Society was instituted; the publication of such curious and useful Papers as may, from time to time, be communicated to them, becomes of course, one material part of their design." The interests of the society evolved over time. By the last half of the nineteenth century, the main areas of inquiry were American paleontology, geology, astronomical and meteorological observations, and Indian ethnology.

A research grant program established in the 1930s has supported such projects as the archeo-

logical excavations of Tikal in Guatemala and the second Byrd Antarctic expedition to measure the depth of the polar ice cap. The society sponsors five research grant and fellowship programs, the most notable being the Franklin Grants in the humanities and a research grant in American history of the early national period. Most grants are used to produce scholarly books and articles, but grant programs have also assisted research in the humanities and social sciences, clinical medicine, North American Indian linguistics, and ethnohistory. There is also a library resident fellowship program for research in the society's collections. One of the recipients of the clinical medicine program, David Fraser, later led the U.S. Public Health Service investigation of Legionnaires disease.

The society's biannual meetings in April and November include the presentation of papers and discussions exploring topics in the sciences and humanities. Topics have included underwater archeology, nuclear magnetic imaging, Shakespeare's writings, and race relations in modern America. The society also supports additional sessions, during which specialists present papers on topics of interest to a more limited audience. Session topics include the complexity of life, American presidential elections, and the protein as a building-block of life.

The *Proceedings* (begun in 1838), a quarterly, publishes papers delivered at the biannual meetings of the society as well as papers submitted independently. The society also publishes larger studies in *Memoirs* (begun in 1935); subjects have ranged from ancient Egyptian science to modern-day Pennsylvania flora.

The society has counted among its membership such notables as George Washington, John Adams, Thomas Jefferson, Alexander Hamilton, Thomas Paine, John Marshall, the Marquis de Lafayette, Baron von Steuben, Thaddeus Kosciuszko, Robert Fulton, Charles Darwin, Alexander von Humboldt, Robert Frost, Louis Pasteur, Elizabeth Cady Agassiz, John James Audubon, Marie Curie, Gerty T. Cori, Albert Einstein, George C. Marshall, Linus Pauling, Margaret Mead, and Thomas Edison. As the society entered the twenty-first century, it counted more than 850 members, 85 percent of whom resided in the United States. The society confers membership based on scholarly and scientific accomplishments in any of five areas: mathematical and physical sciences; biological sciences; social sciences; humanities; and arts, professions, and leaders in public and private affairs. More than 200 members of the society have received the Nobel Prize.

The society recognizes accomplishments not only by conferring membership but by awarding special prizes and medals as well. The Magellanic Premium (1786) is awarded for discoveries "relating to navigation, astronomy, or natural philosophy." The oldest scientific prize given by an American institution, it has acknowledged such discoveries and accomplishments as the circumnavigation of the globe by submarine and the advent of various forms of space technology. The Benjamin Franklin Medal (1906) is awarded for distinguished achievement in the sciences. The Lashley Award (1935) recognizes achievements in neurobiology. The Lewis Award (1935) honors a publication by the society and has been awarded to Enrico Fermi (1946), Millard Meiss (1967), and Kenneth Setton (1984), among others. The Moe Prize (1982) and Phillips Prize (1888) honor papers in the humanities and jurisprudence. The Barzun Prize (1992), named for Jacques Barzun, one of the founders of the discipline of cultural history, recognizes contributions to American or European literature, education, and cultural history. The Jefferson Medal (1993) is awarded for distinguished achievement in the arts, humanities, or social sciences. The Dalland Prize (2001) recognizes outstanding achievement in patient-oriented clinical research.

The society remains headquartered in Philadelphia and occupies two buildings in Independence National Historical Park: the Philosophical Hall (erected 1785–1789) and the Library, a replica of the original home of the Library Company of Philadelphia (1798). The Library houses 200,000 books and bound periodicals, 7,000,000 manuscripts, and thousands of prints and maps, primarily devoted to U.S. history to 1840 and the history of science and technology, especially eighteenth- and nineteenth-century natural history, linguistics, the modern life sciences, physics, and computer technology. The Library also houses the Benjamin Franklin Papers, the papers of artist Charles Willson Peale and family, and the papers

of Franz Boas, founder of modern American anthropology.

Richard M. Edwards

Sources

American Philosophical Society. http://www.amphilsoc.org.

Carter, Edward C. *One Grand Pursuit: A Brief History of the American Philosophical Society's First 250 Years, 1743–1993.* Philadelphia: American Philosophical Society, 1993.

Smith, Murphy D. *Oak from an Acorn: A History of the American Philosophical Society Library, 1770–1803.* Wilmington, DE: Scholarly Resources, 1976.

ARMINIANISM

Arminianism was a movement that sprang from Dutch Reformed theologian Jacobus Arminius (1559–1609), who questioned the Calvinist doctrine of predestination.

Educated at Geneva, Arminius was a respected minister in Amsterdam. He defined predestination in terms of God's foreknowledge. In his system, God foresees that alongside those whom he has decreed for election there are those who freely choose to reject Christ and are damned. Although all are depraved, God has extended "prevenient" grace to all, so that faith is possible for all under Christ's universal atonement. Arminius's views contradicted orthodox Calvinism.

Arminius became a more controversial figure upon his appointment to a professorship at the University of Leyden in 1603. There, he clashed with fellow professor Franciscus Gomarus, a strict Calvinist who had protested his hiring. Critics repeatedly condemned Arminius but failed to dislodge him from his teaching post, where he remained until his death in 1609.

His followers, led by his successor at Leyden, Simon Episcopus, formalized Arminius's theology, thereby giving birth to "Arminianism." When followers of Gomarus (Gomarists) attempted to have them removed from teaching posts, Arminians responded by drafting a Remonstrance (statement of beliefs) in 1610. The Dutch Arminian supporters of the statement henceforward became known as Remonstrants.

The Synod of Dort (November 1618–May 1619) condemned the Remonstrants as heretics and permanently alienated them from the Dutch Reformed Church. Over the next two centuries, Arminianism moved well beyond Arminius and increasingly came to be associated with the complete rejection of Calvinism. It proved to be most influential in Great Britain and the United States.

Controversy over Arminianism arose in colonial New England during the early eighteenth century, as Anglicans gradually established a presence there. Calvinism had been the established orthodoxy among Congregational New Englanders since the arrival of Puritans in the early seventeenth century; they were determined to keep Anglicanism, and the Arminianism that often accompanied it, at bay. The motive was as much political as theological, for they feared a large Anglican presence would lead to conformity with the mother country and the eventual establishment of the Church of England.

In 1722 what came to be called the "great apostasy" shook Yale College (then the bastion of Reformed orthodoxy), and indeed the rest of New England, when it was discovered that Yale rector Timothy Cutler had been meeting regularly with a group of dissident clergy, along with former and current Yale tutors, all of whom had embraced Arminianism. Trustees at Yale quickly fired those involved and enforced strict new guidelines for tutors and rectors. The following summer, a popular young minister and Yale graduate named Jonathan Edwards delivered a commencement address in which he denounced Arminianism in apocalyptic terms and reaffirmed the soundness of strict Calvinist doctrine. At least temporarily, Yale remained a Calvinist stronghold. Later, as a famous revivalist and author, Edwards would continue to do battle with what he called the "great noise" over Arminianism.

In the subsequent Great Awakening of the 1730s, 1740s, and 1750s, Edwards and other Calvinists called themselves New Lights and were opposed by Arminians who called themselves Old Lights. The Old Lights, led by the likes of Charles Chauncy, minister of Boston, promoted a rational Christianity informed by science. They believed that the natural world was an "elder scripture," that the study of nature would reveal knowledge of God, and that said natural theology showed that God is good and just and does not arbitrarily condemn humans to everlasting damnation. The Old Lights believed in a morality based on duty, order, and reason, and they were reluctant revolutionaries during the American Revolution.

By the late eighteenth century, Old Light Arminians and scientists such as Jeremy Belknap of Boston were embracing universal salvation. During the Second Great Awakening of the early nineteenth century, the former New Lights, especially the Methodists and Baptists, embraced the relaxed approach toward Calvinist damnation of Arminianism. By the mid-nineteenth century, Arminianism was firmly established as the new orthodoxy of American Protestantism.

Stephen Peterson

Sources

Gonzales, Justo L. *A History of Christian Thought: From the Protestant Reformation to the Twentieth Century.* Vol. 3. Revised ed. Nashville, TN: Abingdon, 1987.

MacCulloch, Diarmaid. "Prophets Without Honor: Arminius and the Arminians." *History Today,* October 1989, 27–34.

Sell, Alan P.F. *The Great Debate: Calvinism, Arminianism, and Salvation.* Grand Rapids, MI: Baker House, 1983.

Slaatte, Howard A. *The Arminian Arm of Theology: The Theologies of John Fletcher and His Precursor, James Arminius.* Washington, DC: University Press of America, 1977.

DEISM

Deism, the belief in a "natural" or "rational" monotheism, flourished in the eighteenth-century European Enlightenment. Inherently unorthodox and opposed to religious doctrine, deism offered a malleable, and often quite personal, religious system. As such, there was little agreement, then or now, as to its exact tenets. Instead, deism has had numerous variants.

For many educated elites skeptical of Christianity, the new faith offered a means to reform the ancient religion by removing unreasonable doctrines—such as transubstantiation and salvation—while maintaining a divine presence in a designed world. More radical thinkers argued that deism could replace Christianity. For nearly two centuries, debates raged over the sanctity of religious orthodoxy and the religious implications of modern science, as intellectuals struggled to create a framework for a "rational" religion.

Although varied, deism contained a number of general principles. First, most deists were committed to monotheism and the existence of a higher power. At the same time, owing to the religious strife of the seventeenth century, most adherents were vehemently opposed to organized religions, particularly the ritualism of Roman Catholicism and the extremism of English Puritanism. Such religious excesses and intolerance were considered antithetical to true religion; the belief in a vengeful God, or doctrines of asceticism and religious hatred, were considered modern perversions. Ordained priests and other spiritual authorities were equally suspect.

Instead, most deists believed in a single universal religion, often accepting the wisdom of figures from other spiritual and philosophical traditions, such as Socrates, Buddha, and Muhammad. Every tradition had something to offer, in part because each advocated universal principles such as human benevolence and morality. Other universal principles, including empiricism, skepticism, and rationalism—the hallmarks of the Scientific Revolution—soon became a foundation for the new religion.

The scientific discoveries of the seventeenth century deeply influenced the development and contours of deism. Following the publication of Isaac Newton's *Principia* in 1687, which explained mathematically the movements of heavenly bodies by a universal gravitational force, many intellectuals argued that the universe is governed by rational laws. Controversy arose over the idea of divine intervention: Some deists insisted God merely determined the original universal laws (like gravity) and then stepped back, allowing the world to run itself much like a machine; others maintained that God continued as an active presence in the world. Regardless, most deists accepted that the world was designed and operated according to scientific rules that could be understood through human reason. The celebration of human reason led many to dispute the role of revelation and denounce a number of traditional doctrines as mere superstition.

Many deists were prominent and influential members of society. The list of European deists includes such luminaries as Pierre Bayle, Gottfried Wilhelm Leibniz, Jean d'Alembert, Moses Mendelssohn, and Immanuel Kant. Newton, notably, remained a Christian.

Deism also influenced many leaders of the American Revolution. Freemasons arrived in the colonies in the early eighteenth century and quietly established themselves as the primary adherents of deism there, eventually finding favor among the educated colonial elite, including Benjamin Franklin, Thomas Paine, Thomas Jefferson, and John Adams.

Jefferson's deism was evident in the *Jefferson Bible,* a compilation of the New Testament that focused on Jesus the teacher and moralist but omitted passages that suggested divinity and the supernatural. Paine, one of America's foremost deists, described in *The Age of Reason* (1794) a rational approach to religion and included a polemic against Christianity and the New Testament, arguing that science contradicts most stories of Jesus in the Gospels. Franklin, in his *Autobiography,* denied that he was a deist, yet his religious proclivities focused on a rational source of universal goodness—the deist God.

J.G. Whitesides

Sources

Byrne, Peter. *Natural Religion and the Nature of Religion: The Legacy of Deism.* New York: Routledge, 1989.

Jacobs, Margaret C. *Living the Enlightenment: Freemasonry and Politics in Eighteenth-Century Europe.* Oxford, UK: Oxford University Press, 1991.

FIELD MUSEUM OF NATURAL HISTORY

The Field Museum in Chicago has provided natural history and cultural exhibits for the general public as well as materials for scholarly research since it was created to house its core biological and anthropological collections, which had been assembled for the World's Columbian Exposition of 1893. Originally known as the Columbian Museum of Chicago, it was renamed the Field Museum of Natural History in 1905 in honor of Marshall Field, the Chicago department store magnate who was the museum's first major benefactor.

Field, who died in 1906, also bequeathed to the museum its sustaining funds and the lakefront building (1921) that anchors the Chicago Parks District's Museum Campus. This campus also includes the John G. Shedd Aquarium (1929) and the Adler Planetarium (1930). Though the Field Museum was officially renamed the Chicago Natural History Museum in 1943, that name was never commonly used, and the Field name was restored in 1966, both to honor Field and to reflect the name used by most Chicagoans.

The Field Museum is divided into four main departments: Anthropology, Botany, Geology, and Zoology. The curatorial and scientific staff work as a team in conducting interdisciplinary research in anthropology, archeology, ethnography, evolutionary biology, paleontology, and systematic biology; in managing the collections; and in collaborating on public programming with the museum's Departments of Education and Exhibits.

The Harris Loan Program, which began in 1912, circulates artifacts, specimens, audiovisual materials, and activity kits to Chicago area schools. The Field also maintains a 250,000-volume natural history library and many interactive exhibits. One such exhibit, the Underground Adventure, provides a different perspective on life by shrinking visitors to the size of a penny relative to the oversized exhibit components. Two of the museum's laboratories can be viewed by the public: the MacDonald's Prep Lab prepares fossils for study, and the Regenstein Laboratory demonstrates methods of archeological preservation and study.

The museum has more than 20 million specimens, all of which are being photographically digitized for easier scholarly and Internet access. Some of its best-known animal exhibits include Sue, the world's largest and most complete *Tyrannosaurus rex* skeleton; two very large African (Kenyan) elephants that have stood in the museum's Stanley Field Hall since 1921; the two preserved male lions of Tsavo, which killed over 140 workers during the construction of a railroad bridge over the Tsavo River in East Africa in 1898, a story told in the 1996 film *The Ghost and the Darkness*; and many other specimens located in the Nature Walk, Mammals of Asia, Mammals of Africa, and other exhibits. Carl Akeley, the museum's head taxidermist from 1895 to 1909, invented techniques that enabled the Field Museum to display animals in dioramas mimicking their natural habitats, a presentation method that continues in natural history exhibits throughout the world.

The museum houses many cultural exhibits, including traditional clothing from Tibet and China, a nineteenth-century Maori meeting house, and a Native American exhibit that features totem poles and traditional costumes as well as a Pawnee earth lodge. The Inside Ancient Egypt exhibit interactively demonstrates some of the daily activities common to life in Cleopatra's Egypt, such as food preparation, religious practice, burial, dress, business, and family relationships. Twenty-three human and animal mummies are on display, along with a tomb adorned with 5,000-year-old hieroglyphs.

The Grainger Hall of Gems and the Hall of Jades concentrate on diamonds and other gems and their uses. The Evolving Planet, a Life over Time exhibit, traces 4 billion years of history and the evolution of life on Earth and includes a dinosaur hall.

Richard M. Edwards

Sources

Alexander, Edward P. *The Museum in America: Innovators and Pioneers.* American Association for State and Local History Book Series. Lanham, MD: AltaMira, 1997.

Danlioy, Victor J. *Chicago's Museums: A Complete Guide to the City's Cultural Attractions.* Chicago: Chicago Review Press, 1991.

The Field Museum. http://www.fieldmuseum.org.

HARVARD MUSEUM OF NATURAL HISTORY

Formally established in 1995, the Harvard Museum of Natural History is the offspring institution of the Harvard University Herbaria, the Museum of Comparative Zoology, and the Mineralogical and Geological Museum. The primary mission of Harvard's Museum of Natural History is to display the collections of its parent institutions. In addition, it offers educational programs to children, adults, and teachers, as well as public lectures and films dedicated to various aspects of natural history and tours to places of historical and environmental interest around the world. The museum is housed in the same Cambridge, Massachusetts, building as its parent institutions and the institutionally distinct Peabody Museum of Archaeology and Ethnology.

Founded by botanist Asa Gray in 1858, the Harvard University Herbaria is the oldest of the three parent institutions. The nucleus of its collection was donated by William Hooker, director of the Royal Botanic Garden in Kew, England. First called the Museum of Vegetable Products, the institution was originally dedicated to horticulture. With a strong practical focus, the museum's early directors sought practical uses for wild and domesticated plants.

Today, the Herbaria consists of three parts: the Gray Herbarium and the Botanical Museum, both housed in the University Museum building, dedicated in 1891, and the Arnold Arboretum, founded in 1872 and named for its original benefactor, whaling merchant James Arnold. The arboretum is located on 265 acres in the Jamaica Plain section of Boston.

The second-oldest of the Museum of Natural History's parent institutions is the Museum of Comparative Zoology, founded in 1859 by Swiss-born and German-educated zoologist and geologist Louis Agassiz, who emigrated to the United States in 1847 to become a professor in those disciplines at Harvard College. So closely linked is the man and institution that the museum is often called "The Agassiz." Combining his own collection with those already belonging to the college and those donated by benefactors, Agassiz set up the exhibits to highlight the diversity and comparative relationships of Earth's fauna. At the time, most natural history collections were the possessions of amateur zoologists and paleontologists, and usually were displayed in random and haphazard ways.

Harvard's zoological collection is now housed in the University Museum building. While it still collects and preserves specimens, the museum is largely dedicated to education and research efforts in various branches of zoology and paleontology, with departments including Biological Oceanography, Entomology, Herpetology, Ichthyology, Invertebrate Paleontology, Invertebrate Zoology, Mammalogy, Marine Biology, Malacology, Ornithology, Population Genetics, and Vertebrate Paleontology.

Before the opening of the University Museum building in 1891, the collections of the Mineralogical and Geological Museum were maintained by the university's Chemistry Department. Considered one of the finest of its type in the world, the mineralogical collection is noted for the diversity and uniqueness of its specimens.

James Ciment

Sources

Dupree, A. Hunter. *Asa Gray, American Botanist, Friend of Darwin.* Baltimore: Johns Hopkins University Press, 1959, 1988.

Harvard Museum of Natural History. http://www.hmnh.harvard.edu.

Lurie, Edward. *Nature and the American Mind: Louis Agassiz and the Culture of Science.* New York: Science History Publications, 1974.

Pick, Nancy. *The Rarest of the Rare: Stories Behind the Treasures at the Harvard Museum of Natural History.* New York: HarperResource, 2004.

HAZARD, EBENEZER (1744–1817)

Ebenezer Hazard, a collector and editor of historical documents, provided the first major public record of the United States and its colonial past. He also served as postmaster of New York (1775), surveyor of post roads (1776–1782), and postmaster general of the United States (1782–1789). In 1792 and 1794, he published two volumes of *Historical Collections: Consisting of State Papers, and Other Authentic Documents; Intended as Materials for an History of the United States of America.*

Born in Philadelphia in 1744 and educated at Princeton, Hazard began his professional life as a bookseller before becoming involved in the postal service during the Revolutionary War.

Especially interested in history, both human and natural, Hazard engaged his free time in the pursuit of knowledge of both subjects. His friend and collaborator in this antiquarian pursuit was the historian and geographer Jeremy Belknap. The Belknap-Hazard collection of letters, spanning twenty years (1779–1798), includes a host of fascinating epistolary investigations into geography, mineralogy, political science, natural science, and particularly history.

Hazard was also a correspondent with other notable thinkers of the time, such as the geographer Jedidiah Morse, the lexicographer Noah Webster, the historian William Gordon, and the naturalist Thomas Jefferson. Hazard was recognized for his scientific interests with membership in the American Philosophical Society and the American Academy of Arts and Sciences.

During his many years of service with the U.S. Post Office, Hazard had conceived of a collection of historical documents. Spending years on horseback during the Revolutionary War, putting post offices in order and determining the best post roads, he had a chance to see most of the new nation and to interact with public officials. Astonished to discover that most public documents were haphazardly collected and stored, he decided that the best means of preserving America's history was by multiplying the copies of its public documents. To this end, he set about copying as many as he could, spending weeks at a time crouched at a writing desk, transcribing documents. As the years passed, he collected such a vast number of such documents that he decided to publish them, providing Americans with a public record of their brief past.

Hazard's *Historical Collections* focused mostly on the colonial documents of New England, such as the Records of the United Colonies of New England (the New England Confederation). Thomas Jefferson welcomed the project, writing to Hazard that the collected documents "are curious monuments of the infancy of our country.... Time and accident are committing daily havoc on the original [documents] deposited in our public offices. The late war has done the work of centuries in this business. The lost cannot be recovered, but let us save what remains; not by vaults and locks which fence them from the public eye and use in consigning them to the waste of time, but by such a multiplication of copies, as shall place them beyond the reach of accident."

Although Hazard's *Historical Collections* was a financial failure, the two volumes of documents quickly became the standard that other collectors and editors would use in determining what to preserve and how to preserve the past. Peter Force and Jared Sparks, for example, two of the most important documentary editors of the nineteenth century, relied heavily on Hazard's *Historical Collections* in their own historical work.

Russell Lawson

Sources

Lawson, Russell M. *The American Plutarch: Jeremy Belknap and the Historian's Dialogue with the Past.* Westport, CT: Praeger, 1998.

Shelley, Fred. "Ebenezer Hazard: America's First Historical Editor." *William and Mary Quarterly*, 3rd ser., 13 (1955).

HEMPEL, CARL GUSTAV (1905–1997)

The philosopher of science Carl Gustav Hempel is best known for his work on logical positivism (he preferred the term "logical empiricism"), an approach that asserts the primacy of scientific observation over metaphysical or subjective argument

Born in Orianenburg, Germany, on January 8, 1905, Hempel had an eclectic education. He studied mathematics and symbolic logic at the University of Göttingen, and mathematics, physics, and philosophy at the University of Heidelberg and the University of Berlin. He was awarded his doctorate in philosophy from the latter institution in 1934. While there, he became a member of the Berlin Group, an influential circle of philosophers who asserted that experience is the only source of knowledge and that symbolic, or mathematical, logic offers the key to the analysis of philosophical problems.

After Adolf Hitler's rise to power in Germany, Hempel emigrated to Belgium. Although he was not Jewish, his father-in-law was Jewish, and Hempel was an outspoken critic of Nazi anti-Semitism. In 1937, Hempel accepted a position at the University of Chicago as a research associate in philosophy but returned briefly to Belgium. With the Nazis threatening war in Europe, Hempel permanently emigrated to the United States in 1939.

He taught philosophy at a number of institutions of higher learning, including the City College of New York (1939–1940), Queens College (1940–1948), Yale University (1948–1955), Princeton University (1955–1964), the Hebrew University in Jerusalem (1964–1966), and the University of Pittsburgh (1976–1985). Between teaching stints in Israel and Pittsburgh, he taught courses at the Berkeley and Irvine campuses of the University of California.

Hempel was a pioneer in the study of the deductive-nomological model of scientific inquiry, which sees scientific theories as the result of deductive arguments involving two parts: observed facts and natural law. Hempel first explored this model in his work with philosopher Paul Oppenheim, which included a 1936 article, "Studies in the Logic of Explanation," published in the journal *Philosophy of Science.* This work was critical to the relationship between scientific observation or experimentation and scientific theory.

Hempel and Oppenheim's work also helped explain the basic differences between fundamental theory and derived theory. Fundamental theory is universal, like the laws of physics discovered by Isaac Newton, which apply to all objects at every time and in every place. Derived theory is determined from the observation of specific things, such as Johannes Kepler's observations of the motions of the sun and the planets, which apply only to those bodies in that particular space.

Hempel's major published works include *Fundamentals of Concept Formation in Empirical Science* (1952) and the 1988 article "Provisos: A Problem Concerning the Inferential Function of Scientific Theories," published in the philosophical journal *Erkenntnis.* He died in Princeton, New Jersey, on November 9, 1997.

James Ciment

Sources

Fetzer, James H., ed. *Science, Explanation, and Rationality: Aspects of the Philosophy of Carl G. Hempel.* New York: Oxford University Press, 2000.
Scheffler, Isaac. *The Anatomy of Inquiry.* New York: Alfred A. Knopf, 1963.

HERMENEUTICS

Hermeneutics, or the science of interpretation, takes its name from the Greek god Hermes, a messenger of the gods who proclaimed and interpreted the words of the gods to mortals. Hermeneutical principles were applied by ancient Greeks to sacred and legal texts. The medieval Catholic Church's hermeneutics recognized four acceptable interpretations of scripture: the literal, the allegorical, the tropological (moral), and the anagogical (spiritual or mystical).

During the Renaissance, the rediscovery of ancient texts and a new appreciation for the Bible text in its original languages (Hebrew and Greek) led eventually to what became known as the "grammatico-historical" approach. Protestant Reformers were determined to discover a universally applicable set of principles that could be applied by any Christian, not just clergy. A belief in the perspicuity (clarity) of the scriptures was combined with the Protestant doctrine of the priesthood of all believers, leading the Reformers to contend that one can discern the Bible's true meaning through the use of proper hermeneutical principles.

Simultaneously, Protestantism sparked an interest in the physical sciences; hermeneutics, as a "science," played a role in at least three ways. First, the very concept that interpreting the Bible is "science" showed the interplay between special revelation (scripture) and general revelation (nature). Second, the relationship between special and general revelation (on matters such as cosmology) was bound up in the question of hermeneutics and how a particular scriptural text (or scientific phenomenon) should be viewed. Third, Protestants considered natural science to be like a book that can be read inductively. The philosopher Francis Bacon was perhaps the best-known proponent of this "two-book," inductive empirical approach. The American scientist and clergyman Jeremy Belknap in 1792 proclaimed that nature is "elder scripture," holding out answers about God and his creation to the inquisitive mind of the scientist and Christian.

Fueled by the Renaissance and the Enlightenment, hermeneutics developed particularly in three areas in early modern Europe: classical learning, law, and philosophy. A renewed interest in the Greek and Latin classics led scholars on a quest for the authentic text of ancient documents. A revived interest in Roman law prompted scholars to seek universal principles of interpretation, often in terms of exegeting a text grammatically. The third area, philosophy, led ultimately to modern hermeneutics.

Perhaps more than any other eighteenth-century philosopher, the German Christian Wolff enunciated the view that hermeneutics entails universally valid principles—laws that apply for all fields of knowledge requiring interpretation. Wolff emphasized authorial intent, as judged by how effective the author had been in using syntax to convey his or her intended meaning. The theories of another German, Friedrich Schleiermacher, marked a further turning point for hermeneutics. Schleiermacher distinguished between a written text considered linguistically and the same writing as an expression of the author's life experience. For Schleiermacher, the key to understanding was found in the nexus between these two aspects.

Schleiermacher's approach, which reflected the ideas of the Romantic movement, clashed with the earlier, Enlightenment-inspired quest for the rational, and it affected the natural sciences as well as the liberal arts. Modern humans were seen as vacillating between a subjectivistic approach and one that purported to be totally objective. American contributors to this debate included the pragmatists Charles S. Peirce and William James.

Recent European hermeneutical philosophers exerting a strong influence on American philosophy and science include Paul Ricoeur and Michael Polanyi. Hermeneutics continues to play a significant role in the search for a universal and all-encompassing principle of interpretation—whether the object of discussion is a written text or the phenomena of the cosmos.

Frank J. Smith

Sources

Bleicher, Josef. *Contemporary Hermeneutics: Hermeneutics as Method, Philosophy, and Critique.* London: Routledge and Kegan Paul, 1980.

Hanko, Herman C. "Issues in Hermeneutics." *Protestant Reformed Theological Journal,* issues for April 1990, November 1990, April 1991, November 1991.

Kuhn, Thomas S. *The Essential Tension: Selected Studies in Scientific Tradition and Change.* Chicago: University of Chicago Press, 1977.

Mueller-Vollmer, Kurt. *The Hermeneutics Reader: Texts of the German Tradition from the Enlightenment to the Present.* New York: Continuum, 1988.

KUHN, THOMAS S. (1922–1996)

Best known for his landmark book *The Structure of Scientific Revolutions* (1962), Thomas S. Kuhn remains one of the most influential and popular historians and philosophers of science of the past half-century. Since publication, *Structure* has been translated into twenty-five languages, and the English edition alone has sold more than 1 million copies. This work has influenced historians, philosophers, scientists, economists, and sociologists, making it required reading in a variety of disciplines. It has also generated a sizable body of literature that examines, critiques, and develops many of Kuhn's ideas.

Thomas Samuel Kuhn was born on July 18, 1922, in Cincinnati, Ohio. Following an accelerated undergraduate program in physics, and work on developing radar technology for the U.S. government during World War II, he returned to Harvard for graduate study in physics in 1945.

During the dissertation stage of his graduate training in solid state physics, Kuhn convinced his mentor James B. Conant, the president of Harvard University and a chemist, to support his appointment to Harvard's Society of Fellows in order to transform himself into a historian of science. Kuhn hoped the history of science would be an avenue to pursue philosophical questions about the development of scientific ideas. Kuhn received his doctorate in physics in 1949 and, despite being largely self-taught in philosophy and the history of science, became an instructor and then an assistant professor of general education and history of science until 1956. It was during this time that his groundbreaking work on *Structure* began as a series of lectures at the Lowell Institute in Boston.

In 1956, Kuhn accepted a post at the University of California, Berkeley. In addition to teaching history of science and intellectual history from a scientific point of view, Kuhn published his first book, *The Copernican Revolution* (1957). In it, he examined an early example of the type of scientific revolution addressed in his later writings.

Challenging the orthodox understanding of scientific progress, he suggested that the devel-

Thomas Kuhn's view that science progresses in abrupt, periodic revolutions, or "paradigm shifts," rather than by the uniform accumulation of knowledge, was itself revolutionary. *(Bill Pierce/Time & Life Pictures/Getty Images)*

opment of scientific ideas is not a steady and logically driven march toward understanding the truth. He argued that Copernicus had transformed humans' conception of the universe and their relationship to it, overthrowing Aristotelian ideas but still maintaining a degree of continuity with ancient doctrines. Kuhn thus introduced two main points found in his later, more developed philosophy of science. He suggested that a scientific revolution is real and measurable progress, but the process of change is not as simple as replacing old bad beliefs with new good ones.

Logical Positivists, Karl Popper, and the Kuhnian Revolution

In the 1950s, philosophy of science was dominated by two groups: On the one hand were logical positivist philosophers, including Rudolf Carnap, Hans Reichenbach, and Carl Hempel; on the other hand were Karl Popper and his followers.

The logical positivists were primarily concerned with the logical analysis of scientific knowledge, emphasizing that scientific theories are verified by experiment and evidence. They argued that a scientific theory is meaningful only if it can be proved to be true or false by means of experience, at least in principle—an assertion called the "verifiability principle."

Popper and his followers maintained that scientific theory, and human knowledge in general, are generated by the creative imagination to solve problems that have arisen in specific historical and cultural settings. For Popper, scientific progress is the result of experimental testing. Differentiating science from nonscience, Popper argued that a theory should be considered scientific only if there is the possibility that it can be proven false. This made "falsifiability" the criterion of demarcation between what is and is not genuine science.

With the publication of *The Structure of Scientific Revolutions* in 1962, Kuhn provided a new picture of scientific theory. While the logical positivists and Popper's group suggested that science was a gradually growing body of knowledge, Kuhn argued that it changes through dramatic revolutions in thought. This challenged the notion that scientific change was strictly a rational process.

In *Structure*, Kuhn claimed that typical scientists are not objective and independent thinkers; rather, they are generally conservative individuals who accept the theories they have been taught, and who apply their knowledge to solving the problems that their theories dictate. For Kuhn, most scientists are puzzle solvers who aim to discover what they already know in advance based on the shared understanding about how problems are to be understood. This implicit body of intertwined theoretical and procedural beliefs is said to operate as a "paradigm," or framework, guiding the research efforts of scientific communities. For Kuhn, the history of science is not gradual and cumulative, but interrupted by a series of intellectual revolutions in which paradigms are successively replaced.

Structure was initially well received by a variety of audiences in history, philosophy, and the social sciences. In 1964, Kuhn joined the new Program in History and Philosophy of Science at Princeton University, where he was the M. Taylor

Pyne Professor of Philosophy and History of Science. As a rising young historian whose ideas had implications for the philosophy of science, Kuhn attended the International Colloquium in the Philosophy of Science, held at Bedford College, London, in July 1965. Among the major philosophers at the colloquium were Popper, Imre Lakatos, Paul Feyerabend, Stephen Toulmin, and the positivists collectively, including Kuhn's Princeton colleague Carl Hempel.

The published proceedings from the London colloquium, *Criticism and the Growth of Knowledge* (1970), edited by Imre Lakatos and Alan Musgrave, cemented Kuhn's place among the elite philosophers of science. In the book, many of the colloquium contributions were revised to respond to Kuhn's ideas. A revised edition of *Structure* was also published in 1970 and included a postscript in which Kuhn responded to some of the major critiques of his groundbreaking approach.

Kuhn's next major work, *The Essential Tension* (1977), was a collection of historical and philosophical essays. In this book, he argued that an essential tension between tradition and innovation is needed to make progress in an intellectual field. The historical essays he included remained focused on the internal historical development of scientific disciplines, but in his philosophical essays Kuhn revised and extended many of the central ideas from *Structure*. To clarify his argument about the nature of revolutionary change in science, Kuhn expanded the idea of an essential tension between new and old scientific ideas, the nature of paradigms, the relationship between successive theories following paradigm shifts, and the criteria for theory choice by practicing scientists. These arguments were largely developed in response to critiques of Kuhnian philosophy by scholars such as Stephen Toulmin and Ian Hacking.

In 1978, Kuhn provided an unorthodox history of early quantum theory in *Black-Body Theory and the Quantum Discovery: 1894–1912*. He challenged the received view of this critical period in the history of physics by arguing that Max Planck was not the founder of quantum theory in 1900, because he was still working in an older classical tradition. For Kuhn, it was the misreading of Planck's work by Albert Einstein and Paul Ehrenfest and their subsequent attempt at problem-solving that initiated early quantum theory.

Many physicists have strongly resisted this interpretation, while historians of science have frequently critiqued or ignored it. As critics at the time noted, the index of this massive and scholarly work contained no references to earlier Kuhnian ideas of paradigms or scientific revolutions. Historian Steve Fuller argues this represented the climax of Kuhn's attempts to distance himself from his radical interpreters, noting that Kuhn frequently admitted he preferred his critics to his followers.

Later Work and the Kuhnian Legacy

Kuhn's distinction in the fields of history and philosophy of science was sufficiently established by 1979 to earn him a position on the faculty of the Massachusetts Institute of Technology as the Laurence S. Rockefeller Professor of Philosophy, where he remained until he retired in 1991. Although Kuhn intended to write a sequel to *Structure* to provide a definitive statement of his position in response to his critics, he did not finish this project. The University of Chicago Press in 2000 issued a posthumous second collection of Kuhn's essays, *The Road Since Structure: Philosophical Essays, 1970–1993, with an Autobiographical Interview*. These essays reveal how Kuhn spent his last decades defending, developing, and substantially refining many of the basic concepts set forth in *Structure*, including the nature of scientific progress, paradigm shifts, and the relationship between old and new scientific frameworks.

Kuhn was honored with the prestigious George Sarton Medal for lifetime achievement in the History of Science in 1982. After suffering from cancer during the last years of his life, he died on June 17, 1996, at his home in Cambridge, Massachusetts. His work remains a highly influential landmark of twentieth-century intellectual history and continues to stimulate debates about science, culture, and policy across academic disciplines.

Eric Boyle

Sources

Bird, Alexander. *Thomas Kuhn.* Princeton, NJ: Princeton University Press, 2001.

Fuller, Steve. *Thomas Kuhn: A Philosophical History of Our Times.* Chicago: University of Chicago Press, 2000.

Hoyningen-Huen, Paul. *Reconstructing Scientific Revolutions: Thomas S. Kuhn's Philosophy of Science.* Trans. Alexander T. Levine. Foreword by Thomas S. Kuhn. Chicago: University of Chicago Press, 1993.

Kuhn, Thomas S. *The Road Since Structure: Philosophical Essays, 1970–1993, with an Autobiographical Interview.* Ed. James Conant and John Haugeland. Chicago: University of Chicago Press, 2000.

Nickles, Thomas, ed. *Thomas Kuhn.* New York: Cambridge University Press, 2003.

LAWRENCE SCIENTIFIC SCHOOL, HARVARD UNIVERSITY

When it opened in 1847, Harvard University's Lawrence Scientific School was one of only a handful of U.S. institutions offering laboratory education in the sciences on the European model, as well as an emphasis on original research by faculty and students. At the time, no other educational institution came close to the financial resources of the Lawrence School, which was endowed with a $50,000 bequest from textile magnate Amos Lawrence, cited as "the largest single gift" to "a U.S. college before the Civil War." At the same time, Yale provided competition to Harvard with the establishment of the Sheffield Scientific School. In addition, the federal government became involved in funding science and engineering education as part of a growing awareness that the future economic strength of the United States would depend increasingly on scientific and engineering expertise. "Our country abounds in men of action," Lawrence declared in the letter accompanying his bequest to Harvard," and "hard hands are ready to work hard materials." But, he asked, "Where shall sagacious heads be taught to direct those hands?" The Lawrence School, then, represented a concerted effort to build up a national scientific and technical infrastructure to rival and eventually surpass that of Europe, mirroring a similar competition in the economic and cultural spheres.

The first stage in this process was to transfer advanced European laboratory training to the western shores of the Atlantic, in particular the rigorous training methods employed at Justus Liebig's famous school of applied chemistry at Giessen in Germany. Liebig's program—incorporating everything from the basics of how to bend glass for tubes to sophisticated qualitative analysis, as well as an emphasis on the importance of original research—had attracted enthusiastic international students, including many Americans, since its founding in 1824.

One of those American students was Eben Horsford, who studied at Giessen from 1844 to 1846 before being called to the Rumford chair in science at Harvard. Horsford arrived in Cambridge, Massachusetts, determined to build a "Giessen on the Charles," and he was instrumental in convincing Lawrence to underwrite the Scientific School at Harvard (Lawrence helped finance Harvard's geology and engineering departments as well). Horsford developed the Lawrence School's first laboratory course in analytical chemistry, and he supervised the school's chemistry lab for sixteen years.

The Lawrence School developed postgraduate and professional training programs based on the German model, but by the 1860s, Horsford had shifted most of his efforts to the pursuit of wealth in industry, leaving Charles W. Eliot (the future president of Harvard) to take over his teaching responsibilities in the chemistry program. By that time, the bright scientific stars in the Harvard firmament included mathematician Charles Peirce, chemist Oliver Wolcott Gibbs (who was named to the Rumford chair, and thus leadership of the Lawrence School, with Horsford's departure), and pathbreaking botanist Asa Gray. Yet none of the Lawrence School luminaries shone brighter than the internationally renowned, Swiss-born naturalist Louis Agassiz, whose popular U.S. lecture tour in 1846 led to the offer of a chair in zoology and geology at Harvard (also funded by Lawrence), which he accepted in 1848.

Despite his lifelong opposition to Darwinian theories of evolution (which he opposed on both scientific and religious grounds), Agassiz won wide professional acclaim for his original studies on the impacts of glaciation in the northern hemisphere. It was as a teacher, fundraiser, and popularizer of serious scientific study in America, however, that Agassiz had his most lasting impact. His ability to raise money was particularly important at a time when securing scarce outside funding could prove crucial to the development or even the survival of young scientific institutions like the Lawrence School. Not only did Agassiz pry $75,000 from private donors to establish a Museum of Comparative Zoology at Harvard in 1859, but he leveraged another $100,000 from the Massachusetts state legislature at a time when government funding for science was meager.

But it was Charles W. Eliot, Harvard's president from 1869 to 1909, who solidified the university's position in scientific education. Under Eliot, Harvard established its famed "elective system," which greatly expanded the potential for advanced course offerings in the sciences and helped to attract the brightest students and faculty. Following the lead of Yale's Sheffield School, Eliot instituted formalized graduate programs in the sciences at Harvard and the Lawrence School beginning in 1872, and he made a concerted effort to woo the most brilliant scholars to Cambridge. Eliot made faculty advancement contingent upon scholarly production and original research, and he encouraged observance of the sabbatical year to afford faculty the respite from teaching and administrative duties necessary to pursue research and publishing.

To make sure that students were ready to undertake a rigorous college education, Eliot took the lead in promoting nationwide college entrance examinations. He also supervised the compilation of eighty-three chemistry and forty physics experiments that all high school students should complete before applying to Harvard. These "Harvard lists" influenced the teaching of high school science across the United States, as schools began building laboratory facilities and upgrading their curricula to meet the new standards.

Not all of Eliot's changes benefited the Lawrence School, which Eliot actually tried to merge in 1904 with Boston's rival scientific institution, the Massachusetts Institute of Technology (MIT). Court action and the resistance of faculty and students eventually blocked the merger. In the meantime, the Lawrence School suffered from doubts about its future, despite the fact that prominent industrialist Gordon McKay had pledged his substantial estate to the school.

In the end, McKay's money had to be used elsewhere at Harvard as the Lawrence Scientific School officially ceased to exist in 1906. The university's administration, led by Eliot, decided to move Lawrence's graduate programs into a new Harvard Graduate School of Applied Sciences, while the undergraduate courses became part of the regular Harvard curriculum.

Jacob Jones

Sources

Elliott, Clark A., and Margaret W. Rossiter. *Science at Harvard University: Historical Perspectives.* Bethlehem, PA: Lehigh University Press, 1992.

Love, James Lee. *The Lawrence Scientific School in Harvard University, 1847–1906.* Burlington, NC, 1944.

Miller, Howard. *Dollars for Research: Science and Its Patrons in Nineteenth-Century America.* Seattle: University of Washington Press, 1970.

Whitman, Frank P. "The Beginnings of Laboratory Teaching in America." *Science,* new ser. (August 19, 1898): 201–6.

MASSACHUSETTS HISTORICAL SOCIETY

The Massachusetts Historical Society was founded in 1791 by a small group of Bostonians who called it simply "the Historical Society." The organization's mission was to "collect, preserve, and communicate materials for a complete history of this country," not just the state. In 1794, it was chartered and renamed the Massachusetts Historical Society, to distinguish it from other fledgling historical societies.

Patriotic sentiment after the American Revolution inspired the establishment of academies and scholarly societies throughout the new nation. The historian and congregational minister Jeremy Belknap, known as the "American Plutarch," played a pivotal role in founding the society. Recognizing the need for a repository of rare books and historical manuscripts, he sought to establish a storehouse and archive. Belknap's insistence that history must be both factually accurate and based on primary sources contributed to his reputation as the founder of the "scientific history" movement in the United States. His writings helped augment the sense of pride that was developing in American thought. At the time, there was no federal depository or archive for government materials. Public documents were disappearing or poorly preserved. Only a few libraries existed, none of which compared with the renowned university libraries of Great Britain. To be a competitive cultural force, the United States needed to preserve its own history.

Belknap believed that a large collection of historical materials was essential to preserve an accurate view of history. An enthusiastic merchant from New York City, John Pintard, shared Belknap's view and proposed the formation of an American Antiquarian Society. Both men believed that the American Philosophical Society of Philadelphia and the American Academy of Arts and Sciences of Boston were too focused on science instead of history. They envisioned a new association of learned gentlemen, similar to the Society of Antiquaries of London. Belknap and Pintard met in 1789 to discuss their concept, and Belknap launched the Historical Society two years later. It was the first historical society in the United States.

An avid collector himself, Belknap made his personal possessions the cornerstone of the society's holdings. Lacking significant competition for manuscripts, the organization was able to acquire a great deal of articles. Early members donated valuable family papers, books, and artifacts and actively sought other contributors. For example, Belknap convinced Paul Revere to write an account of his famous ride for the society's archive, which survives to this day.

The society focused on documents relating to the history of America after the arrival of European settlers. A 1791 letter on the goals of the society publicized its desire to compile a natural, political, and ecclesiastical history. The organization drafted a constitution that included details on its intention to collect observations in natural history and topography, as well as specimens of "natural and artificial curiosities." For many years, therefore, the society amassed materials on natural history. After 1833, however, it turned over most of these objects to the Boston Society of Natural History and focused instead on political and cultural history.

The Massachusetts Historical Society stood at the vanguard of organized historical research in the United States, and it was a model for other research societies that began collecting primary source materials and preserving them for posterity. John Pintard founded the New York Historical Society in 1804. Individual libraries, the Library of Congress, and the National Archives also began collecting primary source materials. By the late twentieth century, there were approximately 8,000 state and local historical societies across the nation.

Another integral aspect of the society's mission was to disseminate historical information. Few American historians from afar could afford to travel to the collections in various cities, so the society tried to make its sources accessible through publication. The society began publishing historical titles in 1792 and continues to release books and monographs. The earliest published collections were printed by the Apollo Press, owned by Belknap's son Joseph. Today, the society helps to produce a scholarly journal, the *New England Quarterly*, which includes major articles on regional history, literature, and culture. Also, the *Massachusetts Historical Review*, published annually, contains essays, photographs, historical documents, and review articles.

The Massachusetts Historical Society maintains a research library and manuscript repository, along with millions of rare documents and artifacts. Notable holdings include papers from the family of John and Abigail Adams, maps and personal accounts from the Battle of Bunker Hill, and the pen that Abraham Lincoln used to sign the Emancipation Proclamation. The society organizes exhibits and public lectures, produces documentary television programs and films, and lends its materials to other nonprofit and educational institutions.

Robin O'Sullivan

Sources

Massachusetts Historical Society. http://www.masshist.org.
Riley, Stephen T. *The Massachusetts Historical Society 1791–1959.* Boston: Massachusetts Historical Society, 1959.
Tucker, Louis Leonard. *Clio's Consort: Jeremy Belknap and the Founding of the Massachusetts Historical Society.* Boston: Massachusetts Historical Society, 1989.

MORISON, SAMUEL ELIOT (1887–1976)

Samuel Eliot Morison, one of the leading American historians of the twentieth century, was known especially for his lively writing style.

Born on July 9, 1887, to a privileged family in the Beacon Hill area of Boston, Morison was fortunate to be in a milieu conducive to studying American history. He grew up among educated relatives and a city that was itself a historic monument. He received his B.A. at Harvard in 1908 and married Elizabeth Bessie Shaw Greene in 1910.

Eager to see the world, Morison went to Paris in 1913 and then returned to study and earn his Ph.D. in history. His dissertation was a biography of his grandfather, the Massachusetts political leader Harrison Gray Otis. He began teaching history at Harvard in 1915; in addition to his assigned duties, Morison taught adult education classes for the blue-collar, working poor of Cambridge, Massachusetts.

Morison's first publications were maritime histories directed at the less-educated reader. From 1922 to 1925, he took a leave of absence to teach at the University of California at Berkeley and then at Oxford University, where he was named the Harmsworth Professor of American History. He returned to Harvard in 1926, and, in 1941, he was named Jonathan Trumbull Professor of American History.

During World War II, Morison was a lieutenant commander in the U.S. Navy. Promoted to rear admiral, he was commissioned to write a history of American involvement in the Pacific. This resulted in the fifteen-volume *History of U.S. Naval Operations in World War II* (1947–1962). His wife, Elizabeth, died in 1945, and, four years later, Morison married the socially prominent Priscilla Barton, a distant cousin. In 1955, he retired from Harvard.

Morison combined his avocation, sailing, with his professional interests, retracing the voyages of European and American sailors. The resulting books were his best work. These included a two-volume biography of Christopher Columbus, *Admiral of the Ocean Sea* (1942), which won a Pulitzer Prize, and *John Paul Jones* (1959), for which he won another Pulitzer. His journeys

The distinguished historian and Harvard professor Samuel Eliot Morison, who specialized in maritime history, combined research and firsthand experience. A lifelong sailor, he retraced Columbus's voyages before writing a Pulitzer Prize–winning biography of the explorer. *(Dmitri Kessel/Time & Life Pictures/Getty Images)*

also resulted in a two-volume history, *The European Discovery of America* (1971–1974). His final work was *A Concise History of the American Republic* (1976), co-written with Henry Steele Commager and William E. Leuchtenberg.

Morison advocated narrative history that did not sacrifice good scholarship. As he wrote in *History as a Literary Art* (1948), "the quality of imagination, if properly restrained by the conditions of historical discipline, is of great assistance in enabling one to discover problems to be solved, to grasp the significance of facts, to form hypotheses, to discern causes in their first beginnings and, above all, to relate the past creatively to the present." Morison died in Boston on May 15, 1976.

Lana Thompson

Sources

Morison, Samuel Eliot. *Admiral of the Ocean Sea: A Life of Christopher Columbus.* Boston: Little, Brown, 1989.
———. *The Great Explorers: The European Discovery of America.* Oxford, UK: Oxford University Press, 1986.

———. *John Paul Jones: A Sailor's Biography*. Boston: Little, Brown, 1959.

Wilcomb, E. Washburn. "Samuel Eliot Morison, Historian." *William and Mary Quarterly* 36 (1979): 325–52.

NATIONAL ACADEMY OF SCIENCES

On March 3, 1863, in the midst of the Civil War, President Abraham Lincoln approved an act of Congress that established the National Academy of Sciences (NAS). Fifty members were appointed; upon the death or resignation of any member, the remaining members were to appoint a replacement. The legislation stipulated that the academy "shall, whenever called upon by any department of the Government, investigate, examine, experiment, and report upon any subject of science or art." The establishment of such an academy became necessary as the United States was developing into a technological society, and scientists wished to create an institution similar in function to those already established in European nations.

The original membership of fifty grew to approximately 150 by 1916, but even the larger body was unable to handle all the requests for advice from the government pertaining to military preparedness prior to America's entry into World War I. This prompted President Woodrow Wilson to establish the National Research Council (NRC) as part of the academy. The NRC was able to draw on the larger scientific community to aid members of the NAS. Wilson was later persuaded that, with the rapid expansion of American commerce and industry, the NRC should continue its work after the war. Thus, on May 11, 1918, he signed an executive order perpetuating the NRC.

Currently, four organizations comprise what is known collectively as "the Academies": the National Academy of Sciences (NAS); the National Research Council (NRC); the National Academy of Engineering (NAE), established in 1964; and the Institute of Medicine (IOM), established in 1970. Headquartered in Washington, D.C, these are private nonprofit institutions, independent of the federal government. Election to the Academies is considered a high honor and comes with no monetary compensation. Members must be U.S. citizens and are elected annually in recognition of distinguished achievement and scholarly research in their fields. Noncitizens are elected as foreign associates.

The NRC currently serves as the operating agency of the Academies. It draws from thousands of scientists, engineers, and other professionals who volunteer to serve, along with members, on committees formed to study specific scientific concerns and issue reports on their findings. The Academies advise the federal government, state governments, and private organizations. The 1997 Federal Advisory Committee Act requires that, to the best of their ability, the Academies appoint the most qualified individuals for whom there is no conflict of interest relevant to the function of the committee. Committees are expected to produce balanced and objective reports, as these reports often influence government policy.

The official journal of the Academies is the *Proceedings of the National Academy of Sciences* (PNAS). Founded in 1914, this journal publishes research reports by both members and nonmembers.

Mary F. Grosch

Sources

Hilgartner, Stephen. *Science on Stage: Expert Advice as Public Drama*. Stanford, CA: Stanford University Press, 2000.

National Academy of Sciences. http://www.nasonline.org.

National Academy of Sciences. Washington, DC: National Academy of Sciences, 1969.

NATIONAL SCIENCE FOUNDATION

The National Science Foundation (NSF) is an independent federal agency headquartered in Arlington, Virginia. It was officially established on May 10, 1950, when the National Science Foundation Act was approved by President Harry S. Truman. Its continuing mission is "to promote the progress of science; to advance the national health, prosperity, and welfare; to secure the national defense; and for other purposes."

Under President Franklin D. Roosevelt, the Office of Scientific Research and Development

(OSRD), a forerunner of the NSF, was developed for the support of military research during World War II. President Franklin Delano Roosevelt and Vannevar Bush, a leading mathematician, engineer, physicist, and head of OSRD, foresaw a similar agency to continue in times of peace. The first official proposal for what was to become the NSF was a report by Bush entitled *Science: The Endless Frontier*, published in 1945. The report recommended that the federal government accept responsibility for promoting science by funding new research and encouraging children in the study of science.

Legislation introduced from 1945 to 1949 to establish the National Science Foundation included a politician-drafted bill that envisioned the president of the United States as the ultimate authority. A scientist-backed bill (drafted by the OSRD) favored a foundation run by a science board. A compromise was reached in 1949 whereby a dual authority was proposed—a Science Board and director with similar powers—and the NSF became a reality in 1950. The NSF director oversees the staff and management and is responsible for planning, budgeting, and day-to-day operations. The National Science Board establishes NSF policies. The board's twenty-four members serve a six-year term and are appointed by the president of the United States and confirmed by the U.S. Senate.

The mission of the NSF remains the same as when it was created in 1950, which is to initiate and support scientific research and programs to promote scientific education. Other responsibilities have been added to the mission, including the promotion of international collaboration to address global concerns and participation by women, minorities, and persons with disabilities who remain underrepresented in the scientific fields.

The NSF does not conduct research itself but funds fundamental research in science (except medical science) and engineering. This includes fields such as mathematics, computer science, and the social sciences. According to the NSF Web site, its "job is to determine where the frontiers are, identify the leading U.S. pioneers in these fields and provide money and equipment to help them continue." The agency funds both existing and emerging fields of study. Proposals for research, both solicited and unsolicited, are received by the foundation and undergo a rigorous evaluation by independent reviewers before final funding decisions are made.

Science education is supported in the form of fellowships and trainee positions for graduate students, scientists, and science teachers. The NSF also promotes the enhancement of teachers' skills and the improvement of curricula at the elementary through high school levels.

Mary F. Grosch

Sources

Bush, Vannevar. *Science, the Endless Frontier.* A Report to the President by Vannevar Bush, Director of Office of Scientific Research and Development, July 1945. Washington, DC: U.S. Government Printing Office, 1945.

Lomask, Milton. *A Minor Miracle: An Informal History of the National Science Foundation.* Washington, DC: National Science Foundation, 1976.

National Science Foundation. http://www.nsf.gov.

RENSSELAER POLYTECHNIC INSTITUTE

In late 1824, attorney and itinerant lecturer Amos Eaton, with financial support from the region's great landowner, Stephen Van Rensselaer, founded the Rensselaer School in Troy, New York. It was the first technically oriented college in the English-speaking world. Eaton had earned bachelor's and master's degrees at Williams College, had studied science at Yale under Benjamin Silliman, and had lectured on scientific subjects throughout the northeastern United States. Rensselaer was initially staffed by a senior professor (Eaton), a junior professor, and one or two adjunct lecturers. In that era, practical experience or intensive study of a scientific subject sufficed to qualify one as a professor; many of Rensselaer's instructors in subsequent decades were recent graduates of the school. Enrollment was small, and a bachelor's degree could be earned in one or two years. From the beginning, however, Rensselaer's rigorous standards resulted in rates of attrition as high as two-thirds of the student body.

The institution specialized in technical and scientific topics at a time when a liberal arts focus was taken for granted elsewhere. By 1861, it took the name Rensselaer Polytechnic Institute (RPI). At every point, the young school's emphasis was on practical applications of science and technology in everyday life. Students were required to prepare lectures and carry out their own experiments, rather than passively witness an instructor's experimentation. Rensselaer's degree in Civil Engineering (C.E.)—the world's first such degree—and Bachelor of Natural Science (B.N.S) gave way to the more common B.Eng. and B.S. only after other colleges began to compete in the field of scientific education.

The school also stated the intention of educating the sons (and eventually the daughters) of the rising middle class. Financial assistance, or at least credit, was extended to a few well-qualified students of need, and RPI pioneered in opening its doors to foreign and ethnic minority students long before this was the norm in American education.

RPI operated on a shoestring budget until the 1890s, when Palmer Ricketts took over as president. Ricketts actively sought funds from the growing body of RPI alumni, as well as from philanthropists such as Andrew Carnegie and Mrs. Russell Sage. RPI grew in size and endowment, adding students, faculty, buildings, and areas of study during Ricketts's forty-two-year administration. RPI added formal graduate study in 1913, awarding its first doctorate three years later. During the course of the twentieth century, RPI placed increasing emphasis on original research, outside funding, and the acquisition of world-class technology. Once the only school of its kind in North America, RPI faced increasing competition from scientific schools at Yale, Harvard, Columbia, and other major universities—often staffed by RPI graduates. Its previous emphasis on training qualified civil engineers had to give way to more areas of study.

Rensselaer remains a highly successful, financially stable institution of higher learning, with five schools: architecture, engineering, humanities and social sciences, management and technology, and science. Resisting the trend in American education to grow at any cost, RPI has kept its enrollment to just over 7,000 students. Its faculty and alumni include dozens of world-famous scholars, inventors, and entrepreneurs. RPI continues to make important innovations in the development of American technological education.

David Lonergan

Sources

Baker, Ray P. *A Chapter in American Education: Rensselaer Polytechnic Institute 1824–1924.* New York: Charles Scribner's Sons, 1924.

Rezneck, Samuel. *Education for a Technological Society: A Sesquicentennial History of Rensselaer Polytechnic Institute.* Troy, NY: Rensselaer Polytechnic Institute, 1968.

ROYAL SOCIETY OF LONDON

The founding in 1660 of the Royal Society, an all-male, London-based organization with state sponsorship but financial and organizational independence, greatly encouraged the conduct of science in the English colonies in America. The society's initial goal was to compile records of scientific and technological phenomena.

For information from America, the society needed the help of American residents. The first colonial admitted as a fellow of the Royal Society was John Winthrop, Jr., whose description of how Americans made pitch was the first paper read to the society (in 1662) by a colonial. From 1660 to 1783, dozens of colonial Americans were admitted as fellows, and hundreds of letters from the colonies were published in the society's journal, *Philosophical Transactions.* Colonial fellows benefited by not having to pay admission fees or dues until 1753, when the society's financial needs led it to abolish the exemption.

The English leaders of the Royal Society in the seventeenth century viewed colonial fellows more as sources of knowledge and artifacts than as original thinkers. Colonial residents had access to the plants, animals, and minerals of America, and they were particularly valued for their ability to explain technical processes peculiar to the colonies.

Realizing that information about colonial technology and natural history should be gathered systematically, the Royal Society prepared lists of queries that were sent to America with individuals who were journeying to the colonies. In doing so, the society hoped to satisfy British scientific curiosity as well as strengthen the economy of the empire by identifying colonial resources for exploitation. In addition, the Royal Society promoted visits by scientists to America for the purpose of gathering information; it sponsored observations of the eighteenth-century transits of Venus, for example. The society also encouraged natural historians such as Mark Catesby.

By the mid-eighteenth century, Americans were gaining more respect as original scientists, as evidenced by the awarding of the Royal Society's highest honor, the Copley Medal, to Benjamin Franklin in 1753 and the waiving of his fees upon admission as a fellow in 1756. Other prominent colonial scientists who were fellows included Cotton Mather, William Byrd II, and the Harvard astronomer John Winthrop IV. The Royal Society also provided the inspiration and organizational model for the first American scientific societies, the seventeenth-century Boston Philosophical Society and the far more successful American Philosophical Society.

The amateur culture of the Royal Society, as opposed to its far more professionalized French rival, the Royal Academy of Sciences, greatly influenced the development of scientific culture in early America. The close connection between the Royal Society and the American scientific community was weakened by the War of Independence, but, in subsequent years, American scientists continued to be admitted to the Royal Society.

William E. Burns

Sources

Burns, William E. *Science and Technology in Colonial America.* Westport, CT: Greenwood, 2005.

Royal Society. http://www.royalsoc.ac.uk.

Stearns, Raymond Phineas. *Science in the British Colonies of America.* Urbana: University of Illinois Press, 1970.

SARTON, GEORGE (1884–1956)

George Alfred Léon Sarton, the father of poet May Sarton, was a pioneer in establishing the history of science as a distinct discipline. He wrote fifteen books and more than 300 articles in a dynamic career as a scholar and editor.

Born on August 31, 1884, in Ghent, Belgium, Sarton entered the University of Ghent to study philosophy, but after two years he withdrew in disgust for a year before beginning his work in the natural sciences in 1905. Although his work in chemistry earned him a gold medal from the university, he received his doctor of science degree in 1911 for a historical and philosophical thesis on the celestial mechanics of Newton.

In 1912 Sarton founded *Isis,* a scholarly journal devoted to the history and philosophy of science. During the early days, according to Harvard historian of science I. Bernard Cohen, Sarton's wife, Mabel, wrapped and mailed each issue herself.

During his forty years as editor of *Isis,* Sarton compiled an index of thousands of publications dealing with the history of science throughout the world in the form of a critical bibliography, which helped make scholars aware of the resources and growing literature in the field. For *Isis,* he recruited a distinguished editorial board that included mathematician Henri Poincaré, sociologist and philosopher Emile Durkheim, physiologist Jacques Loeb, and chemist Friedrich Wilhelm Ostwald. The range of fields represented by these scholars reflected Sarton's conviction that the history of science was by nature an encyclopedic discipline. The ultimate goal was to create a synthesis of science and the humanities—an ideal he called "the new humanism."

Sarton began to accumulate notes for what would become his *Introduction to the History of Science,* which he first conceived as a relatively short, two- or three-volume history of science to 1900. His devotion to compiling the history was disrupted by the devastation of the German occupation of Belgium in 1914. When his family was forced to abandon their home, Sarton buried the notes for his book in a metal trunk in the garden.

The family moved to England, where Sarton worked for the British War Office. In 1916, he emigrated to the United States, where, in 1918, Robert S. Woodward, the president of the Carnegie Institution in Washington, D.C., created for him the position of research associate in the history of science.

With the support of the Carnegie Institution, the recovery of his notes after the war, and the use of the Widener Library at Harvard University in exchange for honorary teaching responsibilities, Sarton completed the three-volume *Introduction to the History of Science* over a nearly thirty-year period, from 1919 to 1947. In that mammoth work, he reviewed and cataloged the scientific and cultural contributions of nearly every civilization from antiquity through the fourteenth century.

With his emphasis on critical bibliography, his sweeping survey of scientific inquiry, and the journal he created, Sarton helped create the elements required by the new field of history of science. In 1924, when the History of Science Society was founded, *Isis* became its official publication (Sarton continued to assume financial responsibility until 1940). His influence is further evidenced by his numerous honorary degrees (from such institutions as Brown University and Harvard University), the scholarly honor societies to which he was elected (including the American Academy of Arts and Sciences and the Philosophical Society of Philadelphia), and the roles he served as founding member of the International Academy of the History of Science and president of the International Union of the History of Science. The George Sarton Medal, established in his honor, remains the most prestigious prize of the History of Science Society; it has been awarded annually since 1955 to an outstanding historian of science selected from the international scholarly community.

Eric Boyle

Sources

Garfield, Eugene. "The Life and Career of George Sarton: The Father of the History of Science." *Journal of the History of Behavioral Sciences* 21:2 (1985): 107–17.

Thackray, Arnold. "Sarton, Science, and History." *Isis* 75 (1984): 19–20.

Thackray, Arnold, and Robert K. Merton. "On Discipline Building: The Paradoxes of George Sarton." *Isis* 63 (1972): 473–95.

choose to announce their important discoveries to peers and the public. Thomas Alva Edison founded *Science* in 1880, but the first weekly issue did not appear until 1883. Since 1900, it has been sponsored by the American Association for the Advancement of Science (AAAS).

The first issue was an ambitious overview of the sciences, with articles on a diversity of topics and a surprising number of illustrations. In 1923, it was a twenty-page, sparsely illustrated weekly publication, offering detailed essays, information on current discoveries, book reviews, and news items. By the middle of the twentieth century, little had changed: It had grown into a forty-six-page compilation, but it was still sparsely illustrated, with somewhat longer papers, news, announcements, book reviews, technical papers, and footnoted communications. In 1964, it was much more impressive: a weekly with multiple sections, an eye-catching cover picture, and 100 or more pages, many of them illustrated.

Today, *Science* publishes a plethora of different types of items and articles, including brief letters, news notices, technical comments, research articles, reports, and discussions of science policy. Issues contain either an in-depth cover story (such as "HIV/Aids in Asia") or special sections with multiple articles on the same topic (such as "The State of the Planet" or "Genomic Medicine"). Some weekly issues run more than 200 pages and are generously illustrated in color.

Many well-known scholars have been associated with *Science*. Stephen Jay Gould, Edward O. Wilson, and Margaret Mead have been published in it, as have Nobel laureates Gertrude Elion, Joshua Lederberg, Barbara McClintock, Albert Michelson, Robert Millikan, Glenn Seaborg, and Rosalyn Yalow, among others.

Robert Hauptman

Source

Science magazine. http://www.sciencemag.org.

SCIENCE

Science and its British analogue, *Nature*, are two of the most prestigious scientific publications in the world, the journals that most researchers would

SCIENTIFIC AMERICAN

Devoted to disseminating scientific discoveries and theories to the intelligent layperson, *Scientific American* is the oldest continuously published magazine in the United States. Similar publications such as *National Geographic* and *Psychology Today* serve their readers very well, but they are more specialized in their interests and do not maintain the sophisticated intellectual level that is the hallmark of *Scientific American*. Without pandering or condescending to its readers, but concomitantly avoiding the esoteric articulations or incredibly complex mathematics that one might find in *Cell* or *Physical Review Letters*, the periodical offers lucid and concise overviews and explanations to its readers.

Scientific American began in 1845 as a weekly publication. By 1850, it promoted the work of the U.S. Patent Agency, and early issues concluded with a list of recent patents. In 1907, it consisted of twenty oversize, generously illustrated pages and was a cross between *Popular Mechanics* and *National Geographic*. By 1939, it was a monthly publication similar to today's version, but the illustrations continued to appear in black and white. Some color was evident by 1951, and, during the course of that year, book reviewers included I.I. Rabi, I. Bernard Cohen, Ernest Nagel, and Jacob Bronowski.

Over the years, *Scientific American* has changed in size, structure, organization, and emphases, but it has remained constant in its goal: to present state-of-the-art overviews of technology and the sciences (physics, astronomy, chemistry, biology, geology, and mathematics) and their subdisciplines. Great discoveries naturally interest readers, and the magazine has managed to induce 127 Nobel Prize winners to write 213 articles on their specializations. Long before well-known scientists such as Guglielmo Marconi, the Wright brothers, and Robert Goddard succeeded in their endeavors, their efforts were covered in *Scientific American*. Albert Einstein, Jonas Salk, Robert Jarvik, Francis Crick, Linus Pauling, and John Kenneth Galbraith are among the innumerable authors who have offered invaluable perspectives on their work.

As early as 1899, the editors presented special thematic issues (on bicycles and cars), and, for many years, each September number has been devoted to a particular topic. For example, the 1950 special issue was devoted to "The Age of Science," for which J. Robert Oppenheimer, Harlow Shapley, Max Born, Theodosius Dobzhansky, Alfred Kroeber, and Linus Pauling, among

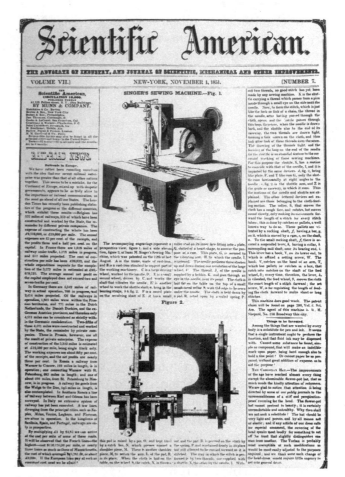

Scientific American, the nation's oldest continuously published magazine, was first issued in August 1845 as a weekly broadsheet. The November 1, 1851, edition featured Isaac Singer's invention of the continuous-stitch sewing machine. *(Mansell/Time & Life Pictures/Getty Images)*

others, contributed overviews of their respective disciplines. The 2003 special issue focused on "Better Brains," and the 2004 special topic was "Beyond Einstein."

Issues are now replete with color images and graphics, diverse brief and more detailed articles, and interviews with notable figures. During the second half of the twentieth century, two authors helped to make *Scientific American* the extraordinary publication that it is: Martin Gardner's column on mathematical puzzles and games was eagerly awaited each month, and Philip Morrison's wide-ranging and incisive book reviews apprised readers of new publications in all disciplines. *Scientific American* is published in sixteen foreign languages and has a circulation of 1 million copies. The entire run is held in hardbound volumes and various microformats by many academic research libraries. A database of digital archives (dating to 1993) is available by subscription.

Robert Hauptman

Sources

Mitchell, Carolyn B. *Life in the Universe: Readings from Scientific American Magazine.* New York: W.H. Freeman, 1994.
Scientific American. http://www.sciam.com.

SHEFFIELD SCIENTIFIC SCHOOL, YALE UNIVERSITY

On August 19, 1847, the governing body of Yale College—the Yale Corporation—approved the establishment of a postgraduate program in the applied sciences, the first of its kind in the United States. The Yale Scientific School emerged in 1853 when Yale's civil engineering program joined the School of Applied Chemistry.

With only two professorships, eight students, limited funding, and a grudging acceptance by the parent college, the school was slow to develop. Only after New Haven railroad developer Joseph Earl Sheffield made a major financial and property bequest to the school in 1858 did the program achieve solvency. In 1860, the school began offering the Doctor of Philosophy degree. The first Ph.D. was awarded in 1861, the same year the organization changed its name to the Sheffield Scientific School to honor its primary benefactor.

Sheffield's bequest indicated an increasing awareness that scientific training and professional research would be crucial to America's future industrial growth. Previously, such training was hard to come by outside of Europe, and aspiring practitioners in applied chemistry or physics usually had to go abroad for their graduate studies. In the first half of the nineteenth century, Yale's small science faculty boasted one of the country's most prominent advocates of American science at the time, Benjamin Silliman, Sr., founder and editor of the eminent *American Journal of Science* (1818). Yet even Silliman found it difficult to secure funds and support to make science education a required part of the Yale curriculum (his popular courses were offered as

electives). In the years before the Sheffield endowment, professors at the Yale Scientific School often had to purchase their own books and equipment, and even pay rent to Yale College for use of a campus building. Moreover, advocates of expanded science and engineering instruction continued to run up against resistance from more traditionally minded educators who argued that combining pedestrian "vocational" training such as engineering with the conventional "classical" curriculum of ancient Latin and Greek literature and philosophy would dilute the very purpose of the college, which they viewed as preparing gentlemen for their future roles as social, governmental, and military leaders.

The situation changed when James Dwight Dana, a Yale professor of geology, called for a new scientific school at Yale—the central theme in his commencement address of 1856—which inspired a number of private donors to contribute funds for the enterprise, including Sheffield. Sheffield's son-in-law, John Addison Porter, became a member of the faculty at the new institution. Porter was an alumnus of the rigorous training program in applied chemistry at the internationally famous Justus von Liebig Laboratory at the University of Giessen in Germany. Liebig's students began with practical laboratory work—learning how to sharpen knives, drill corks for flasks, and bend glass for tubes—before moving on to qualitative analysis and finally an individual research problem directed by Liebig himself. Liebig's fame drew students from around the world, including many from the United States, and most left as disciples or at least disseminators of his methods in their home countries, including both Porter and assistant chemist Samuel W. Johnson at Sheffield.

The Sheffield School attracted other first-rate professors, including Daniel Coit Gilman, future founding president of Johns Hopkins University, who took charge of building up Sheffield's library while also serving as professor of geography. Another Sheffield professor was Francis Amasa Walker, later president of the Massachusetts Institute of Technology. Among the early graduates of Sheffield were such famous engineer-scientists as J. Willard Gibbs, who made his name in fields as disparate as theoretical physics and the development of improved railcar braking systems.

The Sheffield faculty worked on public outreach as well. In 1866, they initiated a highly popular series of public science presentations—originally titled "Public Lectures to Mechanics" but soon referred to simply as the "Sheffield Lectures"—which often drew New Haven audiences numbering in the hundreds. After Gilman's departure in 1872, Sheffield's scientist-directors, mineralogist George Jarvis Brush, and chemist Russell Henry Chittenden strengthened the school's reputation for superior scientific and engineering training, as well as the quality of its faculty research and public outreach programs.

By the early twentieth century, however, it had become obvious that there was great duplication of effort between the Sheffield School and its parent institution, as Yale had added significantly to the science side of its liberal arts curriculum. Thus, when Yale decided upon a general reorganization of the university in 1918–1919, Sheffield was left with responsibility of teaching a four-year course of undergraduate, pre-professional instruction in science and engineering (leading to the Bachelor of Science degree), but its graduate program was transferred to a new graduate school administered by Yale University.

Only in 1945, after twenty-five years as an undergraduate institution, did Sheffield resume its original role of postgraduate scientific training (engineering instruction had moved to a separate School of Engineering in 1932). The reversion proved short-lived, as the Sheffield Scientific School officially ceased to exist in 1956, its faculty and graduate students reclassified under the Division of Science at Yale University.

Jacob Jones

Sources

Baitsall, George A., ed. *The Centennial of the Sheffield Scientific School.* New Haven, CT: Yale University Press, 1950.

Chittenden, Russell H. *History of the Sheffield Scientific School of Yale University.* New Haven, CT: Yale University Press, 1928.

Miller, Howard. *Dollars for Research: Science and Its Patrons in Nineteenth-Century America.* Seattle: University of Washington Press, 1970.

Warren, Charles H. "Sheffield Scientific School—The First Hundred Years." *Scientific Monthly* 67:1 (1948): 58–63.

White, Gerald T. "Benjamin Silliman, Jr., and the Origins of the Sheffield Scientific School at Yale." *Ventures* 8:1 (1968): 19–25.

Smithsonian Institution

The Smithsonian Institution is made up of eighteen museums and nine research centers—most located in Washington, D.C.—devoted to understanding, exploring, and explaining American natural and human history and culture. The Smithsonian was established in 1846 from an estate owned by the British scientist James Smithson, who bequeathed to the United States a trust in the amount of $508,318. Smithson's early vision, an institution dedicated to the growth and diffusion of knowledge, continues to be fulfilled today.

The first building, referred to as "the Castle," was completed in 1855. Its unique architecture—a Gothic Revival design drafted by James Renwick—has eye-catching turrets, spires, parapets, and towers. During the 1880s and 1890s, other Smithsonian buildings were opened: the United States National Museum Building, the Astrophysical Observatory, and the National Zoo. The latter included an "animal house" and an outdoor display area for 839 fauna from all over the world. Smithsonian Park, an area with tree-shrouded winding paths, in time became known as the National Mall.

Visitors to the Natural History Museum can observe preserved fauna from all over the world, a living insect colony, minerals and gems, and carefully prepared skeletons of animals. Exhibits illustrate how Earth came to be, how animals adapted to ecologic niches by either surviving or dying out, and how people evolved and spread throughout the world. What visitors do not see are the areas where scientists, technicians, and artists study, clean, classify, and document the holdings.

The National Air and Space Museum has twenty-two galleries, which house a huge collection of historic aircraft and spacecraft. The history of air and space technology is illustrated with the Wright brothers' 1903 airplane, the *Spirit of St. Louis*, and the *Apollo 11* command module. Planetarium presentations are given throughout the day. In 2001, the space history section began to save artifacts from the *Apollo* space program.

One focus is the spacesuits developed for the moon landings of the late 1960s and early 1970s.

The fabrics provide valuable information as they age under the scrupulous observation of museum scientists. Since both degradable and permanent materials are used to construct a spacesuit, the staff has documented, identified, and photographed every spacesuit. Accessories such as helmets, gloves, and books also have been collected. Because of this project, the museum will be a worldwide authority for spacesuit preservation and conservation.

Extensive art collections are found in many Smithsonian buildings. The Freer Gallery designed by Charles Platt, an Italian Renaissance–style building adjacent to the Smithsonian Castle, opened on the National Mall in 1923. The collection includes Asian and American art: bronzes, jades, screens, scrolls, ceramics, paintings, and metalworks. The Hirshhorn Museum and Sculpture Garden opened in 1974. Architect Gordon Bunshaft designed this building in the shape of a cylindrical drum. The collection focuses on twentieth-century art.

The National Museum of American History and the National Museum of African American History house collections of sociocultural interest. There is a division of cultural history, where exquisite keyboard and stringed instruments are displayed; among the violins, cellos, and violas are creations of Stradivarius.

The National Portrait Gallery and the American Art Museum are housed at the Patent Office Building, which was designed in Greek Revival style by Pierre L'Enfant. Both focus on individual artists rather than styles of art.

Lana Thompson

Sources

Small, Lawrence. "Fanciful and Sublime." *Smithsonian Magazine* (December 2002): 12.
———. "A Pantheon After All." *Smithsonian Magazine* (July 2002): 16.
———. "Pursuing Perfection." *Smithsonian Magazine* (September 2002): 16.
Smithsonian Institution. http://www.si.edu.

SPELMAN COLLEGE

Spelman College, founded in 1881 in Atlanta, Georgia, is one of two surviving black colleges for women and one of the largest undergraduate producers of African American women in science.

Scientific study at Spelman College, as at most historically black colleges and universities, suffered for decades under segregationist policies in higher education. Science programs were underfunded, poorly equipped, and largely rooted in a philosophy of manual and industrial training. Those who did pursue science were limited to practical fields, such as home economics, agriculture, the mechanical arts, and premedicine. The cost of instruction in these areas was less than in research-based disciplines that required equipment and instrumentation. When Spelman College was founded, the concept of a liberal arts education to support the development of black female scientists ran counter to what white society thought black women could or should be.

African American consciousness-raising in the 1960s and the emerging women's movement in the 1970s would help to change attitudes and national policies. Two members of the mathematics faculty at Spelman in particular, Shirley Mathis McBay and Etta Zuber Falconer, questioned the college's commitment to educating African-American women in science. The chemistry curriculum was little more than a service course for students pursuing majors in home economics or physical education, and those with any science interests outside of premedicine had

The Smithsonian Institution in Washington, D.C., was created by an act of Congress in 1846 "for the increase and diffusion of knowledge." It was funded with a bequest by British chemist and mineralogist James Smithson, who never set foot in the United States. *(Karen Bleier/AFP/Getty Images)*

to petition to take the majority of their courses at neighboring colleges. McBay and Falconer, only the ninth and eleventh black women in the United States to earn doctorates in their field, wanted to put in place a structure that would nurture the growth of future black women in science. With the support of Spelman's president, Albert E. Manley, the two began work in the early 1970s to reinvent and build Spelman's science program.

Over the next two decades, departments were added to include biology, chemistry, mathematics, computer science, and physics, as well as a dual degree program in engineering. Under the dual degree program, students spent three years at Spelman taking preengineering coursework and two years at neighboring Georgia Institute of Technology for an engineering specialty, graduating with degrees from both institutions. In addition to curricular changes, academic-year and summer programs were established to increase student recruitment, retention, and graduation. The college actively recruited and hired faculty with doctoral degrees and active research portfolios so that students could gain hands-on experience with research.

By the 1990s, the number of science graduates increased by more than 450 percent, from 28 in 1968 to 132 in 1996. Denise Stephenson-Hawk, a mathematics major, was one of those graduates. After leaving Spelman in 1976, she became the first African American woman to earn a doctoral degree in fluid dynamics, from Princeton University. Greer Lauren Geiger, also a 1976 graduate, earned an M.D., from Harvard Medical School, specializing in ophthalmology. Geiger's fascination with the body, visual arts, and the biology and chemistry of the eye led her to pioneer a form of eye surgery that uses gas inside the eye to repair macular holes that result in the loss of the center of vision.

In 1995, the National Science Foundation named Spelman College a model institution for excellence in undergraduate science and mathematics education. For the period 1997–2001, the NSF ranked Spelman among the top fifteen baccalaureate-origin institutions graduating African Americans in the sciences who went on to earn the doctoral degree in science disciplines. With a second model designation by the NSF in 2000, Spelman continues to rank as one of the top producers of African American women in science.

Olivia A. Scriven

Sources

Barnett, Harold M. "Spelman's Response to the Scientific Challenge." *Spelman Messenger* (Summer/Fall 1993): 12–15.

Falconer, Etta Z. "A Story of Success: The Sciences at Spelman College." *SAGE: A Scholarly Journal on Black Women* 6:2 (1989).

Noble, Jeanne L. *The Negro Woman's College Education.* New York: Columbia University Press, 1956.

Pearson, Willie, Jr., and H. Kenneth Bechtel, eds. *Blacks, Science, and American Education.* New Brunswick, NJ: Rutgers University Press, 1989.

TAYLOR, FREDERICK (1856–1915)

Frederick Winslow Taylor, the founder of modern scientific management and the first industrial efficiency engineer, was born in Germantown, Pennsylvania, on March 20, 1856, to a wealthy Philadelphia Quaker family. At age twelve, Taylor traveled to Europe with his family, where they remained for three years. On returning to the United States, he attended Phillips Exeter Academy, an elite prep school in New Hampshire.

Taylor developed what would become chronic health problems and, in an attempt to combat nightmares, invented a kind of harness to wear during sleep. Because of chronic headaches, insomnia, and poor vision, he chose to forego medical studies at Harvard University for a career in industry. In 1874, he became an apprentice machinist and pattern maker at Enterprise Hydraulic Works, a small Philadelphia pump operation. He completed his apprenticeship in 1878 and, through a family connection, began working in the machine shop of Midvale Steel while enrolled in university studies. He was awarded a mechanical engineering degree from the Stevens Institute of Technology in New Jersey in 1883. The following year, he married Louise Spooner; the couple eventually adopted three children.

Most of Taylor's observations and experimental studies on industrial management were made while he was working at Midvale. Over a twelve-year period, he ascended the corporate hierarchy as a machinist, foreman, master mechanic, and chief engineer. He is best known for developing and introducing time-motion studies, in which workers' movements are measured and controlled to maximize work and boost the speed and volume of production.

Taylor advocated production efficiency based on particular management techniques, such as fitting the best workers and tools to the job, and developing a cooperative approach between workers and management. He vehemently opposed trade unions or any collective bargaining; he believed that workers should cooperate with management above all else. His methods often produced resentment not only among workers but among managers as well. In the Progressive Era of the early twentieth century, with the rise of the American labor movement, "Taylorism" was widely denounced as exploitative, oppressive, dehumanizing, antidemocratic, and harmful to workers. Notwithstanding such criticism, Taylor's philosophy of "the one best way" spread throughout the industrialized world, taking root in shops, offices, and industrial plants, especially steel mills. Taylor's methodology became a source of industrial conflict that continued throughout the twentieth century.

Taylor is credited with launching the movement in scientific management with his best-known text, *The Principles of Scientific Management* (1911). Other works include *A Piece Rate System* (1895), *Shop Management* (1903), and *Concrete Costs* (with S.E. Thompson, 1912).

During his career, Taylor also worked at the Manufacturing Investment Company in Madison, Maine. He acted as a consultant and was a professor at Dartmouth College's Tuck School of Business. He died of pneumonia on March 21, 1915.

Heidi Rimke

Sources

Kakar, Sudhir. *Frederick Taylor: A Study in Personality and Innovation.* Cambridge, MA: MIT Press, 1970.

Kanigel, Robert. *The One Best Way: Frederick Winslow Taylor and the Enigma of Efficiency.* New York: Viking, 1997.

Nelson, Daniel. *Frederick W. Taylor and the Rise of Scientific Management.* Madison: University of Wisconsin Press, 1980.

Sicilia, David B. *The Principles of Scientific Management.* Norwalk, CT: Easton, 1993.

Wrege, Charles D., and Ronald G. Greenwood. *Frederick W. Taylor, the Father of Scientific Management: Myth and Reality.* Homewood, IL: Business One Irwin, 1991.

DOCUMENTS

America's First Historical Editor

Ebenezer Hazard's Historical Collections *(1792) was the first edited compilation of American documents. The following is the preface to the work, in which the author explains his reasons for collecting and publishing the documents.*

When the Conduct of Individuals in a Community is such as to attract public Attention, others are very naturally led to many Inquiries respecting them; so, when Civil States rise into Importance, even their earliest History becomes the Object of Speculation. Secluded from the rest of the World, the Anglo-American Colonies were viewed merely as Dependencies on Great Britain; and little more of them, comparatively, was known, than at what time they were discovered, and by whom: But when they dared to assert their Claims to Freedom, and in Defence of them to oppose the Parent State, whose Power even Europe dreaded—when they compelled her to consent to their Emancipation, and to acknowledge them as Independent States—they were then thought worthy of more respectful Attention, and an Acquaintance with their History was faught for with Avidity. But although the public Mind was anxious for Information, it could not be easily obtained: The Histories which had appeared, relating to a few individual States only, were not sufficient to gratify the inquisitive and were, in general, written so long since, as not now to prove satisfactory; and Materials for furnishing a more comprehensive View of the Subject were much dispersed, and not within the Reach of many. To remove this Obstruction from the Path of Science, and, at the same Time, to lay the Foundation of a good American History, is the Object of the following Compilation. It was the Compiler's original Intention to visit each State in the Union, and to remain there a sufficient Time to form a complete Collection of such Materials for its History as had escaped the Ravages of Time and Accident. His Design was honoured with the Approbation and Patronage of Congress, whose Recommendation of it gained him immediate Access to the Archives of New Hampshire and Massachusetts, including those of the old Colony of Plymouth, and the Province of Maine; but before he could proceed further an Appointment, as Post Master General of the United States, obliged him to reside at the Seat of Federal Government, and prevented his continuing the Work in the Method he at first proposed:—the papers collected since have been picked up just as they happened to fall in his Way: Hence the Compilation, although large, is necessarily very far from being *complete*; but he has, notwithstanding, thought it expedient to publish it in its present State, lest it should be scattered and lost;—he hopes too, that by laying a Foundation, he may induce others to prosecute Work which he conceives is not devoid of either Utility or Entertainment.

Source: Ebenezer Hazard, *Historical Collections: Consisting of State Papers, and Other Authentic Documents, Intended as Materials for an History of the United States of America* (Philadelphia: Thomas Dobson, 1792–1794).

The Virgin and the Dynamo

In his autobiography, The Education of Henry Adams *(1918), the author summarizes his view of history as a perceived contest between human attempts to control the environment, and the subsequent development of technology (the Dynamo), and the inherent human need for security found by accommodating and embracing the pastoral mentality of dependence on nature and God (the Virgin). The following excerpt is taken from Chapter 25, "The Virgin and the Dynamo."*

Then he showed his scholar the great hall of dynamos, and explained how little he knew about electricity or force of any kind, even of his own special sun, which spouted heat in inconceivable volume, but which, as far as he knew, might spout less or more, at any time, for all the certainty he felt in it. To him, the dynamo itself was but an ingenious channel for conveying somewhere the heat latent in a few tons of poor coal hidden in a dirty engine-house carefully kept out of sight; but to Adams the dynamo became a symbol of infinity. As he grew accustomed to the great gallery of machines, he began to feel the forty-foot dynamos as a moral force, much as the early Christians felt the Cross. The planet itself seemed less impressive, in its old-fashioned, deliberate, annual or daily revolution, than this huge wheel, revolving within arm's length at some vertiginous speed, and barely murmuring—scarcely humming an audible warning to stand a hair's-breadth further for respect of power—while it would not wake the baby lying close against its frame. Before the end, one began to pray to it; inherited instinct taught the natural expression of man before silent and infinite force. Among the thousand symbols of ultimate energy the dynamo was not so human as some, but it was the most expressive. . . .

Historians undertake to arrange sequences,—called stories, or histories—assuming in silence a relation of cause and effect. These assumptions, hidden in the depths of dusty libraries, have been astounding, but commonly unconscious and childlike; so much so, that if any captious critic were to drag them to light, historians would probably reply, with one voice, that they had never supposed themselves required to know what they were talking about. Adams, for one, had toiled in vain to find out what he meant. He had even published a dozen volumes of American history for no other purpose than to satisfy himself whether, by severest process of stating, with the least possible comment, such facts as seemed sure, in such order as seemed rigorously consequent, he could fix for a familiar moment a necessary sequence of human movement. The result had satisfied him as little as at Harvard College. Where he saw sequence, other men saw something quite different, and no one saw the same unit of measure. He cared little about his experiments and less about his statesmen, who seemed to him quite as ignorant as himself and, as a rule, no more honest; but he insisted on a relation of sequence. And if he could not reach it by one method, he would try as many methods as science knew. Satisfied that the sequence of men led to nothing and that the sequence of their society could lead no further, while the mere sequence of time was artificial,

and the sequence of thought was chaos, he turned at last to the sequence of force; and thus it happened that, after ten years' pursuit, he found himself lying in the Gallery of Machines at the Great Exposition of 1900, his historical neck broken by the sudden irruption of forces totally new. . . .

The historian was thus reduced to his last resources. Clearly if he was bound to reduce all these forces to a common value, this common value could have no measure but that of their attraction on his own mind. He must treat them as they had been felt; as convertible, reversible, interchangeable attractions on thought. He made up his mind to venture it; he would risk translating rays into faith. Such a reversible process would vastly amuse a chemist, but the chemist could not deny that he, or some of his fellow physicists, could feel the force of both. When Adams was a boy in Boston, the best chemist in the place had probably never heard of Venus except by way of scandal, or of the Virgin except as idolatry; neither had he heard of dynamos or automobiles or radium; yet his mind was ready to feel the force of all, though the rays were unborn and the women were dead. . . .

This problem in dynamics gravely perplexed an American historian. The Woman had once been supreme; in France she still seemed potent, not merely as a sentiment, but as a force. Why was she unknown in America? For evidently America was ashamed of her, and she was ashamed of herself, otherwise they would not have strewn fig-leaves so profusely all over her. When she was a true force, she was ignorant of fig-leaves, but the monthly-magazine-made American female had not a feature that would have been recognized by Adam. The trait was notorious, and often humorous, but any one brought up among Puritans knew that sex was sin. In any previous age, sex was strength. Neither art nor beauty was needed. Every one, even among Puritans, knew that neither Diana of the Ephesians nor any of the Oriental goddesses was worshipped for her beauty. She was goddess because of her force; she was the animated dynamo; she was reproduction—the greatest and most mysterious of all energies; all she needed was to be fecund. . . .

[I]n mechanics, whatever the mechanicians might think, both energies acted as interchangeable force on man, and by action on man all known force may be measured. Indeed, few men of science measured force in any other way. After once admitting that a straight line was the shortest distance between two points, no serious mathematician cared to deny anything that suited his convenience, and rejected no symbol, unproved or unproveable, that helped him to accomplish work. The symbol was force, as a compass-needle or a triangle was force, as the mechanist might prove by losing it, and nothing could be gained by ignoring their value. Symbol or energy, the Virgin had acted as the greatest force the Western world ever felt, and had drawn man's activities to herself more strongly than any other power, natural or supernatural, had ever done; the historian's business was to follow the track of the energy; to find where it came from and where it went to; its complex source and shifting channels; its values, equivalents, conversions. It could scarcely be more complex than radium; it could hardly be deflected, diverted, polarized, absorbed more perplexingly than other radiant matter. Adams knew nothing about any of them, but as a mathematical problem of influence on human progress, though all were occult, all reacted on his mind, and he rather inclined to think the Virgin easiest to handle.

The pursuit turned out to be long and tortuous, leading at last to the vast forests of scholastic science. From Zeno to Descartes, hand in hand with Thomas Aquinas, Montaigne, and Pascal, one stumbled as stupidly as though one were still a German student of 1860. Only with the instinct of despair could one force one's self into this old thicket of ignorance after having been repulsed by a score of entrances more promising and more popular. Thus far, no path had led anywhere, unless perhaps to an exceedingly modest living. Forty-five years of study had proved to be quite Futile for the pursuit of power; one controlled no more force in 1900 than in 1850, although the amount of force controlled by society had enormously increased. The secret of education still hid itself somewhere behind ignorance, and one fumbled over it as feebly as ever. In such labyrinths, the staff is a force almost more necessary than the legs; the pen becomes a sort of blind-man's dog, to keep him from falling into the gutters. The pen works

for itself, and acts like a hand, modelling the plastic material over and over again to the form that suits it best. The form is never arbitrary, but is a sort of growth like crystallization, as any artist knows too well; for often the pencil or pen runs into side-paths and shapelessness, loses its relations, stops or is bogged. Then it has to return on its trail, and recover, if it can, its line of force. The result of a year's work depends more on what is struck out than on what is left in; on the sequence of the main lines of thought, than on their play or variety. Compelled once more to lean heavily on this support, Adams covered more thousands of pages with figures as formal as though they were algebra, laboriously striking out, altering, burning, experimenting, until the year had expired, the Exposition had long been closed, and winter drawing to its end, before he sailed from Cherbourg, on January 19, 1901, for home.

Source: Henry Adams, *The Education of Henry Adams* (1918. Boston: Houghton Mifflin, 1961).

America's First Historical Society

The clergyman, historian, and scientist Jeremy Belknap was chiefly responsible for the founding of the first historical society in America, the Massachusetts Historical Society, in 1791. The following is his "Plan of an Antiquarian Society," penned in August 1790. It became the blueprint for the Massachusetts Historical Society.

A Society to be formed consisting of not more than *seven at first* for the purpose of collecting, preserving, and communicating the Antiquities of America. Admissions to be made in such manner as the associated shall judge proper. The number of members to be limited. A President, Recording and Corresponding Secretary, Treasurer, Librarian, and Cabinet keeper to be appointed. Each Member to pay ___ at his admission and yearly. This and other money to be applied to promoting the objects of the Society. Each Member on his admission shall engage to use his utmost endeavours to collect and communicate to the society Manuscripts, printed books and pamphlets, historical facts, biographical anecdotes, observations in natural history, specimens of natural and artificial Curiosities and any other matters which may elucidate the natural, and political history of America from the earliest times to the present day. All communications which are thought worthy of being preserved shall be entered at large in the books of the Society . . . and the originals kept on file. Letters shall be written to Gentlemen in each of the United States requesting them to form similar societies and a correspondence shall be kept up between them for the purpose of communicating discoveries . . . to each other. Each society through the United States shall be desired from time to time to publish such of their Communications as they may judge proper, and all publications shall be made on paper and in pages of the same size that they may be bound together—and Each Society so publishing shall be desired to send gratuitously to each of the other Societies one dozen Copies at least of each publication. Quarterly meetings to be held for the purpose of communicating—and in this State the quarterly meetings shall be held on the days next following those appointed for the meetings of the American Academy of Arts and Sciences. When the Society's funds can afford it Salaries shall be granted to the Secretaries and other Officers.

Source: Jeremy Belknap Papers, Collections of the Massachusetts Historical Society, ser. 5, vol. 3 (Boston: Massachusetts Historical Society, 1891).

American Nobel Laureates in Science

Physiology or Medicine

1933 **Thomas H. Morgan**, "for his discoveries concerning the role played by the chromosome in heredity."

1934 **George R. Minot**, **William P. Murphy**, and **George H. Whipple**, "for their discoveries concerning liver therapy in cases of anaemia."

1943 **Edward A. Doisy**, "for his discovery of the chemical nature of vitamin K."

1944 **Joseph Erlanger** and **Herbert S. Gasser**, "for their discoveries relating to the highly differentiated functions of single nerve fibres."

1946 **Hermann J. Muller**, "for the discovery of the production of mutations by means of X-ray irradiation."

1947 **Carl F. Cori** and **Gerty T. Cori**, "for their discovery of the course of the catalytic conversion of glycogen."

1950 **Philip S. Hench** and **Edward C. Kendall** (with Tadeus Reichstein of Switzerland), "for their discoveries relating to the hormones of the adrenal cortex, their structure and biological effects."

1952 **Selman A. Waksman**, "for his discovery of streptomycin, the first antibiotic effective against tuberculosis."

1953 **Fritz A. Lipmann**, "for his discovery of co-enzyme A and its importance for intermediary metabolism."

1954 **John F. Enders**, **Frederick C. Robbins**, and **Thomas H. Weller**, "for their discovery of the ability of poliomyelitis viruses to grow in cultures of various types of tissue."

1956 **Andre F. Cournand** and **Dickinson W. Richards** (with Werner Forssman of Germany), "for their discoveries concerning heart catheterization and pathological changes in the circulatory system."

1958 **George W. Beadle** and **Edward L. Tatum**, "for their discovery that genes act by regulating definite chemical events"; **Joshua Lederberg**, "for his discoveries concerning genetic recombination and the organization of the genetic material of bacteria."

1959 **Arthur Kornberg** and **Severo Ochoa**, "for their discovery of the mechanisms in the biological synthesis of ribonucleic acid and deoxyribonucleic acid."

1961 **Georg von Békésy**, "for his discoveries of the physical mechanism of stimulation within the cochlea."

1962 **James D. Watson** (with Francis H.C. Crick and Maurice H.F. Wilkins of the United Kingdom), "for their discoveries concerning the molecular structure of nucleic acids and its significance for information transfer in living material."

1964 **Konrad Bloch** (with Feodor Lynen of Germany), "for their discoveries concerning the mechanism and regulation of the cholesterol and fatty acid metabolism."

1966 **Charles B. Huggins**, "for his discoveries concerning hormonal treatment of prostatic cancer"; **Peyton Rous**, "for his discovery of tumour-inducing viruses."

1967 **Haldan K. Hartline** and **George Wald** (with Ragnar Granit of Sweden), "for their discoveries concerning the primary physiological and chemical visual processes in the eye."

1968 **Robert W. Holley**, **Har G. Khorana**, and **Marshall W. Nirenberg**, "for their interpretation of the genetic code and its function in protein synthesis."

1969 **Max Delbrück**, **Alfred D. Hershey**, and **Salvador E. Luria**, "for their discoveries concerning the replication mechanism and the genetic structure of viruses."

1970 **Julius Axelrod** (with Ulf von Euler of Sweden and Sir Bernard Katz of the United Kingdom), "for their discoveries concerning the humoral transmittors in nerve terminals and the mechanism for their storage, release and inactivation."

1971 **Earl W. Sutherland, Jr.**, "for his discoveries concerning the mechanisms of the action of hormones."

1972 **Gerald M. Edelman** (with Rodney R. Porter of the United Kingdom), "for their discoveries concerning the chemical structure of antibodies."

1974 **George E. Palade** (with Albert Claude and Christian de Duve of Belgium), "for their discoveries concerning the structural and functional organization of the cell."

1975 **David Baltimore**, **Renato Dulbecco**, and **Howard M. Temin**, "for their discoveries concerning the interaction between tumour viruses and the genetic material of the cell."

1976 **Baruch S. Blumberg** and **D. Carleton Gajdusek**, "for their discoveries concerning new mechanisms for the origin and dissemination of infectious diseases."

1977 **Roger Guillemin** and **Andrew V. Schally**, "for their discoveries concerning the peptide hormone production of the brain"; **Rosalyn Yalow**, "for the development of radioimmunoassays of peptide hormones."

1978 **Daniel Nathans** and **Hamilton O. Smith** (with Werner Arber of Switzerland), "for the discovery of restriction enzymes and their application to problems of molecular genetics."

1979 **Allan M. Cormack** (with Godfrey N. Hounsfield of the United Kingdom), "for the development of computer assisted tomography."

1980 **Baruj Benacerraf** and **George D. Snell** (with Jean Dausset of France), "for their discoveries concerning genetically determined structures on the cell surface that regulate immunological reactions."

1981 **David H. Hubel** (with Torsten N. Wiesel of Sweden), "for their discoveries concerning information processing in the visual system"; **Roger W. Sperry**, "for his discoveries concerning the functional specialization of the cerebral hemispheres."

1983 **Barbara McClintock**, "for her discovery of mobile genetic elements."

1985 **Michael S. Brown** and **Joseph L. Goldstein**, "for their discoveries concerning the regulation of cholesterol metabolism."

1986 **Stanley Cohen** and **Rita Levi-Montalcini**, "for their discoveries of growth factors."

1987 **Susumu Tonegawa**, "for his discovery of the genetic principle for generation of antibody diversity."

1988 **Gertrude B. Elion** and **George H. Hitchings** (with Sir James W. Black of the United Kingdom), "for their discoveries of important principles for drug treatment."

1989 **J. Michael Bishop** and **Harold E. Varmus**, "for their discovery of the cellular origin of retroviral oncogenes."

1990 **Joseph E. Murray** and **E. Donnall Thomas**, "for their discoveries concerning organ and cell transplantation in the treatment of human disease."

1992 **Edmond H. Fischer** and **Edwin G. Krebs**, "for their discoveries concerning reversible protein phosphorylation as a biological regulatory mechanism."

1993 **Phillip A. Sharp** (with Richard J. Roberts of the United Kingdom), "for their discoveries of split genes."

1994 **Alfred G. Gilman** and **Martin Rodbell**, "for their discovery of G-proteins and the role of these proteins in signal transduction in cells."

1995 **Edward B. Lewis** and **Eric F. Wieschaus** (with Christiane Nüsslein-Volhard of Germany), "for their discoveries concerning the genetic control of early embryonic development."

1997 **Stanley B. Prusiner**, "for his discovery of Prions—a new biological principle of infection."

1998 **Robert F. Furchgott**, **Louis J. Ignarro**, and **Ferid Murad**, "for their discoveries concerning nitric oxide as a signalling molecule in the cardiovascular system."

1999 **Günter Blobel**, "for the discovery that proteins have intrinsic signals that govern their transport and localization in the cell."

2000 **Paul Greengard** and **Eric R. Kandel** (with Arvid Carlsson of Sweden), "for their discoveries concerning signal transduction in the nervous system."

2001 **Leland H. Hartwell** (with R. Timothy Hunt and Sir Paul M. Nurse of the United Kingdom), "for their discoveries of key regulators of the cell cycle."

2002 **H. Robert Horvitz** (with Sydney Brenner and John E. Sulston of the United Kingdom), "for their discoveries concerning 'genetic regulation of organ development and programmed cell death.'"

2003   **Paul C. Lauterbur** (with Sir Peter Mansfield of the United Kingdom), "for their discoveries concerning magnetic resonance imaging."

2004   **Richard Axel** and **Linda B. Buck**, "for their discoveries of odorant receptors and the organization of the olfactory system."

2006   **Andrew Z. Fire** and **Craig C. Mello**, "for their discovery of RNA interference—gene silencing by double-stranded RNA."

2007   **Mario R. Capecchi** and **Oliver Smithies** (with Sir Martin J. Evans of the United Kingdom) "for their discoveries of principles for introducing specific gene modifications in mice by the use of embryonic stem cells."

Economic Sciences

1970   **Paul A. Samuelson**, "for the scientific work through which he has developed static and dynamic economic theory and actively contributed to raising the level of analysis in economic science."

1971   **Simon Kuznets**, "for his empirically founded interpretation of economic growth which has led to new and deepened insight into the economic and social structure and process of development."

1972   **Kenneth J. Arrow** (with John R. Hicks of the United Kingdom), "for their pioneering contributions to general economic equilibrium theory and welfare theory."

1973   **Wassily Leontief**, "for the development of the input-output method and for its application to important economic problems."

1975   **Tjalling C. Koopmans** (with Leonid V. Kantorovich of the U.S.S.R.), "for their contributions to the theory of optimum allocation of resources."

1976   **Milton Friedman**, "for his achievements in the fields of consumption analysis, monetary history and theory and for his demonstration of the complexity of stabilization policy."

1978   **Herbert A. Simon**, "for his pioneering research into the decision-making process within economic organizations."

1979   **Theodore W. Schultz** (with Sir Arthur Lewis of the United Kingdom), "for their pioneering research into economic development research with particular consideration of the problems of developing countries."

1980   **Lawrence R. Klein**, "for the creation of econometric models and the application to the analysis of economic fluctuations and economic policies."

1981   **James Tobin**, "for his analysis of financial markets and their relations to expenditure decisions, employment, production and prices."

1982   **George J. Stigler**, "for his seminal studies of industrial structures, functioning of markets and causes and effects of public regulation."

1983   **Gerard Debreu**, "for having incorporated new analytical methods into economic theory and for his rigorous reformulation of the theory of general equilibrium."

1985   **Franco Modigliani**, "for his pioneering analyses of saving and of financial markets."

1986   **James M. Buchanan, Jr.**, "for his development of the contractual and constitutional bases for the theory of economic and political decision-making."

1987   **Robert M. Solow**, "for his contributions to the theory of economic growth."

1990   **Harry M. Markowitz**, **Merton H. Miller**, and **William F. Sharpe**, "for their pioneering work in the theory of financial economics."

1992   **Gary S. Becker**, "for having extended the domain of microeconomic analysis to a wide range of human behaviour and interaction, including nonmarket behaviour."

1993   **Robert W. Fogel** and **Douglass C. North**, "for having renewed research in economic history by applying economic theory and quantitative methods in order to explain economic and institutional change."

1994   **John C. Harsanyi** and **John F. Nash, Jr.** (with Reinhard Selten of Germany), "for their pioneering analysis of equilibria in the theory of non-cooperative games."

1995   **Robert E. Lucas, Jr.**, "for having developed and applied the hypothesis of rational expectations, and thereby having transformed macroeconomic analysis and deepened our understanding of economic policy."

1996　**William Vickrey** (with James A. Mirrlees of the United Kingdom), "for their fundamental contributions to the economic theory of incentives under asymmetric information."

1997　**Robert C. Merton** and **Myron S. Scholes**, "for a new method to determine the value of derivatives."

1999　**Robert A. Mundell**, "for his analysis of monetary and fiscal policy under different exchange rate regimes and his analysis of optimum currency areas."

2000　**James J. Heckman**, "for his development of theory and methods for analyzing selective samples"; **Daniel L. McFadden**, "for his development of theory and methods for analyzing discrete choice."

2001　**George A. Akerlof, A. Michael Spence**, and **Joseph E. Stiglitz**, "for their analyses of markets with asymmetric information."

2002　**Daniel Kahneman**, "for having integrated insights from psychological research into economic science, especially concerning human judgment and decision-making under uncertainty"; **Vernon L. Smith**, "for having established laboratory experiments as a tool in empirical economic analysis, especially in the study of alternative market mechanisms."

2003　**Robert F. Engle III**, "for methods of analyzing economic time series with time-varying volatility (ARCH)."

2004　**Edward C. Prescott** (with Finn E. Kydland of Norway), "for their contributions to dynamic microeconomics: the time consistency of economic policy and the driving forces behind business cycles."

2005　**Robert J. Aumann** and **Thomas C. Schelling**, "for having enhanced our understanding of conflict and cooperation through game-theory analysis."

2006　**Edmund S. Phelps**, "for his analysis of intertemporal tradeoffs in macroeconomic policy."

2007　**Leonid Hurwicz, Eric S. Maskin**, and **Roger B. Myerson**, "for having laid the foundations of mechanism design theory."

Physics

1907　**Albert A. Michelson**, "for his optical precision instruments and the spectroscopic and metrological investigations carried out with their aid."

1923　**Robert A. Millikan**, "for his work on the elementary charge of electricity and on the photoelectric effect."

1927　**Arthur H. Compton**, "for his discovery of the effect named after him."

1936　**Carl D. Anderson**, "for his discovery of the positron."

1937　**Clinton J. Davisson** (with George P. Thomson of the United Kingdom), "for their experimental discovery of the diffraction of electrons by crystals."

1939　**Ernest O. Lawrence**, "for the invention and development of the cyclotron and for results obtained with it, especially with regard to artificial radioactive elements."

1943　**Otto Stern**, "for his contribution to the development of the molecular ray method and his discovery of the magnetic moment of the proton."

1944　**Isidor I. Rabi**, "for his resonance method for recording the magnetic properties of atomic nuclei."

1946　**Percy W. Bridgman**, "for the invention of an apparatus to produce extremely high pressures, and for the discoveries he made therewith in the field of high pressure physics."

1952　**Felix Bloch** and **Edward M. Purcell**, "for their development of new methods for nuclear magnetic precision measurements and discoveries in connection therewith."

1955　**Polykarp Kusch**, "for his precision determination of the magnetic moment of the electron"; **Willis E. Lamb**, "for his discoveries concerning the fine structure of the hydrogen spectrum."

1956　**John Bardeen, Walter H. Brattain**, and **William B. Shockley**, "for their researches on semiconductors and their discovery of the transistor effect."

1959　**Owen Chamberlain** and **Emilio G. Segrè**, "for their discovery of the antiproton."

1960 **Donald A. Glaser**, "for the invention of the bubble chamber."

1961 **Robert Hofstadter**, "for his pioneering studies of electron scattering in atomic nuclei and for his thereby achieved discoveries concerning the structure of the nucleons."

1963 **Maria Goeppert-Mayer** (with J. Hans D. Jensen of Germany), "for their discoveries concerning nuclear shell structure"; **Eugene P. Wigner**, "for his contributions to the theory of the atomic nucleus and the elementary particles, particularly through the discovery and application of fundamental symmetry principles."

1964 **Charles H. Townes** (with Nicolay G. Basov and Aleksandr M. Prokhorov of the USSR), "for fundamental work in the field of quantum electronics, which has led to the construction of oscillators and amplifiers based on the maser-laser principle."

1965 **Richard P. Feynman** and **Julian Schwinger** (with Sin-Itiro Tomonaga of Japan), "for their fundamental work in quantum electrodynamics, with deep-ploughing consequences for the physics of elementary particles."

1967 **Hans A. Bethe**, "for his contributions to the theory of nuclear reactions, especially his discoveries concerning the energy production in stars."

1968 **Luis W. Alvarez**, "for his decisive contributions to elementary particle physics, in particular the discovery of a large number of resonance states, made possible through his development of the technique of using hydrogen bubble chamber and data analysis."

1969 **Murray Gell-Mann**, "for his contributions and discoveries concerning the classification of elementary particles and their interactions."

1972 **John Bardeen**, **Leon N. Cooper**, and **J. Robert Schrieffer**, "for their jointly developed theory of superconductivity, usually called the BCS-theory."

1973 **Ivar Giaever** (with Leo Esaki of Japan), "for their experimental discoveries regarding tunneling phenomena in semiconductors and superconductors, respectively."

1975 **L. James Rainwater** (with Aage N. Bohr and Ben R. Mottelson of Denmark), "for the discovery of the connection between collective motion and particle motion in atomic nuclei and the development of the theory of the structure of the atomic nucleus based on this connection."

1976 **Burton Richter** and **Samuel C.C. Ting**, "for their pioneering work in the discovery of a heavy elementary particle of a new kind."

1977 **Philip W. Anderson** and **John H. van Vleck** (with Sir Nevill F. Mott of the United Kingdom), "for their fundamental theoretical investigations of the electronic structure of magnetic and disordered systems."

1978 **Arno A. Penzias** and **Robert W. Wilson**, "for their discovery of cosmic microwave background radiation."

1979 **Sheldon L. Glashow** and **Steven Weinberg** (with Abdus Salam of Pakistan), "for their contributions to the theory of the unified weak and electromagnetic interaction between elementary particles, including, inter alia, the prediction of the weak neutral current."

1980 **James W. Cronin** and **Val L. Fitch**, "for the discovery of violations of fundamental symmetry principles in the decay of neutral K-mesons."

1981 **Nicolaas Bloembergen** and **Arthur L. Schawlow**, "for their contribution to the development of laser spectroscopy."

1982 **Kenneth G. Wilson**, "for his theory for critical phenomena in connection with phase transitions."

1983 **Subramanyan Chandrasekhar**, "for his theoretical studies of the physical processes of importance to the structure and evolution of the stars"; **William A. Fowler**, "for his theoretical and experimental studies of the nuclear reactions of importance in the formation of the chemical elements of the universe."

1988 **Leon M. Lederman**, **Melvin Schwartz**, and **Jack Steinberger**, "for the neutrino beam method and the demonstration of the doublet structure of the leptons through the discovery of the muon neutrino."

1989 **Hans G. Dehmelt** (with Wolfgang Paul of Germany), "for the development of the ion trap technique"; **Norman F.**

Ramsey, "for the invention of the separated oscillatory fields method and its use in the hydrogen maser and other atomic clocks."

1990 **Jerome I. Friedman** and **Henry W. Kendall** (with Richard E. Taylor of Canada), "for their pioneering investigations concerning deep inelastic scattering of electrons on protons and bound neutrons, which have been of essential importance for the development of the quark model in particle physics."

1993 **Russell A. Hulse** and **Joseph H. Taylor, Jr.**, "for the discovery of a new type of pulsar, a discovery that has opened up new possibilities for the study of gravitation."

1994 **Clifford G. Shull**, "for the development of the neutron diffraction technique."

1995 **Martin L. Perl**, "for the discovery of the tau lepton"; **Frederick Reines**, "for the detection of the neutrino."

1996 **David M. Lee, Douglas S. Osheroff**, and **Robert C. Richardson**, "for their discovery of superfluidity in helium-3."

1997 **Steven Chu** and **William D. Phillips** (with Claude Cohen-Tannoudji of France), "for development of methods to cool and trap atoms with laser light."

1998 **Robert B. Laughlin** and **Daniel C. Tsui** (with Horst L. Störmer of Germany), "for their discovery of a new form of quantum fluid with fractionally charged excitations."

2000 **Jack S. Kilby**, "for his part in the invention of the integrated circuit."

2001 **Eric A. Cornell** and **Carl E. Wieman** (with Wolfgang Ketterle of Germany), "for the achievement of Bose-Einstein condensation in dilute gases of alkali atoms, and for early fundamental studies of the properties of the condensates."

2002 **Raymond Davis, Jr.** (with Masatoshi Koshiba of Japan), "for pioneering contributions to astrophysics, in particular for the detection of cosmic neutrinos"; **Riccardo Giacconi**, "for pioneering contributions to astrophysics, which have led to the discovery of cosmic X-ray sources."

2003 **Alexei A. Abrikosov** and **Anthony J. Leggett** (with Vitaly L. Ginzburg of Russia), "for pioneering contributions to

the theory of superconductors and superfluids."

2004 **David J. Gross, H. David Politzer**, and **Frank Wilczek**, "for the discovery of asymptotic freedom in the theory of the strong interaction."

2005 **Roy J. Glauber**, "for his contribution to the quantum theory of optical coherence"; **John L. Hall** (with Theodor W. Hänsch of Germany), "for their contributions to the development of laser-based precision spectroscopy, including the optical frequency comb technique."

2006 **John C. Mather** and **George F. Smoot**, "for their discovery of the blackbody form and anisotropy of the cosmic microwave background radiation."

Chemistry

1914 **Theodore W. Richards**, "in recognition of his accurate determinations of the atomic weight of a large number of chemical elements."

1932 **Irving Langmuir**, "for his discoveries and investigations in surface chemistry."

1934 **Harold C. Urey**, "for his discovery of heavy hydrogen."

1946 **John H. Northrop** and **Wendell M. Stanley**, "for their preparation of enzymes and virus proteins in a pure form"; **James B. Sumner**, "for his discovery that enzymes can be crystallized."

1949 **William F. Giauque**, "for his contributions in the field of chemical thermodynamics, particularly concerning the behaviour of substances at extremely low temperatures."

1951 **Edwin M. McMillan** and **Glenn T. Seaborg**, "for their discoveries in the chemistry of the transuranium elements."

1954 **Linus C. Pauling**, "for his research into the nature of the chemical bond and its application to the elucidation of the structure of complex substances."

1955 **Vincent du Vigneaud**, "for his work on biochemically important sulphur compounds, especially for the first synthesis of a polypeptide hormone."

1960 **Willard F. Libby**, "for his method to use carbon-14 for age determination in archaeology, geology, geophysics, and other branches of science."

1961 **Melvin Calvin**, "for his research on the carbon dioxide assimilation in plants."

1965 **Robert B. Woodward**, "for his outstanding achievements in the art of organic synthesis."

1966 **Robert S. Mulliken**, "for his fundamental work concerning chemical bonds and the electronic structure of molecules by the molecular orbital method."

1968 **Lars Onsager**, "for the discovery of the reciprocal relations bearing his name, which are fundamental for the thermodynamics of irreversible processes."

1972 **Christian B. Anfinsen**, "for his work on ribonuclease, especially concerning the connection between the amino acid sequence and the biologically active conformation"; **Stanford Moore** and **William H. Stein**, "for their contribution to the understanding of the connection between chemical structure and catalytic activity of the active center of the ribonuclease molecule."

1974 **Paul J. Flory**, "for his fundamental achievements, both theoretical and experimental, in the physical chemistry of the macromolecules."

1976 **William N. Lipscomb**, "for his studies on the structure of boranes illuminating problems of chemical bonding."

1979 **Herbert C. Brown** (with Georg Wittig of Germany), "for their development of the use of boron- and phosphorus-containing compounds, respectively, into important reagents in organic synthesis."

1980 **Paul Berg**, "for his fundamental studies of the biochemistry of nucleic acids, with particular regard to recombinant-DNA"; **Walter Gilbert** (with Frederick Sanger of the United Kingdom), "for their contributions concerning the determination of base sequences in nucleic acids."

1981 **Roald Hoffmann** (with Kenichi Fukui of Japan), "for their theories, developed independently, concerning the course of chemical reactions."

1983 **Henry Taube**, "for his work on the mechanisms of electron transfer reactions, especially in metal complexes."

1984 **Robert B. Merrifield**, "for his development of methodology for chemical synthesis on a solid matrix."

1985 **Herbert A. Hauptman** and **Jerome Karle**, "for their outstanding achievements in the development of direct methods for the determination of crystal structures."

1986 **Dudley R. Herschbach** and **Yuan T. Lee** (with John C. Polanyi of Canada), "for their contributions concerning the dynamics of chemical elementary processes."

1987 **Donald J. Cram** and **Charles J. Pedersen** (with Jean-Marie Lehn of France), "for their development and use of molecules with structure-specific interactions of high selectivity."

1989 **Sidney Altman** and **Thomas R. Cech**, "for their discovery of catalytic properties of RNA."

1990 **Elias James Corey**, "for his development of the theory and methodology of organic synthesis."

1992 **Rudolph A. Marcus**, "for his contributions to the theory of electron transfer reactions in chemical systems."

1993 **Kary B. Mullis**, "for his invention of the polymerase chain reaction (PCR) method."

1994 **George A. Olah**, "for his contribution to carbocation chemistry."

1995 **Mario J. Molina** and **F. Sherwood Rowland** (with Paul J. Crutzen of the Netherlands), "for their work in atmospheric chemistry, particularly concerning the formation and decomposition of ozone."

1996 **Robert F. Curl, Jr.**, and **Richard E. Smalley** (with Sir Harold W. Kroto of the United Kingdom), "for their discovery of fullerenes."

1997 **Paul D. Boyer** (with John E. Walker of the United Kingdom), "for their elucidation of the enzymatic mechanism underlying the synthesis of adenosine triphosphate (ATP)."

1998 **Walter Kohn**, "for his development of the density-functional theory."

1999 **Ahmed H. Zewail**, "for his studies of the transition states of chemical reactions using femtosecond spectroscopy."

2000   **Alan J. Heeger** and **Alan G. MacDiarmid** (with Hideki Shirakawa of Japan), "for the discovery and development of conductive polymers."

2001   **William S. Knowles** (with Ryoji Noyori of Japan), "for their work on chirally catalysed hydrogenation reactions"; **K. Barry Sharpless**, "for his work on chirally catalysed oxidation reactions."

2002   **John B. Fenn** (with Koichi Tanaka of Japan), "for their development of soft desorption ionisation methods for mass spectrometric analyses of biological macromolecules."

2003   **Peter Agre**, "for the discovery of water channels [in cell membranes]"; **Roderick MacKinnon**, "for structural and mechanistic studies of ion channels [in cell membranes]."

2004   **Irwin Rose** (with Aaron Ciechanover and Avram Hershko of Israel), "for the discovery of ubiquitin-mediated protein degradation."

2005   **Robert H. Grubbs** and **Richard R. Schrock** (with Yves Chauvin of France), "for the development of the metathesis method in organic synthesis."

2006   **Roger D. Kornberg**, "for his studies of the molecular basis of eukaryotic transcription."

Bibliography

Books

Abram, Ruth. *Send Us a Lady Physician: Women Doctors in America, 1835–1920.* New York: W.W. Norton, 1985.

Abramson, Howard S. *National Geographic: Behind America's Lens on the World.* New York: Crown, 1987.

Ackerman, Carl W. *George Eastman: Founder of Kodak and the Photography Business.* Delaware, OH: Beard, 1930.

Ackoff, Russell L. *The Second Industrial Revolution.* Washington, DC: Alban Institute, 1975.

Acosta, José de. *Natural and Moral History of the Indies.* Ed. Jane E. Mangan; trans. Frances López-Morillas. Durham, NC: Duke University Press, 2002.

Aczel, A.D. *Fermat's Last Theorem: Unlocking the Secret of an Ancient Mathematical Problem.* New York: Penguin, 1996.

Adams, Ansel. *Our National Parks.* New Haven, CT: Bullfinch, 1992.

Adams, Henry. *The Degradation of the Democratic Dogma.* New York: Macmillan, 1919.

———. *The Education of Henry Adams.* 1918. Boston: Houghton Mifflin, 1961.

Ager, Derek. *The New Catastrophism.* Cambridge, UK: Cambridge University Press, 1993.

Agresti, Alan, and Barbara Finlay Agresti. *Statistical Methods for the Social Sciences.* San Francisco: Dellen, 1979.

Aguado, Edward, and James E. Burt. *Understanding Weather and Climate.* 3rd ed. Upper Saddle River, NJ: Prentice Hall, 2003.

Aiken, Howard. *Synthesis of Electronic Computing and Control Circuits.* Cambridge, MA: Harvard University Press, 1952.

Aitken, Martin J. *Science-Based Dating in Archaeology.* London: Longman, 1990.

Akers, Ronald A. *Criminological Theories: Introduction, Evaluation, and Application.* 3rd ed. Los Angeles: Roxbury, 2000.

Aldrich, Lisa J. *Cyrus McCormick and the Mechanical Reaper.* Greensboro, NC: Morgan-Reynolds, 2002.

Aldrin, Edwin E. *Return to Earth.* New York: Random House, 1973.

Aldrin, Edwin E., and Malcolm McConnell. *Men from Earth.* New York: Bantam, 1989.

Alexander, Edward P. *The Museum in America: Innovators and Pioneers.* American Association for State and Local History Book Series. Lanham, MD: AltaMira, 1997.

Alexander, Thomas M., and Larry Hickman, eds. *The Essential Dewey.* Bloomington: Indiana University Press, 1998.

Allan, Roy. *A History of the Personal Computer.* London, Ontario, Canada: Allan, 2001.

Allen, Frederick M. *Total dietary regulation in the treatment of diabetes.* New York: Rockefeller Institute, 1919.

Allen, Garland E. *Thomas Hunt Morgan: The Man and His Science.* Princeton, NJ: Princeton University Press, 1979.

Alliger, Glen, and Irvin Julian Sjothun. *Vulcanization of Elastomers: Principles and Practice of Vulcanization of Commercial Rubbers.* Melbourne, FL: Krieger, 1978.

Alster, Kristine. *The Holistic Health Movement.* Tuscaloosa: University of Alabama Press, 1989.

Alvarez, Luis W. *Adventures of a Physicist.* New York: Basic Books, 1987.

Ambrose, Stephen E. *Nothing Like It in the World: The Men Who Built the Transcontinental Railroad 1863–1869.* New York: Simon and Schuster, 2001.

———. *Undaunted Courage: Meriwether Lewis, Thomas Jefferson, and the Opening of the American West.* New York: Simon and Schuster, 1996.

American Chemical Society. *The Three Mile Island Accident: Diagnosis and Prognosis.* Washington, DC: American Chemical Society, 1986.

American Practical Navigator: An Epitome of Navigation Originally by Nathaniel Bowditch. Bethesda, MD: National Imagery and Mapping Agency, 2002.

American Psychiatric Association. *Diagnostic and Statistical Manual of Mental Disorders.* 4th ed. Washington, DC: American Psychiatric Association, 1994.

American Society of Civil Engineers. *Hydrology Handbook.* 2nd ed. Reston, VA: ASCE Publications, 1996.

Ames, Louise Bates. *Arnold Gesell: Themes of His Work.* New York: Human Sciences, 1989.

Amyes, Sebastian G.B. *Magic Bullets, Lost Horizons: The Rise and Fall of Antibiotics.* New York: Taylor and Francis, 2001.

Anderson, Fred W. *Orders of Magnitude: A History of NACA and NASA, 1915–1980.* Washington, DC: NASA, 1981.

Anderson, William R., and Clay Blair, Jr. *Nautilus 90 North.* Blue Ridge Summit, PA: Tab, 1989.

Andrewartha, H.G. *Introduction to the Study of Animal Populations.* Chicago: University of Chicago Press, 1971.

Annals of Harvard College Observatory. Vols. 1–120. Cambridge, MA: Harvard University Press, 1856–1956.

Anselm. *Proslogium; Monologium; An Appendix in Behalf of the Fool by Gaunilon; and Cur Deus Homo.* Grand Rapids, MI: Christian Classics Ethereal Library, 2000.

Apel, Karl-Otto. *Charles S. Peirce: From Pragmatism to Pragmaticism.* Amherst: University of Massachusetts Press, 1981.

Archibald, R.C. *A Semicentennial History of the American Mathematical Society, 1888–1938.* New York: American Mathematical Society, 1938.

Armstrong, Neil, Michael Collins, and Edwin E. Aldrin. *First on the Moon.* Boston: Little, Brown, 1970; reprint ed., 2002.

Arnheim, Rudolf. *Radio.* New York: Arno, 1971.

Arnold, Eric. *Volcanoes: Mountains of Fire.* New York: Random House, 1997.

Arraj, Tyra. *Tracking the Elusive Human, Volume 2: An Advanced Guide to the Typological Worlds of C.G. Jung, W.H. Sheldon, Their Integration, and the Biochemical Typology of the Future.* Chiloquin, OR: Inner Growth, 1990.

Artemiadis, Nikolaos K. *History of Mathematics: From a Mathematician's Vantage Point.* Providence, RI: American Mathematical Society, 2004.

Asbell, Milton B. *Dentistry: A Historical Perspective.* Bryn Mawr, PA: Dorrance, 1988.

Ashby, W. Ross. *An Introduction to Cybernetics.* London: Chapman and Hall, 1956.

Aspden, Harold. *Modern Aether Science.* Southampton, UK: Sabberton, 1972.

Audubon, John J. *The Audubon Reader.* Ed. Richard Rhodes. New York: Alfred A. Knopf, 2006.

———. *Birds of America.* Washington, DC: National Audubon Society, 2005.

Austrian, Geoffrey D. *Herman Hollerith: Forgotten Giant of Information Processing.* New York: Columbia University Press, 1982.

Axelrod, Julius. *Philosophy of Medicine and Science.* Washington, DC: Institute of History of Medicine and Medical Research, 1972.

Backus, George, et al. *Foundations of Geomagnetism.* Cambridge, UK: Cambridge University Press, 2005.

Baden, Michael. *Unnatural Death: Confessions of a Medical Examiner.* New York: Ballantine, 1990.

Bagley, K., and Ray Douglas Hurt. *Eli Whitney: American Inventor.* Mankato, MN: Capstone, 2003.

Bailey, Liberty H. *The Standard Cyclopedia of Horticulture.* 1900. New York: Macmillan, 2000.

Bailey, Solon I. *The History and Work of Harvard Observatory, 1839–1927.* New York: McGraw-Hill, 1931.

Bain, David Haward. *Empire Express: Building the First Transcontinental Railroad.* New York: Penguin, 2000.

Baitsall, George A., ed. *The Centennial of the Sheffield Scientific School.* New Haven, CT: Yale University Press, 1950.

Baker, David. *The Rocket: The History and Development of Rocket and Missile Technology.* New York: Crown, 1978.

Baker, Rachel. *The First Woman Doctor: The Story of Elizabeth Blackwell, M.D.* New York: Julian Messner, 1944.

Baker, Ray P. *A Chapter in American Education: Rensselaer Polytechnic Institute 1824–1924.* New York: Charles Scribner's Sons, 1924.

Baldwin, Munson. *With Brass and Gas: An Illustrated and Embellished Chronicle of Ballooning in Mid-Nineteenth Century America.* Boston: Beacon, 1967.

Baldwin, William. *Reliquiae Baldwinianae: Selections from the Correspondence of the Late William Baldwin with Occasional Notes, and a Short Biographical Memoir.* Comp. William Darlington. Philadelphia: Kimber and Sharpless, 1843; reprint ed., New York: Hafner, 1969.

Bankston, John. *Robert Jarvik and the First Artificial Heart: Unlocking the Secrets of Science.* Hockessin, DE: Mitchell Lane, 2002.

Barbour, Philip. *The Three Worlds of Captain John Smith.* Boston: Houghton Mifflin, 1964.

Barnard, Christiaan, and Curtis Bill Pepper. *Christiaan Barnard: One Life.* Ontario, Canada: Macmillan, 1969.

Barringer, Mark D. *Selling Yellowstone: Capitalism and the Construction of Nature.* Lawrence: University Press of Kansas, 2002.

Barry, John M. *The Great Influenza: The Epic Story of the Deadliest Plague in History.* New York: Viking, 2004.

Barry, Roger G., and Richard J. Chorley. *Atmosphere, Weather and Climate.* 8th ed. New York: Routledge, 2003.

Bartky, Ian R. *Selling the True Time: Nineteenth-Century Timekeeping in America.* Stanford, CA: Stanford University Press, 2000.

Bartram, John. *The Correspondence of John Bartram, 1734–1777.* Ed. Edmund Berkeley and Dorothy Smith Berkeley. Gainesville: University Press of Florida, 1992.

Bartram, William. *Travels through North and South Carolina, Georgia, East and West Florida, the Cherokee Country, the Extensive Territories of the Muscogulges, or Creek Confederacy, and the Country of the Choctaws.* New York: Penguin, 1988.

Bashe, Charles. "Constructing the IBM ASCC (Harvard Mark I)." In *Makin' Numbers: Howard Aiken and the Computer,* ed. I. Bernard Cohen and Gregory W. Welch. Cambridge, MA: MIT Press, 1999.

Bayne-Jones, Stanhope. *Preventive Medicine in the United States Army, 1607–1939.* Washington, DC: U.S. Government Printing Office, 1968.

Beall, Otto T., Jr., and Richard H. Shryock. *Cotton Mather: First Significant Figure in American Medicine.* Baltimore: Johns Hopkins University Press, 1954.

Bean, William B. *Walter Reed: A Biography.* Charlottesville: University of Virginia Press, 1982.

Beard, George M., and A.D. Rockwell. *A Practical Treatise on the Medical and Surgical uses of Electricity including Localized and General Electrization.* New York: W. Wood, 1871.

Beattie, Donald A. *Taking Science to the Moon: Lunar Experiments and the Apollo Program.* Baltimore: Johns Hopkins University Press, 2001.

Becker, Carl L. *The Heavenly City of the Eighteenth-Century Philosophers.* New Haven, CT: Yale University Press, 2003.

Beddini, Silvio A. *The Life of Benjamin Banneker.* Rancho Cordova, CA: Landmark Enterprises, 1984.

———. *Thinkers and Tinkers: Early American Men of Science.* New York: Charles Scribner's Sons, 1975.

———. *Thomas Jefferson: Statesman of Science.* New York: Macmillan, 1990.

Beebe, William. *Half Mile Down.* Chicago: Cadmus, 1934.

Beecher, Henry K., and Mark D. Altschule. *Medicine at Harvard: The First Three Hundred Years.* Hanover, NH: University Press of New England, 1977.

Begon, Michael, John L. Harper, and Colin R. Townsend. *Ecology: Individuals, Population, and Communities.* Cambridge, MA: Blackwell Science, 1996.

Behe, Michael J. *Darwin's Black Box: The Biochemical Challenge to Evolution.* New York: Free Press, 1996.

Belknap, Jeremy. *The History of New-Hampshire.* 3 vols. Philadelphia and Boston: 1784, 1791, 1792; rev. ed., Westminster, MD: Heritage Books, 1992.

Belknap Papers. *Collections of the Massachusetts Historical Society.* Ser. 5, vol. 2. Boston: Massachusetts Historical Society, 1877.

Bell, Whitfield J., Jr., ed. *The Complete Poor Richard Almanacks, published by Benjamin Franklin.* 2 vols. Barre, MA: Imprint Society, 1970.

Bendall, Sarah. *Dictionary of Land Surveyors and Local Map Makers of Great Britain and Ireland, 1530–1850.* London: British Library, 1997.

Benjamin, Marina. *Rocket Dreams: How the Space Age Shaped Our Vision of a World Beyond.* New York: Free Press, 2003.

Benowitz, Steven. *Cancer.* Berkeley Heights, NJ: Enslow, 1999.

Benson, Maxine, ed. *From Pittsburgh to the Rocky Mountains: Major Stephen Long's Expedition, 1819–1820.* Golden, CO: Fulcrum, 1988.

Bentley, Wilson A., and W.J. Humphreys. *Snow Crystals.* 1931. Reprint, Mineola, NY: Dover, 1962.

Bercovitch, Sacvan. *The Puritan Origins of the American Self.* New Haven, CT: Yale University Press, 1977.

Berk, Joseph. *The Gatling Gun: 19th Century Machine Gun to 21st Century Vulcan.* Boulder, CO: Paladin, 1991.

Berkeley, Edmund, and Dorothy Smith Berkeley. *Dr. Alexander Garden of Charles Town.* Chapel Hill: University of North Carolina Press, 1969.

———. *Dr. John Mitchell: The Man Who Made the Map of North America.* Chapel Hill: University of North Carolina Press, 1974.

———. *John Clayton: Pioneer of American Botany.* Chapel Hill: University of North Carolina Press, 1963.

———. *The Life and Travels of John Bartram: From Lake Ontario to the River St. John.* Tallahassee: University Presses of Florida, 1982.

Berlandier, Jean Louis. *The Indians of Texas in 1830.* Washington, DC: Smithsonian Institution Press, 1969.

Berlin, Isaiah. *Karl Marx: His Life and Environment.* New York: Time, 1963.

Berlinski, David. *A Tour of the Calculus.* New York: Pantheon, 1995.

Bernstein, Jeremy. *Hans Bethe, Prophet of Energy.* New York: Basic Books, 1980.

———. *Oppenheimer: Portrait of an Enigma.* Chicago: Ivan R. Dee, 2004.

Bernstein, Peter L. *Wedding of the Waters: The Erie Canal and the Making of a Great Nation.* New York: W.W. Norton, 2005.

Berra, Tim N. *William Beebe: An Annotated Bibliography.* Hamden, CT: Archon, 1977.

Berry, William B.N. *Growth of the Prehistoric Time Scale.* San Francisco: Freeman, 1968.

Bettelheim, Bruno. *The Uses of Enchantment: The Meaning and Importance of Fairy Tales.* New York: Vintage Books, 1989.

Bettmann, Otto. *A Pictorial History of Medicine.* Springfield, IL: Charles C. Thomas, 1956.

Beverley, Robert. *The History and Present State of Virginia, in Four Parts.* London: 1705. Reprint ed., New York: Bobbs-Merrill, 1971.

Bickel, Lennard. *Deadly Element: The Story of Uranium.* New York: Stein and Day, 1981.

Bieder, Robert E. *Science Encounters the Indian, 1820–1880: Early Years of American Ethnology.* Norman: University of Oklahoma Press, 1986.

Biermann, Carol A., and Ludwig Biermann. "Rosalyn Sussman Yalow." In *Women in Chemistry and Physics: A Bibliographic Sourcebook,* ed. Louise S. Grinstein, Rose K. Rose, and Miriam H. Rafailovich. Westport, CT: Greenwood, 1993.

Bierne, Piers, and James Messerschmit. *Criminology.* San Diego, CA: Harcourt Brace Jovanovich, 1991.

Bigelow, Henry B. *Memories of a Long and Active Life.* Cambridge, MA: Cosmos, 1964.

———. *Oceanography: Its Scope, Problems, and Economic Importance.* Boston: Houghton Mifflin, 1931.

Biggs, Lindy. *The Rational Factory: Architecture, Technology, and Work in America's Age of Mass Production.* Baltimore: Johns Hopkins University Press, 2003.

Bird, Alexander. *Thomas Kuhn.* Princeton, NJ: Princeton University Press, 2001.

Bird, Kai, and Martin J. Sherwin. *American Prometheus: The Triumph and Tragedy of J. Robert Oppenheimer.* New York: Alfred A. Knopf, 2005.

Birkhoff, Garrett. "Mathematics at Harvard, 1836–1944." In *History of Mathematics: A Century of Mathematics in America.* Providence, RI: American Mathematical Society, 1989.

Bissell, Don. *The First Conglomerate: 145 Years of the Singer Sewing Machine Company.* Brunswick, ME: Audenreed, 1999.

Bjork, Daniel. *B.F. Skinner: A Life.* Washington, DC: American Psychological Association, 1997.

Black, Max, ed. *The Social Theories of Talcott Parsons.* Englewood Cliffs, NJ: Prentice Hall, 1961.

Black, Robert C. *The Younger John Winthrop.* New York: Columbia University Press, 1968.

Blake, John B. "Marie Elizabeth Zakrzewska." In *Notable American Women, 1607–1950: A Biographical Dictionary,* ed. Edward T. James, et al. Cambridge, MA: Belknap Press, 1971.

Blakely, Robert, and Judith Harrington. *Bones in the Basement.* Washington, DC: Smithsonian, 1997.

Blanchard, Duncan C. *The Snowflake Man: A Biography of Wilson Bentley.* Blacksburg, VA: McDonald and Woodward, 1998.

Blanton, Wynton Bolling. *Medicine in Virginia in the Eighteenth Century.* Richmond, VA: Garrett and Massie, 1931.

Bleicher, Josef. *Contemporary Hermeneutics: Hermeneutics as Method, Philosophy, and Critique.* London: Routledge and Kegan Paul, 1980.

Bliss, Michael. *Banting: A Biography.* Toronto: McClelland and Stewart, 1984.

———. *The Discovery of Insulin.* Chicago: University of Chicago Press, 1982.

———. *William Osler: A Life in Medicine.* New York: Oxford University Press, 1999.

Bloch, Konrad. *Blondes in Venetian Paintings, the Nine-Banded Armadillo, and Other Essays in Biochemistry.* New Haven, CT: Yale University Press, 1997.

Blochman, Lawrence G. *Doctor Squibb: The Life and Times of a Rugged Idealist.* New York: Simon and Schuster, 1958.

Bloom, Lynn Z. *Dr. Spock: Biography of a Conservative Radical.* Indianapolis, IN: Bobbs-Merrill, 1972.

Boas, Franz. *Race, Language, and Culture.* Chicago: University of Chicago Press, 1982.

Boehme, Sarah E. *John James Audubon in the West: The Last Expedition: Mammals of North America.* New York: Harry N. Abrams, 2000.

Bogdan, Janet Carlisle. "Childbirth in America, 1650–1990." In *Women, Health, and Medicine in America: A Historical Handbook,* ed. Rima D. Apple. New Brunswick, NJ: Rutgers University Press, 1990.

Boland, Frank Kells. *The First Anesthetic: The Story of Crawford Long.* Athens: University of Georgia Press, 1950.

Bolen, Eric G., and William L. Robinson. *Wildlife Ecology and Management.* 5th ed. Englewood Cliffs, NJ: Prentice Hall, 2003.

Bolles, Edmund Blair. *Einstein Defiant: Genius Versus Genius in the Quantum Revolution.* Washington, DC: Joseph Henry, 2004.

Bolt, B.A. *Inside the Earth.* New York: Freeman, 1982.

Bone, Neil. *The Aurora: Sun-Earth Interactions.* 2nd ed. New York: Praxis, 1996.

Bonner, Thomas Neville. *Iconoclast: Abraham Flexner and a Life of Learning.* Baltimore: Johns Hopkins University Press, 2002.

Bonola, Roberto. *Non-Euclidean Geometry: A Critical and Historical Study of Its Developments.* Trans. H.S. Carslaw. New York: Dover, 1955.

Bonta, Marcia Myers. *Women in the Field: America's Pioneering Women Naturalists.* College Station: Texas A&M University Press, 1991.

Boorstin, Daniel. *The Americans.* 3 vols. New York: Vintage Books, 1964–1973.

———. *The Discoverers: A History of Man's Search to Know His World and Himself.* New York: Vintage Books, 1985.

———. *The Lost World of Thomas Jefferson.* Chicago: University of Chicago Press, 1948.

Borst, Charlotte. *Catching Babies: The Professionalization of Childbirth, 1870–1920.* Cambridge, MA: Harvard University Press, 1995.

Bosher, Cecil. *Landmarks in Cardiac Surgery.* London: Taylor and Francis, 1997.

Boston Women's Health Collective. *Our Bodies, Ourselves: A new edition for a new era.* New York: Simon and Schuster, 2005.

Botting, Douglas. *Humboldt and the Cosmos.* New York: Harper and Row, 1973.

Bourricaud, François, ed. *The Sociology of Talcott Parsons.* Trans. Arthur Goldhammer. Chicago: University of Chicago Press, 1981.

Bowditch, Henry I. *The Young Stethoscopist.* 1846. New York: Hafner, 1964.

Bowditch, Nathaniel. *A History of the Massachusetts General Hospital, to August 5, 1851.* New York: Arno, 1972.

———. *The New American Practical Navigator.* 1802. Arcata, CA: Paradise Cay, 2002.

Bowers, Peter M. *Boeing Aircraft Since 1916.* Annapolis, MD: Naval Institute Press, 1989.

Bowler, Peter J. *Evolution: The History of an Idea.* Berkeley: University of California Press, 2003.

Boyd, Thomas Alvin. *Charles F. Kettering: A Biography.* Washington, DC: Beard, 2002.

Boyd, William. *A Text-Book of Pathology.* Philadelphia: Lea and Febiger, 1947.

Bozeman, Theodore Dwight. *Protestants in an Age of Science: The Baconian Ideal and Antebellum American Religious Thought.* Chapel Hill: University of North Carolina Press, 1977.

Bozorgnia, Yousef. *Earthquake Engineering.* Boca Raton, FL: CRC, 2004.

Braasch, William F. *Early Days in the Mayo Clinic.* Springfield, IL: Charles C. Thomas, 1969.

Brackenridge, Henry Marie. "Journal of a Voyage up the River Missouri." In *Early Western Travels, 1748–1846,* vol. 6, ed. Reuben Gold Thwaites. Cleveland, OH: A.H. Clark, 1905.

Brackett, Virginia. *Steve Jobs: Computer Genius of Apple.* Berkeley Heights, NJ: Enslow, 2003.

Bradbury, John. "Travels in the Interior of America, in the Years 1809, 1810, and 1811." In *Early Western Travels, 1748–1846,* vol. 5, ed. Reuben Gold Thwaites. Cleveland, OH: A.H. Clark, 1905.

Brading, D.A. *The First America: The Spanish Monarchy, Creole Patriots, and the Liberal State 1492–1867.* Cambridge, UK: Cambridge University Press, 1991.

Brandon, Ruth. *A Capitalist Romance: Singer and the Sewing Machine.* London: Barrie and Jenkins, 1977.

Brandt, E. Ned. *Growth Company: Dow Chemical's First Century.* Lansing: Michigan State University Press, 2003.

Brandt, John C., and Robert D. Chapman. *Rendezvous in Space: The Science of Comets.* New York: W.H. Freeman, 1992.

Brayer, Elizabeth. *George Eastman: A Biography.* Baltimore: Johns Hopkins University Press, 1996.

Brekke, Asgeir, and Alv Egeland. *The Northern Light: From Mythology to Space Research.* New York: Springer Verlag, 1983.

Brennan, Bernard P. *William James.* New Haven, CT: Twayne, 1968.

Brent, Joseph. *Charles Sanders Peirce: A Life.* 1993. Reprint, Bloomington: Indiana University Press, 1998.

Breslaw, Elaine G., ed. *Records of the Tuesday Club of Annapolis, 1745–56.* Urbana: University of Illinois Press, 1988.

Brian, Denis. *The Enchanted Voyager: The Life of J.B. Rhine.* Englewood Cliffs, NJ: Prentice Hall, 1982.

Bridenbaugh, Carl. *Cities in Revolt: Urban Life in America, 1743–1776.* New York: Oxford University Press, 1955.

Brill, Thomas. *Light: Its Interaction with Art and Antiquities.* New York: Plenum, 1980.

Brinkley, Douglas. *Wheels for the World: Henry Ford, His Company, and a Century of Progress, 1903–2003.* New York: Viking, 2003.

Brodsky, Alyn. *Benjamin Rush: Patriot and Physician.* New York: Truman Talley, 2004.

Broehl, Wayne G. *John Deere's Company.* Hanover, NH: University Press of New England, 1992.

Bromberg, Joan Lisa. *Fusion: Science, Politics, and the Invention of a New Energy Source.* Cambridge, MA: MIT Press, 1982.

Bronson, Po. *The Nudist on the Late Shift: And Other True Tales of Silicon Valley.* New York: Random House, 1999.

Brooks, Paul. *Rachel Carson: The Writer at Work.* San Francisco: Sierra Club, 1998.

Brooks, William Keith. *Biographical Memoir of Alpheus Hyatt, 1838–1902.* Washington, DC: Judd and Detweiler, 1908.

Brown, Chandos Michael. *Benjamin Silliman: A Life in the Young Republic.* Princeton, NJ: Princeton University Press, 1989.

Brown, Ira, ed. *Joseph Priestley: Selections from His Writings.* University Park: Pennsylvania State University Press, 1962.

Brown, Lawrence D. *Politics and Health Care Organization: HMOs as Federal Policy.* Washington, DC: Brookings Institution, 1983.

Brown, Sanborn C. *Benjamin Thompson, Count Rumford.* Cambridge, MA: MIT Press, 1981.

Brown, Stephanie, Finn-Aage Esbensen, and Gibber Geis. *Criminology: Explaining Crime and Its Context.* Cincinnati, OH: Anderson, 1991.

Brown, Thomas J. *Dorothea Dix: New England Reformer.* Cambridge, MA: Harvard University Press, 1998.

Bruce, Robert V. *Alexander Graham Bell and the Conquest of Solitude.* Boston: Little, Brown, 1973.

Brunner, Henry S. *Land-Grant Colleges and Universities, 1862–1962.* Washington, DC: Department of Health, Education, and Welfare, 1962.

Bryan, C.D.B. *The National Geographic Society: 100 Years of Adventure and Discovery.* New York: Harry N. Abrams, 1987.

Brydson, J.A. *Plastics Materials.* 7th ed. Burlington, VT: Butterworth-Heinemann, 1999.

Bryman, Alan, and Duncan Cramer. *Quantitative Analysis with SPSS 12 and 13: A Guide for Social Scientists.* New York: Routledge, 2005.

Buchler, Justus, ed. *Philosophical Writings of Peirce.* New York: Dover, 1955.

Buckley, Kerry W. *Mechanical Man: John Broadus Watson and the Beginnings of Behaviorism.* New York: Guilford, 1989.

Buhle, Paul. *Marxism in the United States: Remapping the History of the American Left.* London: Verson, 1987.

Bullough, Vern, and Martha Voght. "Women, Menstruation, and Nineteenth-Century Medicine." In *Women and Health in America*, ed. Judith Walzer Leavitt. Madison: University of Wisconsin Press, 1984.

Bunton, Robin, and Alan Petersen, eds. *Genetic Governance: Health, Risk, and Ethics in the Biotech Era.* New York: Routledge, 2005.

Burbank, Luther. *The Life and Work of Luther Burbank.* Columbia, MO: Athena University Press, 2004.

———. *Luther Burbank: His Methods and Discoveries and Their Practical Application.* 12 vols. Ed. John Whitson, Robert John, and Henry Smith Williams. New York: Luther Burbank Society, 1914.

Burbank-Beeson, Emma, Effie Young Slusser, and Mary Belle Williams. *Stories of Luther Burbank and His Plant School.* 1920. Park Forest, IL: University Press of the Pacific, 2002.

Burgaleta, Claudio M. *José de Acosta, S.J. (1540–1600): His Life and Thought.* Chicago: Loyola University Press, 1999.

Burkett, Nancy H., and John B. Hench, eds. *Under Its Generous Dome: The Collections and Programs of the American Antiquarian Society.* Worcester, MA: American Antiquarian Society, 1992.

Burks, Alice. *Who Invented the Computer?* New York: Prometheus, 2003.

Burnham, John C. *How Superstition Won and Science Lost: Popularizing Science and Health in the United States.* New Brunswick, NJ: Rutgers University Press, 1987.

———. *Psychoanalysis and American Medicine, 1894–1918.* New York: International Universities Press, 1967.

Burnham, Robert. *Great Comets.* New York: Cambridge University Press, 2000.

Burns, David M. *Gateway: Dr. Thomas Walker and the Opening of Kentucky.* Middlesboro, KY: Bell County Historical Society, 2000.

Burns, William E. *Science and Technology in Colonial America.* Westport, CT: Greenwood, 2005.

Burrow, James G. *AMA: Voice of American Medicine.* Baltimore: Johns Hopkins University Press, 1963.

Burrows, William E. *The New Ocean: The Story of the First Space Age.* New York: Random House, 1998.

Burton, Jean. *Lydia Pinkham Is Her Name.* New York: Farrar, Straus, 1949.

Bush, Vannevar. *Operational Circuit Analysis.* New York: John Wiley and Sons, 1929.

Butcher, Lee. *Accidental Millionaire: The Rise and Fall of Steve Jobs at Apple Computers.* New York: Paragon, 1988.

Butterfield, Herbert. *The Origins of Modern Science.* New York: Free Press, 1997.

Byrd, Richard E. *Alone.* New York: G.P. Putnam's Sons, 1938.

———. *To the Pole: The Diary and Notebook of Richard E. Byrd, 1925–1927.* Richard Byrd Papers, Ohio State University.

Byrne, Peter. *Natural Religion and the Nature of Religion: The Legacy of Deism.* New York: Routledge, 1989.

Caffrey, Margaret M. *Ruth Benedict: Stranger in This Land.* Austin: University of Texas Press, 1989.

Caidin, Martin. *The Astronauts: The Story of Project Mercury, America's Man-in-Space Program.* New York: Dutton, 1961.

Calvin, Melvin. *Following the Trail of Light: A Scientific Odyssey.* Oxford, UK: Oxford University Press, 1998.

Campbell, John F. *History and Bibliography of the New American Practical Navigator and the American Coast Pilot.* Salem, MA: Peabody Museum, 1964.

Campbell-Kelly, Martin, and William Aspray. *Computer: A History of the Information Machine.* New York: Basic Books, 1996.

Campion, Frank D. *The AMA and U.S. Health Policy Since 1940.* Chicago: Chicago Review, 1984.

Cantwell, Robert. *Alexander Wilson: Naturalist and Pioneer.* Philadelphia: Lippincott, 1961.

Caprara, Giovanni. *Living in Space: From Science Fiction to the ISS.* Buffalo, NY: Firefly, 2000.

Carey, James R., and Shripad Taljapurkar, eds. *Life Span: Evolutionary, Ecological, and Demographic Perspectives.* New York: Population Council, 2003.

Carlson, Elof Axel. *The Unfit: A History of a Bad Idea.* Cold Spring Harbor, NY: Cold Spring Harbor Laboratory Press, 2001.

Carmony, Donald F. *Indiana 1816–1850: The Pioneer Era.* Indianapolis: Indiana Historical Bureau and Indiana Historical Society, 1998.

Carroll, Jennifer Lee. *The Speckled Monster: A Historic Tale of Battling Smallpox.* New York: Dutton, 2003.

Carson, Rachel. *Silent Spring.* Boston: Houghton Mifflin; Cambridge, MA: Riverside Press, 1962.

Carter, Bill, and Merri Sue Carter. *Latitude: How American Astronomers Solved the Mystery of Variation.* Annapolis, MD: Naval Institute Press, 2002.

Carter, Edward C. *One Grand Pursuit: A Brief History of the American Philosophical Society's First 250 Years, 1743–1993.* Philadelphia: American Philosophical Society, 1993.

Carver, George Washington. *George Washington Carver: In His Own Words.* Ed. and intro. Gary R. Kremer. Columbia: University of Missouri Press, 1991.

Carwell, Hattie. *Blacks in Science: Astrophysicist to Zoologist.* New York: Exposition, 1977.

Cary, Edward R. *Geodetic Surveying.* New York: John Wiley and Sons, 1916.

Cash, Philip. *Dr. Benjamin Waterhouse: A Life in Medicine and Public Service (1754–1846).* Sagamore Beach, MA: Science History Publications, 2006.

Cashin, Edward J. *William Bartram and the American Revolution on the Southern Frontier.* Columbia: University of South Carolina Press, 2000.

Cassedy, James H. *Charles V. Chapin and the Public Health Movement.* Cambridge, MA: Harvard University Press, 1962.

———. *Medicine in America: A Short History.* Baltimore: Johns Hopkins University Press, 1991.

Cassidy, David C. *J. Robert Oppenheimer and the American Century.* New York: Pi Press, 2005.

Cassidy, John. *Dot.con: The Greatest Story Ever Sold.* New York: HarperCollins, 2002.

Cassidy, Robert. *Margaret Mead: A Voice for the Century.* New York: Universe, 1982.

Castiglioni, Arturo. *A History of Medicine.* New York: Alfred A. Knopf, 1958.

Castleman, Benjamin, David C. Crockett, and S.B. Sutton, eds. *The Massachusetts General Hospital, 1955–1980.* Boston: Little Brown, 1983.

Caulfield, Timothy A., and Bryn Williams-Jones, eds. *The Commercialization of Genetic Research: Ethical, Legal, and Policy Issues.* New York: Kluwer, 1999.

Cerami, Charles. *Benjamin Banneker.* New York: John Wiley and Sons, 2002.

Ceruzzi, Paul. *A History of Modern Computing.* Cambridge, MA: MIT Press, 1998.

Chaikin, Andrew. *A Man on the Moon: The Voyages of the Apollo Astronauts.* New York: Viking, 1994.

Chaisson, Eric. *The Hubble Wars: Astrophysics Meets Astropolitics in the Two-Billion-Dollar Struggle over the Hubble Space Telescope.* New York: HarperCollins, 1994.

Chandler, Alfred D., Jr. *Shaping the Industrial Century: The Remarkable Story of the Evolution of the Modern Chemical and Pharmaceutical Industries.* Cambridge, MA: Harvard University Press, 2005.

———. *The Visible Hand: The Managerial Revolution in American Business.* Cambridge, MA: Belknap Press, 1977.

Chandrasekhar, Subrahmanyan. *The Mathematical Theory of Black Holes.* Oxford, UK: Clarendon, 1998.

Chapman, Arthur H. *Harry Stack Sullivan: His Life and His Work.* New York: G.P. Putnam's Sons, 1976.

Chapman, Brian. *Glow Discharge Processes.* Hoboken, NJ: John Wiley and Sons, 1980.

Chapman, Walker. *Antarctic Conquest.* New York: Bobbs-Merrill, 1965.

Chatterjee, Sankar. *The Rise of Birds: 225 Million Years of Evolution.* Baltimore: Johns Hopkins University Press, 1997.

Chen, Ko Kuei, ed. *The American Society for Pharmacology and Experimental Therapeutics, Incorporated: The First Sixty Years.* Bethesda, MD: American Society for Pharmacology and Experimental Therapeutics, 1969.

Chesler, Ellen. *Woman of Valor: Margaret Sanger and the Birth Control Movement in America.* New York: Anchor, 1992.

Childs, Herbert. *An American Genius: The Life of Ernest Lawrence.* New York: E.P. Dutton, 1968.

Chittenden, Russell H. *History of the Sheffield Scientific School of Yale University.* New Haven, CT: Yale University Press, 1928.

Christensen, Lars Winther. *Gorenstein Dimensions.* New York: Springer, 2000.

Christian, David. *Maps of Time: An Introduction to Big History.* Berkeley: University of California Press, 2004.

Christianson, Gale E. *Edwin Hubble: Mariner of the Nebulae.* New York: Farrar, Straus and Giroux, 1995.

Christie, Maureen. *The Ozone Layer: A Philosophy of Science Perspective.* New York: Cambridge University Press, 2001.

Christopherson, Robert W. *Geosystems.* 4th ed. Upper Saddle River, NJ: Prentice Hall, 2002.

Cichoke, Anthony J. *Secrets of Native American Herbal Remedies: A Comprehensive Guide to the Native American Tradition of Using Herbs and the Mind/Body/Spirit Connection for Improving Health and Well-Being.* New York: Avery, 2001.

Clapesattle, Helen B. *The Doctors Mayo.* Rochester, MN: Mayo Clinic Health Management, 2003.

Clark, Paul F. *Pioneer Microbiologists of America.* Madison: University of Wisconsin Press, 1961.

Clark, Ronald W. *Einstein: The Life and Times.* New York: Avon, 1999.

Clarke, John Henrik. "Lewis Latimer—Bringer of the Light." In *Blacks in Science: Ancient and Modern,* ed. Ivan Van Sertima. Somerset, NJ: Transaction, 1987.

Clary, David A. *Rocket Man: Robert H. Goddard and the Birth of the Space Age.* New York: Hyperion, 2003.

Clendening, Logan. *Source Book of Medical History.* New York: Dover, 1960.

Cliff, Andrew, Peter Haggett, and Matthew Smallman-Raynot. *Measles: An Historical Geography of a Major Human Viral Disease, from Global Expansion to Local Retreat, 1840–1990.* Cambridge, MA: Blackwell, 1993.

Clubb, Jerome M., and Erwin Scheuch, eds. *Historical Social Research: The Use of Historical and Process-Produced Data.* Stuttgart, Germany: Klett-Cotta, 1980.

Cochrane, Dorothy, Von Hardesty, and Russell Lee. *The Aviation Careers of Igor Sikorsky.* Seattle: University of Washington Press, 1989.

Cohen, I. Bernard. *Benjamin Franklin's Science.* Cambridge, MA: Harvard University Press, 1990.

———, ed. *Benjamin Peirce: "Father of Pure Mathematics" in America.* New York: Arno, 1980.

———. *The Birth of a New Physics.* New York: W.W. Norton, 1985.

———. *Howard Aiken: Portrait of a Computer Pioneer.* Cambridge, MA: MIT Press, 1999.

———. *Introduction to Newton's Principia.* Cambridge, MA: Harvard University Press, 1971.

———. *Makin' Numbers: Howard Aiken and the Computer.* Cambridge, MA: MIT Press, 1999.

———. *Some Early Tools of American Science: An Account of the Early Scientific Instruments and Mineralogical and Biological Collections in Harvard University.* Cambridge, MA: Harvard University Press, 1950.

Cohen, Michael P. *The History of the Sierra Club, 1892–1970.* San Francisco: Sierra Club, 1988.

Coleman, William L. *Yellow Fever in the North: The Method of Epidemiology.* Wisconsin Publications in the History of Science and Medicine, no. 6. Madison: University of Wisconsin Press, 2000.

Collette, Bruce B., and Grace Klein-MacPhee, eds. *Bigelow and Schroeder's Fishes of the Gulf of Maine.* Washington, DC: Smithsonian Institution, 2002.

Collins, Michael. *Carrying the Fire: An Astronaut's Journey.* New York: Farrar, Straus and Giroux, 1974.

———. *Liftoff: The Story of America's Adventure in Space.* New York: Grove, 1988.

———. *Mission to Mars: An Astronaut's Vision of Our Future in Space.* New York: Grove Weidenfeld, 1990.

Collins, Randall, and Michael Makosky. *The Discovery of Society.* New York: Random House, 1972.

Columbus, Christopher. *First Voyage to America: From the Log of the "Santa Maria."* New York: Dover, 1991.

Comfort, Nathaniel C. *The Tangled Field: Barbara McClintock's Search for the Patterns of Genetic Control.* Cambridge, MA: Harvard University Press, 2001.

Compton, Arthur H. *Atomic Quest: A Personal Narrative.* New York: Oxford University Press, 1956.

———. *Cosmos of Arthur Holly Compton.* New York: Alfred A. Knopf, 1967.

Conant, James B. *On Understanding Science: An Historical Approach.* New York: Mentor, 1951.

———. *Science and Common Sense.* New Haven, CT: Yale University Press, 1964.

———. *Slums and Suburbs.* New York: McGraw-Hill, 1961.

Conaway, Charles F. *The Petroleum Industry: A Nontechnical Guide.* Tulsa, OK: PennWell, 1999.

Cooley, Charles Horton. *Human Nature and Society.* 1902. Reprint ed., New York: Schocken, 1964.

———. *Social Organization.* 1909. Reprint ed., New York: Schocken, 1962.

———. *Social Process.* 1918. Reprint ed., Carbondale: Southern Illinois University Press, 1966.

Cooley, Denton A. *Reflections and Observations: Essays of Denton A. Cooley.* Collected by Marianne Kneipp. Austin, TX: Eakin, 1984.

Coombs, Jan Gregoire. *The Rise and Fall of the HMO Movement: An American Health Care Revolution.* Madison: University of Wisconsin Press, 2005.

Cooper, Dan. *Enrico Fermi and the Revolutions of Modern Physics.* Oxford, UK: Oxford University Press, 1998.

Cordeschi, Roberto. *The Discovery of the Artificial: Behavior, Mind, and Machines Before and Beyond Cybernetics.* Boston: Kluwer Academic, 2002.

Corner, George Washington. *George Hoyt Whipple and His Friends: The Life Story of a Nobel Prize Pathologist.* Philadelphia: Lippincott, 1963.

Corper, H.J. *A Modern American Tuberculosis Sanatorium.* New York: Modern Hospital, 1924.

Coser, Lewis A. *Masters of Sociological Thought: Ideas in Historical and Social Context.* New York: Harcourt Brace Jovanich, 1971; 2nd ed. Prospect Heights, IL: Waveland, 1977.

Cox, John D. *Storm Watchers: The Turbulent History of Weather Prediction from Franklin's Kite to El Niño.* Hoboken, NJ: John Wiley and Sons, 2002.

Craven, Wesley F., and Walter B. Hayward, eds. *The Journal of Richard Norwood, Surveyor of Bermuda.* New York: Bermuda Historical Monuments Trust, 1945.

Cravens, Hamilton. *The Triumph of Evolution: The Heredity–Environment Controversy, 1900–1941.* Baltimore: Johns Hopkins University Press, 1978.

Cringely, Robert. *Accidental Empires.* New York: Perseus, 1992.

Cronquist, Arthur. *An Integrated System of Classification of Flowering Plants.* New York: Columbia University Press, 1993.

Crosby, Alfred W. *The Columbian Exchange: Biological and Cultural Consequences of 1492.* Westport, CT: Greenwood, 1972.

———. *Epidemic and Peace, 1918.* Westport, CT: Greenwood, 1976.

Crouch, Tom D. *The Bishop's Boys: A Life of Wilbur and Orville Wright.* New York: W.W. Norton, 1989.

———. *The Eagle Aloft: Two Centuries of the Balloon in America.* Washington, DC: Smithsonian, 1983.

Crouse, Nellis Maynard. *Contributions of the Canadian Jesuits to the Geographical Knowledge of New France, 1632–1675.* Ithaca, NY: Cornell Publications, 1924.

Crovisier, Jacques, and Thérèse Encrenaz. *Comet Science: The Study of Remnants from the Birth of the Solar System.* New York: Cambridge University Press, 2000.

Culliford, S.G. *William Strachey 1572–1621.* Charlottesville: University of Virginia Press, 1965.

Cushing, Harvey. *The Life of Sir William Osler.* London: Oxford University Press, 1940.

Cutler, Manasseh. *Life, Journals, and Correspondence of Rev. Manasseh Cutler, LL.D.,* ed. William P. Cutler and Julia P. Cutler. Cincinnati, OH: R. Clarke, 1888; Athens: Ohio University Press, 1987.

Cutright, Paul Russell. *Lewis and Clark, Pioneering Naturalists*. Urbana: University of Illinois Press, 1969.

Dabney, Robert Lewis. *The Sensualistic Philosophy of the Nineteenth Century Considered*. 1875. Reprint ed., Dallas: Naphtali, 2003.

Dain, Norman. *Clifford Beers: Advocate for the Insane*. Pittsburgh, PA: University of Pittsburgh Press, 1980.

Dameshek, William. *George R. Minot Symposium on Hematology*. New York: Grune and Stratton, 1949.

Dana, Edward Salisbury. "The American Journal of Science from 1818 to 1918." In *A Century of Science in America with Special Reference to the American Journal of Science 1818–1918*, ed. Edward Salisbury Dana. New Haven, CT: Yale University Press, 1918.

Daniel, Thomas M. *Pioneers of Medicine and Their Impact on Tuberculosis*. Rochester, NY: University of Rochester Press, 2000.

Daniels, George H. *Science in American Society: A Social History*. New York: Alfred A. Knopf, 1971.

Danlioy, Victor J. *Chicago's Museums: A Complete Guide to the City's Cultural Attractions*. Chicago: Chicago Review Press, 1991.

Darnell, Regna. *And Along Came Boas: Continuity and Revolution in Americanist Anthropology*. Philadelphia: John Benjamins, 1998.

Dassow Walls, Laura. *Seeing New Worlds: Henry David Thoreau and Nineteenth-Century Natural Science*. Madison: University of Wisconsin Press, 1995.

Davidson, Keay. *Carl Sagan: A Life*. New York: John Wiley and Sons, 1999.

Davies, N.J.H., R.S. Atkinson, and G.B. Rushman. *A Short History of Anaesthesia: The First 150 Years*. Burlington, VT: Butterworth-Heinemann Medical, 1996.

Davies, Pete. *Devil's Flu: The World's Deadliest Influenza Epidemic and the Scientific Hunt for the Virus That Caused It*. New York: Henry Holt, 2000.

Davis, Bernard, ed. *The Genetic Revolution: Scientific Prospects and Public Perceptions*. Baltimore: Johns Hopkins University Press, 1991.

Davis, Kenneth S. *The Cautionary Scientists: Priestley, Lavoisier, and the Founding of Modern Chemistry*. New York: G.P. Putnam's Sons, 1966.

Dawkins, Richard. *The Selfish Gene*. New York: Oxford University Press, 1990.

De Bourgeoing, Jacqueline. *The Calendar: History, Lore, and Legend*. New York: Harry N. Abrams, 2000.

De Camp, L. Sprague. *The Great Monkey Trial*. Garden City, NY: Doubleday, 1968.

De Carle, Donald. *Watch and Clock Encyclopedia*. 2nd ed. London: N.A.G. Press, 1976.

De Voto, Bernard. *The Course of Empire*. Boston: Houghton Mifflin, 1952.

DeBakey, Michael E. *General Surgery*. Washington, DC: U.S. Army Medical History Office, 1955.

Delear, Frank J. *Igor Sikorsky: His Three Careers in Aviation*. New York: Dodd, Mead, 1969.

D'Elia, Donald J. *Benjamin Rush, Philosopher of the American Revolution*. Philadelphia: American Philosophical Society, 1974.

Deloria, Vine, Jr. *Red Earth, White Lies: Native Americans and the Myth of Scientific Fact*. New York: Scribner's, 1995

Dembski, William A. *Design Inference: Eliminating Chance through Small Probabilities*. Cambridge, UK: Cambridge University Press, 1998.

DeRosier, Arthur H., Jr. *William Dunbar, Scientific Pioneer of the Old Southwest*. Lexington: University Press of Kentucky, 2007.

Devles, Peter J., ed. *Encyclopedia of Immunology*. Vol. 3, 2nd ed. San Diego, CA: Academic Press, 1998.

Dewey, John. *Experience and Education*. New York: Simon and Schuster, 1997.

Dick, Steven J. *The Biological Universe*. New York: Cambridge University Press, 1996.

———. *Sky and Ocean Joined: The U.S. Naval Observatory, 1830–2000*. Cambridge, UK: Cambridge University Press, 2002.

Dickson, Paul. *Sputnik: The Shock of the Century*. New York: Walker, 2001.

Diggins, John P. *Up from Communism*. New York: Columbia University Press, 1993.

Diggs, Irene. *Black Inventors*. Chicago: Institute of Positive Education, 1975.

Dirac, Paul. *The Principles of Quantum Mechanics*. New York: Oxford University Press, 1982.

Dix, Dorothea L. *On Behalf of the Insane Poor: Selected Reports*. New York: Arno, 1971.

Dixon, Dougal, et al. *Atlas of Life on Earth: The Earth, Its Landscapes, and Lifeforms*. New York: Barnes and Noble, 2001.

Dobbs, David. *The Great Gulf: Fishermen, Scientists, and the Struggle to Revive the World's Greatest Fishery*. Washington, DC: Island, 2000.

Dodge, Robert K., comp. *A Topical Index of Early U.S. Almanacs, 1776–1800*. Westport, CT: Greenwood, 1997.

Dolman, Claude, and Richard Wolfe. *Suppressing the Disease of Animals and Man: Theobald Smith, Microbiologist*. Boston: Boston Medical Library, 2003.

Dolnick, Edward. *Down the Great Unknown: John Wesley Powell's 1869 Journey of Discovery and Tragedy Through the Grand Canyon*. New York: Harper, 2002.

Donegan, Jane B. *Women and Men Midwives: Medicine, Morality, and Misogyny in Early America*. Westport, CT: Greenwood, 1978.

Dormandy, Thomas. *The White Death: A History of Tuberculosis*. London: Hambledon, 2001.

Doskey, John S., ed. *The European Journals of William Maclure.* Memoirs of the American Philosophical Society, vol. 171. Philadelphia: American Philosophical Society, 1988.

Doyle, Jack. *Trespass Against Us: Dow Chemical and the Toxic Century.* Monroe, ME: Common Courage, 2004.

Doyle, Michael P., Larry R. Beuchat, and Thomas J. Montville, eds. *Food Microbiology: Fundamentals and Frontiers.* Washington, DC: ASM, 2001.

Drake, Ellen. *Geologists and Ideas: A History of North American Geology.* Boulder, CO: Geological Society of America, 1985.

Draper, John William. *Collected Works of John William Draper.* Reprint ed., Temecula, CA: Reprint Services, 1999.

Drlica, Karl. *Double-Edged Sword: The Promise and Risks of the Genetic Revolution.* Reading, MA: Helix, 1994.

DuBois, J. Harry. *Plastics History U.S.A.* Boston: Cahners, 1972.

Duffy, John. *The Healers: A History of American Medicine.* Urbana: University of Illinois Press, 1979.

———. *The Sanitarians: A History of American Public Health.* Urbana: University of Illinois Press, 1992.

Duke, Martin. *The Development of Medical Techniques and Treatments: From Leeches to Heart Surgery.* Guilford, CT: International Universities Press, 1991.

Dunar, Andrew J., and Dennis McBride. *Building Hoover Dam: An Oral History of the Great Depression.* New York: Twayne, 1993.

Duncan, Francis. *Atomic Shield: A History of the United States Atomic Energy Commission.* Vol. 2, *1947–1952.* University Park: Pennsylvania State University Press, 1969.

Dupree, A. Hunter. *Asa Gray, 1810–1888.* Cambridge, MA: Harvard University Press, 1959.

———. *Asa Gray, American Botanist, Friend of Darwin.* Baltimore: Johns Hopkins University Press, 1959; 1988.

———. *Science in the Federal Government: A History of Policies and Activities.* Baltimore: Johns Hopkins University Press, 1986.

Duren, Peter, Richard A. Askey, and Uta C. Merzbach, eds. *A Century of Mathematics in America.* 3 vols. Providence, RI: American Mathematical Society, 1988–1989.

Duxbury, Alison B., Alyn C. Duxbury, and Keith A. Sverdrup. *Fundamentals of Oceanography.* New York: McGraw-Hill, 2001.

Dwight, Timothy. *Travels in New-England and New-York,* vol. II. New Haven, CT: 1821.

Earls, Alan R., Nasrin Rohani, and Marie Cosindas. *Polaroid.* Mount Pleasant, SC: Arcadia, 2005.

Earnest, Ernest. *John and William Bartram, Botanists and Explorers, 1699–1777, 1739–1823.* Philadelphia: University of Pennsylvania Press, 1940.

———. *S. Weir Mitchell, Novelist and Physician.* Philadelphia: University of Pennsylvania Press, 1950.

Easterbrook, Gregg. *Surgeon Koop.* Knoxville, TN: Whittle, 1991.

Eccles, W.J. *The Canadian Frontier, 1534–1760.* Rev. ed. Albuquerque: University of New Mexico Press, 1983.

Echols, Harrison, and Carol A. Gross, eds. *Operators and Promoters: The Story of Molecular Biology and Its Creators.* Berkeley: University of California Press, 2001.

Eckel, Edwin. *The Geological Society of America: Life History of a Learned Society.* Boulder, CO: Geological Society of America, 1982.

Eckert, William G. *Introduction to Forensic Science.* Boca Raton, FL: CRC, 1996.

Eddy, Edward D. *Colleges for Our Land and Time: The Land-Grant Idea in American Education.* New York: Harper, 1957.

Eddy, Mary Baker. *Science and Health.* Boston: Christian Science Publishing, 1875. Rev. ed. Boston: Trustees under the Will of Mary Baker Eddy, 1934.

Edwards, Jonathan. "Sinners in the Hands of an Angry God (1741)." In *The Sermons of Jonathan Edwards: A Reader,* ed. Wilson H. Kimnach, Kenneth P. Minkema, and Douglas A. Sweeney. New Haven, CT: Yale University Press, 1999.

Edwards, Terry. *Fiber Optic Systems: Network Applications.* New York: John Wiley and Sons, 1989.

Ehrenreich, Barbara, and Deidre English. *Witches, Midwives, and Nurses: A History of Women Healers.* New York: Feminist Press, 1973.

Ehrlich, Paul. *The Birder's Handbook: A Field Guide to the Natural History of North American Birds.* New York: Simon and Schuster, 1988.

Einstein, Albert. *Essays in Humanism.* New York: Philosophical Library, 1950.

———. *Relativity: The Special and General Theory.* Trans. Robert W. Lawson. New York: Routledge, 2001.

Ekelund, Robert B., Jr., and Robert F. Hebert. *A History of Economic Theory and Method.* New York: McGraw-Hill, 1975.

Elder, Fred Kingsley. *Woodrow: Apostle of Freedom.* Two Harbors, MN: Bunchberry, 1994.

Ellen, Roy, Ernest Gellner, Grazyna Kubica, and Janusz Mucha, eds. *Malinowski Between Two Worlds: The Polish Roots of an Anthropological Tradition.* Cambridge, UK: Cambridge University Press, 1989.

Elliott, Clark A., and Margaret W. Rossiter. *Science at Harvard University: Historical Perspectives.* Bethlehem, PA: Lehigh University Press, 1992.

Elliott, Josephine Mirabella, ed. *Partnership for Posterity: The Correspondence of William Maclure and Marie*

Duclos Fretageot, 1820–1833. Indianapolis: Indiana Historical Society, 1994.

Ellis, Harold. *A History of Surgery.* New York: Cambridge University Press, 2000.

Enz, Charles P. *No Time to Be Brief: A Scientific Biography of Wolfgang Pauli.* Oxford, UK: Oxford University Press, 2002.

Erickson, Jon. *Plate Tectonics: Unraveling the Mysteries of the Earth.* New York: Checkmark, 2001.

Eriksen, Thomas Hylland, and Finn Sivert Nielsen. *A History of Anthropology.* London: Pluto, 2001.

Erikson, Erik H. *Childhood and Society.* 2nd ed. New York: W.W. Norton, 1963.

———. *Young Man Luther: A Study in Psychoanalysis and History.* New York: W.W. Norton, 1958.

Evans, George G., ed. *Illustrated History of the United States Mint.* Philadelphia: George G. Evans, 1892.

Evans, James. *The History and Practice of Ancient Astronomy.* Oxford, UK: Oxford University Press, 1998.

Evans, R.B., V. Staudt Sexton, and T. C. Cadwallader. *The American Psychological Association: A Historical Perspective.* Washington, DC: American Psychological Association, 1992.

Ewan, Joseph, and Nesta Dunn Ewan. *Benjamin Smith Barton, Naturalist and Physician in Jeffersonian America.* St. Louis: Missouri Botanical Garden Press, 2007.

———. *John Banister and His Natural History of Virginia, 1678–1692.* Urbana: University of Illinois Press, 1970.

Fadiman, Anne. *The Spirit Catches You and You Fall Down.* New York: Farrar, Straus and Giroux, 1998.

Fagan, Brian. *The Great Journey: The Peopling of Ancient America.* New York: Thames and Hudson, 1987.

Fancher, Raymond E. *The Intelligence Men: Makers of the I.Q. Controversy.* New York. W.W. Norton, 1985.

Fancher, Robert T. *Health and Suffering in America: The Context and Content of Mental Health Care.* Piscataway, NJ: Transaction, 2003.

Farina, Amerigo. *Abnormal Psychology.* Englewood Cliffs, NJ: Prentice Hall, 1976.

Farragher, John Mack. *Sugar Creek: Life on the Illinois Prairie.* New Haven, CT: Yale University Press, 1986.

Faulkner, Harold. *The Decline of Laissez Faire, 1897–1917.* New York: Rinehart, 1951.

Federoff, Nina, and David Botstein. *The Dynamic Genome: Barbara McClintock's Ideas in the Century of Genetics.* Cold Spring Harbor, NY: Cold Spring Harbor Laboratory Press, 1992.

Felman, Lewis. *Second Industrial Revolution.* Upper Saddle River, NJ: Prentice Hall, 1985.

Fenichell, Stephen. *Plastic: The Making of a Synthetic Century.* New York: HarperBusiness, 1996.

Fenn, John B. *Engines, Energy, and Entropy.* New York: W.H. Freeman, 1982.

Fenner, Frank, et al. *Smallpox and Its Eradication.* Geneva: World Health Organization, 1988.

Fenner, M., and Eleanor C. Fishburn. *Pioneer American Educators.* Port Washington, NY: Kennikat, 1968.

Fenster, Julie. *Ether Day: The Strange Tale of America's Greatest Medical Discovery and the Haunted Men Who Made It.* New York: HarperCollins, 2001.

Fermi, Laura. *Atoms in the Family: My Life with Enrico Fermi.* Albuquerque: University of New Mexico Press, 1988.

Fernandez, Ronald. *Mappers of Society: The Lives, Times, and Legacies of Great Sociologists.* Westport, CT: Praeger, 2003.

Fetzer, James H., ed. *Science, Explanation, and Rationality: Aspects of the Philosophy of Carl G. Hempel.* New York: Oxford University Press, 2000.

Feynman, Richard P., Robert B. Leighton, and Matthew Sands. *The Feynman Lectures of Physics.* Vol. 1. Reading, MA: Addison-Wesley, 1977.

Feynman, Richard P., with Ralph Leighton. *Surely You're Joking Mr. Feynman.* New York: W.W. Norton, 1985.

Fichter, George. *How to Build an Indian Canoe.* New York: McKay, 1977.

Finegold, Sydney M., and W. Lance George, eds. *Anaerobic Infections in Humans.* San Diego, CA: Academic Press, 1984.

Fingleton, Eamonn. *In Praise of Hard Industries.* Boston: Houghton Mifflin, 1999.

Finkelstein, Joseph. *Windows on a New World: The Third Industrial Revolution.* Westport, CT: Greenwood, 1989.

Fishbein, Morris. *A History of the American Medical Association.* Philadelphia: W.B. Saunders, 1947.

Fitzgerald, Deborah. *Every Farm a Factory: The Industrial Ideal in American Agriculture.* New Haven, CT: Yale University Press, 2003.

Fitzmier, John R. *New England's Moral Legislator: Timothy Dwight, 1752–1817.* Bloomington: Indiana University Press, 1998.

Fleming, James R. *Meteorology in America, 1800–1870.* Baltimore: Johns Hopkins University Press, 1990.

Fleming, Thomas. *The Man Who Dared the Lightning: A New Look at Benjamin Franklin.* New York: William Morrow, 1971.

Flexner, Abraham. *Abraham Flexner: An Autobiography: A Revision, Brought up-to-Date, of the Author's I Remember.* New York: Simon and Schuster, 1960.

———. *Medical Education in the United States and Canada.* Boston: Merrymount, 1910.

Florence, Ronald. *The Perfect Machine: Building the Palomar Telescope.* New York: HarperCollins, 1994.

Flores, Dan. *Journal of an Indian Trader: Anthony Glass and the Texas Trading Frontier, 1790–1810.* College Station: Texas A&M University Press, 1985.

Florey, Klaus, ed. *The Collected Papers of Edward Robinson Squibb, M.D. (1819–1900).* Princeton, NJ: Squibb Institute, 1988.

Fluehr-Lobban, Carolyn. *Race and Racism: An Introduction.* Lanham, MD: AltaMira, 2006.

Fogel, Robert W. *Railroads and American Economic Growth: Essays in Econometric History.* Baltimore: Johns Hopkins University Press, 1964.

———. *Without Consent or Contract: The Rise and Fall of American Slavery.* New York: W.W. Norton, 1989.

Fogel, Robert W., and Stanley L. Engerman. *The Reinterpretation of American Economic History.* New York: Harper and Row, 1971.

———. *Time on the Cross: The Economics of American Negro Slavery.* Boston: Little, Brown, 1974.

Foley, William E. *Wilderness Journey: A Life of William Clark.* Columbia: University of Missouri Press, 2004.

Ford, Edward. *David Rittenhouse: Astronomer-Patriot, 1732–1796.* Philadelphia: University of Pennsylvania Press, 1946.

Forrester, J.W. *Industrial Dynamics.* Cambridge, MA: Productivity, 1961.

Fortey, Richard A. *Fossils: The Key to the Past.* Cambridge, MA: Harvard University Press. 1991.

Fouche, Rayvon, and Shelby Davidson. *Black Inventors in the Age of Segregation: Granville T. Woods, Lewis H. Latimer, and Shelby J. Davidson.* Baltimore: Johns Hopkins University Press, 2003.

Fraham, David. *A Cancer Battle Plan.* New York: Putnam, 1992.

Frangsmyr, Tore, ed. *The Nobel Prizes, 1989.* Stockholm, Sweden: Nobel Foundation, 1990.

Frank, G. Goble. *The Third Force: The Psychology of Abraham Maslow.* New York: Viking, 1970.

Franklin, Benjamin. *An Account of the New Invented Pennsylvanian Fire-Place.* Philadelphia, 1744.

———. *The Autobiography of Benjamin Franklin.* New York: Macmillan, 1962.

———. *Experiments and Observations in Electricity, Made at Philadelphia in America.* London: E. Cave, 1751.

Freedheim, Donald K., and Irving Weiner. *History of Psychology.* New York: Wiley, 2003.

Freud, Sigmund. *The Ego and the Id.* Trans. James Strachey. New York: W.W. Norton, 1960.

Frick, George Frederick, and Raymond Phineas Stearns. *Mark Catesby: The Colonial Audubon.* Urbana: University of Illinois Press, 1961.

Friedel, Robert. *Pioneer Plastic: The Making and Selling of Celluloid.* Madison: University of Wisconsin Press, 1983.

Friedenberg, Zachary B. *The Doctor in Colonial America.* Danbury, CT: Rutledge, 1998.

Friedman, Norman, and James L. Christley. *U.S. Submarines Since 1945: An Illustrated Design History.* Washington, DC: Naval Institute Press, 1994.

Fry, C. George. *Congregationalists and Evolution: Asa Gray and Louis Agassiz.* Lanham, MD: University Press of America, 1989.

Fuller, Steve. *Thomas Kuhn: A Philosophical History of Our Times.* Chicago: University of Chicago Press, 2000.

Fulton, John F. *Harvey Cushing: A Biography.* Springfield, IL: Charles C. Thomas, 1946.

Fussner, F. Smith. *The Historical Revolution.* New York: Routledge, 1962.

The Future of Oil as a Source of Energy. Abu Dhabi: Emirates Center for Strategic Studies and Research, 2003.

Gabor, Andrea. *Einstein's Wife.* New York: Viking, 1995.

Galbraith, John Kenneth. *The Affluent Society.* Boston: Houghton Mifflin, 1958.

Galilei, Galileo. *Sidereus Nuncius, or the Sidereal Messenger.* Trans. Albert Van Helden. Chicago: University of Chicago Press, 1989.

Galishoff, Stuart. "John Jones." In *Dictionary of American Medical Biography,* ed. Martin Kaufman, et al. Westport, CT: Greenwood, 1984.

Gallaudet, Edward M. *Life of Thomas Hopkins Gallaudet: Founder of Deaf-mute Instruction in America.* New York: Henry Holt, 1888.

Gardiner, H. Norman. *Selected Sermons of Jonathan Edwards.* New York: Macmillan, 1904.

Garratt, G.R.M. *The Early History of Radio.* London: Institution of Electrical Engineers, 1994.

Garrett, Laurie. *Betrayal of Trust: The Collapse of Global Public Health.* New York: Hyperion, 2001.

Garrison, Fielding H. *An Introduction to the History of Medicine.* Philadelphia: W.B. Saunders, 1960.

Garwin, Richard L., and Georges Charpak. *Megawatts and Megatons: A Turning Point in the Nuclear Age?* New York: Alfred A. Knopf, 2001.

Gately, Iain. *La Diva Nicotina: The Story of How Tobacco Seduced the World.* London: Simon and Schuster, 2001.

Gates, Paul W. *The Economic History of the United States.* Vol. 3, *The Farmer's Age: Agriculture, 1815–1860.* Armonk, NY: M.E. Sharpe, 1960.

Gathorne-Hardy, Jonathan. *Sex, The Measure of All Things: A Life of Alfred C. Kinsey.* Bloomington: Indiana University Press, 2000.

Gatz, Margaret. *The Role of Genes and Environments for Explaining Alzheimer Disease.* Chicago: Archives of General Psychiatry, American Medical Association, 2006.

Gavaghan, Helen. *Something New Under the Sun: Satellites and the Beginning of the Space Age.* New York: Springer-Verlag, 1998.

Gavrilov, Leonid. *The Biology of a Life Span.* New York: Harwood Academic, 1991.

Gay, Evelyn. *The Medical Profession in Georgia, 1733–1983.* Atlanta: Auxiliary to the Medical Association of Georgia, 1983.

Gehlbach, Stephen H. *American Plagues: Lessons from Our Battles with Disease.* New York: McGraw-Hill, 2005.

Geiser, Samuel Wood. *Naturalists of the Frontier.* Dallas: Southern Methodist University, 1937.

Gell-Mann, Murray. *The Quark and the Jaguar.* New York: W.H. Freeman, 1994.

Gerbi, Aneonello. *Nature in the New World: From Christopher Columbus to Gonzalo Fernández de Oviedo.* Trans. Jeremy Moyle. Pittsburgh, PA: University of Pittsburgh Press, 1985.

Gerstner, Patsy Ann. "The Academy of Natural Sciences of Philadelphia, 1812–1850." In *The Pursuit of Knowledge in the Early American Republic: American Scientific and Learned Societies from Colonial Times to the Civil War,* ed. Alexandra Oleson and Sanborn C. Brown. Baltimore: Johns Hopkins University Press, 1976.

Gesell, Arnold, Frances L. Ilg, and Louise Bates Ames. *The First Five Years of Life.* New York: Harper, 1940.

———. *Youth: The Years from Ten to Sixteen.* New York: Harper, 1956.

Gesell, Arnold, Frances L. Ilg, Louise Bates Ames, and Glenna E. Bullis. *The Child from Five to Ten.* New York: Harper, 1946.

Gest, Howard. *Microbes: An Invisible Universe.* Washington, DC: ASM Press, 2003.

Getman, Frederick H. *The Life of Ira Remsen.* Easton, PA: *Journal of Chemical Education,* 1940.

Geyh, Mebus A., and Helmut Schleicher. *Absolute Age Determination: Physical and Chemical Dating Methods and Their Application.* New York: Springer-Verlag, 1990.

Giauque, William F. *Low Temperature, Chemical, and Magneto Thermodynamics: The Scientific Papers of William F. Giauque.* New York: Dover, 1969.

Gibbs, F.W. *Joseph Priestley: Revolutions of the Eighteenth Century.* Garden City, NY: Doubleday, 1967.

Gibian, Peter. *Oliver Wendell Holmes and the Culture of Conversation.* New York: Cambridge University Press, 2001.

Gibson, John M. *Soldier in White.* Durham, NC: Duke University Press, 1958.

Giddens, Paul Henry. *The Birth of the Oil Industry.* New York: Macmillan, 1938.

Gill, Frank B. *Ornithology.* New York: W.H. Freeman, 1995.

Gill, Theodore. *Biographical Memoir of John Edwards Holbrook, 1794–1871.* Washington, DC: National Academy of Sciences, 1905.

Gillcrist, Dan. *Power Shift: The Transition to Nuclear Power in the U.S. Submarine Force As Told by Those Who Did It.* Lincoln, NE: Universe, 2006.

Gilliam, Ann, ed. *Voices for the Earth: A Treasury of the Sierra Club Bulletin, 1893–1977.* San Francisco: Sierra Club, 1979.

Gilman, Daniel C. *The Life of James Dwight Dana: Scientific Explorer, Mineralogist, Geologist, Zoologist, Professor in Yale University.* New York: Harper and Brothers, 1899.

Ginger, Ray. *Six Days or Forever? Tennessee v. John Thomas Scopes.* Chicago: Quadrangle, 1969.

Gittus, John H. *Uranium.* London: Butterworths, 1963.

Glassborow, Francis. *You Can Do It!: A Beginners Introduction to Computer Programming.* Hoboken, NJ: John Wiley and Sons, 2004.

Gleick, James. *Chaos: Making a New Science.* New York: Viking, 1987.

———. *Genius: The Life and Science of Richard Feynman.* New York: Vintage Books. 1992.

Glenn, John. *John Glenn: A Memoir.* New York: Bantam, 1999.

Godwin, Robert, ed. *Mars: The NASA Mission Reports.* Toronto: Apogee, 2000.

Goertzel, Ted, and Ben Goertzel. *Linus Pauling: A Life in Science and Politics.* New York: Basic Books, 1995.

Goetzmann, William H. *New Lands, New Men: America and the Second Great Age of Discovery.* New York: Viking Penguin, 1986.

Goist, Park Dixon. *From Main Street to State Street: Town, City, and Community in America.* Port Washington, NY: Kennikat, 1977.

Goldblith, Samuel A. *Pioneers in Food Science: Samuel Cate Prescott, MIT Dean and Pioneer Food Technologist.* Vol. 1. Trumbull, CT: Food and Nutrition Press, 1993.

Goldfield, David R., and Blaine A. Brownell. *Urban America: From Downtown to No Town.* Boston: Houghton Mifflin, 1979.

Gonzales, Justo L. *A History of Christian Thought: From the Protestant Reformation to the Twentieth Century.* Vol. 3. Revised ed. Nashville, TN: Abingdon, 1987.

Good, Gregory A., ed. *The Earth, the Heavens, and the Carnegie Institution of Washington.* Washington, DC: American Geophysical Union, 1994.

Goodchild, Peter. *Edward Teller: The Real Dr. Strangelove.* Cambridge, MA: Harvard University Press, 2004.

Gookin, Warner F. *Bartholomew Gosnold: Discoverer and Planter.* North Haven, CT: Archon, 1963.

Gorenstein, Daniel. *Finite Groups.* New York: Harper and Row, 1967.

Gorney, Cynthia. *Articles of Faith: A Frontline History of the Abortion Wars.* New York: Simon and Schuster, 1999.

Gosling, Francis. *Before Freud: Neurasthenia and the American Medical Community, 1870–1910.* Urbana: University of Illinois Press, 1987.

Gould, Stephen Jay. *Ever Since Darwin.* New York: W.W. Norton, 1977.

———. *I Have Landed: The End of a Beginning in Natural History.* New York: Harmony, 2002.

———. *The Mismeasure of Man.* New York: W.W. Norton, 1983.

———. *Ontogeny and Phylogeny.* Cambridge, MA: Harvard University Press, 1977.

———. *The Panda's Thumb: More Reflections in Natural History.* New York: W.W. Norton, 1982.

———. *The Structure of Evolutionary Theory.* Cambridge, MA: Harvard University Press, 2002.

———. *Time's Arrow, Time's Cycle.* Cambridge, MA: Harvard University Press, 1987.

Gradstein, Felix M., et al. *A Geologic Time Scale 2004.* Cambridge, UK: Cambridge University Press, 2004.

Graetzer, Hans, and David Anderson. *The Discovery of Nuclear Fission.* New York: Van Nostrand Reinhold, 1971.

Graham, Frank, Jr. *The Audubon Ark: A History of the National Audubon Society.* New York: Alfred A. Knopf, 1990.

Granshaw, Lindsay, and Roy Porter. *The Hospital in History.* New York: Taylor and Francis, 1990.

Graustein, Jeannette E. *Thomas Nuttall, Naturalist: Explorations in America, 1808–1841.* Cambridge, MA: Harvard University Press, 1967.

Graves, Joseph L., Jr. *The Emperor's New Clothes: Biological Theories of Race at the Millennium.* New Brunswick, NJ: Rutgers University Press, 2002.

Gray, Jeremy. *Ideas of Space: Euclidean, Non-Euclidean, and Relativistic.* 2nd ed. Oxford, UK: Clarendon, 1989.

Grebogi, Celso, and James A. Yorke, eds. *The Impact of Chaos on Science and Society.* New York: United Nations University Press, 1997.

Grebstein, Sheldon Norman, ed. *The Monkey Trial: The State of Tennessee vs. John Thomas Scopes.* Boston: Houghton Mifflin, 1960.

Green, Constance. *Eli Whitney and the Birth of American Technology.* Upper Saddle River, NJ: Pearson Education, 1997.

Greenberg, Marvin J. *Euclidean and Non-Euclidean Geometry: Development and History.* New York: W. H. Freeman, 1993.

Greene, John C. *American Science in the Age of Jefferson.* Ames: University of Iowa Press, 1984.

Greene, Kevin. *Archaeology: An Introduction.* London: Routledge, 2002.

Greenwood, Norman Neill, and A. Earnshaw. *Chemistry of the Elements.* 2nd ed. Oxford, UK: Butterworth-Heinemann, 1997.

Greer, Allan, ed. *The Jesuit Relations: Natives and Missionaries in Seventeenth-Century North America.* London: Macmillan, 2000.

Gribbin, John. *The Scientists: A History of Science Told Through the Lives of Its Greatest Inventors.* New York: Random House, 2003.

Grigs, Barbara. *Green Pharmacy: A History of Herbal Medicine.* New York: Viking, 1981.

Grimwood, James M. *Project Mercury: A Chronology.* Washington, DC: U.S. Government Printing Office, 1963.

Grimwood, James M., Barton C. Hacker, and Peter J. Vorzimmer. *Project Gemini Technology and Operations: A Chronology.* Washington, DC: NASA, 1969.

Grissom, Virgil. *Gemini: A Personal Account of Man's Venture into Space.* New York: Macmillan, 1968.

Grob, Bernard. *Basic Electronics.* 8th ed. Westerville, OH: McGraw-Hill, 1997.

Grob, Gerald N. *The Deadly Truth: A History of Disease in America.* Cambridge, MA: Harvard University Press, 2002.

———. *The Mad Among Us: A History of America's Care of the Mentally Ill.* New York: Free Press, 1994.

Groh, Lynn. *Walter Reed, Pioneer in Medicine.* Champaign, IL: Garrard, 1971.

Groom, Martha J., Gary K. Meffe, and Ronald Carrol, eds. *Principles of Conservation Biology.* 3rd ed. Sunderland, MA: Sinauer, 2005.

Groombridge, Brian, and Martin D. Jenkins. *World Atlas of Biodiversity.* Berkeley: University of California Press, 2002.

Grosvenor, Edwin S., and Morgan Wesson. *Alexander Graham Bell: The Life and Times of the Man Who Invented the Telephone.* New York: Harry N. Abrams, 1997.

Grosvenor, Gilbert. *The National Geographic Society and Its Magazine.* Washington, DC: National Geographic Society, 1957.

Groves, Leslie R. *Now It Can Be Told: The Story of the Manhattan Project.* New York: Da Capo, 1983.

Guild, Reuben Aldridge. "Biographical Introduction." In *The Complete Writings of Roger Williams.* New York: Russell and Russell, 1963.

Gusterson, Hugh. *Nuclear Rites: A Weapons Laboratory at the End of the Cold War.* Berkeley: University of California Press, 1996.

Guth, Alan H. *The Inflationary Universe: The Quest for a New Theory of Cosmic Origins.* New York: Perseus, 1997.

Gyles, John. *Memoirs of Odd Adventures, Strange Deliverances, &c. in the Captivity of John Gyles, Esq., Commander of the Garrison on St. George's River.* Boston:

S. Knesland and T. Green, 1736; reprint ed., New York: Garland, 1977.

Haber, Francis C. *The Age of the World: Moses to Darwin.* Baltimore: Johns Hopkins University Press, 1959.

Hacker, Barton C. *Elements of Controversy: The Atomic Energy Commission and Radiation Safety in Nuclear Weapons Testing, 1947–1974.* Berkeley: University of California Press, 1994.

Hacker, Barton C., and James M. Grimwood. *On the Shoulders of Titans: A History of Project Gemini.* Washington, DC: U.S. Government Printing Office, 1977.

Hafner, Katie, and Matthew Lyon. *Where Wizards Stay Up Late: The Origins of the Internet.* New York: Touchstone, 1998.

Hager, Knut. *The Illustrated History of Surgery.* New York: Bell, 1988.

Hager, Thomas. *Linus Pauling and the Chemistry of Life.* New York: Oxford University Press, 1998.

Hahn, Emily. *Animal Gardens.* Garden City, NY: Doubleday, 1967.

Hakluyt, Richard. *The Principal Navigations, Voyages, Traffiques, and Discoveries of the English Nation.* 1588–1590. Reprint ed., ed. Jack Beeching. London: Penguin, 1972.

Hale, Nathan G. *The Beginnings of Psychoanalysis in the United States, 1876–1917.* New York: Oxford University Press, 1971.

———. *The Rise and Crisis of Psychoanalysis in the United States: Freud and the Americans, 1917–1985.* New York: Oxford University Press, 1995.

Hall, A. Rupert. *Philosophers at War: The Quarrel Between Newton and Leibniz.* Cambridge, UK: Cambridge University Press, 1980.

Hall, Michael G. *The Last American Puritan: The Life of Increase Mather 1639–1723.* Middletown, CT: Wesleyan University Press, 1988.

Hallam, A. *Great Geological Controversies.* Oxford, UK: Oxford University Press, 1989.

Hamilton, Alexander. *Gentleman's Progress: The Itinerarium of Dr. Alexander Hamilton, 1744.* 1907. Ed. Carl Bridenbaugh. Chapel Hill: University of North Carolina Press, 1948.

Hammonds, Evelynn Maxine. *Childhood's Deadly Scourge: The Campaign to Control Diphtheria in New York City, 1880–1930.* Baltimore: Johns Hopkins University Press, 1999.

Handley, Susannah. *Nylon: The Story of a Fashion Revolution.* Baltimore: Johns Hopkins University Press, 1999.

Hanson, Elizabeth. *Animal Attractions: Nature on Display in American Zoos.* Princeton, NJ: Princeton University Press, 2004.

Harding, Robert. *Military Foundation of Panamanian Politics.* Piscataway, NJ: Transaction, 2001.

Harding, Sandra, ed. *The "Racial" Economy of Science.* Bloomington: Indiana University Press, 1993.

Hare, Ronald. *The Birth of Penicillin, and the Disarming of Microbes.* London: Allen and Unwin, 1978.

Hariot, Thomas. *A Brief and True Report of a New Found Land of Virginia.* Kila, MT: Kessinger, 2004.

Harland, David M. *The Space Shuttle: Roles, Missions, and Accomplishments.* New York: Wiley, 1998.

Harlow, Alvin F. *Old Wires and New Waves: The History of the Telegraph, Telephone, and Wireless.* New York: D. Appleton-Century, 1936.

Harper, Francis, ed. *The Travels of William Bartram, Naturalist Edition.* Athens: University of Georgia Press, 1998.

Harris, Henry. *Things Come to Life: Spontaneous Generation Revisited.* Oxford, UK: Oxford University Press, 2002.

Harris, Marvin. *Cultural Materialism: The Struggle for a Science of Culture.* New York: Random House, 1979.

Harrison, Paul J., and Timothy R. Parsons. *Fisheries Oceanography: An Integrative Approach to Fisheries Ecology and Management.* Cambridge, MA: Blackwell Science, 2000.

Harrison, Peter. *The Bible, Protestantism, and the Rise of Natural Science.* Cambridge, UK: Cambridge University Press, 1998.

Harrow, Benjamin. *Eminent Chemists of Our Time.* New York: D. Van Nostrand, 1927.

Harshberger, John W. *The Botanists of Philadelphia and Their Works.* Philadelphia: T.C. Davis and Sons, 1899; reprint ed., Mansfield Center, CT: Martino, 1999.

Hartshorne, Henry. *The Household Cyclopedia of General Information.* New York: Thomas Kelly, 1881.

Hartzell, Judith. *I Started All This: The Life of Dr. William Worrall Mayo.* Greenville, SC: Arvi, 2004.

Hawkes, Peter W., ed. *The Beginnings of Electron Microscopy.* Orlando, FL: Academic Press, 1985.

Hawking, Stephen. *A Brief History of Time: From the Big Bang to Black Holes.* New York: Bantam, 1998.

Hawthorne, Peter. *The Transplanted Heart: The Incredible Story of the Epic Heart Transplant Operations by Professor Christiaan Barnard and His Team.* Chicago: Rand McNally, 1968.

Hawthorne, R.M., Jr. "Ira Remsen." In *American Chemists and Chemical Engineers,* ed. Wyndham D. Miles. Washington, DC: American Chemical Society, 1976.

Hayden, Robert. "Black Americans in the Field of Science and Invention." In *Blacks in Science: Ancient and Modern,* ed. Ivan Van Sertima. Somerset, NJ: Transaction, 1987.

———. *Seven Black American Scientists.* Reading, MA: Addison-Wesley, 1970.

Hazard, Ebenezer. *Historical Collections: Consisting of State Papers, and Other Authentic Documents, Intended as Materials for an History of the United States of America.* Philadelphia: Thomas Dobson, 1792–1794.

Hearnshaw, J.B. *The Analysis of Starlight: One Hundred and Fifty Years of Astronomical Spectroscopy.* Cambridge, UK: Cambridge University Press, 1986.

———. *The Measurement of Starlight: Two Centuries of Astronomical Photometry.* Cambridge, UK: Cambridge University Press, 1996.

Hecht, Jeff, and Dick Teresi. *Laser: Light of a Million Uses.* Mineola, NY: Dover, 1998.

Hecthlinger, Adelaide. *The Great Patent Medicine Era, or Without Benefit of Doctor.* New York: Galahad, 1970.

Hediger, Heini. *Wild Animals in Captivity.* New York: Dover, 1964.

Hedrick, U.P. *A History of Horticulture in the United States to 1860.* New York: Oxford University Press, 1950.

Hegner, Robert. *An Introduction to Zoology.* New York: Macmillan, 1910.

Heidenrich, Conrad. *Explorations and Mapping of Samuel de Champlain.* Toronto: University of Toronto, 1976.

Heilbron, J.L. *Electricity in the 17th and 18th Centuries: A Study in Early Modern Physics.* 2nd ed. Mineola, NY: Dover, 1999.

———. *Geometry Civilized: History, Culture, and Technique.* New York: Oxford University Press, 2000.

Heilbron, J.L., and Robert W. Seidel. *Lawrence and His Laboratory: A History of the Lawrence Berkeley Laboratory.* Berkeley: University of California Press, 1989.

Heilbroner, Robert L. *The Worldly Philosophers.* New York: Simon and Schuster, 1953.

Heisenberg, Werner. *The Physical Principles of Quantum Theory.* New York: Dover, 1949.

Helfand, William H. *Quack, Quack, Quack: The Sellers of Nostrums in Prints Posters, Ephemera and Books.* New York: Grolier, 2002.

Helfman, Gene S., Bruce B. Collette, and Douglas E. Facey. *The Diversity of Fishes.* Malden, MA: Blackwell Science, 2000.

Hellman, Geoffrey. *Bankers, Bones, and Beetles: The First Century of the American Museum of Natural History.* Garden City, NY: Natural History Press, 1969.

Hellman, Hal. *Great Feuds in Medicine: Ten of the Liveliest Disputes Ever.* New York: John Wiley and Sons, 2001.

Henderson-Sellers, Ann, and Peter J. Robinson. *Contemporary Climatology.* 2nd ed. Upper Saddle River, NJ: Prentice Hall, 1999.

Hennepin, Louis. *Description of Louisiana.* Minneapolis: University of Minnesota Press, 1938.

Henretta, James A., David Brody, and Lynn Dumenil. *America: A Concise History.* New York: Bedford/St. Martin's, 1999.

Heppenheimer, T.A. *Countdown: A History of Space Flight.* New York: John Wiley and Sons, 1997.

———. *First Flight: The Wright Brothers and the Invention of the Airplane.* Hoboken, NJ: John Wiley and Sons, 2003.

Herbert, Vernon, and Attilio Bisio. *Synthetic Rubber: A Project That Had to Succeed.* Westport, CT: Greenwood, 1985.

Herbert, Wally. *The Noose of Laurels: Robert E. Peary and the Race to the North Pole.* New York: Atheneum, 1989.

Herken, Gregg. *Brotherhood of the Bomb: The Tangled Lives and Loyalties of Robert Oppenheimer, Ernest Lawrence, and Edward Teller.* New York: Henry Holt, 2002.

Herman, Robin. *Fusion: The Search for Endless Energy.* New York: Cambridge University Press, 2006.

Hermes, Matthew E., ed. *Enough for One Lifetime: Wallace Carothers, Inventor of Nylon.* Washington, DC: American Chemical Society, 1996.

Herrick, Francis Hobart. *Audubon the Naturalist: A History of His Life and Time.* 2 vols. New York: Appleton, 1917; 2nd ed., 1938.

Herrnstein, Richard, and Charles Murray. *The Bell Curve: Intelligence and Class Structure in American Life.* New York: Free Press, 1994.

Hersh, Seymour. *Chemical and Biological Warfare.* Indianapolis, IN: Bobbs-Merrill, 1968.

Hershberg, James B. *James B. Conant: Harvard to Hiroshima and the Making of the Nuclear Age.* New York: Alfred A. Knopf, 1993.

Hewitt, William W. *Astrology for Beginners: An Easy Guide to Understanding and Interpreting Your Chart.* St. Paul, MN: Llewellyn, 1992.

Hewlett, Richard G., and Oscar E. Anderson, Jr. *The New World: A History of the United States Atomic Energy Commission.* Vol. 1, *1939–1946.* University Park: Pennsylvania State University Press, 1962.

Hickman, Cleveland. *Integrated Principles of Zoology.* New York: McGraw-Hill, 2000.

Highfield, Roger, and Paul Carter. *The Private Lives of Albert Einstein.* London: Faber and Faber, 1993.

Hilgartner, Stephen. *Science on Stage: Expert Advice as Public Drama.* Stanford, CA: Stanford University Press, 2000.

Hill, George W. *Biographical Memoir of Asaph Hall.* Biographical Memoirs 6. Washington, DC: National Academy of Sciences, 1908.

Hill, Ralph Nading. *Doctors Who Conquered Yellow Fever.* New York: Random Library, 1966.

Hilmes, Michelle, and Jason Jacobs. *The Television History Book.* London: British Film Institute, 2004.

Hindle, Brooke. *David Rittenhouse*. Princeton, NJ: Princeton University Press, 1964.

———, ed. *Early American Science*. New York: Science History, 1976.

———. *The Pursuit of Science in Revolutionary America, 1735–1789*. Chapel Hill: University of North Carolina Press, 1956.

Hindle, Brooke, and Steven Lubar. *Engines of Change: The American Industrial Revolution, 1790–1860*. Washington, DC: Smithsonian, 1986.

Hitchcock, Edward. *The Religion of Geology and Its Connected Sciences*. London: Collins, 1851;. Hicksville, NY: Regina, 1975.

Hixson, Walter L. *Lindbergh, Lone Eagle*. New York: HarperCollins, 1996.

Hobbs, William Herbert. *Peary*. New York: Macmillan, 1936.

Hobby, Gladys L. *Penicillin: Meeting the Challenge*. New Haven, CT: Yale University Press, 1985.

Hoermann, Alfred R. *Cadwallader Colden: A Figure of the American Enlightenment*. Westport, CT: Greenwood, 2002.

Hoffleit, Dorrit. *Women in the History of Variable Star Astronomy*. Cambridge, MA: American Association of Variable Star Observers, 1993.

Hoffman, Edward. *The Right to Be Human: A Biography of Abraham Maslow*. New York: HarperCollins, 1989.

Hoffmann, Nancy, and John C. Van Horne, eds. *John Bartram, King's Botanist: A Tercentennial Reappraisal*. Memoirs of the American Philosophical Society. Vol. 249. Philadelphia: American Philosophical Society, 2003.

Hoffmann-Axthelm, Walter. *A History of Dentistry*. Chicago: Quintessence, 1981.

Hofman, Sigurd. *On Beyond Uranium: Journey to the End of the Periodic Table*. New York: Taylor and Francis, 2002.

Hofmann, Werner. *Rubber Technology Handbook*. New York: Hanser, 1989.

Hogg, Michael A., and Dominic Abrams. *Social Identifications: A Social Psychology of Intergroup Relations and Group Processes*. New York: Routledge, 1988.

Holifield, E. Brooks. *The Gentlemen Theologians*. Durham, NC: Duke University Press, 1978.

Hollon, W. Eugene. *The Lost Pathfinder: Zebulon Montgomery Pike*. Westport, CT: Greenwood, 1981.

Holmes, Oliver Wendell, Jr. *The Common Law and Other Writings*. Reprint ed. New York: Legal Classics Library, 1981.

Holy Bible. Pilgrim Edition. New York: Oxford University Press, 1952.

Honsberger, Ross. *Episodes in Nineteenth and Twentieth Century Euclidean Geometry*. Washington, DC: Mathematical Association of America, 1996.

Hooten, Earnest A. *The American Criminal*. Cambridge, MA: Harvard University Press, 1939.

———. *Apes, Men, and Morons*. New York: G.P. Putman's Sons, 1937.

———. *Crimes and the Man*. Cambridge, MA: Harvard University Press, 1939.

Hopkins, Jack. *The Eradication of Smallpox*. Boulder, CO: Westview, 1989.

Horwitz, Allan. *Creating Mental Illness*. Chicago: University of Chicago Press, 2002.

Hothersall, David. *History of Psychology*. 2nd ed. New York: McGraw-Hill, 1990.

Hounshell, David A., and John Kenly Smith, Jr. *Science and Corporate Strategy: DuPont R & D, 1902—1980*. New York: Cambridge University Press, 1988.

Howell, Joel D. *Technology in the Hospital: Transforming Patient Care in the Early Twentieth Century*. Baltimore: Johns Hopkins University Press, 1996.

Hoyningen-Huen, Paul. *Reconstructing Scientific Revolutions: Thomas S. Kuhn's Philosophy of Science*. Trans. Alexander T. Levine. Foreword by Thomas S. Kuhn. Chicago: University of Chicago Press, 1993.

Hoyt, William Graves. *Lowell and Mars*. Tucson: University of Arizona Press, 1976.

Hubbard, F. Tracy, and Alfred Rehder. *Nomenclatorial Notes on Plants Growing in the Botanical Garden of the Atkins Institution of the Arnold Arboretum at Soledad, Cienfuegos, Cuba*. Cambridge, MA: Harvard University Press, 1936.

Hubbard, Ruth. *The Politics of Women's Biology*. New Brunswick, NJ: Rutgers University Press, 1990.

———. *Profitable Promises: Essays on Women, Science, and Health*. Monroe, ME: Common Courage, 1994.

Hubble, Edwin. *The Realm of the Nebulae*. New Haven, CT: Yale University Press, 1936.

Hughes, Thomas Parke. *Networks of Power: Electrification in Western Society*. Baltimore: Johns Hopkins University Press, 1983.

———. *Science and the Instrument-maker*. Washington, DC: Smithsonian Institution, 1976.

Hugill, Peter J. *Global Communications Since 1844: Geopolitics and Technology*. Baltimore: Johns Hopkins University Press, 1999.

Huizenga, John R. *Cold Fusion: The Scientific Fiasco of the Century*. Rochester, NY: University of Rochester Press, 1992.

Humphreys, Margaret. *Yellow Fever and the South*. Baltimore: Johns Hopkins University Press, 1999.

Hunt, William R. *Stef: A Biography of Vilhjalmur Stefansson*. Vancouver: University of British Columbia, 1986.

Hunter, Clark. *The Life and Letters of Alexander Wilson*. Philadelphia: American Philosophical Society, 1983.

Hunter, Michael, Antonio Clericuzio, and Lawrence M. Principe, eds. *The Correspondence of Robert Boyle.* London: Pickering and Chatto, 2001.

Hunter, Robert J. *The Origin of the Philadelphia General Hospital.* Philadelphia: Rittenhouse, 1955.

Hurt, R. Douglas. *American Agriculture: A Brief History.* Ames, IA: Blackwell, 1994.

Hurt, Raymond. *The History of Cardiothoracic Surgery from Early Times.* New York: Pantheon, 1996.

Huyghe, Patrick. *The Field Guide to Extraterrestrials: A Complete Overview of Alien Lifeforms—Based on Actual Accounts and Sightings.* New York: Avon, 1996.

Hyatt, Alpheus. *Phylogeny of an Acquired Characteristic.* Ed. Stephen J. Gould. Manchester, NH: Ayer, 1980.

Ifrah, Georges. *The Universal History of Computing: From the Abacus to the Quantum Computer.* Trans. E.F. Harding. New York: John Wiley and Sons, 2001.

Inkeles, Alex. *One World Emerging? Convergence and Divergence in Industrial Societies.* Boulder, CO: Westview, 1998.

———. *Public Opinion in Soviet Russia: A Study in Mass Persuasion.* Cambridge, MA: Harvard University Press, 1950.

Inkeles, Alex, and David H. Smith. *Becoming Modern: Individual Change in Six Developing Countries.* Cambridge, MA: Harvard University Press, 1974.

Intergovernmental Panel on Climate Change (IPCC). *Climate Change 2001: The Scientific Basis.* Ed. J.T. Houghton, et al. Cambridge, UK: Cambridge University Press, 2001.

Iserson, Kenneth. *Death to Dust: What Happens to Dead Bodies?* Tucson, AZ: Galen, 2001.

Jackson, Donald. *Thomas Jefferson and the Stony Mountains: Exploring the West from Monticello.* Urbana: University of Illinois Press, 1981.

Jackson, Donald, and Mary Lee Spence, eds. *The Expeditions of John Charles Frémont.* Urbana: University of Illinois Press, 1970–1984.

Jacob, James R. *The Scientific Revolution: Aspirations and Achievements, 1500–1700.* Atlantic Highlands, NJ: Humanities, 1998.

Jacobs, Margaret C. *Living the Enlightenment: Freemasonry and Politics in Eighteenth-Century Europe.* Oxford, UK: Oxford University Press, 1991.

James, William. *Pragmatism and the Meaning of Truth.* Cambridge, MA: Harvard University Press, 1975.

———. *The Varieties of Religious Experience: A Study in Human Nature.* New York: Longmans, Green, 1902.

Jarzombeck, Mark. *Designing MIT: Bosworth's New Tech.* Boston: Northeastern University Press, 2004.

Jeffers, William A., Jr. "William Francis Giauque (1895–1982)." In *Nobel Laureates in Chemistry 1901–1992,* ed. Laylin K. James. Washington, DC: American Chemical Society, 1993.

Jefferson, Thomas. *The Life and Selected Writings.* Ed. Adrienne Koch and William Peden. New York: Modern Library, 1972.

———. *Notes on the State of Virginia.* Richmond, VA: Randolph, 1853.

Jenkins, Edward. *To Fathom More: African American Scientists and Inventors.* Lanham, MD: University Press of America, 2001.

Jennett, Bryan. *High-Technology Medicine: Benefits and Burdens.* New York: Harper and Row, 1987.

Jimenez, Mary Ann. *Changing Faces of Madness: Early American Attitudes and Treatment of the Insane.* Hanover, NH: University Press of New England, 1987.

Johnson, George. *Strange Beauty: Murray Gell-Mann and the Revolution in 20th-Century Physics.* New York: Alfred A. Knopf, 1999.

Johnson, Paul E. *A Shopkeeper's Millennium: Society and Revivals in Rochester, New York, 1815–1837.* New York: Hill and Wang, 1978.

Johnson, Philip E. *Darwin on Trial.* Downers Grove, IL: InterVarsity Press, 1991, 1993.

Jones, Bessie, and Lyle Boyd. *The Harvard College Observatory: The First Four Directorships.* Foreword by Donald H. Menzel. Cambridge, MA: Harvard University Press, 1971.

Jones, Howard Mumford. *O Strange New World: American Culture: The Formative Years.* Westport, CT: Greenwood, 1982.

Jones, Richard Foster. *Ancients and Moderns: A Study of the Rise of the Scientific Movement in Seventeenth Century England.* Berkley: University of California Press, 1965.

Jonnes, Jill. *Empires of Light: Edison, Tesla, Westinghouse, and the Race to Electrify the World.* New York: Random House, 2003.

Jordan, Chester E. *Colonel Joseph B. Whipple.* Concord, NH: Republican Press Association, 1894.

Josephson, Matthew. *Edison: A Biography.* New York: McGraw-Hill, 1959.

Josselyn, John. *New-Englands Rarities Discovered.* London: Widdowes, 1672.

Judson, Clara. *Reaper Man: The Story of Cyrus Hall McCormick.* Boston: Houghton Mifflin, 1948.

Just, Ernest Everett. *Basic Methods for Experiments in Eggs of Marine Animals.* Philadelphia: Blakiston's, 1922.

———. *Biology of the Cell Surface.* Philadelphia: Blakiston's, 1930.

Kahn, Michael. *Basic Freud: Psychoanalytic Thought for the 21st Century.* New York: Basic Books, 2002.

Kakar, Sudhir. *Frederick Taylor: A Study in Personality and Innovation.* Cambridge, MA: MIT Press, 1970.

Kammen, Michael, ed. *The Past Before Us: Contemporary Historical Writing in the United States.* Ithaca, NY: Cornell University Press, 1980.

Kanigel, Robert. *Apprentice to Genius: The Making of a Scientific Dynasty.* New York: Macmillan, 1986.

———. *The One Best Way: Frederick Winslow Taylor and the Enigma of Efficiency.* New York: Viking, 1997.

Kaplan, David. *The Silicon Boys and Their Valley of Dreams.* New York: HarperCollins, 2000.

Kaplan, Philip J. *F'd Companies: Spectacular Dot-Com Flameouts.* New York: Simon and Schuster, 2002.

Kargon, Robert. *The Rise of Robert Millikan: Portrait of a Life in American Science.* Ithaca, NY: Cornell University Press, 1982.

Kaye, Judith. *The Life of Benjamin Spock.* New York: Twenty-First Century Books, 1993.

———. *Life of Daniel Hale Williams.* Minneapolis, MN: Lerner, 1993.

Keating, William H. *Narrative of an Expedition to the Source of St. Peter's River . . . performed in the year 1823.* Vol. 2. Philadelphia: Carey, 1824.

Keller, Evelyn Fox. *A Feeling for the Organism: The Life and Work of Barbara McClintock.* New York: W.H. Freeman, 1983.

Keller, William F. *The Nation's Advocate: Henry Marie Brackenridge and Young America.* Pittsburgh, PA: University of Pittsburgh Press, 1956.

Kendall, Martha E. *Steve Wozniak: Inventor of Apple Computer.* New York: Walker, 1994.

Kennedy, David M. *Birth Control in America: The Career of Margaret Sanger.* New Haven, CT: Yale University Press, 2004.

Kevles, Daniel J. *In the Name of Eugenics: Genetics and the Uses of Human Heredity.* Cambridge, MA: Harvard University Press, 1985.

———. *The Physicists: The History of a Scientific Community in Modern America.* New York: Vintage Books, 1979.

Keys, Thomas E. *The History of Surgical Anesthesia.* New York: Dover, 1963; Malabar, FL: R.E. Krieger, 1978.

Kidwell, Peggy A., and Paul E. Ceruzzi. *Landmarks in Digital Computing: A Smithsonian Pictorial History.* Washington, DC: Smithsonian Institution, 1994.

King, Lester S. *Transformations in American Medicine: From Benjamin Rush to William Osler.* Baltimore: Johns Hopkins University Press, 1991.

Kinsey, Alfred, et al. *New Introduction to Sexual Behavior in the Human Male.* Bloomington: Indiana University Press, 1998.

Kirby-Smith, H.T. *U.S. Observatories: A Directory and Travel Guide.* New York: Van Nostrand Reinhold, 1976.

Kirschenbaum, Howard, and Valerie Land Henderson, eds. *Carl Rogers: Dialogues.* Boston: Houghton Mifflin, 1989.

Kivisto, Peter. *Key Ideas in Sociology.* Thousand Oaks, CA: Pine Forge, 1998.

Kligler, Benjamin, and Roberta Lee. *Integrative Medicine: Principles for Practice.* New York: McGraw-Hill, 2004.

Kline, Meredith. "Genesis." In *New Bible Commentary Revised.* Grand Rapids, MI: Eerdmans, 1970.

Kline, Wendy. *Building a Better Race: Gender, Sexuality, and Eugenics from the Turn of the Century to the Baby Boom.* Berkeley: University of California Press, 2001.

Koch, Sigmund, and David E. Leary. *A Century of Psychology as Science.* Washington, DC: American Psychological Association, 1992.

Kohlstedt, Sally G. *The Formation of the American Scientific Community: The American Association for the Advancement of Science, 1848–1860.* Urbana: University of Illinois Press, 1976.

———. "Maria Mitchell and the Advancement of Women in Science." In *Uneasy Careers and Intimate Lives: Women in Science,* ed. Prina G. Abir-Am and Dorinda Outram. New Brunswick, NJ: Rutgers University Press, 1987.

Kohlstedt, Sally G., Michael M. Sokal, and Bruce V. Lewenstein. *The Establishment of Science in America: 150 Years of the American Association for the Advancement of Science.* New Brunswick, NJ: Rutgers University Press, 1999.

Kolata, Gina. *Flu: The Story of the Great Influenza Pandemic.* New York: Touchstone, 2001.

Koning, Niek. *The Failure of Agrarian Capitalism: Agrarian Politics in the UK, Germany, the Netherlands, and the USA, 1846–1919.* New York: Routledge, 1994.

Konvitz, Josef. *Cartography in France, 1660–1848: Science, Engineering, and Statecraft.* Chicago: University of Chicago Press, 1987.

Koop, C. Everett. *Koop: The Memoirs of America's Family Doctor.* New York: Random House, 1991.

Koplow, David A. *Smallpox: The Fight to Eradicate a Global Scourge.* Berkeley: University of California Press, 2000.

Korman, Richard. *The Goodyear Story: An Inventor's Obsession and the Struggle for a Rubber Monopoly.* San Francisco: Encounter, 2002.

Kornberg, Arthur. *For the Love of Enzymes: The Odyssey of a Biochemist.* Cambridge, MA: Harvard University Press, 1989.

———. *The Golden Helix: Inside Biotech Ventures.* Dulles, VA: University Science Books, 1995.

Koslowsky, Robert. *A World Perspective Through 21st Century Eyes.* Victoria, Canada: Trafford, 2004.

Kras, Sara Louise. *The Steam Engine.* Philadelphia: Chelsea House, 2004.

Kruif, Paul de. *The Fight for Life.* New York: Harcourt, Brace, 1938.

Kuhn, Thomas S. *The Copernican Revolution.* Cambridge, MA: Harvard University Press, 1957.

———. *The Essential Tension: Selected Studies in Scientific Tradition and Change.* Chicago: University of Chicago Press, 1977.

———. *The Road Since Structure: Philosophical Essays, 1970–1993, with an Autobiographical Interview.* Ed. James Conant and John Haugeland. Chicago: University of Chicago Press, 2000.

———. *The Structure of Scientific Revolutions.* 1962. Chicago: University of Chicago Press, 1996.

Kuo, J. David. *dot.bomb: My Days and Nights at an Internet Goliath.* New York: Little, Brown, 2001.

Kurtz, D.W., ed. *Transits of Venus: New Views of the Solar System and Galaxy.* Cambridge, UK: Cambridge University Press, 2005.

Kurzman, Dan. *Disaster! The Great San Francisco Earthquake and Fire of 1906.* New York: William Morrow, 2001.

Kuznets, Simon. *Modern Economic Growth: Rate, Structure, and Spread.* New Haven, CT: Yale University Press, 1966.

Labaree, Benjamin W. *Colonial Massachusetts: A History.* Millwood, NY: Kraus International, 1979.

Labaree, Leonard W., et al., eds. *The Papers of Benjamin Franklin.* New Haven, CT: Yale University Press, 1959–.

Lagler, Karl F., John E. Bardach, Robert R. Miller, and Dora R. May Passino. *Ichthyology.* New York: John Wiley and Sons, 1977.

Lamont, Corliss. *The Philosophy of Humanism.* 6th ed. New York: Frederick Ungar, 1982.

Landes, David S., and Charles Tilly. *History as Social Science.* Englewood Cliffs, NJ: Prentice Hall, 1971.

Lanham, Url. *The Bone Hunters: The Heroic Age of Paleontology in the American West.* New York: Columbia University Press, 1973.

Lankford, John. *American Astronomy: Community, Careers, and Power, 1859–1940.* Chicago: University of Chicago Press, 1997.

Larson, Edward J. *Summer for the Gods: The Scopes Trial and America's Continuing Debate over Science and Religion.* New York: Basic Books, 1997.

Laskin, David. *Braving the Elements: The Stormy History of American Weather.* New York: Doubleday, 1996.

Lassek, Arthur M. *Human Dissection: Its Drama and Struggle.* Springfield, MA: Charles C. Thomas, 1958.

Launius, Roger D. *Space Stations.* Washington, DC: Smithsonian, 2003.

Laurikainen, K.V. *Beyond the Atom: The Philosophical Thought of Wolfgang Pauli.* Berlin, Germany: Springer-Verlag, 1989.

Lawrence, Christopher, and George Weisz, eds. *Greater than the Parts: Holism in Biomedicine, 1920–1950.* New York: Oxford University Press, 1998.

Lawrence, Philip K., and David W. Thornton. *Deep Stall: The Turbulent Story of Boeing Commercial Airplanes.* Burlington, VT: Ashgate, 2005.

Lawson, Andrew C., et al. *The California Earthquake of April 18, 1906: Report of the State Earthquake Investigation Commission.* Publication 87, 2 vols. Washington, DC: Carnegie Institution of Washington, 1908.

Lawson, John. *New Voyage to Carolina.* Chapel Hill: University of North Carolina Press, 1984.

Lawson, Russell M. *The American Plutarch: Jeremy Belknap and the Historian's Dialogue with the Past.* Westport, CT: Praeger, 1998.

———. "Jedidiah Morse, Geographer." In *Encyclopedia of New England.* New Haven, CT: Yale University Press, 2005.

———. *The Land Between the Rivers: Thomas Nuttall's Ascent of the Arkansas in 1819.* Ann Arbor: University of Michigan Press, 2004.

———. *Passaconaway's Realm: Captain John Evans and the Exploration of Mount Washington.* Hanover, NH: University Press of New England, 2002, 2004.

———. "Religion and Science." In *American Eras: The Colonial Era, 1600–1754,* ed. Jessica Kross. Detroit: Gale Research, 1998.

———. "Science." In *Encyclopedia of New England.* New Haven, CT: Yale University Press, 2005.

———. "Science and Medicine." In *American Eras: The Colonial Era, 1600–1754,* ed. Jessica Kross. Detroit: Gale Research, 1998.

Lay, Thorne. *Modern Global Seismology.* Burlington, VT: Academic Press, 1995.

Leake, Chauncey Depew. *Letheon: The Cadenced Story of Anesthesia.* Austin: University of Texas Press, 1947.

Lear, Linda. *Rachel Carson: Witness for Nature.* New York: Owl, 1998.

Leavitt, Judith Walzer. *Brought to Bed: Child-Bearing in America 1750–1950.* New York: Oxford University Press, 1986.

———. *Typhoid Mary: Captive to the Public's Health.* Boston, MA: Beacon, 1996.

Lehmann, Karl. *Thomas Jefferson, American Humanist.* Charlottesville: University of Virginia Press, 1991.

Lehrman, Robert L. *Physics—The Easy Way.* Hauppauge, NY: Barron's, 1998.

Leicester, Henry M. *The Historical Background of Chemistry.* New York: Dover, 1971.

Lemay, J.A. Leo. *Ebenezer Kinnersley: Franklin's Friend.* Philadelphia: University of Pennsylvania Press, 1964.

Lenzen, Victor F. *Benjamin Peirce and the United States Coast Survey.* San Francisco: San Francisco Press, 1968.

Leopold, Aldo. *Game Management.* Madison: University of Wisconsin Press, 1933.

———. *A Sand County Almanac.* Oxford, UK: Oxford University Press, 1949; New York: Ballantine, 1970.

Lerner, Barron. *Breast Cancer Wars.* New York: Oxford University Press, 2001.

Lescarbot, Marc. *History of New France.* 3 vols. Trans. W.L. Grant. Westport, CT: Greenwood, 1968.

Leverington, David. *Babylon to Voyager and Beyond: A History of Planetary Astronomy.* Cambridge, UK: Cambridge University Press, 2003.

Levin, Beatrice. *Women and Medicine.* Lanham, MD: Scarecrow, 2002.

Levine, Donald N. *Visions of the Sociological Tradition.* Chicago: University of Chicago Press, 1995.

Levins, Harold L. *The Earth Through Time.* 2nd ed. Philadelphia: Saunders College Publishing, 1983.

Levinson, Norman. *Selected Papers of Norman Levinson.* 2 vols. Ed. John A. Nohel and David H. Sattinger. Boston, MA: Birkhäuser, 1998.

Levy, David H. *Clyde Tombaugh: Discoverer of Planet Pluto.* Tucson: University of Arizona Press, 1991.

———. *Shoemaker by Levy: The Man Who Made an Impact.* Princeton, NJ: Princeton University Press, 2000.

Levy, Stuart B. *The Antibiotic Paradox: How Miracle Drugs Are Destroying the Miracle.* New York: Plenum, 1992.

———. *The Antibiotic Paradox: How the Misuse of Antibiotics Destroys Their Curative Powers.* New York: HarperCollins, 2002.

Lewis, Oscar. *Five Families.* New York: Basic Books, 1959.

Libby, Willard F. *Radiocarbon Dating.* Chicago: University of Chicago Press, 1952.

Licht, Walter. *Industrializing America.* Baltimore: Johns Hopkins University Press, 1995.

Liebel-Weckowicz, Helen, and Thaddeus Weckowicz. *A History of Great Ideas in Abnormal Psychology.* New York: Elsevier, 1990.

Liebenau, Jonathan. *Medical Science and Medical Industry: The Formation of the American Pharmaceutical Industry.* Basingstoke, UK: Macmillan, 1987.

Life Library of Photography: Light and Film. New York: Time-Life Books, 1975.

Lifton, Robert Jay. *Death in Life: Survivors of Hiroshima.* New York: Random House, 1968.

———. *Destroying the World to Save It: Aum Shinrikyo, Apocalyptic Violence, and the New Global Terrorism.* New York: Owl, 2000.

———. *The Nazi Doctors: Medical Killing and the Psychology of Genocide.* New York: Basic Books, 1986.

———. *Thought Reform and the Psychology of Totalism: A Study of "Brainwashing" in China.* New York: W.W. Norton, 1961.

Lifton, Robert Jay, and Greg Mitchell. *Hiroshima in America: Fifty Years of Denial.* New York: G.P. Putnam's Sons, 1995.

Light, Michael. *Full Moon.* New York: Alfred A. Knopf, 2002.

Lillie, Frank R. *The Woods Hole Marine Biological Laboratory.* Chicago: University of Chicago Press, 1944.

Lindbergh, Charles A. *The Spirit of St. Louis.* New York: Scribner's, 1953.

Lindholdt, Paul, ed. *John Josselyn, Colonial Traveler: A Critical Edition of Two Voyages to New-England.* Hanover, NH: University Press of New England, 1988.

Lindsten, Jan, ed. *From Nobel Lectures, Physiology or Medicine 1971–1980.* Singapore: World Scientific, 1992.

Linzmayer, Owen W. *Apple Confidential: The Real Story of Apple Computer, Inc.* San Francisco: Publishers Group West, 1999.

Livingston, Bernard. *Zoo: Animals, People, Places.* New York: Arbor House, 1974.

Livingstone, David N. *Darwin's Forgotten Defenders: The Encounter Between Evangelical Theology and Evolutionary Thought.* Grand Rapids, MI: Eerdmans, 1987.

Livingstone, David N., D.G. Hart, and Mark A. Noll, eds. *Evangelicals and Science in Historical Perspective.* Religion in America Series. New York: Oxford University Press. 1999.

Locher, David A. *Collective Behavior.* Englewood, NJ: Prentice Hall, 2001.

Lodish, Harvey, Arnold Berk, and S. Lawrence Zipursky. *Molecular Cell Biology.* 4th ed. New York: W.H. Freeman, 1999.

Lomask, Milton. *A Minor Miracle: An Informal History of the National Science Foundation.* Washington, DC: National Science Foundation, 1976.

Loomis, F.A., ed. *As Long as Life: The Memoirs of a Frontier Doctor.* Seattle, WA: Storm Peak, 1994.

Love, James Lee. *The Lawrence Scientific School in Harvard University, 1847–1906.* Burlington, NC, 1944.

Lovejoy, Arthur O. *The Great Chain of Being.* Cambridge, MA: Harvard University Press, 2005.

Lovell, Edith Haroldsen. *Benjamin Bonneville: Soldier of the American Frontier.* Bountiful, UT: Horizon, 1992.

Luchetti, Cathy. *Medicine Women: The Story of Early-American Women Doctors.* New York: Crown, 1998.

Ludmerer, Kenneth M. *Time to Heal: American Medical Education from the Turn of the Century to the Era of Managed Care.* New York: Oxford University Press, 1999.

Lukoff, Herman. *From Dits to Bits: A Personal History of the Electronic Computer.* Portland, OR: Robotics, 1979.

Luminet, Jean-Pierre. *Black Holes.* Cambridge, UK: Cambridge University Press, 1992.

Lunbeck, Elizabeth. *The Psychiatric Persuasion: Knowledge, Gender, and Power in Modern America.* Princeton, NJ: Princeton University Press, 1994.

Lurie, Edward. *Louis Agassiz: A Life in Science.* Baltimore: Johns Hopkins University Press, 1988.

———. *Nature and the American Mind: Louis Agassiz and the Culture of Science.* New York: Science History Publications, 1974.

Lustig, Mary Lou. "Cadwallader Colden." In *American National Biography*, vol. 5. New York: Oxford University Press, 1999.

Lutz, Norma Jean. *Cotton Mather: Clergyman, Author, and Scholar.* New York: Chelsea House, 2000.

———. *Increase Mather: Clergyman and Scholar.* New York: Chelsea House, 2001.

Lutz, Tom. *American Nervousness, 1903: An Anecdotal History.* Ithaca, NY: Cornell University Press, 1991.

Lyons, Albert, and R. Joseph Petrocelli. *Medicine: An Illustrated History.* New York: Abrams, 1992.

MacDonald, Gordon. *Volcanoes.* Englewood Cliffs, NJ: Prentice Hall, 1972.

Macfarlane, Gwyn. *Howard Florey: The Making of a Great Scientist.* Oxford, UK: Oxford University Press, 1979.

Mack, Pamela E. "Straying from Their Orbits: Women in Astronomy in America." In *Women of Science: Righting the Record*, ed. Gabriele Kass-Simon and Patricia Farnes. Bloomington: Indiana University Press, 1990.

Mackintosh, Ray, ed. *Nucleus: A Trip into the Heart of Matter.* Baltimore: Johns Hopkins University Press, 2002.

Macrae, Norman. *John von Neumann.* New York: Pantheon, 1992.

Madden, Edward H. *Chauncey Wright.* New York: Washington Square, 1964.

———. *Chauncey Wright and the Foundations of Pragmatism.* Seattle: University of Washington Press, 1963.

Magner, Lois. *A History of Medicine.* New York: Marcel Dekker, 1992.

Magueijo, Joao. *Faster than the Speed of Light.* Cambridge, MA: Perseus, 2003.

Mahoney, Michael S. "Cybernetics and Information Technology." In *Companion to the History of Modern Science*, ed. R.C. Olby, et al. New York: Routledge, 1990.

Maienschein, Jane. *100 Years Exploring Life, 1888–1988: The Marine Biological Laboratory at Woods Hole.* Boston: Jones and Bartlett, 1989.

Maier, Thomas. *Dr. Spock: An American Life.* New York: Harcourt Brace, 1998.

Malaurie, Jean. *Ultima Thule: Explorers and Natives in the Polar North.* New York: W.W. Norton, 2003.

Malinowski, Bronislaw. *Argonauts of the Western Pacific.* 1922. Reprint ed., Long Grove, IL: Waveland, 1984.

———. *The Dynamics of Culture Change.* Ed. P.M. Kaberry. Oxford, UK: Oxford University Press, 1945.

Malmstrom, Vincent H. *Cycles of the Sun, Mysteries of the Moon.* Austin: University of Texas Press, 1997.

Malphrus, Benjamin K. *The History of Radio Astronomy and the National Radio Astronomy Observatory: Evolution Toward Big Science.* Melbourne, FL: Krieger, 1996.

Mandelbrot, Benoit B. *The Fractal Geometry of Nature.* New York: W.H. Freeman, 1982.

Manes, Stephen, and Paul Andrews. *Gates: How Microsoft's Mogul Reinvented an Industry and Made Himself the Richest Man in America.* New York: Simon and Schuster, 1994.

Manning, Kenneth R. *Black Apollo of Science: The Life of Ernest Everett Just.* New York: Oxford University Press, 1983.

Manning, Robert E., ed. *Mountain Passages: An Appalachia Anthology.* Boston: Appalachian Mountain Club, 1982.

March, Francis A., in collaboration with Richard J. Beamish. *History of the World War: An Authentic Narrative of the World's Greatest War.* Philadelphia: United Publishers of the United States and Canada, 1919.

Marco, Gino J., Robert M. Hollingworth, and William Durham, eds. *Silent Spring Revisited.* Washington, DC: American Chemical Society, 1986.

Marcus, Alan I. *Agricultural Science and the Quest for Legitimacy: Farmers, Agricultural Colleges, and Experimental Stations, 1870–1890.* Ames, IA: Blackwell, 1985.

Marcus, R. Kenneth, and José A.C. Broekaert, eds. *Glow Discharge Plasmas in Analytical Spectroscopy.* Chichester, UK: John Wiley and Sons, 2003.

Marinacci, Barbara, ed. *Linus Pauling in His Own Words.* New York: Touchstone, 1995.

Marine Biological Laboratory Communications Office. *2004 Guide to Research and Education.* Woods Hole, MA, 2004.

Marion, John Francis. *Philadelphia Medica.* Philadelphia: SmithKline, 1975.

Mark, H., and G.S. Whitby, eds. *Collected Papers of Wallace Hume Carothers on High Polymeric Substances.* New York: Interscience, 1940.

Mark, Hans, and Sidney Fernbach. *Properties of Matter Under Unusual Conditions (In Honor of Edward Teller's 60th Birthday).* New York: John Wiley and Sons, 1969.

Marsh, Othniel Charles. *The Life and Scientific Work of Othniel Charles Marsh.* New York: Arno, 1980.

Marshak, Robert E., ed. *Perspectives in Modern Physics: Essays in Honor of Hans A. Bethe on the Occasion of His 60th Birthday, July 1966.* New York: Interscience, 1966.

Marshall, Alfred. *Principles of Economics.* London: Macmillan, 1959.

Martin, Christopher. *Thomas Aquinas: God and Explanations*. Edinburgh, UK: Edinburgh University Press, 1997.

Marx, Karl, and Friedrich Engels. *The Communist Manifesto*. Trans. Samuel Moore. Harmondsworth, UK: Penguin, 1967.

Maslow, Abraham. *The Journals of A.H. Maslow*. 2 vols. Belmont, CA: Thomson Brooks/Cole, 1979.

Mather, Cotton. *Magnalia Christi Americana, or the Ecclesiastical History of New England*. Ed. Raymond J. Cunningham. New York: Frederick Ungar, 1970.

Mathes, W. Michael. *Vizcaíno and Spanish Expansion in the Pacific Ocean, 1580–1630*. San Francisco: California Historical Society, 1968.

Matricon, Jean, Georges Waysand, and Charles Glashausser. *The Cold Wars: A History of Superconductivity*. Piscataway, NJ: Rutgers University Press, 2003.

Matt, Ridley. *Genome: The Autobiography of a Species in 23 Chapters*. New York: HarperCollins, 1999.

May, George. *Charles E. Duryea: Automaker*. Ann Arbor, MI: Edwards Brothers, 1973.

Mazo, Robert. *Brownian Motion: Fluctuations, Dynamics, and Applications*. New York: Oxford University Press, 2002.

Mazumdar, Pauline M.H. *Species and Specificity: An Interpretation of the History of Immunology*. Cambridge, UK: Cambridge University Press, 1995.

McAllister, Ethel M. *Amos Eaton, Scientist and Educator*. Philadelphia: University of Pennsylvania Press, 1941.

McCartney, Scott. *ENIAC: The Triumphs and Tragedies of the World's First Computer*. New York: Berkley, 2001.

McCaslin, John C. *Petroleum Exploration Worldwide*. Tulsa, OK: PennWell, 1983.

McClintock, Barbara. *The Discovery and Characterization of Transposable Elements: The Collected Papers of Barbara McClintock*. New York: Garland, 1987.

McCluskey, Stephen C. "Historical Archaeoastronomy: The Hopi Example." In *Archaeoastronomy in the New World: American Primitive Astronomy*, ed. A.F. Aveni. Cambridge, UK: Cambridge University Press, 1982.

McCullough, David. *The Great Bridge: The Epic Story of the Building of the Brooklyn Bridge*. New York: Touchstone, 1972.

———. *Mornings on Horseback: The Story of an Extraordinary Family, a Vanished Way of Life, and the Unique Child Who Became Theodore Roosevelt*. New York: Simon and Schuster, 1981.

McCurdy, David W., ed. *Conformity and Conflict: Readings in Cultural Anthropology*. Boston: Allyn and Bacon, 2005.

McDermott, John J., ed. *The Philosophy of John Dewey*. Chicago: University of Chicago Press, 1973.

———. *The Writings of William James: A Comprehensive Edition*. Chicago: University of Chicago Press, 1977.

McElheny, Victor K. *Insisting on the Impossible: The Life of Edwin Land*. New York: Perseus, 1998.

———. *Watson and DNA: Making a Scientific Revolution*. Cambridge, MA: Perseus, 2003.

McGrayne, Sharon Bertsch. *Nobel Prize Women in Science*. New York: Birch Lane, 1993.

McGregor, Deborah Kuhn. *Sexual Surgery and the Origins of Gynecology*. New York: Garland, 1989.

McGucken, William. *Nineteenth-Century Spectroscopy: Development of the Understanding of Spectra, 1802–1897*. Baltimore: Johns Hopkins University Press, 1969.

McHale, Thomas R., and Mary C. McHale. *Early American–Philippine Trade: The Journal of Nathaniel Bowditch in Manila, 1796*. New Haven, CT: Yale University Press, 1962.

McMurtry, John. *The Structure of Marx's World-View*. Princeton, NJ: Princeton University Press, 1978.

Mead, Margaret. *Coming of Age in Samoa: A Psychological Study of Primitive Youth for Western Civilization*. 1928. Reprint ed., New York: HarperCollins, 2001.

———. *Ruth Benedict*. New York: Columbia University Press, 1974.

Meikle, Jeffrey L. *American Plastic: A Cultural History*. New Brunswick, NJ: Rutgers University Press, 1995.

Menand, Louis. *The Metaphysical Club*. New York: Farrar, Straus and Giroux, 2002.

Merrill, James. *The Indians' New World: Catawbas and Their Neighbors from European Contact Through the Era of Removal*. Chapel Hill: University of North Carolina Press, 1989.

Metropolis, Nicholas, Jack Howlett, and Gian-Carlo Rota, eds. *A History of Computing in the Twentieth Century: A Collection of Essays*. New York: Academic Press, 1980.

Meyers, Amy, and Margaret Beck Pritchard. *Empire's Nature: Mark Catesby's New World Vision*. Chapel Hill: University of North Carolina Press, for the Omohundro Institute of Early American History and Culture, 1998.

Micklethwait, David. *Noah Webster and the American Dictionary*. Jefferson, NC: McFarland, 2000.

Micklus, Robert. *The Comic Genius of Dr. Alexander Hamilton*. Knoxville: University of Tennessee Press, 1990.

Middlekauff, Robert. *The Mathers: Three Generations of Puritan Intellectuals 1596–1728*. New York: Oxford University Press, 1971.

Middleton, William Shainline. "Caspar Wistar, Junior." In *Annals of Medical History*, vol. 4. New York: Paul B. Hoeber, 1922.

Miller, Arthur I. *Empire of the Stars: Obsession, Friendship, and Betrayal in the Quest for Black Holes.* New York: Houghton Mifflin, 2005.

Miller, Charles B. *Biological Oceanography.* Cambridge, MA: Blackwell Science, 2004.

Miller, Donald L. *Lewis Mumford: A Life.* New York: Weidenfeld and Nicolson, 1989.

Miller, Genevieve. *Bibliography of the History of Medicine in the United States.* Baltimore: Johns Hopkins University Press, 1964.

Miller, Howard. *Dollars for Research: Science and Its Patrons in Nineteenth-Century America.* Seattle: University of Washington Press, 1970.

Miller, Nathan. *Theodore Roosevelt: A Life.* New York: William Morrow, 1992.

Millikan, Robert A., and I. Bernard Cohen. *Autobiography of Robert A. Millikan.* North Stratford, NH: Ayer, 1980.

Milne, John. *Seismology.* London: Read Books, 2006.

Minetree, Harry. *Cooley: The Career of a Great Heart Surgeon.* New York: HarperCollins, 1973.

Minton, R.B. "Friedrich Bessel and the Companion of Sirius." In *Cosmic Horizons: Astronomy at the Cutting Edge,* ed. Steven Soter and Neil deGrasse Tyson. New York: New Press, 2001.

Mitchell, Carolyn B. *Life in the Universe: Readings from Scientific American Magazine.* New York: W.H. Freeman, 1994.

Mitchell, S. Weir. *Wear and Tear (or Hints for the Overworked).* Philadelphia: J.B. Lippincott, 1887; reprint, New York: Arno, 1973.

Mitford, Jessica. *The Trial of Dr. Spock, the Rev. William Sloane Coffin, Jr., Michael Ferber, Mitchell Goodman, and Marcus Raskin.* New York: Alfred A. Knopf, 1969.

Mitton, Jacqueline. *Informania: Aliens.* Cambridge, MA: Candlewick, 2000.

Monmonier, Mark. *Air Apparent: How Meteorologists Learned to Map, Predict, and Dramatize Weather.* Chicago: University of Chicago Press, 1999.

Moore, Jerry D. *Visions of Culture.* Walnut Creek, CA: AltaMira, 1997.

Moores, Eldridge, ed. *Shaping the Earth: Tectonics of Continents and Oceans.* New York: W.H. Freeman, 1990.

Morange, Michael. *A History of Molecular Biology.* Cambridge, MA: Harvard University Press, 1998.

Morantz-Sanchez, Regina. *Conduct Unbecoming a Woman.* New York: Oxford University Press, 1999.

More, Ellen. *Restoring the Balance.* Cambridge, MA: Harvard University Press, 1999.

Morganstern, Oskar, and John von Neumann. *The Theory of Games and Economic Behavior.* Princeton, NJ: Princeton University Press, 1944.

Morison, Samuel Eliot. *Admiral of the Ocean Sea: A Life of Christopher Columbus.* Boston: Little, Brown, 1989.

———. *Christopher Columbus, Mariner.* New York: New American Library, 1983.

———. *The Great Explorers: The European Discovery of America.* New York: Oxford University Press, 1986.

———. *John Paul Jones: A Sailor's Biography.* Boston: Little, Brown, 1959.

———. *Maritime History of Massachusetts, 1783–1860.* Boston: Houghton Mifflin, 1961.

———. *Samuel de Champlain: Father of New France.* Boston: Little, Brown, 1972.

Moritz, Michael. *The Little Kingdom: The Private Story of Apple Computer.* New York: William Morrow, 1984.

Morris, Henry M. *History of Modern Creationism.* 2nd ed. Santee, CA: Institute for Creation Research, 1993.

Morris, Robin. *A Cognitive Neuropsychology of Alzheimer's Disease.* New York: Oxford University Press, 2004.

Mortenson, Terry. *The Great Turning Point: The Church's Catastrophic Mistake on Geology—Before Darwin.* Green Forest, AR: Master, 2004.

Morton, A.G. *History of Botanical Science.* London: Academic Press, 1981.

Mott, Frank Luther. *A History of American Magazines.* Vol. 3, *1865–1885.* Cambridge, MA: Harvard University Press, 1938.

Moyer, Albert E. *A Scientist's Voice in American Culture: Simon Newcomb and the Rhetoric of Scientific Method.* Berkeley: University of California Press, 1992.

Moyle, Peter B., and Joseph J. Cech, Jr. *Fishes: An Introduction to Ichthyology.* Englewood Cliffs, NJ: Prentice Hall, 1988.

Mueller-Vollmer, Kurt. *The Hermeneutics Reader: Texts of the German Tradition from the Enlightenment to the Present.* New York: Continuum, 1988.

Muir, John. *Nature Writings: The Story of My Boyhood and Youth; My First Summer in the Sierra; The Mountains of California; Stickeen; Essays.* New York: Library of America, 1997.

Mulholland, James A. *A History of Metals in Colonial America.* University: University of Alabama Press, 1981.

Multhauf, Robert P., and Gregory Good. *A Brief History of Geomagnetism and a Catalog of the Collections of the National Museum of American History.* Washington, DC: Smithsonian Institution, 1987.

Mumford, Lewis. *The Culture of Cities.* New York: Harcourt Brace, 1938.

———. *Sketches from Life: The Autobiography of Lewis Mumford: The Early Years.* New York: Dial, 1982.

Murdock, Kenneth B. *Increase Mather: The Foremost American Puritan.* New York: Russell and Russell, 1966.

Murphy, Jim. *An American Plague: The True and Terrifying Story of the Yellow Fever Epidemic of 1793.* New York: Clarion, 2003.

Murray, Charles, and Catherine B. Cox. *Apollo: The Race to the Moon.* New York: Simon and Schuster, 1989.

Murray, Robert K., and Daryl K. Granner. *Harper's Biochemistry.* 25th ed. New York: McGraw-Hill, 2000.

Murtagh, William J. *Keeping Time: The History and Theory of Preservation in America.* New York: Wiley, 2005.

Naef, Andreas P. *The Story of Thoracic Surgery: Milestones and Pioneers.* Lewiston, ME: Hogrefe and Huber, 1990.

National Academy of Sciences. Washington, DC: National Academy of Sciences, 1969.

National Research Council. *Climate Change Science: An Analysis of Some Key Questions.* Washington, DC: National Academy Press, 2001.

National Research Council Institute of Medicine. *Enhancing the Vitality of the National Institutes of Health.* Washington, DC: National Academies Press, 2003.

Neachtain, Ted. *Yellow Fever.* Oakton, VA: Ravensyard, 1999.

Nebeker, Fred. *Calculating the Weather: Meteorology in the 20th Century.* San Diego, CA: Academic Press, 1995.

Neimark, A.E. *A Deaf Child Listened: Thomas Gallaudet, Pioneer in American Education.* New York: William Morrow, 1983.

Nelson, Daniel. *Frederick W. Taylor and the Rise of Scientific Management.* Madison: University of Wisconsin Press, 1980.

Nevins, Allan, and Frank Ernest Hill. *Ford.* New York: Arno, 1976.

Newberry, Sterling P. *EMSA and Its People: The First Fifty Years.* Milwaukee, WI: Electron Microscopy Society of America, 1992.

Newhall, Beaumont. *History of Photography: From 1839 to the Present.* Lebanon, IN: Bulfinch, 1982.

Newman, William R. *Gehennical Fire: The Lives of George Starkey, an American Alchemist in the Scientific Revolution.* Cambridge, MA: Harvard University Press, 1994.

Newman, William R., and Lawrence M. Principe. *Alchemy Tried in the Fire: Starkey, Boyle, and the Fate of Helmontian Chymistry.* Chicago: University of Chicago Press, 2002.

Newton, Isaac. *Newton's Principia: The Central Argument: Translation, Notes, and Expanded Proofs.* 1687. Santa Fe, NM: Green Lion, 1995.

Nichols, Roger L., and Patrick L. Halley. *Stephen Long and American Frontier Exploration.* Newark: University of Delaware Press, 1980.

Nickles, Thomas, ed. *Thomas Kuhn.* New York: Cambridge University Press, 2003.

Nisbet, Robert A. *The Sociological Tradition.* New York: Basic Books, 1966.

Noble, Jeanne L. *The Negro Woman's College Education.* New York: Columbia University Press, 1956.

Norberg, Arthur L. *Computers and Commerce: A Study of Technology and Management at Eckert-Mauchly Computer Company, Engineering Research Associates, and Remington Rand, 1946–1857.* Cambridge, MA: MIT Press, 2005.

Norman, J.R., and P.H. Greenwood. *A History of Fishes.* New York: Halstead, 1975.

North, Michael. *Camera Works: Photography and the Twentieth-Century Word.* New York: Oxford University Press, 2005.

Nuland, Sherwin. *Doctors.* New York: Alfred A. Knopf, 1989.

Nuttall, Thomas. "Journal of Travels into the Arkansas Territory During the Year 1819," in *Early Western Travels, 1748–1846,* vol. 13, ed. Reuben Gold Thwaites. Cleveland, OH: A.H. Clark, 1905.

———. *A Journal of Travels into the Arkansas Territory During the Year 1819.* Philadelphia: T.H. Palmer, 1821; Fayetteville: University of Arkansas Press, 1999.

Nyce, James M., and Paul Kahn, eds. *From Memex to Hypertext: Vannevar Bush and the Mind's Machine.* Boston, MA: Academic Press, 1992.

Nye, David E. *Electrifying America: Social Meanings of a New Technology, 1880–1940.* Cambridge, MA: MIT Press, 1990.

O'Donnell, John M. *The Origins of Behaviorism: American Psychology, 1987–1920.* New York: New York University Press, 1985.

Olasky, Marvin. *Abortion Rites: A Social History of Abortion in America.* Washington, DC: Regnery, 1995.

Olby, Robert. *The Path to the Double Helix.* Seattle: University of Washington Press, 1974; reprint ed., New York: Dover, 1994.

Oleson, Alexandra, and Sanborn C. Brown, eds. *Pursuit of Knowledge in the Early American Republic: American Scientific and Learned Societies from Colonial Times to the Civil War.* Baltimore: Johns Hopkins University Press, 1976.

Olshaker, Mark. *The Instant Image: Edwin Land and the Polaroid Experience.* Briarcliff Manor, NY: Stein and Day, 1978.

O'Malley, Michael. *Keeping Watch: A History of American Time.* New York: Viking, 1990.

Onsager, Lars. *The Collected Work of Lars Onsager.* Hackensack, NJ: World Scientific, 1996.

Orr, Oliver H. *Saving American Birds: T. Gilbert Pearson and the Founding of the Audubon Movement.* Gainesville: University Press of Florida, 1992.

Orton, Vrest. *The Forgotten Art of Building a Good Fireplace: The Story of Benjamin Thompson, Count Rumford, an American Genius, and His Principles of Fireplace Designs Which Have Remained Unchanged for 174 years.* Collingdale, PA: Diane Publishing, 1999.

Oshinsky, David M. *Polio: An American Story.* Oxford, UK: Oxford University Press, 2005.

Oslin, George P. *The Story of Telecommunications.* Macon, GA: Mercer University Press, 1992.

Osterbrock, Donald E. *James E. Keeler: Pioneer American Astrophysicist.* New York: Cambridge University Press, 1984.

Overton, Mark. *Agricultural Revolution in England: The Transformation of the Agrarian Economy 1500–1850.* Cambridge Studies in Historical Geography. Cambridge, UK: Cambridge University Press, 1996.

Pais, Abraham. *Subtle Is the Lord: The Science and the Life of Albert Einstein.* Oxford, UK: Oxford University Press, 1982.

Palladino, Paolo. *Entomology, Ecology, and Agriculture: The Making of Scientific Careers in North America, 1885–1985.* Amsterdam, The Netherlands: Harwood Academic, 1996.

Parascandola, John. *The Development of American Pharmacology: John J. Abel and the Shaping of a Discipline.* Baltimore: Johns Hopkins University Press, 1992.

Paris, Bernard. *Karen Horney: A Psychoanalyst's Search for Self-Understanding.* New Haven, CT: Yale University Press. 1994.

Parkman, Francis. *The Pioneers of France in the New World.* 1865. Lincoln: University of Nebraska Press, 1996.

Parsons, Talcott. *Essays in Sociological Theory.* New York: Free Press, 1949.

———. *The Social System.* New York: Free Press, 1951.

———. *Social Systems and the Evolution of Action Theory.* Glencoe, IL: Free Press, 1977.

———. *The Structure of Social Action.* New York: McGraw-Hill, 1937.

Partington, James Riddick. *Historical Studies on the Phlogiston Theory.* New York: Arno, 1981.

Paschal, Herbert R., Jr. *A History of Colonial Bath.* Raleigh, NC: Edwards and Broughton, 1955.

Passmore, John A., ed. *Priestley's Writings on Philosophy, Science, and Politics.* New York: Collier, 1965.

Patterson, James T. *The Dread Disease: Cancer and Modern American Culture.* Cambridge, MA: Harvard University Press, 1987.

Patterson, Thomas C. *A Social History of Anthropology in the United States.* New York: Berg, 2001.

Paul, Diane B. *Controlling Human Heredity: 1865 to the Present.* Atlantic Highlands, NJ: Humanities Press, 1995.

Paul, John R. *A History of Poliomyelitis.* New Haven, CT: Yale University Press, 1971.

Paul, Maureen, et al. *A Clinician's Guide to Medical and Surgical Abortion.* New York: Churchill Livingstone, 1999.

Pauling, Linus. *General Chemistry.* North Chelmsford, MA: Courier Dover, 1988.

———. *Selected Scientific Papers.* New York: World Scientific, 1999.

Payne-Gaposchkin, Cecilia. *An Autobiography and Other Recollections.* Ed. Katharine Haramundanis. 2nd ed. Cambridge, UK: Cambridge University Press, 1996.

———. *Stars and Clusters.* Cambridge, MA: Harvard University Press, 1979.

Pearcey, Nancy R., and Charles B. Thaxton. *The Soul of Science: Christian Faith and Natural Philosophy.* Wheaton, IL: Crossway, 1994.

Pearson, Thomas Gilbert. *Adventures in Bird Protection.* New York: D. Appleton-Century, 1937.

Pearson, Willie, Jr., and H. Kenneth Bechtel, eds. *Blacks, Science, and American Education.* New Brunswick, NJ: Rutgers University Press, 1989.

Peebles, Curtis. *Asteroids: A History.* Washington, DC: Smithsonian Institution, 2000.

———. *The Corona Project: America's First Spy Satellites.* Annapolis, MD: Naval Institute Press, 1997.

Peirce, Bradford. *Trials of an Inventor: Life and Discoveries of Charles Goodyear.* Seattle, WA: University Press of the Pacific, 2003.

Peirce, Charles. *Writings of Charles S. Peirce: A Chronological Edition.* Bloomington: Indiana University Press, 1982.

Perkowitz, Sidney. *Empire of Light.* New York: Henry Holt, 1996.

Pernick, Martin. *A Calculus of Suffering: Pain, Professionalism, and Anesthesia in Nineteenth-Century America.* New York: Columbia University Press, 1985.

Perry, Helen Swick. *Psychiatrist of America: The Life of Harry Stack Sullivan.* Cambridge, MA: Harvard University Press, 1982.

Petersen, Carolyn Collins, and John C. Brandt. *Hubble Vision: Astronomy with the Hubble Space Telescope.* New York: Cambridge University Press, 1995.

Peterson, Roger Tory. *A Field Guide to the Birds.* Boston and New York: Houghton Mifflin Company, 1934.

Pettijohn, Francis John. *Sedimentary Rocks.* 3rd ed. New York: HarperCollins, 1983.

Philbrick, Nathaniel. *Sea of Glory: America's Voyage of Discovery: The U.S. Exploring Expedition, 1838–1842.* New York: Viking Penguin, 2003.

Pick, Nancy. *The Rarest of the Rare: Stories Behind the Treasures at the Harvard Museum of Natural History.* New York: HarperResource, 2004.

Pickett, Steward T.A., Jurek Kolasa, and Clive G. Jones. *Ecological Understanding.* San Diego, CA: Academic Press, 1994.

Pike, Zebulon. "Arkansaw Journal." In *The Expeditions of Zebulon Montgomery Pike,* vol. II. New York: F.P. Harper, 1895.

———. *The Expeditions of Zebulon Montgomery Pike.* 3 vols. New York: Harper, 1895; reprint ed., 2 vols., Mineola, NY: Dover, 1985.

Pinckney, Elise, ed. *The Letterbook of Eliza Lucas Pinckney.* Columbia: University of South Carolina Press, 1997.

Pirsson, Louis V. *Biographical Memoir of James Dwight Dana, 1813–1895.* Washington, DC: National Academy of Sciences, 1919.

Piszkiewicz, Dennis. *Wernher von Braun: The Man Who Sold the Moon.* Westport, CT: Praeger, 1998.

Pitcher, Everett. *A History of the Second Fifty Years, American Mathematical Society, 1939–1988.* Providence, RI: American Mathematical Society, 1988.

Plotkin, Stanley A., and Edward A. Mortimer, eds. *Vaccines.* Philadelphia: W.B. Saunders, 1988.

Plummer, Charles C., and David McGeary. *Physical Geology.* Dubuque, IA: W.C. Brown, 1979.

Poundstone, William. *Carl Sagan: A Life in the Cosmos.* New York: Henry Holt, 1999.

Powell, John Wesley. *The Exploration of the Colorado River and Its Canyons.* New York: Penguin, 2003.

Powers, Michael D., ed. *Children with Autism: A Parents' Guide.* Rockville, MD: Woodbine House, 1989.

Preble, Edward. "John Jones." In *Dictionary of American Biography.* Vol. 5. Ed. Dumas Malone. New York: Charles Scribner's Sons, 1961.

Prescott, Samuel C. *When MIT Was "Boston Tech."* Cambridge, MA: Technology Press, 1954.

Preston, Richard. *First Light: The Search for the Edge of the Universe.* New York: Random House, 1987.

Pretzer, William S., ed. *Working at Inventing: Thomas Edison and the Menlo Park Experience.* Dearborn, MI: Henry Ford Museum and Greenfield Village, 1989.

Price, Robert. *Johnny Appleseed: Man and Myth.* Bloomington: Indiana University Press, 1954.

Priestley, Joseph. *Considerations on the Doctrine of Phlogiston and the Decomposition of Water.* Philadelphia: Thomas Dobson, 1796.

Pruette, Lorine G. *Stanley Hall: A Biography of a Mind.* New York: D. Appleton, 1926.

Quandt, Jean. *From Small Town to the Great Community: The Social Thought of Progressive Intellectuals.* New Brunswick, NJ: Rutgers University Press, 1970.

Quincy, Josiah. *The History of Harvard University.* 2 vols. Cambridge, MA, 1840.

Quinn, Susan. *A Mind of Her Own: The Life of Karen Horney.* New York: Summit, 1987.

Rabi, I.I. *My Life and Times as a Physicist.* Claremont, CA: Friends of the Colleges at Claremont, 1960.

Rachlin, Howard. *Introduction to Modern Behaviorism.* San Francisco: W.H. Freeman, 1970.

Rackeman, Francis Minot. *The Inquisitive Physician: The Life and Times of George Richards Minot.* Cambridge, MA: Harvard University Press, 1956.

Rainger, Ronald. *An Agenda for Antiquity: Henry Fairfield Osborn and Vertebrate Paleontology at the American Museum of Natural History, 1890–1935.* Tuscaloosa: University of Alabama Press, 1991.

Rall, J.E. *Solomon Berson: Biographical Memoirs.* Washington, DC: National Academy of Sciences, 1990.

Ramsey, Norman. *Molecular Beams.* Wotton-under-Edge, UK: Clarendon, 1956.

Randier, Jean. *Marine Navigation Instruments.* London: John Murray, 1980.

Randolph, Jacob. *A Memoir on the Life and Character of Philip Syng Physick.* Philadelphia: Collins, 1839.

Redmond, Kent C., and Thomas M. Smith. *From Whirlwind to MITRE: The R & D Story of the SAGE Air Defense Computer.* Cambridge, MA: MIT Press, 2000.

———. *Project Whirlwind: The History of a Pioneer Computer.* Bedford, MA: Digital, 1980.

Reeves, Robert. *The Superpower Space Race: An Explosive Rivalry Through the Solar System.* New York and London: Plenum, 1994.

Regal, Brian. *Henry Fairfield Osborn: Race and the Search for the Origins of Man.* Burlington, VT: Ashgate, 2002.

Regis, Pamela. *Describing Early America: Bartram, Jefferson, Crèvecoeur, and the Rhetoric of Natural History.* Dekalb: Northern Illinois University Press, 1992.

Reilly, Philip. *The Surgical Solution: A History of Involuntary Sterilization in the United States.* Baltimore: Johns Hopkins University Press, 1991.

Reinhold, Nathan, ed. *Science in the Nineteenth Century: A Documentary History.* Chicago: University of Chicago Press, 1985.

Reisman, David. *Thorstein Veblen: A Critical Interpretation.* New York: Charles Scriber's Sons, 1953.

Reisman, Judith, and Edward W. Eichel. *Kinsey, Sex, and Fraud: The Indoctrination of a People.* Lafayette, LA: Huntington House, 1993.

Reiss, Oscar. *Medicine in Colonial America.* Lanham, MD: University Press of America, 2000.

Repcheck, Jack. *The Man Who Found Time: James Hutton and the Discovery of Earth's Antiquity.* Cambridge, MA: Perseus, 2003.

Reverby, Susan M., ed. *Tuskegee's Truths: Rethinking the Tuskegee Syphilis Study.* Chapel Hill: University of North Carolina Press, 2000.

Rexer, Lyle, and Rachel Klein. *American Museum of Natural History: 125 Years of Expedition and Discovery.* New York: Harry N. Abrams, 1995.

Reynolds, Terry S. *The Engineer in America: A Historical Anthology from Technology and Culture.* Chicago: University of Chicago Press, 1991.

Rezek, Philipp R. *Autopsy Pathology: A Guide for Pathologists and Clinicians.* Springfield, IL: Charles C. Thomas, 1963.

Rezneck, Samuel. *Education for a Technological Society: A Sesquicentennial History of Rensselaer Polytechnic Institute.* Troy, NY: Rensselaer Polytechnic Institute, 1968.

Rhine, Louisa E. *Something Hidden*. Jefferson, NC: McFarland, 1983.

Rhoades, Lawrence. *A History of the American Sociological Association 1905–1980*. Washington, DC: American Sociological Association, 1981.

Rhodes, Richard. *Dark Sun: The Making of the Hydrogen Bomb*. New York: Simon and Schuster, 1995.

———. *The Making of the Atomic Bomb*. New York: Simon and Schuster, 1995.

Rhonda, James P. *Lewis and Clark Among the Indians*. Lincoln: University of Nebraska Press, 1984.

Rice, Howard C. *The Rittenhouse Orrery: Princeton's Eighteenth-Century Planetarium, 1767–1954*. Princeton, NJ: Princeton University Press, 1954.

Richards, E.G. *Mapping Time: The Calendar and Its History*. New York: Oxford University Press, 2000.

Richardson, Ruth. *Death, Dissection, and the Destitute*. Chicago: Chicago University Press, 1987.

Richelson, Jeffrey T. *America's Secret Eyes in Space: The U.S. Keyhole Spy Satellite Program*. New York: Harper and Row, 1990.

Richter, Brigitte Zoeller. *Alzheimer's Disease: A Physician's Guide to Practical Management*. Totowa, NJ: Humana, 2004.

Richter, Charles F. *Elementary Seismology*. New York: W.H. Freeman, 1995.

Ricklefs, Robert E. *Ecology*. New York: W.H. Freeman, 1990.

Rigden, John S. *Rabi: Scientist and Citizen*. Cambridge, MA: Harvard University Press, 2000.

Riley, James. *Rising Life Expectancy: A Global History*. New York: Cambridge University Press, 2001.

Riley, Stephen T. *The Massachusetts Historical Society 1791–1959*. Boston: Massachusetts Historical Society, 1959.

Ring, Malvin. *Dentistry*. New York: Abrams. 1987.

Rink, Paul. *To Steer by the Stars: The Story of Nathaniel Bowditch*. New York: Doubleday, 1969.

Robbins, Michael W. *The Principio Company: Iron-Making in Colonial Maryland 1720–1781*. New York: Garland, 1986.

Roberts, David. *A Newer World: Kit Carson, John C. Frémont, and the Claiming of the American West*. New York: Simon and Schuster, 2000.

Roberts, J. Timmons, and Amy Hite. *From Modernization to Globalization: Perspectives on Development and Social Change*. Ames, IA: Blackwell, 2000.

Roberts, Jon H. *Darwinism and the Divine in America: Protestant Intellectuals and Organic Evolution, 1859–1900*. Notre Dame, IN: University of Notre Dame Press, 2001.

Robinson, Judith. *Noble Conspirator: Florence S. Mahoney and the Rise of the National Institutes of Health*. Washington, DC: Francis, 2001.

Rodgers, Eugene. *Beyond the Barrier: The Story of Byrd's First Expedition to Antarctica*. Annapolis, MD: Naval Institute Press, 1990.

Rogers, Carl R. *On Becoming a Person*. Boston: Houghton Mifflin, 1961.

———. *On Personal Power*. New York: Delacorte, 1977.

———. *A Way of Being*. Boston: Houghton Mifflin, 1980.

Rogers, George A., and R. Frank Saunders, Jr. "Stephen Elliott: Early Botanist of Coastal Georgia." In *Swamp Water and Wiregrass: Historical Sketches of Coastal Georgia*. Mercer, GA: Mercer University Press, 1984.

Ronda, James P. *Thomas Jefferson and the Rocky Mountains: Exploring the West from Monticello*. Norman: University of Oklahoma Press, 1993.

Roosevelt, Theodore. *An Autobiography (1858–1919)*. New York: Macmillan, 1913.

Rose, Frank. *West of Eden: The End of Innocence at Apple Computer*. New York: Viking, 1989.

Rose, John. *The Cybernetic Revolution*. New York: Barnes and Noble, 1974.

Rosen, George. *A History of Public Health*. MD Monographs on Medical History, No. 1. New York: MD Publications, 1958; Baltimore: Johns Hopkins University Press, 1993.

Rosenberg, Alex. *The Philosophy of Science: A Contemporary Introduction*. New York: Routledge, 2000.

Rosenburg, Charles E. *The Care of Strangers: The Rise of America's Hospital System*. New York: Basic Books, 1997.

———. *The Cholera Years: The United States in 1832, 1849, and 1866*. Chicago: University of Chicago Press, 1962; rev. ed. 1987.

———. *The Trail of the Assassin Guiteau: Psychiatry and Law in the Gilded Age*. Chicago: University of Chicago Press, 1968.

Rosenfeld, Boris A. *A History of Non-Euclidean Geometry: Evolution of the Concept of a Geometric Space*. Trans. Hardy Grant, with Abe Shenitzer. New York: Springer-Verlag, 1988.

Ross, Dorothy. *G. Stanley Hall: The Psychologist as Prophet*. Chicago: University of Chicago Press, 1972.

Ross, Ishbel. *Child of Destiny: The Life Story of the First Woman Doctor*. New York: Harper, 1949.

Ross, Walter S. *The Last Hero: Charles A. Lindbergh*. New York: Harper and Row, 1967.

Rothman, Hal K., and Sara Dant Ewert, eds. *Encyclopedia of American National Parks*. Armonk, NY: M.E. Sharpe, 2004.

Rothstein, W.G. *American Medical Schools and the Practice of Medicine: A History*. New York: Oxford University Press, 1987.

Rowland, Eron, ed. *Life, Letters, and Papers of William Dunbar of Elgin, Morayshire, Scotland, and*

Natchez, Mississippi: Pioneer Scientist of the Southern United States. Jackson: Mississippi Historical Society, 1930.

Ruark, Arthur Edward, and Harold Clayton Urey. *Atoms, Molecules, and Quanta.* New York: McGraw-Hill, 1930.

Rubertone, Patricia E. *Grave Undertakings: An Archaeology of Roger Williams and the Narragansett Indians.* Washington, DC: Smithsonian Institution, 2001.

Rukeyser, Muriel. *Willard Gibbs.* Garden City, NY: Doubleday, Doran, 1942.

Ruse, Michael. *The Evolution Wars: A Guide to the Debates.* Santa Barbara, CA: ABC-CLIO, 2000.

Russell, Colin A. *Michael Faraday: Physics and Faith.* Oxford, UK: Oxford University Press, 2000.

Russell, Howard. *Indian New England before the Mayflower.* Hanover, NH: University Press of New England, 1980.

Rutledge, John, and Deborah Allen. *Rust to Riches: The Coming of the Second Industrial Revolution.* New York: HarperCollins, 1989.

Ryan, Patrick J. *Euclidean and Non-Euclidean Geometry.* New York: Cambridge University Press, 1986.

Sagan, Carl. *Broca's Brain: Reflections on the Romance of Science.* 1979. Reprint ed., New York: Ballantine, 1986.

———. *Cosmos.* 1980. Reprint ed., New York: Ballantine, 1985.

———. *The Dragons of Eden: Speculations on the Evolution of Human Intelligence.* 1977. Reprint ed., New York: Ballantine, 1986.

Sagendorph, Robb Hansell. *America and Her Almanacs: Wit, Wisdom, and Weather, 1639–1970.* Boston: Yankee, 1970.

Salerno, Roger D. *Louis Wirth: A Bio-bibliography.* New York: Greenwood, 1987.

Salsburg, David. *The Lady Tasting Tea: How Statistics Revolutionized Science in the Twentieth Century.* New York: Henry Holt, 2001.

Sammons, Vivian. *Blacks in Science and Medicine.* Philadelphia: Taylor and Francis, 1989.

Sanders, Brad. *Guide to William Bartram's Travels: Following the Trail of America's First Great Naturalist.* Athens, GA: Fevertree, 2002.

Sandler, Martin W. *Photography: An Illustrated History.* Oxford, UK: Oxford University Press, 2002.

Sanger, Margaret. *The Autobiography of Margaret Sanger.* New York: Dover, 2004.

Santiago, Diego de. "Diary of Sebastian Vizcaino, 1602–1603." In *Spanish Exploration in the Southwest, 1542–1706,* ed. Herbert Eugene Bolton. New York: Barnes and Noble, 1946.

Sappol, Michael. *A Traffic in Dead Bodies.* Princeton, NJ: Princeton University Press, 2002.

Sarkar, Sahotra. *Genetics and Reductionism.* Cambridge, UK: Cambridge University Press, 1998.

Sarton, George. *The History of Science and the New Humanism.* Bloomington: Indiana University Press, 1937.

Satter, Beryl. *Each Mind a Kingdom: American Women, Sexual Purity, and the New Thought Movement, 1875–1920.* Berkeley: University of California Press, 1999.

Saunders, John Richard. *The World of Natural History as Revealed in the American Museum of Natural History.* New York: Sheridan House, 1952.

Savitt, T. "Thomas Hopkins Gallaudet." In *Dictionary of American Medical Biography.* Westport, CT: Greenwood, 1984.

Schatzkin, Paul. *The Boy Who Invented Television: A Story of Inspiration, Persistence, and Quiet Passion.* Burtonsville, MD: Teamcom, 2002.

Scheffler, Isaac. *The Anatomy of Inquiry.* New York: Alfred A. Knopf, 1963.

Schubert, Frank N. *The Nation Builders: A Sesquicentennial History of the Corps of Topographical Engineers 1838–1863.* Washington, DC: U.S. Government Printing Office, 1980.

———. *Vanguard of Expansion: Army Engineers in the Trans-Mississippi West, 1819–1879.* Washington, DC: Army Corps of Engineers, 1980.

Schuh, Randall T. *Biological Systematics: Principles and Applications.* Ithaca, NY: Cornell University Press, 2000.

Schutz, Bernard. *Gravity from the Ground Up: An Introductory Guide to Gravity and General Relativity.* Cambridge, UK: Cambridge University Press, 2003.

Schwarz, Ted. *A History of United States Coinage.* New York: A.S. Barnes, 1980.

Seaborg, Glenn T. *Adventures in the Atomic Age: From Watts to Washington.* New York: Farrar, Straus and Giroux, 2001.

———. *A Chemist in the White House: From the Manhattan Project to the End of the Cold War.* Washington, DC: American Chemical Society, 1998.

———. *The Transuranium People: The Inside Story.* Singapore: World Scientific, 1999.

Seaborg, Glenn T., et al. *The Plutonium Story: The Journals of Professor Glenn T. Seaborg, 1939–1946.* Columbus, OH: Battelle, 1994.

Seitz, Frederick, Erich Vogt, and Alvin M. Weinberg. "Eugene Paul Wigner." In *Biographical Memoirs,* National Academy of Sciences, vol. 74. Washington, DC: National Academies Press, 1998.

Seligman, Edwin R.A., and Alvin Johnson, eds. *Encyclopedia of the Social Sciences.* New York: Macmillan, 1930–1935.

Sell, Alan P.F. *The Great Debate: Calvinism, Arminianism, and Salvation.* Grand Rapids, MI: Baker House, 1983.

Serafini, Anthony. *Linus Pauling: A Man and His Science.* New York: Paragon House, 1989.

Serber, Robert. *The Los Alamos Primer: The First Lectures on How to Build an Atomic Bomb.* Berkeley: University of California Press, 1992.

Seroussi, Karyn. *Unraveling the Mystery of Autism and Pervasive Developmental Disorder: A Mother's Story of Research and Recovery.* New York: Broadway, 2002.

Service, Elman Rogers. *Profiles in Ethnology.* New York: Harper and Row, 1963.

Shafer, Henry B. *The American Medical Profession 1783–1850.* New York: AMS, 1936.

Shallat, Todd. *Structures in the Stream: Water, Science, and the Rise of the U.S. Army Corps of Engineers.* Austin: University of Texas Press, 1994.

Shands, Alfred Rives. *William Edmonds Horner, 1793–1853.* Transactions and Studies of the College of Physicians of Philadelphia, vol. 22, 4th ser. Baltimore: Waverly, 1954–1955.

Shapin, Steven. *The Scientific Revolution.* Chicago: University of Chicago Press, 1996.

Sharchburg, Richard P. *Carriages Without Horses: J. Frank Duryea and the Birth of the American Automobile Industry.* Warrendale, PA: Society of Automotive Engineers, 1993.

Shattuck, George Cheever. *Diseases of the Tropics.* New York: Appleton-Century-Crofts, 1951.

Sheehan, William. *The Transits of Venus.* New York: Prometheus, 2004.

Sheets, Bob, and Jack Williams. *Hurricane Watch: Forecasting the Deadliest Storms on Earth.* New York: Vintage Books, 2001.

Sheldon, William H. *Varieties of Delinquent Youth.* New York: Harper, 1949.

———. *Varieties of Temperament: A Psychology of Constitutional Differences.* New York: Harper and Brothers, 1942.

Shiers, George, ed. *The Development of Wireless to 1920.* New York: Arno, 1977.

———. *The Electric Telegraph: An Historical Anthology.* New York: Arno, 1977.

———. *The Telephone: An Historical Anthology.* New York: Arno, 1977.

Shilts, Randy. *And the Band Played On: Politics, People, and the AIDS Epidemic.* New York: Stonewall Inn, 2000.

Shine, Ian, and Sylvia Wrobel. *Thomas Hunt Morgan: Pioneer of Genetics.* Lexington: University Press of Kentucky, 1976.

Shipton, Clifford K. *Biographical Sketches of Graduates of Harvard University.* Vol. 6. Boston: Massachusetts Historical Society, 1937.

Shnayerson, Michael, and Mark J. Plotkin. *The Killers Within: The Deadly Rise of Drug-Resistant Bacteria.* New York: Little, Brown, 2002.

Shorter, Edward. *From Paralysis to Fatigue: A History of Psychosomatic Illness in the Modern Era.* New York: Free Press, 1992.

———. *A History of Psychiatry from the Era of the Asylum to the Age of Prozac.* New York: John Wiley and Sons, 1997.

Shumacker, Harris B. *The Evolution of Cardiac Surgery.* Bloomington: Indiana University Press, 1992.

Shurkin, Joel N. *Engines of the Mind: A History of the Computer.* New York: W.W. Norton, 1984.

Shute, Michael, ed. *The Scientific Work of John Winthrop.* New York: Arno, 1980.

Sibley, David. *The Sibley Guide to Birds.* New York: Alfred A. Knopf, 2000.

Sicilia, David B. *The Principles of Scientific Management.* Norwalk, CT: Easton, 1993.

Sieden, Lloyd S. *Buckminster Fuller's Universe: His Life and Work.* New York: Perseus, 2000.

Sigman, Marian. *Children with Autism: A Developmental Perspective.* Cambridge, MA: Harvard University Press, 1997.

Sigurdsson, Haraldur. *Encyclopedia of Volcanoes.* London: Academic Press, 2000.

Sikorsky, Igor I. *The Story of the Winged-S: An Autobiography.* New York: Dodd, Mead, 1939.

Silfvast, William T. *Laser Fundamentals.* New York: Cambridge University Press, 1996.

Silk, Joseph. *A Short History of the Universe.* New York: Scientific American Library, 1994.

Sills, David L., ed. *International Encyclopedia of the Social Sciences.* New York: Macmillan and Free Press, 1968.

Silverberg, Robert. *Stormy Voyager: The Story of Charles Wilkes.* Philadelphia: Lippincott, 1968.

Silverman, Kenneth. *The Life and Times of Cotton Mather.* New York: Harper and Row, 1984.

———. *Lightning Man: The Accursed Life of Samuel F.B. Morse.* New York: Alfred A. Knopf, 2003.

Silverstein, Arthur M. *A History of Immunology.* San Diego, CA: Academic Press, 1989.

Simmons, John. *Doctors and Discoveries: Lives That Created Today's Medicine.* Boston: Houghton Mifflin, 2002.

Singer, Charles. *A History of Biology to About the Year 1900: A General Introduction to the Study of Living Things.* History of Science and Technology Reprint Series. Iowa City: University of Iowa Press, 1989.

Singh, Simon. *Big Bang: The Origin of the Universe.* New York: Fourth Estate, 2005.

Skidmore-Roth, Linda. *Mosby's Handbook of Herbs and Natural Supplements.* St. Louis, MO: Mosby, 2001.

Skinner, B.F. *About Behaviorism.* New York: Alfred A. Knopf, 1974; New York: Vintage Books, 1976.

———. *Beyond Freedom and Dignity.* Indianapolis, IN: Hackett, 2002.

———. *The Treatment Techniques of Harry Stack Sullivan.* London. Brunner-Routledge, 1978.

———. *Walden Two.* Englewood Cliffs, NJ: Prentice Hall, 1948, 1976.

Sklansky, Jeffrey. *The Soul's Economy: Market Society and Selfhood in American Thought, 1820–1920.* Chapel Hill: University of North Carolina Press, 2002.

Sklar, Kathryn Kish. *Catharine Beecher: A Study in American Domesticity.* New York: W.W. Norton, 1976.

Slaatte, Howard A. *The Arminian Arm of Theology: The Theologies of John Fletcher and His Precursor, James Arminius.* Washington, DC: University Press of America, 1977.

Slack, Charles. *Noble Obsession: Charles Goodyear, Thomas Hancock, and the Race to Unlock the Greatest Industrial Secret of the Nineteenth Century.* New York: Hyperion, 2003.

Slaughter, Thomas P. *Exploring Lewis and Clark: Reflections on Men and Wilderness.* New York: Alfred A. Knopf, 2003.

———. *The Natures of John and William Bartram.* New York: Alfred A. Knopf, 1996.

Small, Miriam R. *Oliver Wendell Holmes.* New York: Twayne, 1962.

Smelser, Neil J., and Paul B. Baltes, eds. *International Encyclopedia of the Social and Behavioral Sciences.* Amsterdam, The Netherlands: Elsevier, 2001.

Smith, Asa. *Smith's Illustrated Astronomy.* New York: Daniel Burgess, 1855.

Smith, Bruce D., ed. *Rivers of Change: Essays on Early Agriculture in Eastern North America.* Washington, DC: Smithsonian Institution, 1992.

Smith, Edward H. *The Entomological Society of America: The First Hundred Years, 1889–1989.* Lanham, MD: Entomological Society of America, 1989.

Smith, Jane S. *Patenting the Sun: Polio and the Salk Vaccine.* New York: William Morrow, 1990.

Smith, John. *The Complete Works of Captain John Smith.* Ed. Philip Barbour. 3 vols. Chapel Hill: University of North Carolina Press, 1986.

Smith, Lawrence D., and William Ray Woodward. *B.F. Skinner and Behaviorism in American Culture.* Cranbury, NJ: Lehigh University Press, 1996.

Smith, Murphy D. *Oak from an Acorn: A History of the American Philosophical Society Library, 1770–1803.* Wilmington, DE: Scholarly Resources, 1976.

Smith, Neil. *American Empire: Roosevelt's Geographer and the Prelude to Globalization.* Berkeley: University of California Press, 2003.

Smith, Peter. *A Way of Life and Selected Writings of Sir William Osler.* New York: Dover, 1958.

Smith, Raymond A. *Encyclopedia of AIDS: A Social, Political, Cultural, and Scientific Record of the HIV Epidemic.* New York: Penguin, 2001.

Smith, Robert Leo, and Thomas M. Smith. *Ecology and Field Biology.* 6th ed. San Francisco: Benjamin Cummings, 2001.

Sobel, Michael I. *Light.* Chicago: University of Chicago Press, 1989.

Socolow, Arthur A. *The State Geological Surveys: A History.* Grand Forks, ND: Association of State Geologists, 1988.

Spink, Wesley W. *Infectious Diseases: Prevention and Treatment in the Nineteenth and Twentieth Centuries.* Minneapolis: University of Minnesota Press, 1978.

Squire, Larry R., ed. *The History of Neuroscience in Autobiography.* Vol. 1. St. Louis, MO: Academic Press, 1998.

Stachel, John. *Einstein's Miraculous Year: Five Papers That Changed the Face of Physics.* Princeton, NJ: Princeton University Press, 1998.

Stage, Sarah. *Female Complaints: Lydia Pinkham and the Business of Women's Medicine.* New York: W.W. Norton, 1979.

Stanford, Alfred. *Navigator: The Story of Nathaniel Bowditch.* Kila, MT: Kessinger, 2004.

Stanton, William. *Great United States Exploring Expedition of 1838–1842.* Berkeley: University of California Press, 1975.

Starr, Paul. *The Social Transformation of American Medicine.* New York: Basic Books, 1982.

Stearns, Peter N. *The Industrial Revolution in World History.* 2nd ed. Boulder, CO: Westview, 1998.

Stearns, Raymond Phineas. *Science in the British Colonies of America.* Urbana: University of Illinois Press, 1970.

Steele, Volney. *Bleed, Blister, and Purge: A History of Medicine on the American Frontier.* Missoula, MT: Mountain Press, 2005.

Steen, Harold K. *The U.S. Forest Service: A History.* Seattle: University of Washington Press, 2004.

Stefansson, Vilhjalmur. *Discovery.* New York: McGraw-Hill, 1964.

Steffen, Jerome O. *William Clark: Jeffersonian Man on the Frontier.* Norman: University of Oklahoma Press, 1977.

Stehle, Philip. *Physics: The Behavior of Particles.* New York: Harper and Row, 1971.

Stephens, Carlene E. *Inventing Standard Time.* Washington, DC: National Museum of American History, Smithsonian Institution, 1983.

Sterling, Christopher H., and John M. Kitross. *Stay Tuned: A History of American Broadcasting.* Mahwah, NJ: Lawrence Erlbaum, 2001.

Stern, Heinrich. *Theory and Practice of Bloodletting.* New York: Rebman, 1915.

Stern, Nancy. *From ENIAC to UNIVAC: An Appraisal of the Eckert-Mauchly Computers.* Bedford, MA: Digital, 1981.

Sternberg, George Miller. *Malaria and Malarial Diseases.* New York: W. Wood, 1884.

———. *Sanitary Lessons of the War, and Other Papers.* New York: Beaufort, 1999.

———. *A Textbook of Bacteriology.* London: J. and A. Churchill, 1896.

Stevens, Joseph E. *Hoover Dam: An American Adventure.* Norman: University of Oklahoma Press, 1988.

Stevens, Rosemary. *In Sickness and Wealth: American Hospitals in the Twentieth Century.* Baltimore: Johns Hopkins University Press, 1999.

Stewart, Chris, and Mike Torrey, eds. *A Century of Hospitality in High Places: The Appalachian Mountain Club Hut System 1888–1988.* Littleton, NH: Appalachian Mountain Club, 1988.

Stewart, Frank H. *History of the First United States Mint: Its People and Its Operations.* Philadelphia: Frank H. Stewart Electric Co., 1924.

Stewart, Robert E. *Seven Decades That Changed America: A History of the American Society of Agricultural Engineers, 1907–1977.* St. Joseph, MI: American Society of Agricultural Engineers, 1970.

Stigler, Stephen M. *Statistics on the Table: The History of Statistical Concepts and Methods.* Cambridge, MA: Harvard University Press, 1999.

Stites, Daniel P., John D. Stobo, and J. Vivian Wells, eds. *Basic and Clinical Immunology.* 6th ed. Norwalk, CT: Appleton and Lange, 1987.

Stocking, George W. *The Ethnographer's Magic and Other Essays in the History of Anthropology.* Madison: University of Wisconsin Press, 1992.

Stoll, Clifford. *Cuckoo's Egg: Tracking a Spy Through the Maze of Computer Espionage.* New York: Pocket Books, 2000.

———. *Silicon Snake Oil: Second Thoughts on the Information Highway.* New York: Anchor, 1996.

Stowell, Marion Barber. *Early American Almanacs: The Colonial Weekday Bible.* New York: Burt Franklin, 1977.

Stratton, Julius A., and Loretta H. Mannix. *Mind and Hand: The Birth of MIT.* Cambridge, MA: MIT Press, 2005.

Straus, Eugene. *Rosalyn Yalow, Nobel Laureate: Her Life and Work in Medicine.* Cambridge, MA: Perseus, 2000.

Strauss, Anselm, ed. *The Social Psychology of George Herbert Mead.* Chicago: University of Chicago Press. 1956.

Strauss, David. *Percival Lowell: The Culture and Science of a Boston Brahmin.* Cambridge, MA: Harvard University Press, 2001.

Streshinsky, Shirley. *Audubon: Life and Art in the American Wilderness.* Athens: University of Georgia Press, 1998.

Strick, James Edgar. *Sparks of Life: Darwinism and the Victorian Debates over Spontaneous Generation.* Cambridge, MA: Harvard University Press, 2000.

Stroud, Patricia Tyson. *Thomas Say: New World Naturalist.* Philadelphia: University of Pennsylvania Press, 1992.

Sturtevant, A.H. *A History of Genetics.* Cold Spring Harbor, NY: Cold Spring Harbor Laboratory Press, 2001.

Sullivan, Woodruff Turner, ed. *Early Years of Radio Astronomy: Reflections Fifty Years After Jansky's Discovery.* New York: Cambridge University Press, 1984.

Sultz, H.A., and K.M. Young. *Health Care USA: Understanding Its Organization and Delivery.* 2nd ed. Gaithersburg, MD: Aspen, 1999.

Suppe, Frederick, ed. *The Structure of Scientific Theories.* 2nd ed. Urbana: University of Illinois Press, 1977.

Sutcliffe, Andrea. *Steam: The Untold Story of America's First Great Invention.* New York: Palgrave Macmillan, 2004.

Sutherland, Edwin H. *The Professional Thief.* Chicago: University of Chicago Press, 1990.

———. *White Collar Crime: The Uncut Version.* New Haven, CT: Yale University Press, 1990.

Swenson, Lloyd. *The Ethereal Aether: A History of the Michelson-Morley-Miller Aether-Drift Experiments.* Austin: University of Texas Press, 1972.

Swenson, Loyd S., Jr., James M. Grimwood, and Charles C. Alexander. *This New Ocean: A History of Project Mercury.* Washington, DC: NASA, 1966.

Swift, David W. *SETI Pioneers: Scientists Talk About Their Search for Extraterrestrial Intelligence.* Tucson: University of Arizona Press, 1990.

Szajnberg, Nathan M., ed. *Educating the Emotions: Bruno Bettelheim and Psychoanalytic Development.* New York: Plenum, 1992.

Taft, Robert. *Photography and the American Scene: A Social History, 1839–1889.* New York: Dover, 1964.

Takeda, Masatoshi, et al. *Molecular Neurobiology of Alzheimer Disease and Related Disorders.* New York: Karger, 2004.

Talbott, John. *A Biographical History of Medicine.* New York: Grune and Stratton, 1970.

Taton, Rene, and Curtis Wilson, eds. *The General History of Astronomy.* Vol. 2, Part B, *Planetary Astronomy from the Renaissance to the Rise of Astrophysics.* Cambridge, UK: Cambridge University Press, 1995.

Taubes, Gary. *Bad Science: The Short Life and Weird Times of Cold Fusion.* New York: Random House, 1993.

Taussig, Helen B. *Congenital Malformations of the Heart.* New York: Commonwealth Fund, 1947.

Taylor, Edwin F., and John Wheeler. *Spacetime Physics.* New York: W.H. Freeman, 1992.

Taylor, John. *When the Clock Struck Zero.* New York: St. Martin's, 1993.

Taylor, R.E., and Martin J. Aitken, eds. *Chronometric Dating in Archaeology: Advances in Archaeological and Museum Science.* Vol. 2. Oxford, UK: Oxford University Press, 1997.

Teller, Edward, with Judith L. Shoolery. *Memoirs: A Twentieth-Century Journey in Science and Politics.* Cambridge, MA: Perseus, 2001.

Terkel, Susan Neiburg. *Colonial American Medicine.* Danbury, CT: Franklin Watts, 1993.

Thacker, Christopher. *The History of Gardens.* Berkeley: University of California Press, 1985.

Thomas, David H. *Skull Wars: Kennewick Man, Archeology, and the Battle for Native American Identity.* New York: HarperCollins, 2000.

Thomas, Robert. *"With Bleeding Footsteps": Mary Baker Eddy's Path to Religious Leadership.* New York: Alfred A. Knopf, 1994.

Thompson, Benjamin (Count Rumford). *Collected Works of Count Rumford.* Cambridge, MA: Harvard University Press, 1968.

Thompson, Lana. *The Wandering Womb: A Cultural History of Outrageous Beliefs About Women.* Amherst, NY: Prometheus. 1999.

Thompson, Neal. *Light This Candle: The Life and Times of Alan Shepard, America's First Spaceman.* New York: Crown, 2004.

Thomson, E.H. "Harvey Williams Cushing." In *Dictionary of Scientific Biography,* vol. 3. New York: Scribner's Sons, 1971.

Thomson, Robert. *The Pelican History of Psychology.* Baltimore: Penguin, 1968.

Thoney, D.A., Paul V. Loiselle, and Neil Schlager, eds. *Grzimek's Animal Life Encyclopedia.* Vol. 4, *Fishes.* Detroit: Gale, 2003.

Thoreau, Henry David. *The Portable Thoreau.* Ed. Carl Bode. New York: Penguin, 1976.

———. *A Week on the Concord and Merrimack Rivers.* Boston: Houghton Mifflin, 1906.

Thorne, Brian. *Carl Rogers.* London: Sage, 1992.

Tilden, Freeman. *The National Parks.* New York: Random House, 1970.

Tinkham, Michael. *Introduction to Superconductivity.* New York: McGraw-Hill, 1996.

Tobin, James. *To Conquer the Air: The Wright Brothers and the Great Race for Flight.* New York: Free Press, 2003.

Tobyn, Graeme. *Culpepper's Medicine: A Practice of Western Holistic Medicine.* Rockport, MA: Element, 1997.

Todd, Edgeley W. *The Adventures of Captain Bonneville, U.S.A., in the Rocky Mountains and the Far West, Di-* gested from His Journal by Washington Irving. Norman: University of Oklahoma Press, 1961.

Todd, James T., and Edward K. Morris, eds. *Modern Perspectives on John B. Watson and Classical Behaviorism.* Westport, CT: Greenwood, 1994.

Tolles, Frederick. *Meeting-House and Counting-House: The Quaker Merchants of Colonial Philadelphia, 1682–1763.* New York: W.W. Norton, 1948.

Tombaugh, Clyde, and Patrick Moore. *Out of the Darkness: The Planet Pluto.* Harrisburg, PA: Stackpole, 1980.

Tomes, Nancy. *The Art of Asylum Keeping: Thomas Story Kirkbride and the Origins of American Psychiatry.* Philadelphia: University of Pennsylvania Press, 1994.

———. *The Gospel of Germs: Men, Women, and the Microbe in American Life.* Cambridge, MA: Harvard University Press, 1999.

Tooker, Elisabeth. *Lewis H. Morgan on Iroquois Material Culture.* Tucson: University of Arizona Press, 1994.

Tooley, R.V. *Map Making in France from the Sixteenth Century to the Eighteenth Century.* London: Butler and Tanner, 1952.

Tops, Franklin H., and Paul F. Wehrle, eds. *Communicable and Infectious Diseases.* 6th ed. St. Louis, MO: Mosby, 1972.

Townsend, John Kirk. "Narrative of a Journey Across the Rocky Mountains to the Columbia River." In *Early Western Travels, 1748–1846,* vol. 21, ed. Reuben Gold Thwaites. Cleveland, OH: A.H. Clark, 1905.

———. *Narrative of a Journey Across the Rocky Mountains to the Columbia River.* Lincoln: University of Nebraska Press, 1978.

———. *Ornithology of the United States of North America.* Philadelphia: Chevalier, 1839.

Trowbridge, Carol. *Andrew Taylor Still, 1828–1917.* Kirksville, MO: Truman State University Press, 1991.

Trower, Peter, ed. *Discovering Alvarez: Selected Works of Luis W. Alvarez with Commentary by His Students and Colleagues.* Chicago: University of Chicago Press, 1987.

Tucker, Louis Leonard. *Clio's Consort: Jeremy Belknap and the Founding of the Massachusetts Historical Society.* Boston: Massachusetts Historical Society, 1989.

Tucker, William H. *The Science and Politics of Racial Research.* Chicago: University of Illinois Press, 1994.

Turner, Gerard L'Estrange. *Scientific Instruments 1500–1900: An Introduction.* Berkeley: University of California Press, 1998.

Turner, Tom. *Sierra Club: 100 Years of Protecting Nature.* New York: Harry N. Abrams, 1991.

Udias, Augustin. *Principles of Seismology.* New York: Cambridge University Press, 2000.

Ulrich, Laura Thatcher. *A Midwife's Tale: The Life of Martha Ballard, Based on Her Diary, 1785–1812.* New York: Alfred A. Knopf, 1990; New York: Vintage Books, 1991.

Unger, Harlow G. *Noah Webster: The Life and Times of an American Patriot.* New York: John Wiley, 1998.

Urey, Harold Clayton. *The Planets: Their Origin and Development.* New Haven, CT: Yale University Press, 1952.

U.S. Agency for Health Care Research. *National Health Care Quality Report.* Washington, DC: Health and Human Services Administration, 2003.

U.S. Centers for Disease Control and Prevention. *Measles: Epidemiology and Prevention of Vaccine Preventable Diseases.* Atlanta: CDC, 2000.

U.S. Congress. Senate Committee on Finance. *Trends in U.S. Life Expectancy.* Washington, DC: U.S. Government Printing Office, 1983.

U.S. Department of Commerce. National Oceanic and Atmospheric Administration. *The NOAA Story.* Washington, DC: U.S. Government Printing Office, 1973.

U.S. House of Representatives. *Advanced Submarine Technology and Antisubmarine Warfare.* Park Forest, IL: University Press of the Pacific, 2005.

Valiela, Ivan. *Marine Ecological Processes.* New York: Springer-Verlag, 1995.

Van Doren, Carl. *Benjamin Franklin.* New York: Viking, 1938.

Van Sertima, Ivan, ed. *Blacks in Science: Ancient and Modern.* Somerset, NJ: Transaction, 1987.

Van Wylen, Gordon J., and Richard E. Sonntag. *Fundamentals of Classical Thermodynamics.* New York: John Wiley and Sons, 1976.

Van Zwoll, Wayne. *America's Great Gunmakers.* South Hackensack, NJ: Stoeger, 1992.

Vasquez, Tim. *Weather Forecasting Handbook.* 5th ed. Garland, TX: Weather Graphics Technologies, 2002.

Veblen, Thorstein. *The Higher Learning in America.* 1924. Reprint ed., New Brunswick, NJ: Transaction, 1993.

———. *The Theory of the Leisure Class.* 1899. Reprint ed., New York: Penguin, 1994.

Verschuur, Gerrit L. *Hidden Attraction: The Mystery and History of Magnetism.* Oxford, UK: Oxford University Press, 1993.

Vogel, Virgil. *American Indian Medicine.* Norman: University of Oklahoma Press, 1990.

von Braun, Wernher. *Conquest of the Moon.* New York: Viking, 1953.

———. *The Mars Project.* Chicago: University of Illinois, 1991.

von Hagen, Victor. *South America Called Them: Explorations of the Great Naturalists: La Condamine, Humboldt, Darwin, and Spruce.* New York: Duell, Sloan, and Pearce, 1955.

Wachhorst, Wyn. *Thomas Alva Edison, an American Myth.* Cambridge, MA: Harvard University Press, 1981.

Wagner, David. *Light and Color.* New York: Wiley, 1982.

Wahl, Paul. *The Gatling Gun.* New York: Arco, 1965.

Waksman, Selman A. *My Life with Microbes.* New York: Simon and Schuster, 1954.

Wald, Robert M. *Space, Time, and Gravity: The Theory of the Big Bang and Black Holes.* Chicago: University of Chicago Press, 1981.

Waldman, Gary. *Introduction to Light: The Physics of Light, Vision, and Color.* New York: Dover, 2002.

Wallace, David Rains. *The Bonehunters' Revenge: Dinosaurs, Greed, and the Greatest Scientific Feud of the Gilded Age.* Boston: Houghton Mifflin, 1999.

Wallace, James, and Jim Erickson. *Hard Drive: Bill Gates and the Making of the Microsoft Empire.* New York: HarperBusiness, 1993.

Wallace, Paul A. *The Muhlenbergs of Pennsylvania.* Philadelphia: University of Pennsylvania Press, 1950.

Walter, William J. *Space Age.* New York: Random House, 1992.

Walters, Michael. *A Concise History of Ornithology.* New Haven, CT: Yale University Press, 2005.

Ward, Bob. *Dr. Space: The Life of Wernher von Braun.* Annapolis, MD: Naval Institute Press, 2005.

Ward, Darrell E. *The Amfar AIDS Handbook: The Complete Guide to Understanding HIV and AIDS.* New York: W.W. Norton, 1998.

Wardell, Walter I. *Chiropractic: History and Evolution of a New Profession.* St. Louis, MO: Mosby Year Book, 1992.

Warner, Deborah Jean. *Alvan Clark and Sons, Artists in Optics.* Washington, DC: Smithsonian, 1968.

Warren, Leonard. *Joseph Leidy: The Last Man Who Knew Everything.* New Haven, CT: Yale University Press, 1998.

Waselkov, Gregory A., and Kathryn E. Holland Braund, eds. *William Bartram on the Southeastern Indians.* Lincoln: University of Nebraska Press, 1995.

Washburn, Emory. *Sketches of the Judicial History of Massachusetts from 1630 to the Revolution in 1775.* New York: Da Capo, 1974.

Watson, James D. *The Double Helix: A Personal Account of the Discovery of the Structure of DNA.* New York: Scribner's, 1998.

———. *Genes, Girls, and Gamow: After the Double Helix.* New York: Random House, 2002.

Watts, Steven. *The People's Tycoon: Henry Ford and the American Century.* New York: Alfred A. Knopf, 2005.

Webb, George E. *The Evolution Controversy in America.* Lexington: University Press of Kentucky, 1994.

Webb, Stephen. *If the Universe Is Teeming with Aliens . . . Where Is Everybody? Fifty Solutions to Fermi's Paradox and the Problem of Extraterrestrial Life.* New York: Copernicus, 2002.

Wechsler, David. *The Measurement and Appraisal of Adult Intelligence.* Baltimore: Williams and Wilkins, 1985.

Weems, John Edward. *Peary: The Explorer and the Man.* Cambridge, MA: Riverside, 1967.

Weiss, Harry B., and Grace M. Ziegler. *Thomas Say: Early American Naturalist.* Springfield, IL: Charles C. Thomas, 1931.

Welchman, Kit. *Erik Erikson: His Life, Work, and Significance.* Philadelphia: Open University Press, 2000.

Welker, Robert Henry. *Natural Man: The Life of William Beebe.* Bloomington: Indiana University Press, 1975.

Welling, William B. *Photography in America: The Formative Years, 1839–1900.* Albuquerque: University of New Mexico Press, 1987.

Wells, Susan. *Out of the Dead House.* Madison: University of Wisconsin Press, 1994.

Wensberg, Peter C. *Land's Polaroid: A Company and the Man Who Invented It.* Boston: Houghton Mifflin, 1987.

Wertz, Richard W., and Dorothy Wertz. *Lying-In: A History of Childbirth in America.* New Haven, CT: Yale University Press, 1989.

Westaby, Stephen. *Landmarks in Cardiac Surgery.* Oxford, UK: Isis Medical Media, 2000.

Westheimer, Michael. *A Brief History of Psychology.* New York: Holt, Rinehart and Winston, 1970.

Wheeler, John. *Geons, Black Holes, and Quantum Foam: A Life in Physics.* New York: W.W. Norton, 1998.

Wheeler, Lynde Phelps. *Josiah Willard Gibbs: The History of a Great Mind.* New Haven, CT: Yale University Press, 1952.

Whipple, Joseph. *The History of Acadie, Penobscot Bay and River, with a More Particular Geographical and Statistical View of the District of Maine.* Bangor, ME: Peter Edes, 1816.

Whisker, James B. *The Gunsmith's Trade.* Lewiston, NY: Edwin Mellen, 1992.

White, Edward A. *Science and Religion in American Thought: The Impact of Naturalism.* Stanford, CA: Stanford University Press, 1952.

White, Leslie A. *The Indian Journals 1859–1862.* Ann Arbor: University of Michigan Press, 1859; reprint ed., New York: Dover, 1993.

Whitehead, Don. *Dow Story: The History of the Dow Chemical Company.* New York: McGraw-Hill, 1968.

Whitfield, Peter. *Astrology: A History.* New York: Harry N. Abrams, 2001.

Whitnah, Donald R. *A History of the United States Weather Bureau.* Urbana: University of Illinois Press, 1961.

Whorton, James C. *Nature Cures: The History of Alternative Medicine in America.* New York: Oxford University Press, 2002.

Wiener, Norbert. *Cybernetics, or Control and Communication in the Animal and the Machine.* Cambridge, MA: MIT Press, 1948.

Wigner, Eugene P. *Symmetries and Reflections.* Woodbridge, CT: Ox Bow, 1979.

Williams, Brian K., and Stacey C. Sawyer. *Using Information Technology: A Practical Introduction to Computers and Communications.* 5th ed. New York: McGraw-Hill, 2003.

Williams, Frances Leigh. *Matthew Fontaine Maury, Scientist of the Sea.* New Brunswick, NJ: Rutgers University Press, 1963.

Williams, Guy. *The Age of Agony: The Art of Healing, c. 1700–1800.* Chicago: American Academy, 1986.

Williams, Vernon J., Jr. *Rethinking Race: Franz Boas and His Contemporaries.* Lexington: University Press of Kentucky, 1996.

Williams, William H. *America's First Hospital: The Pennsylvania Hospital, 1751–1841.* Wayne, PA: Haverford House, 1976.

Wilson, David Scofield. *In the Presence of Nature.* Amherst: University of Massachusetts Press, 1978.

Wilson, Edmund. *An Introduction to Particle Accelerators.* Oxford, UK: Oxford University Press, 2001.

Wilson, Edward O. *Biophilia.* Cambridge, MA: Harvard University Press, 1984.

———. *The Diversity of Life.* Cambridge, MA: Harvard University Press, 1992; New York: W.W. Norton, 1999.

———. *Sociobiology: The New Synthesis.* Cambridge, MA: Harvard University Press, 2000.

Wilson, Leonard G., ed. *Benjamin Silliman and His Circle: Studies on the Influence of Benjamin Silliman on Science in America.* New York: Science History, 1979.

Wilson, Louis N. *G. Stanley Hall: A Sketch.* New York: G.E. Stechert, 1914.

Wilson, Philip K. "Jared Eliot." In *American National Biography,* vol. 7. New York: Oxford University Press, 1999.

Wilson, Robert. *Astronomy Through the Ages: The Story of the Human Attempt to Understand the Universe.* Princeton, NJ: Princeton University Press, 1997.

Winchester, Simon. *A Crack in the Edge of the World: America and the Great California Earthquake of 1906.* New York: HarperCollins, 2005.

Winslow, Charles-Edward Armory. *The History of American Epidemiology.* St. Louis, MO: C.V. Mosby, 1952.

———. *The Life of Hermann M. Biggs, M.D.* Philadelphia: Lea and Febiger, 1929.

Winslow, Ola Elizabeth. *A Destroying Angel: The Conquest of Smallpox in Colonial Boston.* Boston: Houghton Mifflin, 1974.

Wirth, Louis. *The Ghetto.* Chicago: University of Chicago Press, 1928.

Wolfe, Tom. *The Right Stuff.* New York: Farrar, Straus and Giroux, 1979.

Wood, Laura Newbold. *Walter Reed: Doctor in Uniform.* New York: Julian Messner, 1943.

Wood, Richard G. *Stephen Harriman Long, 1784–1864: Army Engineer, Explorer, Inventor.* Glendale, CA: Arthur H. Clark, 1966.

Wood, W. Barry. *From Miasmas to Molecules.* New York: Columbia University Press, 1961.

Woodman, Richard. *The History of the Ship.* London: Conway Maritime, 1997.

Wozniak, Robert H. *Classics in Psychology, 1855–1914: Historical Essays.* Tokyo: Thoemmes, 1999.

Wraight, A. Joseph, and Elliot B. Roberts. *The Coast and Geodetic Survey 1807–1957: 150 Years of History.* Washington, DC: U.S. Government Printing Office, 1957.

Wrege, Charles D., and Ronald G. Greenwood. *Frederick W. Taylor, the Father of Scientific Management: Myth and Reality.* Homewood, IL: Business One Irwin, 1991.

Wrenshall, G.A., G. Hetenyi, and W.R. Feasby. *The Story of Insulin.* Bloomington: Indiana University Press, 1962.

Wright, Helen. *Explorer of the Universe: A Biography of George Ellery Hale.* New York: E.P. Dutton, 1966; Woodbury, NY: American Institute of Physics Press, 1994.

Wright, Helen, Joan N. Warnow, and Charles Weiner, eds. *The Legacy of George Ellery Hale: Evolution of Astronomy and Scientific Institutions in Pictures and Documents.* Cambridge, MA: MIT Press, 1972.

Wright, John Kirtland. *Geography in the Making: The American Geographical Society, 1851–1951.* New York: American Geographical Society, 1952.

Wright, Susan, ed. *Biological Warfare and Disarmament: New Problems/New Perspectives.* New York: Rowman and Littlefield, 2002.

Wylie, Francis. *M.I.T. in Perspective.* Boston: Little, Brown, 1975.

Wynbrandt, James. *The Excruciating History of Dentistry: Toothsome Tales and Oral Oddities from Babylon to Braces.* New York: St. Martin's, 1998.

Yan, Song Y. *Number Theory for Computing.* Berlin: Springer-Verlag, 2000.

Yeats, R.S. *Geology of Earthquakes.* New York: Oxford University Press, 1997.

York, Herbert. *The Advisors: Oppenheimer, Teller, and the Superbomb.* Palo Alto, CA: Stanford University Press, 1989.

Young, James Harvey. *The Toadstool Millionaires: A Social History of Patent Medicines in America Before Federal Regulation.* Princeton, NJ: Princeton University Press, 1961.

Zachary, G. Pascal. *Endless Frontier: Vannevar Bush, Engineer of the American Century.* New York: Free Press, 1997.

Zee, Anthony. *Quantum Field Theory in a Nutshell.* Princeton NJ: Princeton University Press, 2003.

Zehnpfennig, Gladys. *Charles F. Kettering: Inventor and Idealist; a Biographical Sketch of a Man Who Refused to Recognize the Impossible.* Men of Achievement Series. Minneapolis, MN: T.S. Denison, 1962.

Zetterberg, J. Peter, ed. *Evolution Versus Creationism: The Public Education Controversy.* Phoenix, AZ: Oryz, 1983.

Zimmerman, Leo. *Great Ideas in the History of Surgery.* New York: Dover, 1967.

Zinsser, William. *Peterson's Birds: The Art and Photography of Roger Tory Peterson.* New York: Universe, 2002.

Ziporyn, Teresa. *Disease in the Popular American Press: The Case of Diphtheria, Typhoid Fever, and Syphilis, 1870–1920.* New York: Greenwood, 1988.

Zirker, Jack B. *Journey from the Center of the Sun.* Princeton, NJ: Princeton University Press, 2001.

Zung, Thomas T.K. *Buckminster Fuller: Anthology for a New Millennium.* New York: St. Martin's, 2002.

Web Sites

American Academy of Arts and Sciences. http://www.amacad.org.

American Antiquarian Society. http://www.americanantiquarian.org.

American Association for the Advancement of Science. http://www.aaas.org.

American Cancer Society. http://www.cancer.org.

American Dental Association. http://www.ada.org.

American Historical Association. http://www.historians.org.

American Journal of Psychology. http://www.press.uillinois.edu/journals/ajp.html.

American Mathematical Society. http://www.ams.org.

American Medical Association. http://www.ama-assn.org.

American Museum of Natural History. http://www.amnh.org.

American Philosophical Society. http://www.amphilsoc.org.

American Psychological Association. http://www.apa.org.

American Society of Agricultural and Biological Engineers. http://www.asabe.org.

American Society of Agronomy. http://www.agronomy.org.

American Society of Civil Engineers. http://www.asce.org.

American Society of Gene Therapy. http://www.asgt.org.

American Sociological Association. http://www.asanet.org.

American Statistical Association. http://www.asstat.org.

Appalachian Mountain Club. http://www.outdoors.org.

Association of American Medical Colleges. http://www.aamc.org.

Autism Society of America. http://www.autism-society.org.

Black Inventor Online Museum. http://www.blackinventor.com.

Centers for Disease Control. National Center for Infectious Disease. Flu Home Page. http://www.cdc.gov/flu.

Deere and Company. http://www.deere.com.

Discovering Lewis and Clark. http://www.lewis-clark.org.

The Field Museum. http://www.fieldmuseum.org.

Geological Society of America. http://www.geosociety.org.

Harvard Medical School. http://hms.harvard.edu.

Harvard Museum of Natural History. http://www.hmnh.harvard.edu.

International Commission on Stratigraphy. http://www.stratigraphy.org.

Jarvik Heart. http://www.jarvikheart.com.

The Journals of the Lewis and Clark Expedition. http://lewisandclarkjournals.unl.edu.

NASA-Jet Propulsion Laboratory, California Institute of Technology. http://www.jpl.nasa.gov/missions.

Kinsey Institute. http://www.indiana.edu/~Kinsey.

Lawrence Livermore National Laboratory. http://www.llnl.gov.

Marine Biological Laboratory. http://www.mbl.edu.

Massachusetts General Hospital. http://www.mgh.harvard.edu.

Massachusetts Historical Society. http://www.masshist.org.

Mayo Clinic. http://www.mayoclinic.org.

Mount Washington Observatory. http://www.mountwashington.org.

Mount Wilson Observatory. http://www.mtwilson.edu.

National Academy of Sciences. http://www.nasonline.org.

National Aeronautics and Space Administration. http://www.nasa.gov.

National Aeronautics and Space Administration Hubble Space Telescope. http://hubble.nasa.gov.

National Audubon Society. http://www.audubon.org/nas.

National Climatic Data Center. http://www.ncdc.noaa.gov.

National Geographic Society. http://www.nationalgeographic.com.

National Hurricane Center. http://www.nhc.noaa.gov.

National Institutes of Health. http://www.nih.gov.

National Oceanic and Atmospheric Administration. http://www.noaa.gov.

National Optical Astronomy Observatory. http://www.noao.edu.

National Park Service. http://www.nps.gov.

National Park Service Archeology Program. http://www.cr.nps.gov/archeology.

National Science Foundation. http://www.nsf.gov.

National Weather Service. http://www.nws.noaa.gov.

Nobel Foundation. http://nobelprize.org/nobel_prizes.

Pennsylvania Hospital. http://www.pennhealth.com/pahosp.

Planned Parenthood Federation of America. http://www.plannedparenthood.org.

Plastics Historical Society. http://www.plastiquarian.com.

Royal Society. http://www.royalsoc.ac.uk.

Science magazine. http://www.sciencemag.org.

Scientific American. http://www.sciam.com.

Search for Extraterrestrial Intelligence Institute. http://setiathome.ssl.berkeley.edu.

Sierra Club. http://www.sierraclub.org.

Slater Mill. http://www.slatermill.org.

Smithsonian Institution. http://www.si.edu.

System Dynamics Society. http://www.systemdynamics.org.

United Nations Framework Convention on Climate Change. http://unfccc.int/2860.php.

U.S. Department of Agriculture. http://www.usda.gov.

U.S. Forest Service. http://www.fs.fed.us.

U.S. Geological Survey. http://www.usgs.gov.

U.S. Mint. http://www.usmint.gov.

U.S. Naval Observatory. http://www.usno.navy.mil.

U.S. Navy Submarine Force Museum. http://www.ussnautilus.org.

Index